VOLUME ONE HUNDRED AND NINETY FOUR

PROGRESS IN MOLECULAR BIOLOGY AND TRANSLATIONAL SCIENCE

Receptor Endocytosis and Signalling in Health and Disease - Part A

VOLUME ONE HUNDRED AND NINETY FOUR

PROGRESS IN MOLECULAR BIOLOGY AND TRANSLATIONAL SCIENCE

Receptor Endocytosis and Signalling in Health and Disease - Part A

Edited by

INDRA MANI

Department of Microbiology, Gargi College, University of Delhi, New Delhi, India

VIJAI SINGH

Department of Biosciences, School of Science, Indrashil University, Rajpur, Mehsana, India

ELSEVIER

ACADEMIC PRESS

An imprint of Elsevier

Academic Press is an imprint of Elsevier
50 Hampshire Street, 5th Floor, Cambridge, MA 02139, United States
525 B Street, Suite 1650, San Diego, CA 92101, United States
The Boulevard, Langford Lane, Kidlington, Oxford OX5 1GB, United Kingdom
125 London Wall, London EC2Y 5AS, United Kingdom

First edition 2023

ISBN: 978-0-443-19185-5
ISSN: 1877-1173

For information on all Academic Press publications
visit our website at https://www.elsevier.com/books-and-journals

Publisher: Zoe Kruze
Acquisitions Editor: Leticia M. Lima
Developmental Editor: Jhon Michael Peñano
Production Project Manager:
Sudharshini Renganathan
Cover Designer: Matthew Limbert

Typeset by STRAIVE, India

Contents

Contributors

Areej Alzahrani
School of Molecular & Cellular Biology, University of Leeds, Leeds, United Kingdom

Vandana Anang
International Center for Genetic Engineering and Biotechnology (ICGEB), New Delhi, India

Rhea Bhalla
Faculty of Medicine, Department of Infectious Disease, Imperial College London, London, United Kingdom

Aaradhya Bhatnagar
Department of Biosciences, School of Science, Indrashil University, Rajpur, Mehsana, Gujarat, India

Gargi Bhattacharjee
Department of Biosciences, School of Science, Indrashil University, Rajpur, Mehsana, Gujarat, India

Mettle Brahma
Department of Biochemistry, Central University of Haryana, Jant-Pali, Mahendragarh, Haryana, India

Nhat-Le Bui
Center for Biomedicine and Community Health; Faculty of Applied Sciences, International School, Vietnam National University, Hanoi, Vietnam

Eunhee Choi
Department of Pathology and Cell Biology, Vagelos College of Physicians and Surgeons, Columbia University, New York, NY, United States

Dinh-Toi Chu
Center for Biomedicine and Community Health; Faculty of Applied Sciences, International School, Vietnam National University, Hanoi, Vietnam

William R. Critchley
School of Molecular & Cellular Biology, University of Leeds, Leeds, United Kingdom

Gyaltsen Dakpa
National Chung Hsing University, Taichung City, Taiwan

Aysha Divan
School of Molecular & Cellular Biology, University of Leeds, Leeds, United Kingdom

Soumika Ghosal
Amity Stem cell Institute, Amity Medical School, Amity University Haryana, Amity Education Valley Pachgaon, Manesar, Gurugram, Haryana, India

Mariela M. Gironacci
From Universidad de Buenos Aires, Facultad de Farmacia y Bioquímica, Dpto. Química Biológica, IQUIFIB (UBA-CONICET), Buenos Aires, Argentina

Nisarg Gohil
Department of Biosciences, School of Science, Indrashil University, Rajpur, Mehsana, Gujarat, India

Michael A. Harrison
School of Biomedical Sciences, University of Leeds, Leeds, United Kingdom

Suresh K. Kalangi
Stem Cell and Drug Discovery Division, INOVAUGMET Research Pvt Ltd., Bengaluru, Karnataka, India

Khushal Khambhati
Department of Biosciences, School of Science, Indrashil University, Rajpur, Mehsana, Gujarat, India

Sarat Kumar Kottarath
Department of Experimental Therapeutics, MD Anderson Cancer Center, Huston, TX, United States

Komal Kumari
Department of Biotechnology, Central University of South Bihar, Bihar, India

Angela Lalnunthangi
School of Biotechnology, Jawaharlal Nehru University, New Delhi, India

Ngoc Hoan Le
Faculty of Biology, Hanoi National University of Education, Hanoi, Vietnam

Indra Mani
Department of Microbiology, Gargi College, University of Delhi, New Delhi, India

Mulaka Maruthi
Department of Biochemistry, Central University of Haryana, Jant-Pali, Mahendragarh, Haryana, India

Rupesh Maurya
Department of Biosciences, School of Science, Indrashil University, Rajpur, Mehsana, Gujarat, India

Shivranjani C. Moharir
Centre for Cellular and Molecular Biology, Council for Scientific and Industrial Research, Hyderabad, India

Thanh-Lam Nguyen
Center for Biomedicine and Community Health, International School, Vietnam National University, Hanoi, Vietnam

Pratiksha Pandita
ICMR-National Institute of Virology, Pune, Maharashtra, India

Stella Hayoung Park
Department of Pathology and Cell Biology, Vagelos College of Physicians and Surgeons, Columbia University, New York, NY, United States

Sreenivasan Ponnambalam
School of Molecular & Cellular Biology, University of Leeds, Leeds, United Kingdom

Suresh Ramakrishna
College of Medicine; Graduate School of Biomedical Science and Engineering, Hanyang University, Seoul, South Korea

Hannah Reeve
School of Molecular & Cellular Biology, University of Leeds, Leeds, United Kingdom

Aleah D. Roberts
Biochemistry and Biophysics Center, National Heart, Lung, and Blood Institute, National Institutes of Health, Bethesda, MD, United States

Natalia L. Rukavina Mikusic
From Universidad de Buenos Aires, Facultad de Farmacia y Bioquímica, Dpto. Química Biológica, IQUIFIB (UBA-CONICET), Buenos Aires, Argentina

Queen Saikia
School of Molecular & Cellular Biology, University of Leeds, Leeds, United Kingdom

Ashok Saini
Department of Microbiology, Institute of Home Economics, University of Delhi, New Delhi, India

Shreya Shakhreliya
Department of Science and Technology, Nottingham Trent University, Nottingham, United Kingdom

Aayushi Singh
Invictus Oncology Pvt. Ltd, New Delhi, India

Priyanka Singh
Department of Biosciences, School of Science, Indrashil University, Rajpur, Mehsana, Gujarat, India

Vijai Singh
Department of Biosciences, School of Science, Indrashil University, Rajpur, Mehsana, Gujarat, India

Kapil Sirohi
Centre for Cellular and Molecular Biology, Council for Scientific and Industrial Research, Hyderabad, India

Ghanshyam Swarup
Centre for Cellular and Molecular Biology, Council for Scientific and Industrial Research, Hyderabad, India

Justin W. Taraska
Biochemistry and Biophysics Center, National Heart, Lung, and Blood Institute, National Institutes of Health, Bethesda, MD, United States

Swati Tiwari
School of Biotechnology, Jawaharlal Nehru University, New Delhi, India

Chaitenya Verma
Department of Pathology, Wexner Medical Center, Ohio State University, Columbus, OH, United States

Jiayi Wu
Department of Pathology and Cell Biology, Vagelos College of Physicians and Surgeons, Columbia University, New York, NY, United States

Elton Zeqiraj
School of Molecular & Cellular Biology, University of Leeds, Leeds, United Kingdom

Preface

Receptor-mediated endocytosis (RME) is a mechanism to uptake ligands and agonists and deliver them into a target site. RME plays a vital role in cell growth, survival, differentiation, and homeostasis. All of these depend on specific membrane receptor proteins. Moreover, RME plays a vital role in signal generation, duration, specificity, and amplitude. After ligand binding, ligand–receptor complexes traffic through the various subcellular organelles, such as early endosomes, recycling endosomes, and lysosomes. In addition, some receptors also traffic through other organelles, such as the nucleus, mitochondria, and many others. During receptor internalization, various proteins are involved, including clathrin, adaptor proteins, dynamin, and others. Under normal physiological conditions, all these factors play a significant role during RME and signaling.

However, with any anomaly in the receptor internalization, subcellular trafficking, and signaling consequences, disease conditions may occur. Numerous receptors are associated with different organs, which play a vital role in normal functioning in the body, such as the Mas receptor, transferrin, insulin, VEGFR, AT1, BCR, LXRs, Frizzled receptors, and many more. However, the overexpression of the receptor gene in disease conditions and underexpression in normal conditions can be a major challenge in physiological conditions. This volume covers several important receptors in various parts of the body, examining the emphasis in biomedical research on their function for the understanding of receptor internalization, subcellular trafficking, metabolic degradation, and signaling in healthy as well as disease conditions.

The structure of this book covers fundamental to applied aspects. Chapter 1 offers an overview of receptor endocytosis and signaling. Chapter 2 discusses emerging tools for studying receptor endocytosis and signaling. Chapter 3 explores Mas receptor signaling and trafficking in health and disease. Chapter 4 covers regulation of transferrin receptor trafficking by optineurin and its disease-associated mutants. Chapter 5 looks at endocytosis and signaling of the insulin receptor in the regulation of glucose homeostasis. Chapters 6–8 discuss endocytosis of VEGFR, angiotensin II type 1 receptor, and B cell receptor. Chapter 9 explores the multifunctional role of the ubiquitin–proteasome pathway in phagocytosis. Chapters 10 and 11 discuss receptors of immune cells that mediate the recognition of tumors and the role of their receptors in mediating the suppression of tumor immunity. Chapter 12 describes adrenoceptors and SCD1 in adipocytes/adipose tissues: their expression and variation in health and obesity. Chapter 13 explores low-density lipoprotein receptor endocytosis in cardiovascular disease and several factors affecting LDL levels. Chapter 14 discusses endocytosis and signaling of LXRs in the liver. Finally, Chapter 15 examines applications of Frizzled receptors and SFRP5 in lipid metabolism.

This volume thus offers a wide range of topics, with a major focus on various important receptors having interest for biomedical applications, especially receptor endocytosis and signaling. We believe that this volume will enjoy a wide readership from research scientists, clinicians, students, and researchers, and that it will be an excellent contribution to the growth and widening of scientific knowledge in the field of receptor biology in health and disease.

INDRA MANI
VIJAI SINGH

CHAPTER ONE

An overview of receptor endocytosis and signaling

Indra Mani[a,*] and Vijai Singh[b]

[a]Department of Microbiology, Gargi College, University of Delhi, New Delhi, India
[b]Department of Biosciences, School of Science, Indrashil University, Rajpur, Mehsana, Gujarat, India
*Corresponding author: e-mail addresses: indramanibhu@gmail.com; indra.mani@gargi.du.ac.in

Contents

Abstract

Endocytosis is a cellular process which mediates receptor internalization, nutrient uptake, and the regulation of cell signaling. Microorganisms (many bacteria and viruses) and toxins also use the same process and enter the cells. Generally, endocytosis is considered in the three forms such as phagocytosis (cell eating), pinocytosis (cell drinking), and highly selective receptor-mediated endocytosis (clathrin-dependent and independent). Several endocytic routes exist in an analogous, achieving diverse functions. Most studies on endocytosis have used transformed cells in culture. To visualize the receptor internalization, trafficking, and signaling in subcellular organelles, a green fluorescent protein-tagged receptor has been utilized. It also helps to visualize the endocytosis effects in live-cell imaging. Confocal laser microscopy increases our understanding of receptor endocytosis and signaling. Site-directed mutagenesis studies demonstrated that many short-sequence motifs of the cytoplasmic domain of receptors significantly play a vital role in receptor internalization, subcellular trafficking, and signaling. However, other factors also regulate receptor internalization through clathrin-coated vesicles. Receptor endocytosis can occur through clathrin-dependent and

Progress in Molecular Biology and Translational Science, Volume 194
ISSN 1877-1173
https://doi.org/10.1016/bs.pmbts.2022.06.018

clathrin-independent pathways. This chapter briefly discusses the internalization, trafficking, and signaling of various receptors in normal conditions. In addition, it also highlights the malfunction of the receptor in disease conditions.

Abbreviations

Ang II/ROS	angiotensin II/reactive oxygen species
APs	adaptor proteins
Arf	ADP-ribosylation factor
CALM	clathrin assembly lymphoid myeloid leukaemia
CIE	clathrin-independent endocytosis
CPZ	chlorpromazine
Dab2	disabled-2
GABA	gamma-aminobutyric acid receptors
GPCRs	G-protein-coupled receptors
LAMP 1 and LAMP 2	lysosome-associated membrane proteins 1 and 2
LDLR	low-density lipoprotein receptor
MDC	monodansylcadaverine
NF-kB	Nuclear factor kappa-light-chain-enhancer of activated B cells
PDGF	Platelet-derived growth factor receptor
PKC/ERK	protein kinase C/extracellular signal-related kinase
Rab	Ras-associated binding
RTKs	receptor tyrosine kinases
TNF	tumor necrosis factor

1. Introduction

Endocytosis is a cellular process that is generally responsible for membrane receptor internalization, nutrient uptake, and the regulation of cell signaling. Microorganisms (many bacteria and viruses) and toxins also internalize and enter the cells.[1] Receptor endocytosis occurs in various stages of cells, such as in proliferating, quiescent, and terminally differentiated cells. In addition, endocytosis varies broadly reliant on cell types and functions.[1] Receptor-mediated endocytosis has been involved in various functions such as delivery of iron into the cells by transferrin receptor,[2] transport of hormones across cells through Insulin receptor,[3] clearance of the growth factor by Platelet-derived growth factor (PDGF) receptor,[4] transport of epidermal growth factor (EGF) across cells by EGF receptor (EGFR),[5] delivery of tumor necrosis factor (TNF) by tumor necrosis factor receptor,[6] delivery of interleukin-1 and 2 by interleukin receptor 1 and 2,[7,8] delivery of fibronectin by fibronectin receptor,[9] transports newly synthesized lysosomal enzymes to primary lysosomes by a mannose-6-phosphate receptor,[10] delivery of cholesterol by low-density lipoprotein receptor (LDLR),[11] etc.

Internalization of the membrane receptor occurs when ligands bind with the receptor and other molecules like clathrin, adapter, etc., are recruited. Then the plasma membrane initiates the invagination and forms vesicles, which are cut by dynamin molecules. The vesicles carry ligand-receptor complexes and traffic through different subcellular organelles such as early endosome, late endosome, lysosome, etc. Some receptors skip the trafficking into lysosomes and recycle back on plasma membranes with the help of recycling endosomes. These recycled receptors are involved in re-internalization.[12–17] Due to this, it maintains cell homeostasis.

Endosomal trafficking is regulated by various proteins, which belong to the small GTPase superfamily like members of the Ras-associated binding (Rab) and ADP-ribosylation factor (Arf) families.[17] Further, the endocytic system is linked with exosome vesicles, and various subcellular organelles such as the endoplasmic reticulum (ER) and Golgi complex are responsible for the processing and export of proteins. Subcellular organelles include ER, vesicles from the trans-Golgi network, and some endosomes are localized around the nucleus to form perinuclear clouds, but the mechanism is unclear.[18] Some receptors are also trafficked through mitochondria and nucleus, but detailed mechanisms are not well understood.[19,20] Apart from these subcellular organelles, EGFR traffics from the cell membrane to the nucleus, endoplasmic reticulum, Gargi apparatus, and mitochondria.[21] It seems to be a receptor involved in these subcellular organelles for diverse biological and physiological roles. This chapter highlights an overview of receptor internalization, subcellular trafficking, metabolic degradation, and signaling.

2. Role of clathrin in endocytosis

Clathrin coated vesicles are formed through the interactions of adaptor proteins, CALM (clathrin assembly lymphoid myeloid leukemia), Epsin, dynamin, the epidermal growth factor receptor pathway substrate 15 (EPS15), endophilin, and other cytosolic proteins.[22] However, other factors also regulate receptor internalization through clathrin-coated vesicles, such as receptor autophosphorylation, inhibitory signal sequence intrinsic to the receptor, and threonine and serine phosphorylation.[23–25] Often serotonin 5-HT(2B) receptors co-expressed with 5-HT(1B) and cross talk with each other. When 5-HT(1B) expressed alone, then internalization of this occurs through clathrin-independent and caveolin1-dependent. On the other hand, 5-HT(2B) was clathrin-dependent and caveolin1-independent.[26] Two types of endocytic pathways occur in eukaryotic cells such as clathrin-dependent and clathrin-independent pathway.

2.1 Clathrin-dependent endocytosis

In the endocytosis pathways, different populations of coated vesicles involve, such as clathrin-coated vesicles bud from the trans-Golgi network and the plasma membrane.[27] The size of clathrin-coated vesicles ranges from 80 to 120 nm in diameter and vesicles containing ligand-receptor complex traffic through different subcellular organelles.[28,29] Clathrin dependent pathway has been well characterized in the various receptor such as LDL,[30] complement receptor 1 (CR1),[31] insulin receptor,[23] beta(2)-adrenergic receptor (beta(2)AR),[32] G-protein-coupled receptors (GPCRs),[33] vasopressin V1B receptor,[34] angiotensin domain type 1 receptor-associated proteins (APJ) receptor,[35] muscarinic m2 Receptor (m2R),[36] EGFR,[37] lymphotoxin β receptor (LTβR),[38] insulin-like growth factor 1 (IGF-1) receptor,[39] alpha-amino-3-hydroxy-5-methyl-4-isoxazolepropionic acid (AMPA) receptor,[40] etc.

2.1.1 Role of dynamin in internalization of receptor

For the receptor-mediated internalization, dynamin (100 kDa GTPase) is an important component of vesicle formation. It plays a vital role in membrane fission,[41] an important regulator of the clathrin pathway and chemical inhibitors of dynamin are usually used to evaluate the role of the clathrin route in receptor signaling.[42] For the generation of clathrin-coated vesicles in the clathrin-mediated internalization, dynamin forms a ring structure after the recruitment of coated pits and then pinches off coated vesicles into the cytoplasm and the phenomenon known as pinchase by its intrinsic GTPase activity.[43–46] Chlorpromazine (CPZ) and monodansylcadaverine (MDC) have been utilized as dynamin GTPase blockers to inhibit the function of the dynamin. Somanna et al. has shown that after treatment with CPZ and MDC significantly blocked the clathrin-mediated ANP-NPRA (Atrial Natriuretic peptide-Nattriuretic peptide receptor-A) internalization.[15] Further, an expression of a dominant-negative mutant of dynamin (K44A/dynamin) has been utilized in many studies, which inhibit the creation of coated vesicles during the internalization of ligand-receptor complexes.[15,47–49] A study has shown that the after coexpression of a dominant negative form of dynamin (DN-dynamin, dynamin K44A) significantly diminished the agonist-dependent internalization of Muscarinic M4 receptor, but not of M2 receptors.[50] In addition, it has shown that interaction of dynamin with AP2 plays a critical role in the clathrin-mediated internalization.[51]

2.2 Clathrin-independent endocytosis

A study has demonstrated the clathrin-independent endocytosis (CIE) pathway for EGFR employed at low ligand concentration, and the CIE pathway operated down-regulate signaling at higher concentration of ligand.[52] In addition, its significant role has been shown in the Wingless/Wnt signaling in developing Drosophila wing discs,[53] the chemerin receptor CMKLR1,[54] GluK2 receptor,[55] etc. A schematic representation of clathrin-dependent and clathrin-independent endocytosis is given in Fig. 1.[56] Our knowledge

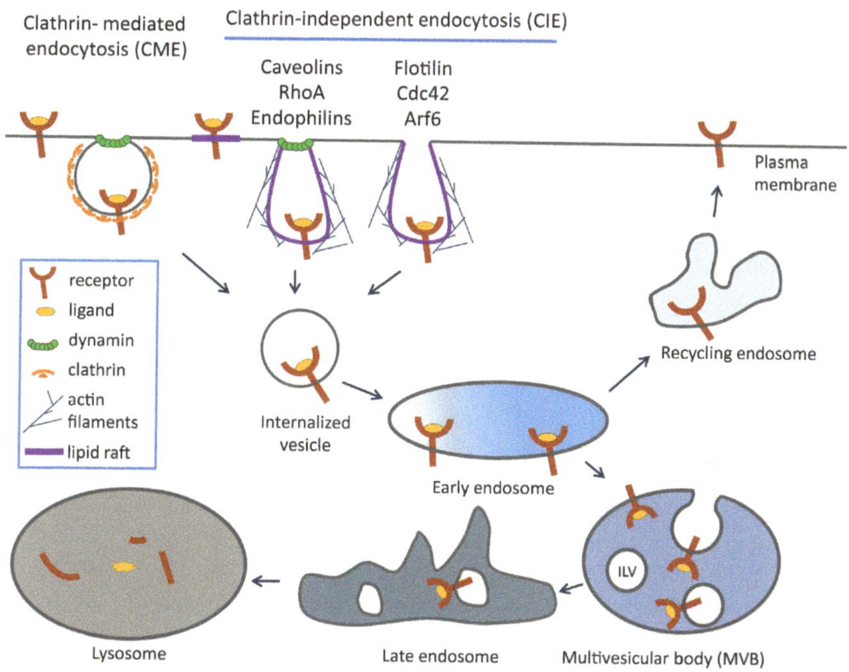

Fig. 1 Schematic representation of receptor-mediated internalization and subcellular trafficking. Receptors can be internalized either clathrin-mediated endocytosis (CME) or clathrin-independent endocytosis (CIE). CIE often occurs from the plasma membrane microdomains (lipid rafts). Different regulatory molecules are involved along with dynamin during pinchase. After internalization, receptor traffic to different subcellular organelles such as early endosome, recycling endosome, multivesicular body (MVB), late endosome, and lysosome. Receptor sorted in intraluminal vesicles (ILV) and changes in color from blue to grey indicates increased acidification of endosomal lumen. *Adapted with permission from J. Cendrowski, A. Mamińska, M. Miaczynska, Endocytic regulation of cytokine receptor signaling, Cytokine Growth Factor Rev 32 (2016) 63–73.*

about clathrin-independent pathways is still far from detailed understanding. However, a comprehensive study needs to know the kinetic pathway of clathrin-independent receptor internalization.

3. Role of adaptor protein in internalization of the receptor

In endocytosis, various adaptor proteins (APs) are involved. Aromatic amino acids, such as tyrosine-based Yxxphi sorting signals present in the carboxyl-terminus of the receptor, which interact with mu 1, mu 2, mu 3, and mu 4 subunits of APs (AP-1, AP-2, AP-3, and AP-4) and plays a vital role in the internalization and trafficking.[57–60] During internalization, a different subunits of APs interacted with different receptor, such as μ2-adaptin in Frizzled receptor internalization,[61] adaptor protein Disabled-2 (Dab2) in transforming growth factor-beta (TGF-β) receptor endocytosis,[62] roles of the adaptin AP2 complex in constitutive endocytosis of the gamma-aminobutyric acid (GABA) receptors,[63,64] beta-arrestin acts as a clathrin adaptor in internalization of the beta2-adrenergic receptor and G-protein-coupled receptors (GPCRs),[33,65] and AP complex-1 and 2 involved in trafficking of transferrin receptor 1.[66] Various studies have demonstrated that the endocytic adaptor proteins play a vital role in health and disease conditions.

4. Subcellular trafficking

Subcellular trafficking is one of the important steps after coated vesicle formation. Subcellular trafficking has been extensively studied in many receptors. Chemokine receptor internalization is routed in two different ways, such as clathrin-mediated and lipid rafts/caveolae-dependent. Further chemokine receptors modified in different ways such as glycosylation, ubiquitination, sulfation, and palmitoylation, may also affect subcellular trafficking, signaling, and chemotaxis.[67] Transferrin receptor-mediated internalization and recycling of transferrin (Tf) has been shown in rat reticulocytes. Through the receptor-mediated endocytosis, Tf protein binds to circulating iron and delivers to cells.[2,68] A recent study has shown that iron-induced the unconventional internalization and trafficking of TfR1 in cultured HepG2 hepatoma cells exposed to ferric ammonium citrate (FAC). Knockdown experiments have shown that

TfR1 lysosomal degradation is independent of Tf, clathrin adaptor AP2, H-ferritin, and MARCH8.[69] This study suggested that iron is a key regulator of the trafficking of TfR1. Similarly, unconventional mechanisms of endocytosis exist in parallel to CME to internalize specific cargoes and support several cellular functions.[70]

Many lysosomal membrane proteins (LMPs) are present in mammalian lysosomes. It is involved in numerous functions like acidification of the lysosomal lumen, control of membrane fusion and fission events, and import of sorting and recycling of degraded products. Some of the LMPs, such as lysosome-associated membrane proteins 1 and 2 (LAMP1 and LAMP2) are the most abundant and extensively studied.[71] These protein markers (LAMP1 and LAMP2) are being utilized to visualize the receptor localization into lysosomal compartments. Various receptors are internalized and degraded into lysosomes such as fibroblast growth factor receptor,[72] Interleukin-1 receptors, [73] TGF beta receptor,[74] etc. Some receptors skip the lysosomal compartment and traffic with other subcellular organelles such as Mannose 6-phosphate receptors because they contain TGN-sorting signals.[75]

Earlier to understand subcellular trafficking, various methods were used, such as a radioisotope, coimmunoprecipitation, and immunofluorescence. Further, the green fluorescent protein (GFP) tagged receptor has been extensively used to demonstrate the several receptor internalizations and subcellular trafficking processes. GFP-tagged receptor used to visualize the various receptor endocytosis, such as gamma-aminobutyric acid subtype A (GABAA) receptor,[76] LDL receptor,[77] endothelin B receptor (ETB),[78] h5-HT(4a) and h5-HT(4b) receptors,[79] galanin R2 receptor,[80] beta2-adrenergic receptor (beta2AR),[81] neuropeptide Y (NPY) receptors,[82] transferrin receptor,[83] NPRA,[14] EGF receptor,[84] dopamine D2 autoreceptors,[85] NPRA,[15] VEGFR2 (vascular endothelial growth factor receptor 2),[86] etc.

5. Receptor trafficking in non-traditional subcellular organelles

Receptor internalization and subcellular trafficking from other than early endosome, late endosome, recycling endosome, and lysosome, it also trafficked through the nucleus, mitochondria, endoplasmic reticulum, and Golgi complex. Apart from the endo-lysosomal trafficking, EGFR also traffics from the cell surface to various subcellular organelles such as in the nucleus, mitochondria, endoplasmic reticulum, and Golgi apparatus after endocytosis, where it appears to be involved in several biological and physiological activities.[21] Different receptor tyrosine kinases (RTKs) trafficking has been reported in the nucleus and mitochondria.[20] Subcellular trafficking of ErbB-1,[87] TGF-β,[88] IL-15R2,[89] Met,[90] growth hormone (GH) receptor,[91] TrKA,[92] etc., has been shown in the nucleus while ErbB-1 in mitochondria.[93] Similarly, a GH receptor is also trafficked in mitochondria.[94] Recently, it has been shown that MasR trafficked in the nucleus after treatment with Ang-(1-7).[95]

Receptor trafficking in non-traditional organelles suggests that the receptor itself is a signaling element between the cell surface and the nucleus. There is less understanding of the receptor trafficking in the nucleus. A further comprehensive study is required to mechanistically understand the receptor translocation in the nucleus and mitochondria. A study has demonstrated that AT1R and AT2R are endocytosed as heterodimers in the endoplasmic reticulum in cultured LLC-PK_1 cells.[96] Further investigations are needed to understand the varied pathways of the internalization and trafficking of different ligand-receptor complexes in various subcellular organelles in both health and disease conditions.

6. Role of short-sequence motif in the internalization and trafficking of the receptors

Various short-sequence signal motifs in receptor play an important role in the receptor internalization, trafficking, and signaling. It has been extensively studied in the different receptors. The utmost common small-signal sequence motifs are found in several membrane receptors such as YATL in LDL-related receptors,[97] YSKV in CI-Mannose-6-phosphate receptor,[98] FDNPVY and NPXY in LDL receptors, [99,100] YKKL in Protease-activated receptor-1,[101] NPxY in integrin receptor,[102] YQPL in T-cell receptor,[98] YWL in glutamate receptor,[103] GTALL in luteinizing hormone (LH) receptor,[104] YTRF/Q in transferrin receptor,[98] YxxPhi in Insulin-like growth factor receptor,[98] YEQGL in P2X receptor,[105] [YF]xNPx[YF] in platelet-activating factor receptor,[106] GDAY and FQQI in natriuretic peptide receptor-A (NPRA), [13,59,107] etc. Delineate the role of a short sequence in the receptor can provide significant evidence to understand the malfunction of receptor endocytosis and signaling.

7. Signaling

Receptor endocytosis and signaling play a vital role in the duration, strength, degradation, diversification, retroendocytosis, and spatial restriction of any receptors.[108,109] Remarkably, several analyses suggest that internalization and subcellular trafficking can induce prolonged receptor signaling. It is needed for some biological effects of the hormones, such as angiotensin II, TSH, LH, and adrenalin.[110,111] Multiple biological and physiological responses are stimulated by the various receptor signaling. EGFR elicits responses such as cell survival, proliferation, and migration. Abnormal receptor internalization and trafficking affect signaling, consequently, disease conditions occur. There are various receptors and their unusual signaling has contributed in diverse disease conditions, such as heart failure,[112–114] kidney failure, [115,116] neuropathy, [117,118] neuropsychiatric disorders,[119] lung disease, [120,121] cancers, [109,122,123] skin disease, [124,125] etc. More details about the disease occurrence due to abnormalities in receptor internalization, subcellular trafficking, and signaling are given in Table 1. In chronic kidney disease (CKD), several signaling pathways are involved in hypoxia-induced renal fibrosis, including pathways mediated by Notch, angiotensin II/reactive oxygen species (Ang II/ROS), Nuclear factor kappa-light-chain-enhancer of activated B cells (NF-kB), protein kinase C/extracellular signal-related kinase (PKC/ERK),[140] etc. Further,

Table 1 List of disease occurrence due to abnormalities in receptor endocytosis and signaling

S. no	Receptor	Abnormalities	Disease	Reference
1	EGFR	Signaling	Cancer	126
2	LDLR	Trafficking	Familial hyperlipidemia	11
3	Notch 1	Signaling	Pulmonary arterial hypertension (PAH)	127
4	Adrenergic receptors (ARs)	Signaling	Cancer, heart failure, obesity, etc.	128
5	Neurotrophin receptor	Signaling	Huntington's disease	129
6	Ang II type 1 (AT1) receptor	Signaling	Aging related diseases, cardiovascular diseases and Albuminuria	130,131

Continued

Table 1 List of disease occurrence due to abnormalities in receptor endocytosis and signaling—cont'd

S. no	Receptor	Abnormalities	Disease	Reference
7	γ-Aminobutyric acid type-A receptor ($GABA_AR$)	Signaling	Depression and epilepsy	132
8	Oxytocin receptor (OTR)	Signaling	Psychosocial diseases	133
9	*N*-methyl-D-aspartate (NMDA) receptor	Signaling	Neurodegenerative diseases	134
10	G-protein-coupled receptor	Signaling	Heart failure	114
11	Insulin/IGF-1 receptor	Signaling	Kidney failure	115
12	TrkA receptor	Signaling	Neuropathy	118
13	NMDA receptor	Trafficking	Neuropsychiatric disorders	119
14	P2 receptor	Signaling	Lung disease	120
15	CB1 and CB2 receptors	Signaling	skin disease	124
16	Glycine receptors	Trafficking	Neurological disease	135
17	AMPA receptor	Internalization	Depression of synapse strength, and forgetting.	136
18	Mutant P86S glucagon receptor (GCGR)	Internalization	Mahvash disease	137
19	Gonadotropin receptor	Trafficking	Reproductive failure and/or abnormal or absence of development of secondary sex characteristics	138
20	Integrin receptor	Trafficking	Cancer	139

serotonin signaling is responsible for various types of cancer, such as ovary cancer, melanoma, lung cancer, urinary bladder cancer, placenta cancer, and glioblastoma.[141] Indeed, an error in receptor endocytosis and signaling can cause diverse diseases.

8. Conclusions and future perspectives

Receptor internalization, subcellular trafficking, and signaling play a vital role in normal physiology. Any malfunction that occurs in these processes, is consequently responsible for disease conditions. Confocal laser microscopy was extensively used to visualize these processes in cultured fixed cells and live cells. It also helps to enhance our understanding of endocytosis and signaling pathways. Receptor-mediated endocytosis occurs through clathrin-dependent and independent ways. Traditionality, receptor trafficking occurs through the endo-lysosomal organelles, and some receptors skip the lysosome and recycle back to the plasma membrane via recycling endosome. How is the recycling receptor separated from the down-regulated receptor not well understood? However, in non-traditional ways receptor traffics/translocate apart from endo-lysosomal compartments, such as in the nucleus, mitochondria, endoplasmic reticulum, and Golgi complex. Various studies have been performed on different receptors to understand receptor endocytosis and signaling processes. Some receptors that concurrently generate signaling molecules during ligand-induced receptor internalization and trafficking are not well established. Further use of the emerging genomic editing tools like the clustered regularly interspaced short palindromic repeats (CRISPR)-CRISPR associated (Cas) (CRISPR-CAS) system may provide a deeper understanding of the role of endocytosis and signaling in physiology and pathophysiological conditions.

Conflict of interest

None.

Reference

1. Hinze C, Boucrot E. Endocytosis in proliferating, quiescent and terminally differentiated cells. *J Cell Sci.* 2018;131(23):jcs216804.
2. Harding C, Heuser J, Stahl P. Receptor-mediated endocytosis of transferrin and recycling of the transferrin receptor in rat reticulocytes. *J Cell Biol.* 1983;97(2):329–339.
3. King GL, Johnson SM. Receptor-mediated transport of insulin across endothelial cells. *Science.* 1985;227(4694):1583–1586.
4. Keating MT, Williams LT. Processing of the platelet-derived growth factor receptor. Biosynthetic and degradation studies using anti-receptor antibodies. *J Biol Chem.* 1987;262(16):7932–7937.
5. Maratos-Flier E, Kao CY, Verdin EM, King GL. Receptor-mediated vectorial transcytosis of epidermal growth factor by Madin-Darby canine kidney cells. *J Cell Biol.* 1987;105(4):1595–1601.

6. Imamura K, Spriggs D, Kufe D. Expression of tumor necrosis factor receptors on human monocytes and internalization of receptor bound ligand. *J Immunol.* 1987;139(9):2989–2992.
7. Grenfell S, Smithers N, Miller K, Solari R. Receptor-mediated endocytosis and nuclear transport of human interleukin 1 alpha. *Biochem J.* 1989;264(3):813–822.
8. Smith KA, Cantrell DA. Interleukin 2 regulates its own receptors. *Proc Natl Acad Sci USA.* 1985;82(3):864–868.
9. Bretscher MS. Endocytosis and recycling of the fibronectin receptor in CHO cells. *EMBO J.* 1989;8(5):1341–1348. Erratum in: EMBO J. 8(7):2122.
10. Lobel P, Fujimoto K, Ye RD, Griffiths G, Kornfeld S. Mutations in the cytoplasmic domain of the 275 kd mannose 6-phosphate receptor differentially alter lysosomal enzyme sorting and endocytosis. *Cell.* 1989;57(5):787–796.
11. Go GW, Mani A. Low-density lipoprotein receptor (LDLR) family orchestrates cholesterol homeostasis. *Yale J Biol Med.* 2012;85(1):19–28.
12. Signoret N, Pelchen-Matthews A, Mack M, Proudfoot AE, Marsh M. Endocytosis and recycling of the HIV coreceptor CCR5. *J Cell Biol.* 2000;151(6):1281–1294.
13. Pandey KN, Nguyen HT, Garg R, Khurana ML, Fink J. Internalization and trafficking of guanylyl (guanylate) cyclase/natriuretic peptide receptor A is regulated by an acidic tyrosine-based cytoplasmic motif GDAY. *Biochem J.* 2005;388(pt 1):103–113.
14. Mani I, Garg R, Tripathi S, Pandey KN. Subcellular trafficking of guanylyl cyclase/natriuretic peptide receptor-A with concurrent generation of intracellular cGMP. *Biosci Rep.* 2015;35(5):e00260.
15. Somanna NK, Mani I, Tripathi S, Pandey KN. Clathrin-dependent internalization, signaling, and metabolic processing of guanylyl cyclase/natriuretic peptide receptor-A. *Mol Cell Biochem.* 2018;441(1-2):135–150.
16. Hinaux H, Bachem K, Battistara M, et al. Revisiting the developmental and cellular role of the pigmentation gene yellow in Drosophila using a tagged allele. *Dev Biol.* 2018;438(2):111–123.
17. Iraburu MJ, Garner T, Montiel-Duarte C. Revising endosomal trafficking under insulin receptor activation. *Int J Mol Sci.* 2021;22(13):6978.
18. Naslavsky N, Caplan S. The enigmatic endosome—sorting the ins and outs of endocytic trafficking. *J Cell Sci.* 2018;131(13):jcs216499.
19. Grewal T, Enrich C. Annexins—modulators of EGF receptor signalling and trafficking. *Cell Signal.* 2009;21(6):847–858.
20. Carpenter G, Liao HJ. Trafficking of receptor tyrosine kinases to the nucleus. *Exp Cell Res.* 2009;315(9):1556–1566.
21. Wang YN, Hung MC. Nuclear functions and subcellular trafficking mechanisms of the epidermal growth factor receptor family. *Cell Biosci.* 2012;2(1):13.
22. Mousavi SA, Malerød L, Berg T, Kjeken R. Clathrin-dependent endocytosis. *Biochem J.* 2004;377(Pt 1):1–16.
23. Carpentier JL, Paccaud JP, Gorden P, Rutter WJ, Orci L. Insulin-induced surface redistribution regulates internalization of the insulin receptor and requires its autophosphorylation. *Proc Natl Acad Sci USA.* 1992;89(1):162–166.
24. Miettinen HM, Matter K, Hunziker W, Rose JK, Mellman I. Fc receptor endocytosis is controlled by a cytoplasmic domain determinant that actively prevents coated pit localization. *J Cell Biol.* 1992;116(4):875–888.
25. Carpentier JL, Lew DP, Paccaud JP, et al. Internalization pathway of C3b receptors in human neutrophils and its transmodulation by chemoattractant receptors stimulation. *Cell Regul.* 1991;2(1):41–55.
26. Janoshazi A, Deraet M, Callebert J, et al. Modified receptor internalization upon coexpression of 5-HT1B receptor and 5-HT2B receptors. *Mol Pharmacol.* 2007;71(6):1463–1474.

27. Robinson MS. The role of clathrin, adaptors and dynamin in endocytosis. *Curr Opin Cell Biol.* 1994;6(4):538–544.
28. Conner SD, Schmid SL. Regulated portals of entry into the cell. *Nature.* 2003;422 (6927):37–44.
29. Gong Q, Huntsman C, Ma D. Clathrin-independent internalization and recycling. *J Cell Mol Med.* 2008;12(1):126–144.
30. Goldstein JL, Brown MS, Anderson RG, Russell DW, Schneider WJ. Receptor-mediated endocytosis: concepts emerging from the LDL receptor system. *Annu Rev Cell Biol.* 1985;1:1–39.
31. Paccaud JP, Reith W, Johansson B, Magnusson KE, Mach B, Carpentier JL. Clathrin-coated pit-mediated receptor internalization. Role of internalization signals and receptor mobility. *J Biol Chem.* 1993;268(31):23191–23196.
32. Liang W, Curran PK, Hoang Q, Moreland RT, Fishman PH. Differences in endosomal targeting of human (beta)1- and (beta)2-adrenergic receptors following clathrin-mediated endocytosis. *J Cell Sci.* 2004;117(Pt 5):723–734.
33. Fessart D, Simaan M, Zimmerman B, et al. Src-dependent phosphorylation of beta2-adaptin dissociates the beta-arrestin-AP-2 complex. *J Cell Sci.* 2007;120 (pt 10):1723–1732.
34. Kashiwazaki A, Fujiwara Y, Tsuchiya H, Sakai N, Shibata K, Koshimizu TA. Subcellular localization and internalization of the vasopressin V1B receptor. *Eur J Pharmacol.* 2015;765:291–299.
35. He L, Chen L, Li L. The mechanosensitive APJ internalization via clathrin-mediated endocytosis: a new molecular mechanism of cardiac hypertrophy. *Med Hypotheses.* 2016;90:6–10.
36. Lambert L, Dubayle D, Fafouri A, et al. Endocytosis of activated muscarinic m2 receptor (m2R) in live mouse hippocampal neurons occurs via a clathrin-dependent pathway. *Front Cell Neurosci.* 2018;12:450.
37. Caballero-Díaz D, Bertran E, Peñuelas-Haro I, et al. Clathrin switches transforming growth factor-β role to pro-tumorigenic in liver cancer. *J Hepatol.* 2020; 72(1):125–134.
38. Maksymowicz M, Miączyńska M, Banach-Orłowska M. Clathrin- and dynamin-dependent endocytosis limits canonical NF-κB signaling triggered by lymphotoxin β receptor. *Cell Commun Signal.* 2020;18(1):176.
39. Dam DHM, Jelsma SA, Yu JM, Liu H, Kong B, Paller AS. Flotillin and AP2A1/2 promote IGF-1 receptor association with clathrin and internalization in primary human keratinocytes. *J Invest Dermatol.* 2020;140(9):1743–1752.e4.
40. Sathler MF, Khatri L, Roberts JP, et al. Phosphorylation of the AMPA receptor subunit GluA1 regulates clathrin-mediated receptor internalization. *J Cell Sci.* 2021;134(17):jcs257972.
41. Hinshaw JE. Dynamin and its role in membrane fission. *Annu Rev Cell Dev Biol.* 2000;16:483–519.
42. Basagiannis D, Zografou S, Goula E, Gkeka D, Kolettas E, Christoforidis S. Chemical inhibitors of dynamin exert differential effects in VEGF signaling. *Cells.* 2021;10(5):997.
43. Warnock DE, Schmid SL. Dynamin GTPase, a force-generating molecular switch. *Bioessays.* 1996;18(11):885–893.
44. Marks B, Stowell MH, Vallis Y, et al. GTPase activity of dynamin and resulting conformation change are essential for endocytosis. *Nature.* 2001;410(6825):231–235.
45. Rappoport JZ, Benmerah A, Simon SM. Analysis of the AP-2 adaptor complex and cargo during clathrin-mediated endocytosis. *Traffic.* 2005;6(7):539–547.
46. McMahon HT, Boucrot E. Molecular mechanism and physiological functions of clathrin-mediated endocytosis. *Nat Rev Mol Cell Biol.* 2011;12(8):517–533.

47. Zhang J, Ferguson SS, Barak LS, Ménard L, Caron MG. Dynamin and beta-arrestin reveal distinct mechanisms for G protein-coupled receptor internalization. *J Biol Chem*. 1996;271(31):18302–18305.
48. Ceresa BP, Kao AW, Santeler SR, Pessin JE. Inhibition of clathrin-mediated endocytosis selectively attenuates specific insulin receptor signal transduction pathways. *Mol Cell Biol*. 1998;18(7):3862–3870.
49. Johannessen LE, Pedersen NM, Pedersen KW, Madshus IH, Stang E. Activation of the epidermal growth factor (EGF) receptor induces formation of EGF receptor- and Grb2-containing clathrin-coated pits. *Mol Cell Biol*. 2006;26(2):389–401.
50. Vögler O, Bogatkewitsch GS, Wriske C, Krummenerl P, Jakobs KH, van Koppen CJ. Receptor subtype-specific regulation of muscarinic acetylcholine receptor sequestration by dynamin. Distinct sequestration of m2 receptors. *J Biol Chem*. 1998;273 (20):12155–12160.
51. Wang LH, Südhof TC, Anderson RG. The appendage domain of alpha-adaptin is a high affinity binding site for dynamin. *J Biol Chem*. 1995;270(17):10079–10083.
52. Sigismund S, Woelk T, Puri C, et al. Clathrin-independent endocytosis of ubiquitinated cargos. *Proc Natl Acad Sci USA*. 2005;102(8):2760–2765.
53. Hemalatha A, Prabhakara C, Mayor S. Endocytosis of Wingless via a dynamin-independent pathway is necessary for signaling in *Drosophila* wing discs. *Proc Natl Acad Sci USA*. 2016;113(45):E6993–E7002.
54. Zhou JX, Liao D, Zhang S, Cheng N, He HQ, Ye RD. Chemerin C9 peptide induces receptor internalization through a clathrin-independent pathway. *Acta Pharmacol Sin*. 2014;35(5):653–663.
55. Du JJ, Yan L, Zhang W, Xu H, Zhu QJ. Clathrin-independent but dynamin-dependent mechanisms mediate Ca2+-triggered endocytosis of the glutamate GluK2 receptor upon excitotoxicity. *J Integr Neurosci*. 2020;19(3):449–458.
56. Cendrowski J, Mamińska A, Miaczynska M. Endocytic regulation of cytokine receptor signaling. *Cytokine Growth Factor Rev*. 2016;32:63–73.
57. Hirst J, Lindsay MR, Robinson MS. GGAs: roles of the different domains and comparison with AP-1 and clathrin. *Mol Biol Cell*. 2001;12(11):3573–3588.
58. Parent JL, Labrecque P, Driss Rochdi M, Benovic JL. Role of the differentially spliced carboxyl terminus in thromboxane A2 receptor trafficking: identification of a distinct motif for tonic internalization. *J Biol Chem*. 2001;276(10):7079–7085.
59. Mani I, Garg R, Pandey KN. Role of FQQI motif in the internalization, trafficking, and signaling of guanylyl-cyclase/natriuretic peptide receptor-A in cultured murine mesangial cells. *Am J Physiol Renal Physiol*. 2016;310(1):F68–F84.
60. Yoneyama Y, Lanzerstorfer P, Niwa H, et al. IRS-1 acts as an endocytic regulator of IGF-I receptor to facilitate sustained IGF signaling. *Elife*. 2018;7:e32893.
61. Yu A, Rual JF, Tamai K, et al. Association of dishevelled with the clathrin AP-2 adaptor is required for frizzled endocytosis and planar cell polarity signaling. *Dev Cell*. 2007;12(1):129–141.
62. Penheiter SG, Singh RD, Repellin CE, et al. Type II transforming growth factor-beta receptor recycling is dependent upon the clathrin adaptor protein Dab2. *Mol Biol Cell*. 2010;21(22):4009–4019.
63. Kittler JT, Delmas P, Jovanovic JN, Brown DA, Smart TG, Moss SJ. Constitutive endocytosis of GABAA receptors by an association with the adaptin AP2 complex modulates inhibitory synaptic currents in hippocampal neurons. *J Neurosci*. 2000;20 (21):7972–7977.
64. Kanematsu T, Fujii M, Mizokami A, et al. Phospholipase C-related inactive protein is implicated in the constitutive internalization of GABAA receptors mediated by clathrin and AP2 adaptor complex. *J Neurochem*. 2007;101(4):898–905.

65. Goodman Jr OB, Krupnick JG, Santini F, et al. Beta-arrestin acts as a clathrin adaptor in endocytosis of the beta2-adrenergic receptor. *Nature.* 1996;383(6599):447–450.
66. Promchan K, Natarajan V. Leucine zipper transcription factor-like 1 binds adaptor protein complex-1 and 2 and participates in trafficking of transferrin receptor 1. *PLoS One.* 2020;15(1):e0226298.
67. Neel NF, Schutyser E, Sai J, Fan GH, Richmond A. Chemokine receptor internalization and intracellular trafficking. *Cytokine Growth Factor Rev.* 2005;16(6):637–658.
68. Brandsma ME, Jevnikar AM, Ma S. Recombinant human transferrin: beyond iron binding and transport. *Biotechnol Adv.* 2011;29(2):230–238.
69. Gammella E, Lomoriello IS, Conte A, et al. Unconventional endocytosis and trafficking of transferrin receptor induced by iron. *Mol Biol Cell.* 2021;32(2):98–108.
70. Renard HF, Boucrot E. Unconventional endocytic mechanisms. *Curr Opin Cell Biol.* 2021;71:120–129.
71. Bajaj L, Lotfi P, Pal R, Ronza AD, Sharma J, Sardiello M. Lysosome biogenesis in health and disease. *J Neurochem.* 2019;148(5):573–589.
72. Sorokin A, Mohammadi M, Huang J, Schlessinger J. Internalization of fibroblast growth factor receptor is inhibited by apoint mutation at tyrosine 766. *J Biol Chem.* 1994;269:17056–17061.
73. Solari R, Smithers N, Kennard N, Ray K, Grenfell S. Receptor mediated endocytosis and intracellular fate of IL-1. *Biochem Pharmcol.* 1994;47:93–101.
74. Massagué J, Kelly B. Internalization of transforming growth factor-beta and its receptor in BALB/c 3T3 fibroblasts. *J Cell Physiol.* 1986;128(2):216–222.
75. Díaz E, Pfeffer SR. TIP47: a cargo selection device for mannose 6-phosphate receptor trafficking. *Cell.* 1998;93(3):433–443.
76. Kittler JT, Wang J, Connolly CN, Vicini S, Smart TG, Moss SJ. Analysis of GABAA receptor assembly in mammalian cell lines and hippocampal neurons using gamma 2 subunit green fluorescent protein chimeras. *Mol Cell Neurosci.* 2000;16(4):440–452.
77. Holst HU, Dagnaes-Hansen F, Corydon TJ, et al. LDL receptor-GFP fusion proteins: new tools for the characterisation of disease-causing mutations in the LDL receptor gene. *Eur J Hum Genet.* 2001;9(11):815–822.
78. Foster N, Loi TH, Owe-Young R, Stanley KK. Lysosomal traffic of liganded endothelin B receptor. *Biochim Biophys Acta.* 2003;1642(1-2):45–52.
79. Pindon A, Van Hecke G, Josson K, et al. Internalization of human 5-HT4a and 5-HT4b receptors is splice variant dependent. *Biosci Rep.* 2004;24(3):215–223.
80. Xia S, Kjaer S, Zheng K, et al. Visualization of a functionally enhanced GFP-tagged galanin R2 receptor in PC12 cells: constitutive and ligand-induced internalization. *Proc Natl Acad Sci USA.* 2004;101(42):15207–15212.
81. Prenner L, Sieben A, Zeller K, Weiser D, Häberlein H. Reduction of high-affinity beta2-adrenergic receptor binding by hyperforin and hyperoside on rat C6 glioblastoma cells measured by fluorescence correlation spectroscopy. *Biochemistry.* 2007;46(17):5106–5113.
82. Kilpatrick LE, Briddon SJ, Holliday ND. Fluorescence correlation spectroscopy, combined with bimolecular fluorescence complementation, reveals the effects of β-arrestin complexes and endocytic targeting on the membrane mobility of neuropeptide Y receptors. *Biochim Biophys Acta.* 2012;1823(6):1068–1081.
83. He Q, Sun X, Chu C, et al. Endocytosis of a functionally enhanced GFP-tagged transferrin receptor in CHO cells. *PLoS One.* 2015;10(3):e0122452.
84. Pinilla-Macua I, Grassart A, Duvvuri U, Watkins SC, Sorkin A. EGF receptor signaling, phosphorylation, ubiquitylation and endocytosis in tumors in vivo. *eLife.* 2017;6:e31993. https://doi.org/10.7554/eLife.31993.

85. Robinson BG, Bunzow JR, Grimm JB, et al. Desensitized D2 autoreceptors are resistant to trafficking. *Sci Rep.* 2017;7(1):4379.
86. Bayliss AL, Sundararaman A, Granet C, Mellor H. Raftlin is recruited by neuropilin-1 to the activated VEGFR2 complex to control proangiogenic signaling. *Angiogenesis.* 2020;23(3):371–383.
87. Lin SY, Makino K, Xia W, et al. Nuclear localization of EGF receptor and its potential new role as a transcription factor. *Nat Cell Biol.* 2001;3(9):802–808.
88. Zwaagstra JC, Guimond A, O'Connor-McCourt MD. Predominant intracellular localization of the type I transforming growth factor-beta receptor and increased nuclear accumulation after growth arrest. *Exp Cell Res.* 2000;258(1):121–134.
89. Pereno R, Giron-Michel J, Gaggero A, et al. IL-15/IL-15Ralpha intracellular trafficking in human melanoma cells and signal transduction through the IL-15Ralpha. *Oncogene.* 2000;19(45):5153–5162.
90. Gomes DA, Rodrigues MA, Leite MF, et al. c-Met must translocate to the nucleus to initiate calcium signals. *J Biol Chem.* 2008;283(7):4344–4351.
91. Lobie PE, Wood TJ, Chen CM, Waters MJ, Norstedt G. Nuclear translocation and anchorage of the growth hormone receptor. *J Biol Chem.* 1994;269(50):31735–31746.
92. Bonacchi A, Taddei ML, Petrai I, et al. Nuclear localization of TRK-A in liver cells. *Histol Histopathol.* 2008;23(3):327–340.
93. Boerner JL, Demory ML, Silva C, Parsons SJ. Phosphorylation of Y845 on the epidermal growth factor receptor mediates binding to the mitochondrial protein cytochrome c oxidase subunit II. *Mol Cell Biol.* 2004;24(16):7059–7071.
94. Ardail D, Debon A, Perret-Vivancos C, et al. Growth hormone internalization in mitochondria decreases respiratory chain activity. *Neuroendocrinology.* 2010;91(1):16–26.
95. Cerniello FM, Silva MG, Carretero OA, Gironacci MM. Mas receptor is translocated to the nucleus upon agonist stimulation in brainstem neurons from spontaneously hypertensive rats but not normotensive rats. *Cardiovasc Res.* 2020;116(12):1995–2008.
96. Ferrão FM, Cardoso LHD, Drummond HA, et al. Luminal ANG II is internalized as a complex with AT_1R/AT_2R heterodimers to target endoplasmic reticulum in LLC-PK_1 cells. *Am J Physiol Renal Physiol.* 2017;313(2):F440–F449.
97. Thies RS, Webster NJ, McClain DA. A domain of the insulin receptor required for endocytosis in rat fibroblasts. *J Biol Chem.* 1990;265(17):10132–10137.
98. Ktistakis NT, Thomas D, Roth MG. Characteristics of the tyrosine recognition signal for internalization of transmembrane surface glycoproteins. *J Cell Biol.* 1990;111(4):1393–1407.
99. Davis CG, Lehrman MA, Russell DW, Anderson RG, Brown MS, Goldstein JL. The J.D. mutation in familial hypercholesterolemia: amino acid substitution in cytoplasmic domain impedes internalization of LDL receptors. *Cell.* 1986;45(1):15–24.
100. Bansal A, Gierasch LM. The NPXY internalization signal of the LDL receptor adopts a reverse-turn conformation. *Cell.* 1991;67(6):1195–1201.
101. Traub LM. Sorting it out: AP-2 and alternate clathrin adaptors in endocytic cargo selection. *J Cell Biol.* 2003;163(2):203–208.
102. Bogdanović O, Delfino-Machín M, Nicolás-Pérez M, et al. Numb/Numbl-Opo antagonism controls retinal epithelium morphogenesis by regulating integrin endocytosis. *Dev Cell.* 2012;23(4):782–795.
103. Colgan L, Liu H, Huang SY, Liu YJ. Dileucine motif is sufficient for internalization and synaptic vesicle targeting of vesicular acetylcholine transporter. *Traffic.* 2007; 8(5):512–522.
104. Kishi M, Liu X, Hirakawa T, Reczek D, Bretscher A, Ascoli M. Identification of two distinct structural motifs that, when added to the C-terminal tail of the rat LH receptor, redirect the internalized hormone-receptor complex from a degradation to a recycling pathway. *Mol Endocrinol.* 2001;15(9):1624–1635.

105. Levitan ES, Takimoto K. Surface expression of Kv1 voltage-gated K+ channels is governed by a C-terminal motif. *Trends Cardiovasc Med*. 2000;10(7):317–320.
106. Mulkearns EE, Cooper JA. FCH domain only-2 organizes clathrin-coated structures and interacts with disabled-2 for low-density lipoprotein receptor endocytosis. *Mol Biol Cell*. 2012;23(7):1330–1342.
107. Mani I, Pandey KN. Emerging concepts of receptor endocytosis and concurrent intracellular signaling: mechanisms of guanylyl cyclase/natriuretic peptide receptor-A activation and trafficking. *Cell Signal*. 2019;60:17–30.
108. Levy JR, Olefsky JM. Retroendocytosis of insulin in rat adipocytes. *Endocrinology*. 1986;119(2):572–579.
109. Caldieri G, Malabarba MG, Di Fiore PP, Sigismund S. EGFR trafficking in physiology and cancer. *Prog Mol Subcell Biol*. 2018;57:235–272.
110. Zhuo JL, Imig JD, Hammond TG, Orengo S, Benes E, Navar LG. Ang II accumulation in rat renal endosomes during Ang II-induced hypertension: role of AT(1) receptor. *Hypertension*. 2002;39(1):116–121.
111. Calebiro D, Godbole A. Internalization of G-protein-coupled receptors: Implication in receptor function, physiology and diseases. *Best Pract Res Clin Endocrinol Metab*. 2018;32(2):83–91.
112. Lamba S, Abraham WT. Alterations in adrenergic receptor signaling in heart failure. *Heart Fail Rev*. 2000;5(1):7–16.
113. Wasilewski MA, Myers VD, Recchia FA, Feldman AM, Tilley DG. Arginine vasopressin receptor signaling and functional outcomes in heart failure. *Cell Signal*. 2016;28(3):224–233.
114. Altamish M, Samuel VP, Dahiya R, et al. Molecular signaling of G-protein-coupled receptor in chronic heart failure and associated complications. *Drug Dev Res*. 2020;81(1):23–31.
115. Ising C, Koehler S, Brähler S, et al. Inhibition of insulin/IGF-1 receptor signaling protects from mitochondria-mediated kidney failure. *EMBO Mol Med*. 2015;7(3):275–287.
116. Kramer J, Schwanbeck R, Pagel H, Cakiroglu F, Rohwedel J, Just U. Inhibition of notch signaling ameliorates acute kidney failure and downregulates platelet-derived growth factor receptor β in the mouse model. *Cells Tissues Organs*. 2016;201(2):109–117.
117. Lee SM, Chin LS, Li L. Dysregulation of ErbB receptor trafficking and signaling in demyelinating charcot-marie-tooth disease. *Mol Neurobiol*. 2017;54(1):87–100.
118. Li L, Gruner K, Tourtellotte WG. Retrograde nerve growth factor signaling abnormalities in familial dysautonomia. *J Clin Invest*. 2020;130(5):2478–2487.
119. Lau CG, Zukin RS. NMDA receptor trafficking in synaptic plasticity and neuropsychiatric disorders. *Nat Rev Neurosci*. 2007;8(6):413–426.
120. Wirsching E, Fauler M, Fois G, Frick M. P2 purinergic signaling in the distal lung in health and disease. *Int J Mol Sci*. 2020;21(14):4973.
121. Kiyokawa H, Morimoto M. Notch signaling in the mammalian respiratory system, specifically the trachea and lungs, in development, homeostasis, regeneration, and disease. *Dev Growth Differ*. 2020;62(1):67–79.
122. Asai N, Jijiwa M, Enomoto A, et al. RET receptor signaling: dysfunction in thyroid cancer and Hirschsprung's disease. *Pathol Int*. 2006;56(4):164–172.
123. Butti R, Das S, Gunasekaran VP, Yadav AS, Kumar D, Kundu GC. Receptor tyrosine kinases (RTKs) in breast cancer: signaling, therapeutic implications and challenges. *Mol Cancer*. 2018;17(1):34.
124. Tóth KF, Ádám D, Bíró T, Oláh A. Cannabinoid signaling in the skin: therapeutic potential of the "c(ut)annabinoid" system. *Molecules*. 2019;24(5):918.

125. Brandão LAC, Tricarico PM, Gratton R, et al. Multiomics integration in skin diseases with alterations in notch signaling pathway: PlatOMICs phase 1 deployment. *Int J Mol Sci*. 2021;22(4):1523.
126. Tomas A, Futter CE, Eden ER. EGF receptor trafficking: consequences for signaling and cancer. *Trends Cell Biol*. 2014;24(1):26–34.
127. Siebel C, Lendahl U. Notch signaling in development, tissue homeostasis, and disease. *Physiol Rev*. 2017;97(4):1235–1294.
128. Wang JJ, Li DJ, Du MR. Research advances on adrenergic receptor signaling involved in disease microenvironment through regulation of macrophages. *Sheng Li Xue Bao*. 2020;72(2):227–234.
129. Simmons DA, Longo FM, Massa SM. Neurotrophin receptor signaling as a therapeutic target for Huntington's disease. *CNS Neurol Disord Drug Targets*. 2017;16(3):291–302.
130. Kamo T, Akazawa H, Komuro I. Pleiotropic effects of angiotensin II receptor signaling in cardiovascular homeostasis and aging. *Int Heart J*. 2015;56(3):249–254.
131. Königshausen E, Zierhut UM, Ruetze M, et al. Angiotensin II increases glomerular permeability by β-arrestin mediated nephrin endocytosis. *Sci Rep*. 2016;6:39513.
132. MacKenzie G, Maguire J. Neurosteroids and GABAergic signaling in health and disease. *Biomol Concepts*. 2013;4(1):29–42.
133. Grinevich V, Knobloch-Bollmann HS, Eliava M, Busnelli M, Chini B. Assembling the puzzle: pathways of oxytocin signaling in the brain. *Biol Psychiatry*. 2016;79(3):155–164.
134. Bading H. Therapeutic targeting of the pathological triad of extrasynaptic NMDA receptor signaling in neurodegenerations. *J Exp Med*. 2017;214(3):569–578.
135. Schaefer N, Roemer V, Janzen D, Villmann C. Impaired glycine receptor trafficking in neurological diseases. *Front Mol Neurosci*. 2018;11:291.
136. Awasthi A, Ramachandran B, Ahmed S, et al. Synaptotagmin-3 drives AMPA receptor endocytosis, depression of synapse strength, and forgetting. *Science*. 2019;363(6422):eaav1483.
137. Yu R, Wawrowsky K, Zhou C. A natural inactivating mutant of human glucagon receptor exhibits multiple abnormalities in processing and signaling. *Endocrinol Nutr*. 2011;58(6):258–266.
138. Ulloa-Aguirre A, Zariñán T, Gutiérrez-Sagal R, Dias JA. Intracellular trafficking of gonadotropin receptors in health and disease. *Handb Exp Pharmacol*. 2018;245:1–39.
139. Pellinen T, Tuomi S, Arjonen A, et al. Integrin trafficking regulated by Rab21 is necessary for cytokinesis. *Dev Cell*. 2008;15(3):371–385.
140. Liu M, Ning X, Li R, et al. Signalling pathways involved in hypoxia-induced renal fibrosis. *J Cell Mol Med*. 2017;21(7):1248–1259.
141. Karmakar S, Lal G. Role of serotonin receptor signaling in cancer cells and anti-tumor immunity. *Theranostics*. 2021;11(11):5296–5312.

CHAPTER TWO

Emerging tools for studying receptor endocytosis and signaling

Pratiksha Pandita[a,†], Rhea Bhalla[b,†], Ashok Saini[c,†], and Indra Mani[d,*]
[a]ICMR-National Institute of Virology, Pune, Maharashtra, India
[b]Faculty of Medicine, Department of Infectious Disease, Imperial College London, London, United Kingdom
[c]Department of Microbiology, Institute of Home Economics, University of Delhi, New Delhi, India
[d]Department of Microbiology, Gargi College, University of Delhi, New Delhi, India
*Corresponding author: e-mail addresses: indramanibhu@gmail.com; indra.mani@gargi.du.ac.in

Contents

Abstract

Ligands, agonists, or antagonists use receptor-mediated endocytosis (RME) to reach their intracellular targets. After the internalization of ligand-receptor complexes, it traffics through different subcellular organelles such as early endosome, recycling endosome, lysosome, etc. Further, after the ligand binding to the receptor, different second messengers are generated, such as cGMP, cAMP, IP_3, etc. Several methods have been used, such as radioligand binding assay, western blotting, co-immunoprecipitation (co-IP), qRT-PCR, immunofluorescence and confocal microscopy, microRNA/siRNA, and bioassays to understand the various events, such as internalization, subcellular trafficking, signaling, metabolic degradation, etc. This chapter briefly discusses the key

† Contributed equally the first authors.

Progress in Molecular Biology and Translational Science, Volume 194
ISSN 1877-1173
https://doi.org/10.1016/bs.pmbts.2022.10.003

principles and methods used to study internalization, subcellular trafficking, signaling, and metabolic degradation of numerous receptors.

Abbreviations

AMPA receptor	α-amino-3-hydroxy-5-methyl-4-isoxazolepropionic acid receptor
cAMP	cyclic adenosine monophosphate
cGMP	cyclic guanosine monophosphate
co-IP	co-immunoprecipitation
GABAA receptor	gamma-aminobutyric acid subtype A receptor
IP3	inositol 1,4,5-trisphosphate
miRNAs	microRNAs
PDGF	platelet-derived growth factor
qRT-PCR	real-time reverse transcription-PCR
RME	receptor-mediated endocytosis
siRNA	small interfering RNA
VEGFR	vascular endothelial growth factor

1. Introduction

Receptor-mediated endocytosis (RME) is a critical step in the process by which ligand, agonist, or antagonist reach their intracellular targets. In addition, after the internalization of ligand-receptor complexes, it traffics through diverse subcellular organelles such as early endosomes, recycling endosomes, lysosomes, etc.[1–7] After ligand binding to the receptor, it induces the generation of various second messengers such as cyclic nucleotides (cGMP, cAMP), Inositol 1,4,5-trisphosphate (IP_3), etc.[8–10] Several investigators had studied RME of various ligands. After binding the ligand, receptors usually undergo an endocytotic pathway. In such a study, RME of vascular endothelial growth factor (VEGF-A) was studied in which radiolabeled VEGF-A was used which binds with its receptors, vascular endothelial growth factor receptors (VEGFR1 and VEGFR 2) in high affinity in rat liver sinusoidal endothelial cells. The expression level of VEGF-A was determined by immunoblotting.[11] Very recently, Wang et al. studied the effect of Dcbld 2 deletion on the platelet-derived growth factor (PDGF) signaling pathway. DCBLD2 mainly controls vascular hyperplasia by modulating the PDGF receptor-β endocytosis by caveolin-1 in vascular smooth muscle cells.[12] They had observed an increase in PDGF signaling when Dcbld 2 had been deleted. The deletion was confirmed by western blotting analysis.

In menopausal women, risk of cardiovascular diseases has been increased due to the raised level of low-density-lipoprotein cholesterol (LDL-C). The mechanism and influence of high postmenopausal follicle-stimulating hormone (FSH) levels on lipid profiles had been investigated by Song et al.[13] in more than 400 Chinese postmenopausal women. The FSH and LDL-C binds to their respective receptors—FSHR and LDLR in the cells and plays significant role in lipid metabolism. The levels of FSHR and LDLR were examined using western blotting, and it was observed that LDLR levels decreases which further decreases the endocytosis of LDL resulting in increased its level in blood.[13] Many researchers had studied the trafficking events of various neurotransmitters using a recycling assay in which they attached the receptor with a biotinylated reactive linker group via amino-terminal (sulfo-NHS-SS-biotin).[14] After the addition of the reducing agent, such as glutathione, the biotin gets separated, and the receptor is internalized with the ligand. After endocytosis, the internalized biotinylated receptor is captured by solubilizing the cells and streptavidin affinity chromatography was done. The amount and presence of the receptor then determined by sodium dodecyl sulfate-polyacrylamide gel electrophoresis (SDS-PAGE) followed by Western blotting.[15] This approach has been used to study the kinetics and control mechanisms of various receptors such as α-amino-3-hydroxy-5-methyl-4-isoxazolepropionic acid (AMPA) receptor internalization and *N*-methyl-D-aspartate (NMDA) receptors[16] as well as to study the internalization of various neurotransmitter receptors including gamma-aminobutyric acid subtype A (GABAA) receptors,[17] and GABAB receptors[18] in neuronal cells.

The leucine-rich repeat kinase 2 (LRRK2) is defective gene at the PARK8 locus causing the autosomal dominant form of Parkinson's disease (PD). Identification of Rab5b as an LRRK2-interacting protein was done. With the help of co-immunoprecipitation (co-IP) assays, it has been shown that it precisely interacts with LRRK2. Finally, it was hypothesized that the interaction of LRRK2 with Rab5b plays a significant role in synaptic function by controlling the endocytosis of synaptic vesicles.[19] During endocytosis, vesicular transport from early to late endosomes is mainly regulated by Rab7 protein (RabGTPase). Yang et al.[20] characterizes one novel Rab7-like GTPase (Rab7b), which was found to be homologous to Rab7 protein. The tissue specific level of Rab7b mRNA was analyzed in various tissues like lungs, muscles, heart cells and peripheral blood leukocytes. Moreover, Rab7b selectively expressed in some types of cells, such as monocytes, monocyte-derived immature dendritic cells (DCs) as determined by reverse-transcription PCR (RT-PCR) or western blotting.[20] The data suggests the role of Rab7b in phorbol myristate acetate (PMA)-induced differentiation of acute promyelocytic leukemia cells and regulation of functioning of monocytes.

In mammalian brain cells, it has been found that one microtubule activated GTPase (Dynamin) is essential for endocytosis. In this study, another dynamin gene (*Dyn 2*) was isolated from cDNA library using 500-bp cDNA probe generated from enzymatic region of the dynamin-encoding gene. Further, the expression of *Dyn2* gene was analyzed in all the tissues of rat and found that it is expressed in all tissues but at different levels, highest in testis cells. Hence, the ubiquitous distribution of dynamin in rat tissues is due to Dyn2 not brain dynamin.[21] Equine infectious anemia virus (EIAV) infects various cells, such as macrophages, fibroblast cells of equine, etc. It is a type of lentivirus and its entry mechanism still not well understood. Thus, an engineered mouse cell line was developed which expresses equine lentivirus receptor-1 to study and characterize the specific lentiviral entry mechanism. The effects of various drugs specifically affect the entry of EIAV were also assessed using qRT-PCR assay of early reverse transcripts formed inside the cell and by measuring the production of virion.[22] Taken together, the results had shown that EIAV enters the cell via its receptor in a low-pH-dependent endocytic pathway.

B-cell receptor (BCR) signaling and endocytosis are simultaneously initiated after binding of antigen with BCR. The spatio-temporal dynamics of signaling enzymes and/or proteins involved in BCR signaling pathway can provide the in-depth knowledge. Hence, the study of BCR signaling by confocal microscopy could talk about the distribution of phosphorylated kinases present in activated B cells in spatio-temporal manner.[23] Natriuretic peptide receptor-A (NPRA) is a receptor for both atrial natriuretic peptide (ANP) and brain natriuretic peptide (BNP).[24] After binding with ligands, NPRA-ligand complex gets internalized into cells, where it is redistributed to various intracellular compartments and finally degraded in lysosomes. Sometimes these complexes are escaped from lysosomal degradation and recycled back to the cell membrane.[25] Moreover, NPRA is endocytosed by dynamin-dependently clathrin-coated vesicles pathway.[26] The whole pathway of NPRA-ligand internalization and recycling was studied using confocal microscopy.[27] Similar kind of study was done in which NPRA was tagged with enhanced green fluorescence protein (eGFP) in HEK-293 cells and murine mesangial cells (MMCs) to analyses its localization and other assays.[3,4,9,26] Graphical illustration of endocytosis and subcellular trafficking of glucagon-like peptide receptor (GLP-1R) is shown in Fig. 1.[5]

To study the receptor endocytosis and signaling, various methods are used, such as radioligand binding assay, western blotting, co-immunoprecipitation, qRT-PCR, immunofluorescence and confocal microscopy, microRNA/

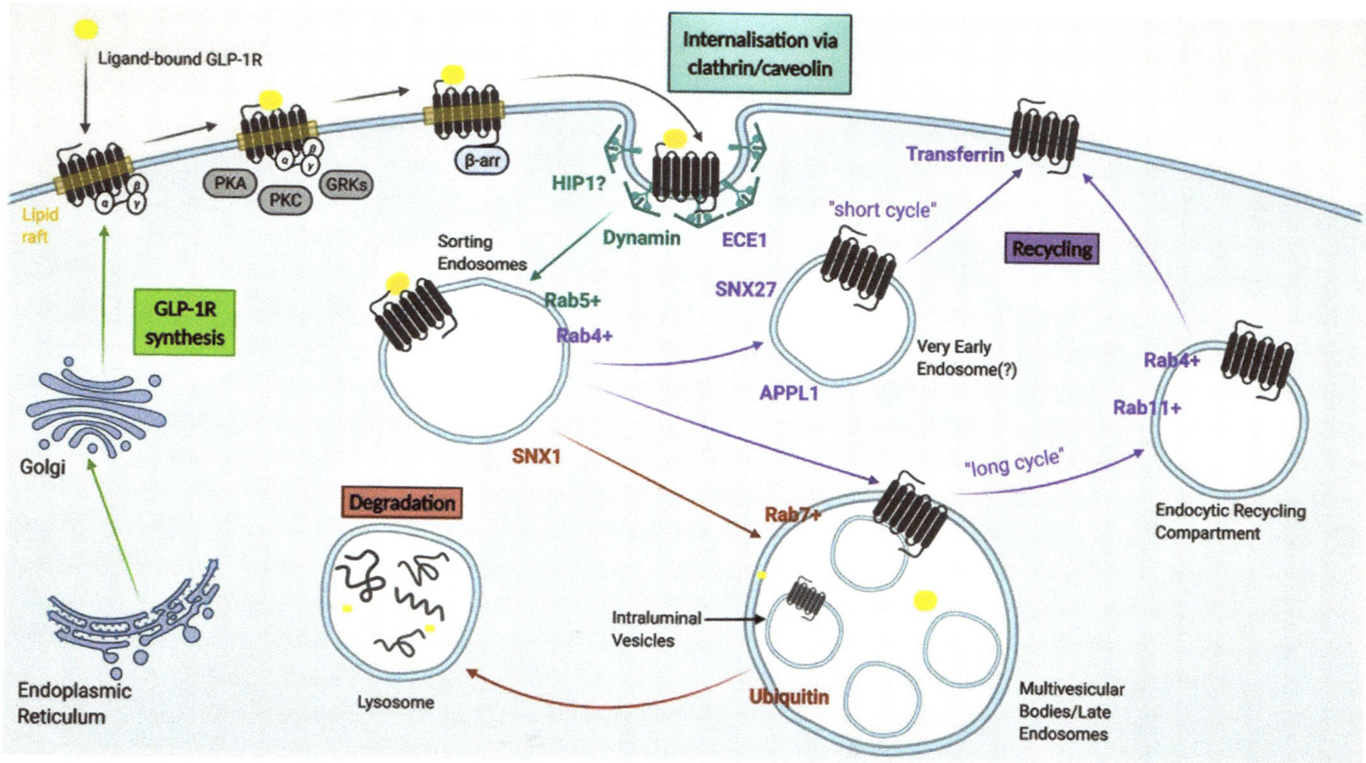

Fig. 1 Graphical representation of endocytosis and subcellular trafficking of glucagon-like peptide receptor (GLP-1R). After internalization, the receptor traffics through the early endosome, recycling endosome and lysosome. PKA, Protein Kinase A; PKC, Protein Kinase C; GRKs, G protein-coupled receptor kinases; HIP1, Huntingtin-interactin protein 1; ECE1, endothelin-converting enzyme-1; APPL1, PTB domain and leucine zipper motif 1; SNX1, sorting nexin 1. *Adapted from Marzook A, Tomas A, Jones B. The interplay of glucagon-like peptide-1 receptor trafficking and signalling in pancreatic beta cells.* Front Endocrinol (Lausanne). *2021;12:678055. https://doi.org/10.3389/fendo.2021.678055.*

siRNA, and bioassays for second messengers. This chapter briefly discusses the basic principle and application of these methods, which are used to study the internalization, subcellular trafficking, signaling and metabolic degradation of numerous receptors.

2. Radioligand binding assay

Radioligand binding assay is very sensitive and provides quantitative data about receptor internalization. Different radiolabeled ligands, such as H^3, F^{18}, C^{11}, I^{125}, etc., have been used to analyze receptor internalization, recycling, and metabolic degradation.[3,28–30] Most of the cases, radioactive isotope I^{125} has been used in the ligands. Various steps are involved in the study of radioligand-based internalization, recycling, and metabolic degradation of the receptor. The major steps are the following; binding of the radioligand to the receptor, allowing for internalization, lysis of cells, and finally counting of the radioactivity in the cells.[3,9] It has been used for the study of various receptors such as cannabinoid receptors,[28] 5-HT(2A) receptors,[31] Sigma-2 receptors,[29] EGFR,[32] NPRA,[3] GPCR,[33] LDL,[34] etc.

3. Western blot

Western blotting or immunoblotting or protein immunoblot is an important method used in molecular biology and other fields of biology that was developed in 1979.[35] A need arose in the 1970s for the development of a technique that could determine the specific antigen using antisera. This need led to the invention of a technique which could produce a replica of proteins onto a nitrocellulose membrane post their separation by electrophoresis on polyacrylamide gels. The separated proteins were then transferred to the nitrocellulose membrane and the fragments could be probed with antisera which aids in the detection of specific proteins. It was William Burnette in 1981 that carried out the process of SDS-PAGE for separation of the sample and coined the term "Western Blotting."[36] Western blotting is a widespread technique for detection and measuring concentration of the specific proteins in a biological sample .[37] The western blot depends on four main principles—(1) equal loading of protein samples, (2) separation of the samples based on molecular size, (3) transfer of these separated proteins on a suitable membrane and (4) probing of the transferred proteins with labeled antibodies.[38]

The process starts with the preparation of the samples to extract proteins from them. Cell lysates are taken and treated with a lysis buffer to extract proteins. The lysis buffer usually contains protease inhibitors and detergents. The sample containing proteins is mixed with a gel-loading buffer containing glycerol which make it easier to load it into the wells. Dyes such as bromophenol blue are also added, which helps in as tracking the movement of the sample through the gel.[39] The loading buffer also contains SDS, the anionic detergent agent. The sample is then heated for a few minutes, this aids in the denaturation and unfolding of the protein and surrounds it with a negative charge imparted by the SDS so that the proteins have a uniform negative charge and all proteins move toward the positive terminal (anode) during polyacrylamide gel electrophoresis. Before denaturation of the protein sample by heat, it is usually treated with reducing agent like 2-mercaptoethanol that disrupts inter and intra disulfide bonds, hence further helping in the unfolding of different types of secondary and tertiary protein structures.[40]

The rate of movement of protein samples through the polyacrylamide gel is mainly depends on numerous factors such as concentration of polyacrylamide used for the preparation of gel, type of buffers used in the system, size of the protein and voltage applied. The gel system of PAGE consists of two types of gels—stacking gel and resolving gel. The ready protein samples are loaded into wells of the stacking gels. Stacking gel layer is formed at the top (~3–5 cm) having pH ~6.8 and lower percentage of acrylamide (~3–5%) while the bottom or resolving gel layer is made up of higher acrylamide concentration (~6–15%) in buffer with higher pH (~8.8). SDS-PAGE is therefore stated to be done in a discontinuous buffer system. As soon as current is applied, the loaded samples start moving through the pores present in the resolving gel and the migrated proteins are separated on the basis of size as it is proportional to mass-to-charge ratio (i.e. size).[41] Then, the proteins resolved on the gel are transferred and immobilized to either a polyvinyliden fluoride (PVDF) membrane or a nitrocellulose membrane. This is followed by washing, blocking, and incubating the membrane with a target protein-specific primary antibody.[42] The most used ways of incubation include—either the use labeled primary antibody directly or labeled secondary antibody that is meant against the primary antibody. The antibodies used can be labeled by attaching them either with an enzyme which can produce a colored product or a light signal at the site where the antibody interact with its specific target antigen (protein) when incubated with substrate or with fluorochrome which can produce a fluorescent signal after excitation with a particular wavelength.[43]

Immunoblotting has been widely used to study the receptor kinetics during endocytosis. The expression level or amount of the receptor or ligand present on the membrane or inside the cell is usually detected using this technique. In one of such study of macrophage phagocytosis and decrease in inflammation by all-trans retinoic acid—the protein amount of toll-like receptor 4 (TLR 4) and the cluster of differentiation (CD14) are analyzed, which are mainly located on the membrane of macrophages with the help of western blotting.[44]

In another study, western blotting has used to analyze the effect of Eps15 homology domain 1 (EHD1) on β2-adrenoceptor (β2AR) signaling—Mammalian EHD1 plays an integral part in the progression of tumors, however, its role in cancer angiogenesis remains unknown.[45]

It has been observed that excitotoxicity has been induced in spiral ganglion neurons (SGNs) in the presence of sodium salicylate (SS), which mainly affects dopamine D1-like receptors (D1Rs) signaling pathway by inhibiting GABAA receptors (GABAARs) and increases their internalization involving. Qin et al.[46] expose GABAARs to various activators and inhibitors such as sodium salicylate, dynasore (inhibitor of D1R) Bis-indolylmaleimide-I (inhibitor of protein kinase C), SKF38393 (agonist), SCH23390 (antagonist) to investigate the role of these receptors and their signaling pathway proteins and the expression of receptor and other proteins was confirmed using immunofluorescence and western blotting.[46] The data showed the increase in internalization when the receptors were exposed to SS, while the inhibiting the D1Rs and protein kinase C (PKC) decreased the internalization.

Influenza A virus is an enveloped, ssRNA(−) virus belongs to orthomyxoviridae family. During infection it derives its envelope from the lipid bilayer membrane of the host. To enter a cell, influenza virus binds to sialic acid residues on the target cell via its hemagglutinin (HA) protein and is then internalized by endocytosis.[47] During the endocytotic pathway, HA undergone conformational changes due to the low pH (~5.0–5.7) of endosome and further fuses to the membrane of late endosome consequential releasing of viral genome in the cytoplasm to initiate replication.[48] It has been observed that viral proteins-HA and NA localization to the membrane has been decreased in the cells which are deficient in sphingomyelin synthase. Drews et al.[49] had shown the role of glucosylceramidase (GBA) enzyme which metabolize glycosphingolipid in trafficking of influenza virus to late endosomes which is required for fusion of membranes, entry of virus and further infection. They investigated the function of GBA in viral entry and further infection by deleting it out using CRISPR-Cas9 technology. The levels of GBA and other receptors or proteins (EGFR and Cathepsin B) in the knockout cell lines had been confirmed using western blotting and found that cells lacking GBA showed influenza virus infection.[49]

Cerebral cortical disinhibition is usually shown by the patients of generalized epilepsy which is caused due to mutations in GABAARs which are major inhibitory ligand-gated ion channels found in the brain cells of mammals consisting of five subunits which are expressed from 8 gene families which comprise of multiple isoforms (α1–6, β1–3, γ1–3, δ, ϵ, θ, π, and ρ1–3). To create functional receptors, these 19 subunits can be assembled in a variety of ways, however most synaptic receptors in the cortex comprise two α-subunits, two β-subunits, and one γ-subunit.[50] In heterozygous knock-out (Hetα1KO) of the human epilepsy gene (GABAAR α1-subunit), the epilepsy was found absent in mouse.[51] The Hetα1KO affects the expression as well as the physiology of GABAA receptors in cortex by reducing the expression of β2 subunit while did not affect the expression of β1 or β3 subunits resulting in small reduction of GABAARs. The expression levels of various proteins—neutravidin, Na^+/K^+ ATPase, β2, β3 and GABAAR α1-subunit were quantified using western blotting and the localization was confirmed by immunofluorescence. Also, the interaction of α1 subunit with other subunits α2 and α3 was checked using co-immunoprecipitation. Moreover, relative abundances of mRNAs of α1 and α3 subunits mRNA was measured in wild-type and Hetα1KO mouse cortex cells using qRT-PCR.[52] The study had explained the various effects of Hetα1KO on cortical GABAAR-mediated inhibitory neurotransmission.

Recently, Ma et al.[53] had evaluated the effect of morphine tolerance on [D-Ala(2), N-Me-Phe(4), Gly(5)-ol]-enkephalin (DAMGO)-induced internalization of micro-opioid receptor (MOR). Morphine and DAMGO was administered in male Sprague-Dawley rats to induce antinociceptive tolerance and the latency in response generation was measured. The expression and localization of membrane-bound and cytosolic MOR was checked by western blotting and immunofluorescence respectively. They found that DAMGO stimulates internalization and recycling of MOR from membrane, but the expression remains unchanged on membrane as well as in cytoplasm whereas morphine inhibits internalization of MOR and thus give rise to tolerance.[53]

Chronic kidney disease (CKD) is mainly caused due to hypertensive nephropathy (HN).[54] Due to the development of renal fibrosis that causes interstitial fibrosis, glomerular sclerosis, and tubular atrophy[55] affecting renal functions.[56] But the particular mechanism causing renal fibrosis is enigmatic although some strategies are known which could stop or limit the development of fibrosis.[57] Also, various clinical data and animal experiments have revealed that higher serum gastrin levels are linked with CKD.[58] Gu et al.[59] had studied the mechanism involved in this pathway. They had reported that angiotensin II (Ang II)-induced hypertensive nephropathy is mainly responsible for enhanced the expression of renal cholecystokinin B receptor (CCKBR). Also, after the treatment with gastrin, Ang II-induced apoptosis of renal tubule cells was found to be decreased, Ang II-mediated inhibition of efferocytosis of macrophages was reversed, and inflammation of kidney was reduced. Moreover, peroxisome proliferator-activated receptor α (PPAR-α) regulates efferocytosis and its deletion leads to reversed the effects of gastrin. The levels of various proteins such as interleukines (IL-6 and IL-1β), caspase-3, CCKBR, tumor necrosis factor α, proliferating cell nuclear antigen (PCNA) and glyceraldehyde-3-phosphate dehydrogenase (GAPDH) involved in this pathway was determined by immunoblotting and qPCR.[59] They conclude that gastrin may be acts as a regulator of PPAR-α.

More than 120 million people are affected by the major depressive disorder, mainly caused by environmental stress-related factors.[60] It has been found that glutamate-mediated excitatory system is critically involved in the development of depressive disorders,[61] but the underlying mechanisms is still unknown. The neuroplastic alterations of stress-induced depression-like activity in the ventrolateral periaqueductal gray and the link between stress and AMPA receptors were examined. They had analyzed the membrane bound and cytosolic expression of glutamate receptor-1 (GluR1) and AMPA receptor using western blotting.[62] Their study demonstrates that depression-like activities are mainly linked with decrease in glutamatergic, but not GABAergic, transmission in the ventrolateral periaqueductal gray leading to chronic stress-induced neuroplastic changes.

Many other studies had used western blotting as a tool to measure the amount of various proteins in the cells involved in endocytosis, such as huntington associated protein 1 (HAP1),[63] VSTM2A,[64] sulfated syndecan-1,[65] cAMP response element binding protein (CREB),[66] etc.

4. Co-immunoprecipitation and immunoblotting

Co-immunoprecipitation (co-IP) is the most useful tool to find out interactions between two or more proteins *in vivo*, using antibodies against the one of the interacting protein partners. In principle, if the proteins of interest are interacting with each other or associated with each other in cell then both the protein will be precipitated out when the antibody against only one protein is used for precipitation. These studies mainly rely on the protein-specific antibody, which is usually immobilized on a solid support or resin to allow precipitation of target protein from the solution. The utilization of co-IP is to study interactions between proteins which are present in the solution and/or to identify unknown interacting partners with that particular protein.[67]

To perform co-IP assay, an antibody against a target protein usually coupled to sepharose beads either through protein A or G is taken and incubated with the cell extract. Then the beads were washed to remove the non-specific and unbound partners. Further, the beads linked with antibody and protein complexes comprising the target protein and binding partners are precipitated by centrifugation and the protein components in the complexes are recovered by increasing the salt concentration and followed by visualization on SDS-PAGE and western blotting using antibodies specific to the other interacting partners.[68] Co-immunoprecipitation studies are widely used to find out the interacting partners in a signaling or endocytotic pathways where receptor initially interacts with ligands and then get internalize and interact with other molecules in the cell. Also, various ligands in cell interact with many other proteins involved in regulation of various activities of cell. Few examples of such applications are discussed below.

The growth hormone (GH) regulates the growth of its target cells by binding with its receptor, GH receptor (GHR) and forms a complex [hormone-(receptor)$_2$] and thus initiates the signaling pathway. After binding, the complex is endocytosed and transported to the lysosome in a regulated manner via the ubiquitin dependent endocytotic pathway (UbE). Gent et al.[69] had demonstrated the GHR dimerization in ER and at the surface of cell in a ligand-independent manner using co-IP of an epitope-tagged truncated GHR with non-truncated GHR.[69] It is postulated that dimerization of GHR molecules is critical for ubiquitin system-dependent endocytosis. Rice blast disease is caused by *Pyricularia oryzae*. Autophagy plays a significant role in development of disease and infection by *P. oryzae*. But the linkage between autophagy and endocytosis was remains unexplored in *P. oryzae*. This study identified three *P. oryzae* VPS9 domain-containing proteins, PoVps9, PoMuk1 and PoVrl1 involved in the process.[70] The interaction between these domains with each other was determined by co-IP assays which suggests that PoVps9 directly interacts with the PoVps21 thus involved in endocytosis and autophagy.

Proteins of synaptic vesicles are retrieved with the involvement of AP180. In this study researchers have shown that in *Drosophila* recycling of synaptical vesicles has been severely affected at the neuromuscular synapse of larva after inactivation of AP180. Moreover, integrity of proteins of synaptical vesicles is dependent on AP180 at the cell membrane during endocytosis. It has been shown that various synaptical vesicles proteins interact directly with AP180 which is confirmed by co-IP. Also, Like-AP180 (LAP) as a monomeric clathrin assembly protein which is involved in sorting and retrieval of synaptical vesicles proteins to form endocytotic synaptical vesicles. These synaptical vesicle proteins clustered in the cell membrane after fusion thus in this study the direct interaction of LAP with synaptical vesicles has been examined using co-IP.[71] Yeast two hybrid studies have shown that a protein involved in cell signaling and endocytosis Disabled 2 (Dab2) interacts directly with myosin-VI (actin-based motor protein). Their interaction was further confirmed by co-IP and other *in vitro* binding assays suggesting the inter-linkage of actin cytoskeleton with endocytotic pathway of any receptor.[72]

Internalization of an endocytic receptor is a complicated and active process. In this study, they attempted to create and test ligand-bound quantum dot (NGF-QD) bioconjugates for imaging discrete receptor endocytic processes inside of living NGF-responsive cells (PC12). They demonstrated and validated the use of QD-receptor complexes for imaging receptor trafficking at synchronized time points following QD-ligand receptor binding and internalization ($t = 15$–150 min) using various techniques such as single particle tracking, QD hybrid gel co-immunoprecipitation and immunocolocalization. New dynamic findings demonstrating (1) endocytosis occurs in a stunningly controlled manner and (2) diffusive and active modes of transport within cells are quick and effective serve to highlight the particular significance of these probes.[73] E-cadherin forms a continuous layer of adhesive dimers at cell-cell junctions. The dynamics of adhesive dimer assembly have been studied using mixed culture cross-linking and co-IP assays. Researchers had discovered that the quantity of these dimers increased substantially within minutes after endocytosis inhibition by ATP depletion or hypertonic sucrose. This increase was accompanied by efficient E-cadherin recruitment into adherens junctions. These findings suggest that endocytosis is the primary mechanism for the dissociation of E-cadherin adhesive dimers. Its inhibition prevents the replacement of the monomeric cadherin pool, thereby preventing the formation of new dimers.[74]

The connexion-36 (Cx36) is gap junctional protein was co-purified with the Cav-1 protein, which is present in lipid rafts. But, the application of this interaction was unknown. Kotova et al. had explored the importance of Cav-1 protein interaction with Cx36 in the context of intracellular and membrane transport. Forster resonance energy transfer (FRET) and co-IP assays were used to check the interaction between these proteins, and found that the interaction of Cx36 and Cav-1 is dependent on Ca^{2+} levels.[75] Spinocerebellar Ataxia type-2 (SCA2) disease (a type of neurodegenerative disease) is caused due to the unstable increase of an internal polyglutamine domain of Ataxin-2 protein. Nonis et al. in 2008 determined the function of ataxin-2 and find out its interacting partners (endophilin A1 and A3) involved in synaptic vesicle endocytosis.[76] Their data suggests the role of ataxin-2 in endocytic receptor cycling.

5. Northern blotting

The northern blotting is a major molecular biology technique used for measuring the amount and size of specific mRNA and to estimate their abundance.[77] The amount of mRNA of a particular gene mainly provides two information—(i) tissue-specific or time-specific expression of that particular gene which indicates its physiological function of its encoded protein and (ii) factors which regulates the expression of that gene in particular tissue or time. Other techniques which can be used to measure mRNA in a sample are reverse transcriptase polymerase chain reaction (RT-PCR) and RNase protection assay. The basic principle of this technique is that it separates mRNA based on size then it is fixed onto a nylon membrane keeping the same position as on the gel on which they are separated. The immobilized RNA is then allowed to hybridize with a labeled probe complementary to the sequence (or gene of interest) and further after washing the hybridization signal is recorded and analyzed. This technique can determine the abundance and sizes of the desired transcripts due to the usage of size-dependent separation. It is a very powerful approach for detecting transcript variations of genes.

The basic steps involve the extraction of total RNA from a cell or tissue, by means of chaotropic agents such as guanidinium isothiocyanate (GITC) which helps in the tissue solubilization and digestion of proteins like RNases and solubilizes RNA. As the quantity of mRNA is usually less so some additional steps may be included to isolate mRNA. The extracted mRNA is then separated by agarose-gel electrophoresis, followed by immobilization onto a nylon membrane. Nylon membranes have higher binding capacity and robustness due to the presence of positive-charges on it, thus these membranes are more preferred for immobilization of DNA/RNA. A northern blot provides an exact reflection on nylon membrane of the RNA molecules separated in an agarose or polyacrylamide gel. After blotting, the membrane was either baked at high temperature or exposed to UV which cross-links the RNA with the binding groups present on membrane making it resistant to further washing steps. Then, a labeled hybridization probe is prepared followed by its hybridization with the RNA immobilized on the membrane. There are two main forms of hybridization probes used—(i) complementary DNA (cDNA) probes and (ii) antisense oligonucleotides designed against the target sequence.

The unbound and non-specific binding probes is removed by washing it buffer with increasing salt concentration. The membrane is further dried and the hybridization signals can be detected by various means usually on X-ray film and quantified using various detecting methods using either radioactivity (Radioactive labeled probes having ^{32}P) or by non-radioactive strategies such as chemiluminescence. Molecular markers of known molecular weight are usually run in-gel to determine the exact sizes of RNA molecules in the sample. Four major types of markers are used: (1) Commercially available RNA standards, (2) Commercially available DNA standards, (3) Tracking dyes like, bromophenol blue and (4) Highly abundant rRNAs (such as 28S and 18S) already present within the prepared RNA sample. Northern blotting can be used to monitor the expression of a particular gene in different cells or tissues or organs or in different development stages or under environment stress levels, etc. It can also be used to compare the expression level of oncogenes and tumor suppressor genes between normal and cancerous cells. It can also be used to study mRNA splicing and screening of recombinants obtained after transformation. Other applications of the northern blotting in the field of signaling and receptor endocytosis are discussed below.

In a study, researchers had investigated endocytotic role of annexin VI (Calcium-dependent phospholipid-binding protein) protein in human squamous carcinoma cell line A431. They had found out that endogenous annexin VI is not expressed in A431 cells. The results were confirmed using northern blotting, western blotting, and PCR/Southern blotting. The annexin VI protein and mRNA could not be detected in western blotting and northern blotting respectively.[78] During the advanced stages of prostate cancer, growth of cancer became androgen-independent and more dependent on growth factors such as epidermal growth factor (EGF).

Moreover, changes in growth factor RME have been observed during cancer progression. Using western blotting and northern blotting, researchers investigated the internalization and signaling of EGF receptors (EGFRs) in various prostate cancer cell lines in relation to the expression level of the endocytosis-related REPS2 gene. The results had shown that REPS2 gene is expressed at very high levels which is further correlated with decreased internalization and signaling of EGFR. As a result, decreased REPS2 expression may result in increased EGFR expression and signaling, leading to androgen independence in advanced prostate tumors.[79]

Another study focused of iron delivery into on the Sertoli cells of the rat seminiferous epithelium after the secretion of diferric testicular transferrin (tTf). High affinity transferrin binding sites are present on the surface of rat spermatids which binds with tTf and initiates its internalization processes. RNA was isolated from various spermatocyte cells to compare the expression level of Tf receptor and ferritin using northern blotting assay. The expression level of these proteins was found to be inversely proportional to the maturation stages of spermatocytes. Thus, they conclude that diferric Tf binds to its receptor on the cell membrane of germ cells, get internalized via RME and then recycled back to the membrane by forming apotTf-Tf receptor complex releasing apotTf into the adluminal fluid.[80]

Recently, it has been found that Cbl-interacting protein (CIN85) is an adaptor protein which is involved in the endocytosis of many tyrosine kinase receptors. In this study, researchers had isolated one novel interacting protein, CAMGAP1 (a type of RhoGAP) which interacts with SH3 domain of CIN85. Additionally, basic local alignment search tool (BLAST) search analysis suggests the presence of three homologs of CAMGAP protein in human and mouse. Thus, they examined the expression level of these three CAMGAPs in a tissue specific manner using northern blot analysis.[81] It was reported that the human monocytic cell line (Mono Mac6sr) degrades acetylated LDL (acLDL) and oxidized LDL via receptor-specific pathways. Binding site of acLDL was characterized in untreated and LPS-differentiated Mono Mac 6sr cells and found out that these cells have more than one binding site but RT-PCR and northern blot analysis do not detect any mRNA specific to acLDL receptor in any type of cells.[82]

6. Real-time reverse transcription PCR (qRT PCR)

Real-time reverse transcription-PCR (qRT-PCR) is one of the best techniques for detecting and quantifying targeted RNA and is increasingly used in diagnostic assays. Quantitative results are more meaningful than qualitative data, and it simplifies assay standardization and quality control. Fundamental basis of this technique have remained the same since its start in which targets are detected by fluorescence-based methods and the signal is amplified at every step of PCR. The reason for such mainstream usage of this technique is as following (i) post-PCR processing is not required (ii) RNAs present in different abundance can be compared easily and (iii) qualitative as well as a quantitative analysis can be done.[83]

The basic principle is that RNA is isolated and then converted into cDNA by the process of reverse transcription. Then this cDNA is further used as a template for DNA amplification using PCR. As this technique not only amplifies DNA but also quantitate it in real-time using various methods such as SYBR Green dyes, hybridization probes giving fluorescence after removal of quenchers, etc. SYBR Green binds to ds DNA and, when excited by a light source, fluoresces about 1000 times brighter than the unbound type. Therefore, the fluorescence signal generated during the reaction is directly proportional to the amount of dsDNA formed during the PCR reaction. While other hybridization probes (such as TaqMan) binds to the template and do not produce any fluorescence due to presence of quencher molecule at 3′ end. During amplification, the TaqMan degrades due to the 5′ exonuclease activity of Taq polymerase which releases the fluorophore (attached to 5′ end) giving fluorescence and the signal is directly proportional to the amount of DNA formed.

Amplification mainly consists of four different interrelated phases: baseline phase, geometric phase, linear phase, and plateau phase. For the first few cycles of amplification, the intensity of fluorescence is usually low, i.e., below the detection threshold known as baseline phase, then it steadily increases as the product accumulates marks the geometric phase. Finally, the amount of product became constant in later phases showing a plateau at later cycles of amplification. The number of cycles in which the intensity curve enters the logarithmic phase is called the cycle threshold (Ct) and is inversely proportional to the amount of template. Ct value talks about the relative abundance of the different RNA (or cDNA) in the sample. Lesser the Ct value more is the abundance of that particular RNA in the sample and vice-versa. To quantitate the expression usually a standard RNA (or template) is used whose expression remains constant in cell.[84] RT-PCR has been used to study the expression level of various genes involved in endocytic pathways. Some of the examples are discussed below.

DNA origami structures can be easily developed into nanometer-sized functional structures which can be used for targeted delivery of drugs and biosensor development. But these DNA nanostructures are not efficiently up taken by cells. In a study, investigators have coupled the DNA origami structure with variable number of transferrin (Tf) protein which is important for iron uptake in cells and observed very high uptake of the assembly. The rate of uptake of modified and unmodified structures was analyzed using confocal fluorescence microscopy and quantitative (qPCR) techniques which showed that the cytoplasmic uptake of transferring bound DNA structure is 22-fold higher than the unmodified structure.[85] Degradation of filtered albumin and proximal tubular endocytosis is mainly depending on the megalin/cubilin receptor complex. In this study, megalin/cubilin genes were inactivated by knockout technique to assess their effects on albumin uptake. The inactivation of these genes (Megalin/cubilin/podocin) was assessed by qRT-PCR and immunohistochemistry assays by comparing their mRNA levels in podocin knockout (KO), inducible megalin/cubilin KO mice and megalin/cubilin/podocin KO mice and compared with control mice in whole kidney sample.[86]

In another study, mechanism of internalization of human IgG into epithelium cells of intestine was studied. The mRNA level of neonatal Fc receptors (FcRn) was measured using qRT-PCR and the uptake was determined by using fluorescein isothiocyanate (FITC) labeled IgG via immunofluorescence. mRNA level of FcRn was found higher during differentiation process of cells. Results suggest the presence of another mechanism of endocytosis of human IgG via FcRn.[87] Many times, the infection of HIV is not productive and HIV particles got degraded in lysosomes. In this study, role of endocytosis in the productive entry of HIV has been studied. They used trans dominant-negative mutants of dynamin and Eps15 to explain the mechanism of entry into HeLa cells. When the expression of dominant-negative mutant of dynamin was induced in $CD4^+$ HeLa cells, the infection of HIV was found to be reduced to 40–80% independently of infectious dose and isolates. This reduction was further confirmed using qRT-PCR analysis of a fusion protein of virion suggesting that infection is blocked at early stages. Similar results were observed for Eps15 mutant confirming the important role of endocytosis in productive infection.[88]

During viral infections, some viruses (e.g., influenza virus) initially get attached to the membrane receptor and then get internalize into the host cell via fusion of viral envelop to the cell membrane and uses endocytotic pathway for further infection. It has been observed that influenza virus escaped from the endocytic compartment to the cytoplasm and nucleus. The presence of influenza virus in various compartments can be traced out by detecting its RNA genome in that particular compartment. In this study, investigators have developed a density gradient ultracentrifugation-based method for subcellular fractionation and further detection of viral RNA in them using qRT-PCR.[89]

In another similar kind of study, the propagation of yellow head virus (YHV) in *Penaeus monodon* was studied. Earlier, it has been observed that YHV inflection up-regulates clathrin coat assembly protein-17 (AP17) levels in the infected cells.[90] AP17 protein is mainly plays important role in clathrin-mediated endocytosis. Jatuyosporn et al.[91] had observed that the levels of AP17 proteins have been increased at least threefold after post-infection by YHV as analyzed using qRT-PCR suggesting that YHV get internalized via clathrin-mediated endocytosis.[91]

7. Immunofluorescence and confocal microscopy

Light microscopy is used to view the samples under white light but sometimes due to non-uniform thickness of sample creates blurred images as light coming from different depth of focal planes. This issue has been resolved by the confocal microscopy in which there is provision (pinhole mechanism) to obstruct the light coming from out of focal planes and thus no blurring of image is observed. Using confocal microscopy, imaging of thick samples can be done by moving the pinhole to observe the light coming from different focal planes of the sample.[92,93] Confocal microscope was designed in 1955 in which two pinholes were placed in front of light source (zirconium arc lamp or lasers) and detector (photomultiplier tubes) respectively.[94]

Now a days, it is combined with the fluorescence microscopy application and provides very high-resolution images of the sample labeled with fluorophores. The images were taken from various focal planes of the sample and combined to produce a 3D image. This is known as optical sectioning.[94] Confocal microscopy is most widely used microscopy technique to view the localization and changes occurring in live cells of the target molecules-proteins and organelles in the sample. In confocal microscopy, different target molecules can be viewed at the same time by labeling them with different fluorochromes. It is one of the best techniques to study the receptor dynamics during cell signaling. Few of the studies which have used confocal microscopy to study endocytosis and cell signaling pathways have been mentioned below.

During innate immune response, phagocytes and other cells detect the invading pathogens through their immunoregulatory receptor proteins on the membrane and eliminate them. One of the examples of this type of receptor is channel catfish (*Ictalurus punctatus*) leukocyte immune-type receptors (IpLITRs) which regulates various processes during innate immune response including phagocytosis. But, the detailed mechanism of target internalization is still not clear. In this study, researchers had developed one method which can detect F-actin and phosphoinositide dynamics occurring during phagocytosis via IpLITRs using confocal microscopy and fluorescence-activated cell sorting (FACS). Results suggests that phagocytosis induced by IpLITR redistributes PI(4,5)P2 and changes the polymerization of F-actin.[95]

G-protein coupled receptors (GPCRs) are major kind of receptors involved in cell signaling pathways. A related kind of receptors-adhesion GPCR (aGPCR) is also known which are structurally and functionally similar to GPCR. The GPR125 (ADGRA3) receptor is belongs to aGPCR family which also have role in endocytosis, Wnt signaling, cell polarity and act as stem cell marker. In this study, Spiess et al. have found that GPR125 is present on the membrane and get internalized via β-arrestin independent pathway but usually colocalizes with transferring receptor-1. The localization of GPR125 during endocytosis and its interaction with other ligands has been determined by confocal microscopy.[96]

Intracellular adhesion molecule-1 (ICAM-1) is a cell surface receptor present on many cells (leucocytes, fibroblasts, endothelial cells, and macrophages) and belongs to immunoglobulin superfamily proteins and plays a major role in cell-signaling during immune response. The role of ICAM-1 in macrophages is still not well understood. In this study, role of lipopolysaccharide (LPS) layer of bacteria has been determined in regulating the activity of ICAM-1. The expression of ICAM-1 on the surface as well as cytoplasm of macrophages was measured using FACS and confocal microscopy.[97] The results suggest that LPS stimulates the expression of ICAM-1 on surface as well as inside the cells, thus enhance the phagocytosis by macrophages.

Receptor tyrosine kinases (RTKs) are one of the types of receptors which are present in plasma membranes of various cells mainly in monomeric and non-phosphorylated forms. Upon the binding of ligand, they dimerize and phosphorylate their cytosolic domain and initiate a signaling cascade inside the cell.[98] Insulin-like growth factor-1 receptor (IGF1R) is one kind of receptor tyrosine kinase which mainly involves in the signaling cascade to keratinocyte proliferation, their migration and survival. But its endocytotic mechanism in keratinocytes is unclear. Dam et al.[99] had demonstrated the IGF1R internalization pathway using various techniques such as confocal microscopy and co-immunoprecipitation, etc., and shown that IGF1R interacts with flotillin-1 and clathrin during internalization.[99]

Signal transducers and activators of transcription (STATs) are one kind of transcription factors which controls the gene expression of those genes which are mainly involved in differentiation, development, apoptosis, and proliferation.[100,101] These transcription factors (as STAT3) are usually activated by binding with their cell surface receptors which are activated after binding by various ligands such as cytokines, growth factors, etc., after binding of these ligands to the cell surface receptors. Post-activation STAT3 dimerizes and transported to the nucleus where it binds to the promoter of the target genes and regulates their expression.[100] But, mechanism of movement of STAT3 from cytoplasm to the nucleus is yet to be clear and the role of receptor endocytosis in STAT3 translocalization has been explained by colocalization and expression inhibition studies using confocal microscopy and immunoblotting techniques.[102]

In other studies, confocal microscopy is used to determine the expression and localization of various proteins during endocytosis and cell signaling such as GPCR,[103] PID-1,[104] serotonin 1A receptor,[105] etc. Furthermore, GFP-tagged receptor used to visualize the several receptor endocytosis, such as LDL receptor,[106] endothelin B receptor (ETB),[107] galanin R2 receptor,[108] beta2-adrenergic receptor (beta2AR),[109] neuropeptide Y (NPY) receptors,[110] transferrin receptor,[111] dopamine D2 autoreceptors,[112] EGF receptor,[113] VEGFR2,[114] etc.

8. MicroRNA/small interference RNA

MicroRNA (miRNA) and small interference (siRNA) directly repress the expression of targeted genes. There are different mechanisms by which miRNA binds to the target region based on their complementarity.[115] miRNAs bind to two complementarity regions, such as the seed region (2–8 bases of the miRNA) and the central regions (~9–12 bases), to cause mRNA cleavage and subsequent degradation. mRNA degradation and translational repression depend on the degree of complementarity of miRNA with the target region.[116] To knockdown or suppress the expression of receptor genes, RNA interference used as a valuable method to understand the ligands and nanomaterials endocytosis.[117] miRNA and siRNA have been used to understand the endocytosis pathway in the various receptor.[118,119] A study has reported that LDL endocytic trafficking through LDLR is regulated by miR-17. Further, it has been shown that mir-17 directly targets the LDLR gene and downregulates its expression.[120] Similarly, miR-145 inhibited EGFR internalization in human colon cancer cells SW1116.[121] In addition, it has been emphasized that miRNA can use in prognosis, diagnosis, and therapy of cancer.[122] miRNA based study needs to more explore to understand receptor internalization, subcellular trafficking, signaling, metabolic degradation, etc.

9. Bioassay

Enzyme-linked or catalytic receptor produces cyclic nucleotides (cGMP, cAMP) as the second messenger after ligand interaction. Signaling molecules regulate various biological processes like secretary signaling, cellular growth and contractility, neuroplasticity and learning, cardiovascular homeostasis, sensory transduction, and inflammation.[123] It has been shown that the generation of the cAMP was continued after the internalization of GPCR with their ligands.[8] cGMP interacts with various intracellular effector proteins such as cGMP-dependent protein kinases (PKG), cGMP-activated phosphodiesterases (PDEs) and cGMP-regulated ion channels (CNGs).[124] Rybalkin et al. have described two methods used to measure cGMP and cAMP hydrolytic activities. These methods are based on radioactive substrates and nonradioactive based on isothermal titration calorimetry.[125] Recently, Yu et al. have developed a rapid and simple bioassay for real-time detection of cGMP generation during signaling in live cells. It is a flexible luciferase-based biosensor.[126] Apart from cyclic nucleotides signaling molecules, different proteins involved in endocytosis and signaling such as β-arrestins,[127,128] GRB2,[129,130] APPL1 and APPL2,[131,132] Epsin,[133,134] Disheveled,[135,136] p38 MAPK,[137,138] Intersectin,[139,140] etc. Different methods are being used to measure their expression, interaction and downregulation during receptor endocytosis and signaling.

10. Conclusion and future perspectives

Receptor internalization, trafficking, signaling and metabolic degradation are key events of receptor function and their regulation. These phases play a vital role in physiological and pathophysiological conditions. To mechanistically understand all these events during receptor activation and attenuation, we need to explore genetics, physiology and biochemistry level understanding of the ligand-receptor and associated proteins. There are various methods such as radioligand binding assay, western blotting, co-immunoprecipitation, qRT-PCR, immunofluorescence and confocal microscopy, microRNA/siRNA, and bioassays have been discussed. Further, some recent tools need to explore to understand the different events in *in vitro* (cell culture) and *in vivo* (animal model). Real-time quantitative analysis and live-cell imaging are being used to visualize the receptor endocytosis and signaling processes in the various receptors. Moreover, development in the technology/methods will be helpful to understand the internalization, subcellular trafficking, signaling and metabolic degradation of several receptors.

Conflict of interest

None.

References

1. Barford K, Deppmann C, Winckler B. The neurotrophin receptor signaling endosome: where trafficking meets signaling. *Dev Neurobiol*. 2017;77(4):405–418. https://doi.org/10.1002/dneu.22427.
2. Lombardi JP, Kinzlmaier DA, Jacob TC. Visualizing GABA a receptor trafficking dynamics with fluorogenic protein labeling. *Curr Protoc Neurosci*. 2020;92(1):e97. https://doi.org/10.1002/cpns.97.
3. Mani I, Garg R, Tripathi S, Pandey KN. Subcellular trafficking of guanylyl cyclase/natriuretic peptide receptor-A with concurrent generation of intracellular cGMP. *Biosci Rep*. 2015;35(5):e00260. https://doi.org/10.1042/BSR20150136.
4. Mani I, Pandey KN. Emerging concepts of receptor endocytosis and concurrent intracellular signaling: mechanisms of guanylyl cyclase/natriuretic peptide receptor-A activation and trafficking. *Cell Signal*. 2019;60:17–30. https://doi.org/10.1016/j.cellsig.2019.03.022.
5. Marzook A, Tomas A, Jones B. The interplay of glucagon-like peptide-1 receptor trafficking and signalling in pancreatic beta cells. *Front Endocrinol (Lausanne)*. 2021;12, 678055. https://doi.org/10.3389/fendo.2021.678055.
6. Roney JC, Cheng XT, Sheng ZH. Neuronal endolysosomal transport and lysosomal functionality in maintaining axonostasis. *J Cell Biol*. 2022;221(3):e202111077. https://doi.org/10.1083/jcb.202111077.
7. Wang J, Fedoseienko A, Chen B, Burstein E, Jia D, Billadeau DD. Endosomal receptor trafficking: retromer and beyond. *Traffic*. 2018;19(8):578–590. https://doi.org/10.1111/tra.12574.
8. Calebiro D, Nikolaev VO, Persani L, Lohse MJ. Signaling by internalized G-protein-coupled receptors. *Trends Pharmacol Sci*. 2010;31(5):221–228. https://doi.org/10.1016/j.tips.2010.02.002.
9. Mani I, Garg R, Pandey KN. Role of FQQI motif in the internalization, trafficking, and signaling of guanylyl-cyclase/natriuretic peptide receptor-A in cultured murine mesangial cells. *Am J Physiol Ren Physiol*. 2016;310(1):F68–F84. https://doi.org/10.1152/ajprenal.00205.2015.

10. Mikoshiba K. IP3 receptor/Ca2+ channel: from discovery to new signaling concepts. *J Neurochem*. 2007;102(5):1426–1446. https://doi.org/10.1111/j.1471-4159.2007.04825.x.
11. Mousavi SA, Skjeldal F, Fønhus MS, et al. Receptor-mediated endocytosis of VEGF-A in rat liver sinusoidal endothelial cells. *Biomed Res Int*. 2019;2019:5496197. https://doi.org/10.1155/2019/5496197.
12. Wang S, Liu X, Meng Z, et al. DCBLD2 regulates vascular hyperplasia by modulating the platelet derived growth factor receptor-β endocytosis through Caveolin-1 in vascular smooth muscle cells. *FASEB J*. 2022;36(9), e22488. https://doi.org/10.1096/fj.202200156RR.
13. Song Y, Wang ES, Xing LL, et al. Follicle-stimulating hormone induces postmenopausal dyslipidemia through inhibiting hepatic cholesterol metabolism. *J Clin Endocrinol Metab*. 2016;101(1):254–263. https://doi.org/10.1210/jc.2015-2724.
14. Arancibia-Cárcamo IL, Fairfax BP, Moss SJ, et al. Studying the localization, surface stability and endocytosis of neurotransmitter receptors by antibody labeling and biotinylation approaches. In: Kittler JT, Moss SJ, eds. *The Dynamic Synapse: Molecular Methods in Ionotropic Receptor Biology*. Boca Raton (FL): CRC Press/Taylor & Francis; 2006.
15. Snyder EM, Nong Y, Almeida CG, et al. Regulation of NMDA receptor trafficking by amyloid-beta. *Nat Neurosci*. 2005;8(8):1051–1058.
16. Ehlers MD. Reinsertion or degradation of AMPA receptors determined by activity-dependent endocytic sorting. *Neuron*. 2000;28(2):511–525.
17. Kittler JT, Thomas P, Tretter V, et al. Huntingtin-associated protein 1 regulates inhibitory synaptic transmission by modulating gamma-aminobutyric acid type A receptor membrane trafficking. *Proc Natl Acad Sci U S A*. 2004;101(34):12736–12741.
18. Fairfax BP, Pitcher JA, Scott MG, et al. Phosphorylation and chronic agonist treatment atypically modulate $GABA_B$ receptor cell surface stability. *J Biol Chem*. 2004;279(13):12565–12573.
19. Shin N, Jeong H, Kwon J, et al. LRRK2 regulates synaptic vesicle endocytosis. *Exp Cell Res*. 2008;314(10):2055–2065. https://doi.org/10.1016/j.yexcr.2008.02.015.
20. Yang M, Chen T, Han C, Li N, Wan T, Cao X. Rab7b, a novel lysosome-associated small GTPase, is involved in monocytic differentiation of human acute promyelocytic leukemia cells. *Biochem Biophys Res Commun*. 2004;318(3):792–799. https://doi.org/10.1016/j.bbrc.2004.04.115.
21. Cook TA, Urrutia R, McNiven MA. Identification of dynamin 2, an isoform ubiquitously expressed in rat tissues. *Proc Natl Acad Sci*. 1994;91(2):644–648.
22. Jin S, Zhang B, Weisz OA, Montelaro RC. Receptor-mediated entry by equine infectious anemia virus utilizes a pH-dependent endocytic pathway. *J Virol*. 2005;79(23):14489–14497. https://doi.org/10.1128/JVI.79.23.14489-14497.2005.
23. Mishra AR, Chaturvedi A. B cell receptor signaling and compartmentalization by confocal microscopy. *Methods Mol Biol*. 2018;1707:121–129. https://doi.org/10.1007/978-1-4939-7474-0_9.
24. Koller KJ, Goeddel DV. Molecular biology of the natriuretic peptides and their receptors. *Circulation*. 1992;86(4):1081–1088.
25. Pandey KN. Endocytosis and trafficking of natriuretic peptide receptor-A: potential role of short sequence motifs. *Membranes (Basel)*. 2015;5(3):253–287.
26. Somanna NK, Mani I, Tripathi S, Pandey KN. Clathrin-dependent internalization, signaling, and metabolic processing of guanylyl cyclase/natriuretic peptide receptor-a. *Mol Cell Biochem*. 2018;441(1–2):135–150.
27. Dickey DM, Flora DR, Potter LR. Antibody tracking demonstrates cell type-specific and ligand-independent internalization of guanylyl cyclase A and natriuretic peptide receptor C. *Mol Pharmacol*. 2011;80(1):155–162.

28. Pertwee RG. Pharmacology of cannabinoid receptor ligands. *Curr Med Chem.* 1999;6 (8):635–664. 10469884.
29. Tu Z, Xu J, Jones LA, et al. Radiosynthesis and biological evaluation of a promising sigma(2)-receptor ligand radiolabeled with fluorine-18 or iodine-125 as a PET/SPECT probe for imaging breast cancer. *Appl Radiat Isot.* 2010;68(12):2268–2273. https://doi.org/10.1016/j.apradiso.2010.06.004.
30. Volkow ND, Wang GJ, Fowler JS, et al. Relationship between blockade of dopamine transporters by oral methylphenidate and the increases in extracellular dopamine: therapeutic implications. *Synapse.* 2002;43(3):181–187. https://doi.org/10.1002/syn.10038.
31. Adams KH, Pinborg LH, Svarer C, et al. A database of [(18)F]-altanserin binding to 5-HT(2A) receptors in normal volunteers: normative data and relationship to physiological and demographic variables. *NeuroImage.* 2004;21(3):1105–1113. https://doi.org/10.1016/j.neuroimage.2003.10.046.
32. Henriksen L, Grandal MV, Knudsen SL, van Deurs B, Grøvdal LM. Internalization mechanisms of the epidermal growth factor receptor after activation with different ligands. *PLoS One.* 2013;8(3), e58148. https://doi.org/10.1371/journal.pone.0058148.
33. Flanagan CA. GPCR-radioligand binding assays. *Methods Cell Biol.* 2016;132:191–215. https://doi.org/10.1016/bs.mcb.2015.11.004.
34. Riad A, Zeng C, Weng CC, et al. Sigma-2 receptor/TMEM97 and PGRMC-1 increase the rate of internalization of LDL by LDL receptor through the formation of a ternary complex. *Sci Rep.* 2018;8(1):16845. https://doi.org/10.1038/s41598-018-35430-3.
35. Taylor SC, Berkelman T, Yadav G, Hammond M. A defined methodology for reliable quantification of Western Blot data. *Mol Biotechnol.* 2013;55:217–226. https://doi.org/10.1007/s12033-013-9672-6.
36. MacPhee DJ. Methodological considerations for improving Western blot analysis. *J Pharmacol Toxicol Methods.* 2010;61(2):171–177. https://doi.org/10.1016/j.vascn.2009.12.001.
37. Gürtler A, Kunz N, Gomolka M, et al. Stain-free technology as a normalization tool in Western blot analysis. *Anal Biochem.* 2013;433(2):105–111. https://doi.org/10.1016/j.ab.2012.10.010.
38. Gavini K, Parameshwaran K. Western Blot. In: *StatPearls [Internet].* Treasure Island (FL): StatPearls Publishing; 2022 Apr 28. 2022 Jan 31194430.
39. Mahmood T, Yang PC. Western blot: technique, theory, and trouble shooting. *N Am J Med Sci.* 2012;4(9):429–434. https://doi.org/10.4103/1947-2714.100998.
40. Hnasko TS, Hnasko RM. The Western blot. *Methods Mol Biol.* 2015;1318:87–96. https://doi.org/10.1007/978-1-4939-2742-5_9.
41. Kurien BT, Scofield RH. Western blotting. *Methods.* 2006;38(4):283–293. https://doi.org/10.1016/j.ymeth.2005.11.007.
42. Fowler SJ. Use of monoclonal antibodies for western blotting with enhanced chemiluminescent detection. *Methods Mol Biol.* 1995;45:115–127. https://doi.org/10.1385/0-89603-308-2:115.
43. Alegria-Schaffer A, Lodge A, Vattem K. Performing and optimizing Western blots with an emphasis on chemiluminescent detection. *Methods Enzymol.* 2009;463:573–599. https://doi.org/10.1016/S0076-6879(09)63033-0.
44. Li S, Lei Y, Lei J, Li H. All-trans retinoic acid promotes macrophage phagocytosis and decreases inflammation via inhibiting CD14/TLR4 in acute lung injury. *Mol Med Rep.* 2021;24(6):868. https://doi.org/10.3892/mmr.2021.12508.
45. Wang T, Xing Y, Meng Q, et al. Mammalian Eps15 homology domain 1 potentiates angiogenesis of non-small cell lung cancer by regulating β2AR signaling. *J Exp Clin Cancer Res.* 2019;38(1):174. https://doi.org/10.1186/s13046-019-1162-7.

46. Qin J, Wei T, Chen H, et al. Salicylate induced GABAAR internalization by dopamine D1-like receptors involving protein kinase C (PKC) in spiral ganglion neurons. *Med Sci Monit.* 2021;27:e933278. https://doi.org/10.12659/MSM.933278.
47. Imai M, Kawaoka Y. The role of receptor binding specificity in interspecies transmission of influenza viruses. *Curr Opin Virol.* 2012;2:160–167. https://doi.org/10.1016/j.coviro.2012.03.003.
48. White JM, Delos SE, Brecher M, Schornberg K. Structures and mechanisms of viral membrane fusion proteins. *Crit Rev Biochem Mol Biol.* 2008;43:189–219. https://doi.org/10.1080/10409230802058320.
49. Drews K, Calgi MP, Harrison WC, et al. Glucosylceramidase maintains influenza virus infection by regulating endocytosis. *J Virol.* 2019;93(12):e00017–e00019. https://doi.org/10.1128/JVI.00017-19.
50. Sieghart W. Structure, pharmacology, and function of GABAA receptor subtypes. *Adv Pharmacol.* 2006;54:231–263.
51. Arain FM, Boyd KL, Gallagher MJ. Decreased viability and absence-like epilepsy in mice lacking or deficient in the GABAA receptor α1 subunit. *Epilepsia.* 2012;53:e161–e165.
52. Zhou C, Huang Z, Ding L, et al. Altered cortical GABAA receptor composition, physiology, and endocytosis in a mouse model of a human genetic absence epilepsy syndrome. *J Biol Chem.* 2013;288(29):21458–21472. https://doi.org/10.1074/jbc.M112.444372.
53. Ma X, Chen R, Huang M, et al. DAMGO-induced μ opioid receptor internalization and recycling restore morphine sensitivity in tolerant rat. *Eur J Pharmacol.* 2020;878, 173118. https://doi.org/10.1016/j.ejphar.2020.173118.
54. Seccia TM, Caroccia B, Calo LA. Hypertensive nephropathy. Moving from classic to emerging pathogenetic mechanisms. *J Hypertens.* 2017;35:205–212.
55. Meguid El Nahas A, Bello AK. Chronic kidney disease: the global challenge. *Lancet.* 2005;365:331–340. https://doi.org/10.1016/S0140-6736(05)17789-7.
56. Sun YB, Qu X, Caruana G, Li J. The origin of renal fibroblasts/myofibroblasts and the signals that trigger fibrosis. *Differentiation.* 2016;92:102–107. https://doi.org/10.1016/j.diff.2016.05.008.
57. Kawai T, Masaki T, Doi S, et al. PPAR-gamma agonist attenuates renal interstitial fibrosis and inflammation through reduction of TGF-beta. *Lab Investig.* 2009;89:47–58. https://doi.org/10.1038/labinvest.2008.104.
58. McLeland SM, Lunn KF, Duncan CG, Refsal KR, Quimby JM. Relationship among serum creatinine, serum gastrin, calcium-phosphorus product, and uremic gastropathy in cats with chronic kidney disease. *J Vet Intern Med.* 2014;28:827–837. https://doi.org/10.1111/jvim.12342.
59. Gu D, Fang D, Zhang M, et al. Gastrin, via activation of PPARα, protects the kidney against hypertensive injury. *Clin Sci.* 2021;135(2):409–427. https://doi.org/10.1042/CS20201340.
60. Duman RS, Aghajanian GK, Sanacora G, Krystal JH. Synaptic plasticity and depression: new insights from stress and rapid-acting antidepressants. *Nat Med.* 2016; 22:238–249.
61. Auer DP, Putz B, Kraft E, Lipinski B, Schill J, Holsboer F. Reduced glutamate in the anterior cingulate cortex in depression: an in vivo proton magnetic resonance spectroscopy study. *Biol Psychiatry.* 2000;47:305–313.
62. Ho YC, Lin TB, Hsieh MC, et al. Periaqueductal gray glutamatergic transmission governs chronic stress-induced depression. *Neuropsychopharmacology.* 2018;43(2):302–312. https://doi.org/10.1038/npp.2017.199.
63. Lim Y, Wu LLY, Chen S, et al. HAP1 is required for endocytosis and signalling of BDNF and its receptors in neurons. *Mol Neurobiol.* 2018;55:1815–1830. https://doi.org/10.1007/s12035-016-0379-0.

64. Dong Y, Zhang Y, Kang W, et al. VSTM2A suppresses colorectal cancer and antagonizes Wnt signaling receptor LRP6. *Theranostics.* 2019;9(22):6517.
65. Kang D, Jung SH, Lee GH, et al. Sulfated syndecan 1 is critical to preventing cellular senescence by modulating fibroblast growth factor receptor endocytosis. *FASEB J.* 2020;34(8):10316–10328.
66. Ashourpour F, Jafari A, Babaei P. Co-treatment of AMPA endocytosis inhibitor and GluN2B antagonist facilitate consolidation and retrieval of memory impaired by β amyloid peptide. *Int J Neurosci.* 2022;132(7):714–723. https://doi.org/10.1080/00207454.2020.1837800.
67. Iqbal H, Akins DR, Kenedy MR. Co-immunoprecipitation for identifying protein-protein interactions in Borrelia burgdorferi. *Methods Mol Biol.* 2018;1690:47–55. https://doi.org/10.1007/978-1-4939-7383-5_4.
68. Takahashi Y. Co-immunoprecipitation from transfected cells. *Methods Mol Biol.* 2015;1278:381–389. https://doi.org/10.1007/978-1-4939-2425-7_25.
69. Gent J, van Kerkhof P, Roza M, Bu G, Strous GJ. Ligand-independent growth hormone receptor dimerization occurs in the endoplasmic reticulum and is required for ubiquitin system-dependent endocytosis. *Proc Natl Acad Sci U S A.* 2002;99(15):9858–9863. https://doi.org/10.1073/pnas.152294299.
70. Zhu XM, Liang S, Shi HB, et al. VPS9 domain-containing proteins are essential for autophagy and endocytosis in Pyricularia oryzae. *Environ Microbiol.* 2018;20(4):1516–1530. https://doi.org/10.1111/1462-2920.14076.
71. Vanlandingham PA, Barmchi MP, Royer S, et al. AP180 couples protein retrieval to clathrin-mediated endocytosis of synaptic vesicles. *Traffic.* 2014;15(4):433–450. https://doi.org/10.1111/tra.12153.
72. Morris SM, Arden SD, Roberts RC, et al. Myosin VI binds to and localises with Dab2, potentially linking receptor-mediated endocytosis and the actin cytoskeleton. *Traffic.* 2002;3(5):331–341. https://doi.org/10.1034/j.1600-0854.2002.30503.x.
73. Rajan SS, Liu HY, Vu TQ. Ligand-bound quantum dot probes for studying the molecular scale dynamics of receptor endocytic trafficking in live cells. *ACS Nano.* 2008;2(6):1153–1166.
74. Troyanovsky RB, Sokolov EP, Troyanovsky SM. Endocytosis of cadherin from intracellular junctions is the driving force for cadherin adhesive dimer disassembly. *Mol Biol Cell.* 2006;17(8):3484–3493.
75. Kotova A, Timonina K, Zoidl GR. Endocytosis of connexin 36 is mediated by interaction with caveolin-1. *Int J Mol Sci.* 2020;21(15):5401.
76. Nonis D, et al. Ataxin-2 associates with the endocytosis complex and affects EGF receptor trafficking. *Cell Signal.* 2008;20(10):1725–1739.
77. Josefsen K, Nielsen H. Northern blotting analysis. *Methods Mol Biol.* 2011;703:87–105. https://doi.org/10.1007/978-1-59745-248-9_7.
78. Smythe E, Smith PD, Jacob SM, Theobald J, Moss SE. Endocytosis occurs independently of annexin VI in human A431 cells. *J Cell Biol.* 1994;124(3):301–306. https://doi.org/10.1083/jcb.124.3.301.
79. Oosterhoff JK, Kühne LC, Grootegoed JA, Blok LJ. EGF signalling in prostate cancer cell lines is inhibited by a high expression level of the endocytosis protein REPS2. *Int J Cancer.* 2005;113(4):561–567. https://doi.org/10.1002/ijc.20612.
80. Petrie Jr RG, Morales CR. Receptor-mediated endocytosis of testicular transferrin by germinal cells of the rat testis. *Cell Tissue Res.* 1992;267(1):45–55. https://doi.org/10.1007/BF00318690.
81. Sakakibara T, et al. Identification and characterization of a novel rho GTPase activating protein implicated in receptor-mediated endocytosis. *FEBS Lett.* 2004;566(1–3):294–300.

82. Scheithe R, et al. Acetylated LDL endocytosis by the human monocytic Mono Mac 6sr cells is not mediated by the macrophage type I and II scavenger receptors. *Arterioscler Thromb Vasc Biol.* 1997;17(6):1024–1031.
83. Arya M, Shergill IS, Williamson M, Gommersall L, Arya N, Patel HR. Basic principles of real-time quantitative PCR. *Expert Rev Mol Diagn.* 2005;5(2):209–219. https://doi.org/10.1586/14737159.5.2.209.
84. Heid CA, Stevens J, Livak KJ, Williams PM. Quantitative real time PCR. *Genome Res.* 1996;6:986–994.
85. Schaffert DH, Okholm AH, Sørensen RS, et al. Intracellular delivery of a planar DNA origami structure by the transferrin-receptor internalization pathway. *Small.* 2016;12 (19):2634–2640. https://doi.org/10.1002/smll.201503934.
86. Weyer K, Andersen PK, Schmidt K, et al. Abolishment of proximal tubule albumin endocytosis does not affect plasma albumin during nephrotic syndrome in mice. *Kidney Int.* 2018;93(2):335–342. https://doi.org/10.1016/j.kint.2017.07.024.
87. Sato K, et al. Effects of endocytosis inhibitors on internalization of human IgG by Caco-2 human intestinal epithelial cells. *Life Sci.* 2009;85(23–26):800–807.
88. Daecke J, et al. Involvement of clathrin-mediated endocytosis in human immunodeficiency virus type 1 entry. *J Virol.* 2005;79(3):1581–1594.
89. Su WC, Lai MMC. Quantitative RT-PCR analysis of influenza virus endocytic escape. *Methods Mol Biol.* 2018;1836:185–194. https://doi.org/10.1007/978-1-4939-8678-1_9.
90. Pongsomboon S, Tang S, Boonda S, Aoki T, Hirono I, Tassanakajon A. A cDNA microarray approach for analyzing transcriptional changes in Penaeus monodon after infection by pathogens. *Fish Shellfish Immunol.* 2011;30(1):439–446. https://doi.org/10.1016/j.fsi.2010.10.015.
91. Jatuyosporn T, et al. The essential role of clathrin-mediated endocytosis in yellow head virus propagation in the black tiger shrimp Penaeus monodon. *Dev Comp Immunol.* 2014;44(1):100–110.
92. Dean PN. Confocal microscopy: principles and practices. *Curr Protoc Cytom.* 1998; 2.7.1–2.8.12. Unit 2.8 (Chapter 2).
93. Nwaneshiudu A, Kuschal C, Sakamoto FH, Anderson RR, Schwarzenberger K, Young RC. Introduction to confocal microscopy. *J Invest Dermatol.* 2012;132(12): e3. https://doi.org/10.1038/jid.2012.429.
94. Elliott AD. Confocal microscopy: principles and modern practices. *Curr Protoc Cytom.* 2020;92(1):e68. https://doi.org/10.1002/cpcy.68.
95. Jones LO, Stafford JL. Imaging flow cytometry and confocal microscopy-based examination of F-actin and phosphoinositide dynamics during leukocyte immune-type receptor-mediated phagocytic events. *Dev Comp Immunol.* 2019;92:199–211. https://doi.org/10.1016/j.dci.2018.11.018.
96. Spiess K, Bagger SO, Torz LJ, et al. Arrestin-independent constitutive endocytosis of GPR125/ADGRA3. *Ann N Y Acad Sci.* 2019;1456(1):186–199. https://doi.org/10.1111/nyas.14263.
97. Zhong H, Lin H, Pang Q, et al. Macrophage ICAM-1 functions as a regulator of phagocytosis in LPS induced endotoxemia. *Inflamm Res.* 2021;70(2):193–203. https://doi.org/10.1007/s00011-021-01437-2.
98. Goh LK, Sorkin A. Endocytosis of receptor tyrosine kinases. *Cold Spring Harb Perspect Biol.* 2013;5(5), a017459.
99. Dam DHM, Jelsma SA, Yu JM, Liu H, Kong B, Paller AS. Flotillin and AP2A1/2 promote IGF-1 receptor association with Clathrin and internalization in primary human keratinocytes. *J Invest Dermatol.* 2020;140(9):1743–1752.e4. https://doi.org/10.1016/j.jid.2020.01.015.

100. Darnell Jr JE. STATs and gene regulation. *Science*. 1997;277(5332):1630–1635. https://doi.org/10.1126/science.277.5332.
101. Schindler C, Darnell Jr JE. Transcriptional responses to polypeptide ligands: the JAK-STAT pathway. *Annu Rev Biochem*. 1995;64:621–651. https://doi.org/10.1146/annurev.bi.64.070195.003201.
102. Bild AH, Turkson J, Jove R. Cytoplasmic transport of Stat3 by receptor-mediated endocytosis. *EMBO J*. 2002;21(13):3255–3263.
103. Kumar GA, Chattopadhyay A. Statin-induced chronic cholesterol depletion switches GPCR endocytosis and trafficking: insights from the serotonin1A receptor. *ACS Chem Neurosci*. 2019;11(3):453–465.
104. Fischer AW, Albers K, Krott LM, et al. The adaptor protein PID1 regulates receptor-dependent endocytosis of postprandial triglyceride-rich lipoproteins. *Mol Metab*. 2018;16:88–99.
105. Kumar GA, Chattopadhyay A. Membrane cholesterol regulates endocytosis and trafficking of the serotonin1A receptor: insights from acute cholesterol depletion. *Biochim Biophys Acta Mol Cell Biol Lipids*. 2021;1866(4):158882.
106. Holst HU, Dagnaes-Hansen F, Corydon TJ, et al. LDL receptor-GFP fusion proteins: new tools for the characterisation of disease-causing mutations in the LDL receptor gene. *Eur J Hum Genet*. 2001;9(11):815–822.
107. Foster N, Loi TH, Owe-Young R, Stanley KK. Lysosomal traffic of liganded endothelin B receptor. *Biochim Biophys Acta*. 2003;1642(1–2):45–52.
108. Xia S, Kjaer S, Zheng K, et al. Visualization of a functionally enhanced GFP-tagged galanin R2 receptor in PC12 cells: constitutive and ligand-induced internalization. *Proc Natl Acad Sci U S A*. 2004;101(42):15207–15212.
109. Prenner L, Sieben A, Zeller K, Weiser D, Häberlein H. Reduction of high-affinity beta2-adrenergic receptor binding by hyperforin and hyperoside on rat C6 glioblastoma cells measured by fluorescence correlation spectroscopy. *Biochemistry*. 2007;46(17):5106–5113.
110. Kilpatrick LE, Briddon SJ, Holliday ND. Fluorescence correlation spectroscopy, combined with bimolecular fluorescence complementation, reveals the effects of β-arrestin complexes and endocytic targeting on the membrane mobility of neuropeptide Y receptors. *Biochim Biophys Acta*. 2012;1823(6):1068–1081.
111. He Q, Sun X, Chu C, et al. Endocytosis of a functionally enhanced GFP-tagged transferrin receptor in CHO cells. *PLoS One*. 2015;10(3):e0122452.
112. Robinson BG, Bunzow JR, Grimm JB, et al. Desensitized D2 autoreceptors are resistant to trafficking. *Sci Rep*. 2017;7(1):4379.
113. Pinilla-Macua I, Grassart A, Duvvuri U, Watkins SC, Sorkin A. EGF receptor signaling, phosphorylation, ubiquitylation and endocytosis in tumors in vivo. *elife*. 2017;6:e31993. https://doi.org/10.7554/eLife.31993.
114. Bayliss AL, Sundararaman A, Granet C, Mellor H. Raftlin is recruited by neuropilin-1 to the activated VEGFR2 complex to control proangiogenic signaling. *Angiogenesis*. 2020;23(3):371–383.
115. Bartel DP. MicroRNAs: target recognition and regulatory functions. *Cell*. 2009;136(2):215–233. https://doi.org/10.1016/j.cell.2009.01.002.
116. Martinez J, Tuschl T. RISC is a 5′ phosphomonoester-producing RNA endonuclease. *Genes Dev*. 2004;18(9):975–980. https://doi.org/10.1101/gad.1187904.
117. Gilleron J, Querbes W, Zeigerer A, et al. Image-based analysis of lipid nanoparticle-mediated siRNA delivery, intracellular trafficking and endosomal escape. *Nat Biotechnol*. 2013;31(7):638–646. https://doi.org/10.1038/nbt.2612.
118. Gerasymchuk D, Hubiernatorova A, Domanskyi A. MicroRNAs regulating cytoskeleton dynamics, endocytosis, and cell motility-a link between neurodegeneration and cancer? *Front Neurol*. 2020;11:549006. https://doi.org/10.3389/fneur.2020.549006.

119. Sundaramurthy V, Barsacchi R, Samusik N, et al. Integration of chemical and RNAi multiparametric profiles identifies triggers of intracellular mycobacterial killing. *Cell Host Microbe*. 2013;13(2):129–142.
120. Serva A, Knapp B, Tsai YT, et al. miR-17-5p regulates endocytic trafficking through targeting TBC1D2/Armus. *PLoS One*. 2012;7(12), e52555. https://doi.org/10.1371/journal.pone.0052555.
121. Zhang L, Zhang Y, Wong SH, et al. Common deregulation of seven biological processes by MicroRNAs in gastrointestinal cancers. *Sci Rep*. 2018;8(1):3287. https://doi.org/10.1038/s41598-018-21573-w.
122. Mani I, Vasdev K. MicroRNA in prognosis, diagnosis and therapy of cancer. *Cell Cell Life Sci J*. 2018;3:000134.
123. Feil R, Kemp-Harper B. cGMP signalling: from bench to bedside. Conference on cGMP generators, effectors and therapeutic implications. *EMBO Rep*. 2006;7 (2):149–153. https://doi.org/10.1038/sj.embor.7400627.
124. Lincoln TM, Cornwell TL. Intracellular cyclic GMP receptor proteins. *FASEB J*. 1993;7(2):328–338. https://doi.org/10.1096/fasebj.7.2.7680013.
125. Rybalkin SD, Hinds TR, Beavo JA. Enzyme assays for cGMP hydrolyzing phosphodiesterases. *Methods Mol Biol*. 2013;1020:51–62. https://doi.org/10.1007/978-1-62703-459-3_3.
126. Yu L, Shi X, Han C, Rao C, Wang J. A rapid reporter assay for recombinant human brain natriuretic peptide (rhBNP) by GloSensor technology. *J Pharm Anal*. 2018;8 (5):297–301. https://doi.org/10.1016/j.jpha.2018.04.003.
127. Lefkowitz RJ, Pitcher J, Krueger K, Daaka Y. Mechanisms of beta-adrenergic receptor desensitization and resensitization. *Adv Pharmacol*. 1998;42:416–420. https://doi.org/10.1016/s1054-3589(08)60777-2.
128. Moore CA, Milano SK, Benovic JL. Regulation of receptor trafficking by GRKs and arrestins. *Annu Rev Physiol*. 2007;69:451–482. https://doi.org/10.1146/annurev.physiol.69.022405.
129. Rozakis-Adcock M, Fernley R, Wade J, Pawson T, Bowtell D. The SH2 and SH3 domains of mammalian Grb2 couple the EGF receptor to the Ras activator mSos1. *Nature*. 1993;363(6424):83–85. https://doi.org/10.1038/363083a0.
130. Waterman H, Katz M, Rubin C, et al. A mutant EGF-receptor defective in ubiquitylation and endocytosis unveils a role for Grb2 in negative signaling. *EMBO J*. 2002;21(3):303–313. https://doi.org/10.1093/emboj/21.3.303.
131. Lin DC, Quevedo C, Brewer NE, et al. APPL1 associates with TrkA and GIPC1 and is required for nerve growth factor-mediated signal transduction. *Mol Cell Biol*. 2006;26 (23):8928–8941. https://doi.org/10.1128/MCB.00228-06.
132. Varsano T, Dong MQ, Niesman I, et al. GIPC is recruited by APPL to peripheral TrkA endosomes and regulates TrkA trafficking and signaling. *Mol Cell Biol*. 2006;26 (23):8942–8952. https://doi.org/10.1128/MCB.00305-06.
133. Aguilar RC, Longhi SA, Shaw JD, et al. Epsin N-terminal homology domains perform an essential function regulating Cdc42 through binding Cdc42 GTPase-activating proteins. *Proc Natl Acad Sci U S A*. 2006;103(11):4116–4121. https://doi.org/10.1073/pnas.0510513103.
134. Kazazic M, Bertelsen V, Pedersen KW, et al. Epsin 1 is involved in recruitment of ubiquitinated EGF receptors into clathrin-coated pits. *Traffic*. 2009;10(2):235–245. https://doi.org/10.1111/j.1600-0854.2008.00858.x.
135. Schulte G, Bryja V. The frizzled family of unconventional G-protein-coupled receptors. *Trends Pharmacol Sci*. 2007;28(10):518–525. https://doi.org/10.1016/j.tips.2007.09.001.
136. Yu A, Rual JF, Tamai K, et al. Association of dishevelled with the clathrin AP-2 adaptor is required for frizzled endocytosis and planar cell polarity signaling. *Dev Cell*. 2007;12 (1):129–141. https://doi.org/10.1016/j.devcel.2006.10.015.

137. Cavalli V, Vilbois F, Corti M, et al. The stress-induced MAP kinase p38 regulates endocytic trafficking via the GDI:Rab5 complex. *Mol Cell.* 2001;7(2):421–432. https://doi.org/10.1016/s1097-2765(01)00189-7.
138. Loesch M, Chen G. The p38 MAPK stress pathway as a tumor suppressor or more? *Front Biosci.* 2008;13:3581–3593. https://doi.org/10.2741/2951.
139. Hussain NK, Jenna S, Glogauer M, et al. Endocytic protein intersectin-l regulates actin assembly via Cdc42 and N-WASP. *Nat Cell Biol.* 2001;3(10):927–932. https://doi.org/10.1038/ncb1001-927.
140. Martin NP, Mohney RP, Dunn S, Das M, Scappini E, O'Bryan JP. Intersectin regulates epidermal growth factor receptor endocytosis, ubiquitylation, and signaling. *Mol Pharmacol.* 2006;70(5):1643–1653.

CHAPTER THREE

Mas receptor endocytosis and signaling in health and disease

Natalia L. Rukavina Mikusic and Mariela M. Gironacci*
From Universidad de Buenos Aires, Facultad de Farmacia y Bioquímica, Dpto. Química Biológica, IQUIFIB (UBA-CONICET), Buenos Aires, Argentina
*Corresponding author: e-mail addresses: mariela@qb.ffyb.uba.ar; marielagironacci@gmail.com

Contents

Abstract

The renin angiotensin system (RAS) plays a major role in blood pressure regulation and electrolyte homeostasis and is mainly composed by two axes mediating opposite effects. The pressor axis, constituted by angiotensin (Ang) II and the Ang II type 1 receptor (AT1R), exerts vasoconstrictor, proliferative, hypertensive, oxidative and pro-inflammatory actions, while the depressor/protective axis, represented by Ang-(1-7), its Mas receptor (MasR) and the Ang II type 2 receptor (AT2R), opposes the actions elicited by the pressor arm. The MasR belongs to the G protein-coupled receptor (GPCR) family. To avoid receptor overstimulation, GPCRs undergo internalization and trafficking into the cell after being stimulated. Then, the receptor may induce other signaling cascades or it may even interact with other receptors, generating distinct biological responses. Thus, control of a GPCR regarding space and time affects the specificity of the signals transduced by the receptor and the ultimate cellular response. The present chapter is focused on the signaling and trafficking pathways of MasR under physiological conditions and its participation in the pathogenesis of numerous brain diseases.

Progress in Molecular Biology and Translational Science, Volume 194
ISSN 1877-1173
https://doi.org/10.1016/bs.pmbts.2022.09.001

1. Introduction

The renin-angiotensin system (RAS) is essential for blood pressure regulation and electrolyte and water homeostasis. The RAS is a complex interconnected system composed of two axes.[1,2] The pressor axis, represented by angiotensin (Ang) II and the Ang II type 1 receptor (AT1R), exerts vasoconstrictive, trophic and pro-inflammatory actions. The protective axis counteracts AT1R-mediated actions and exerts vasodilatory, anti-inflammatory and anti-fibrotic actions via Ang-(1-7) and its receptor Mas (MasR).[1,2] The Ang II type 2 receptor (AT2R) that opposes the responses mediated by AT1R also belongs to this protective axis of the RAS. Ang-(1-7) differs from the Ang II sequence by the lack of the C-terminal phenylalanine residue. Ang-(1-7) is synthesized from Ang II by the mono-carboxypeptidase angiotensin converting enzyme (ACE) 2.[3] With less efficiency, ACE2 generates Ang-(1-9) from Ang I, which then is cleaved by ACE, prolyl oligopeptidase or the enzyme neprilysin generating Ang-(1-7).[4,5] In addition, Ang-(1-7) is also synthesized from Ang I. Neutral endopeptidase, thimet oligopeptidase and prolyl oligopeptidase cleave the bond between Pro^7-Phe^8 residues of the Ang I sequence, hence producing Ang-(1-7) (Fig. 1).[2,6]

RAS expression in several tissues has been identified including the kidney, heart, adipose, arterial wall, and brain. Tissue RAS does not depend on systemic RAS regulation.[7] All the constituents of the Ang-(1-7)/MasR axis have been recognized to be present in almost all cell type of the brain.[8] Centrally, Ang-(1-7) is involved in the regulation of cardiovascular function, influencing blood pressure.[9,10] Acting on MasR, Ang-(1-7) acts as a neuroprotective agent by counteracting the detrimental effects of the Ang II/AT1R axis.[10,11] Thus, Ang-(1-7), via MasR activation, induces vasodilation and angiogenesis and reduced oxidative stress and inflammation.[12] Furthermore, MasR activation prevented neurodegeneration, demyelination of axons, synapse changes and injury of astrocytes and oligodendrocytes elicited by Shigha toxin 2.[13] In this sense, alterations of the centrally Ang-(1-7)/MasR axis have been involved in the pathogenesis of several cardiovascular and neurodegenerative diseases.[14–17] The present chapter is focused on the signaling and trafficking pathways of MasR under physiological conditions and in the pathogenesis of numerous diseases at central level.

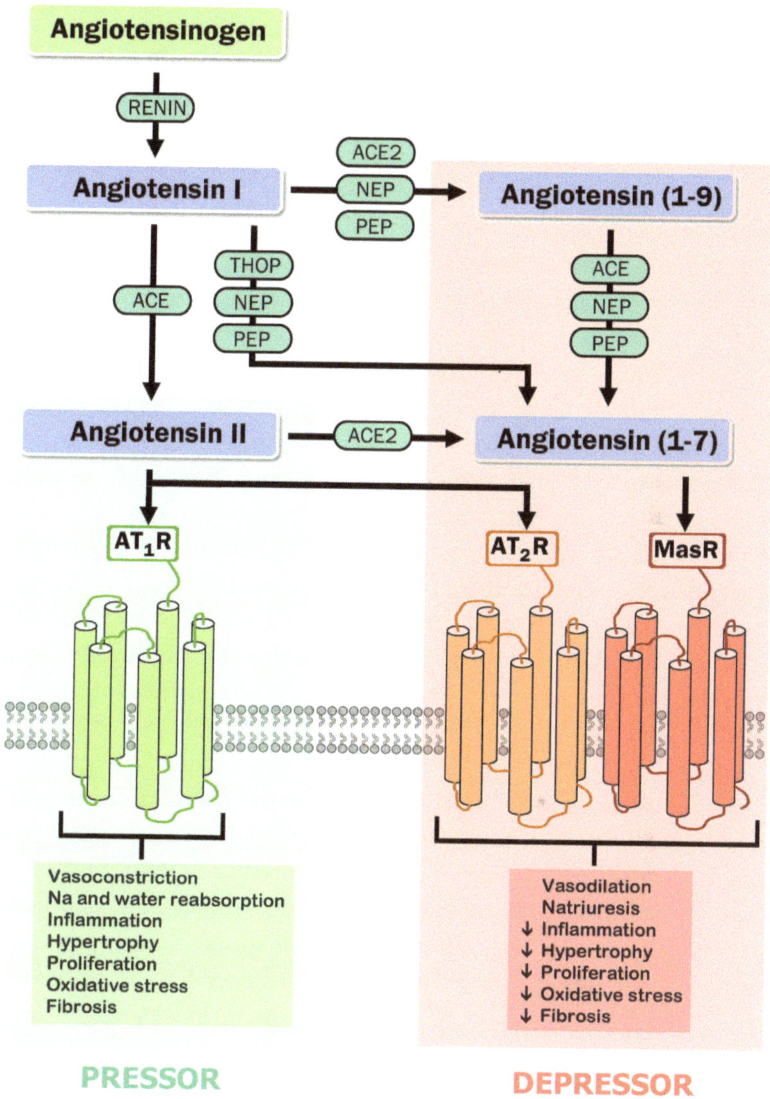

Fig. 1 The pressor and protective axis of the renin-angiotensin system. Generation of angiotensin II and angiotensin-(1-7) is presented. The pressor axis represented by angiotensin II, AT1R and ACE and the depressor axis represented by angiotensin-(1-7), MasR, AT2R and ACE2 are depicted. The main biological responses coupled to AT1R, AT2R and MasR stimulation are included under each receptor. Abbreviations: ACE, angiotensin-converting enzyme; ACE2, angiotensin-converting enzyme 2; Ang II, Angiotensin II; AT1R, Ang II type 1 receptor; AT2R, Ang II type 2 receptor; Mas R, Mas receptor; NEP, neutral endopeptidase (neprilysin); THOP, thimet oligopeptidase; PEP, prolyl endopeptidase;

2. Gene structure and expression of MasR

MasR is a member of the G-protein-coupled receptors (GPCRs) family. In 1986, Young et al. identified the gene encoding the MasR, the Mas proto-oncogene.[18] MasR is currently catalogued under a family of Mas-related G protein–coupled receptors (Mrgprs).[19] MasR is the prototype and the best studied member of the Mrgpr family, mainly constituted by GPCRs whose main ligands are neuropeptides and exert major physiological actions.[20] MasR is ubiquitously distributed in many tissues and cells. It is primarily expressed in the brain and testes and, less strongly, in the kidney, heart, vessels, adipose tissue, lungs and skeletal muscle.[21–24] The MasR coding region has been isolated and sequenced in humans[18], rat[25], and mouse.[26] The gene is positioned on chromosome 17 in mouse[27], on chromosome 1 in rat and on chromosome 6 in humans.[28] The complete mouse gene of MasR is formed by 12 exons and spans over 27 kb. The presence of alternative splicing and at least four promoters results in at least 12 MasR different transcripts. Termination of MasR transcription is controlled by two different polyadenylation signals. This evidence shows that the MasR gene structure is one of the most intricate inside the group of coding genes for the GPCRs identified to date.[24]

MasR has been reported to be present in hippocampus, amygdala, cortex and hypoglossal nucleus, along with regions of medulla and forebrain associated with cardiovascular functions.[22,29] MasR is predominantly present in nuclear and non-nuclear compartments of neurons[30–32], but it has also been detected in astrocytes and microglia.[30,32] Additionally, the expression of MasR was higher in hypothalamic neurons of spontaneously hypertensive rats (SHR) than from Wistar-Kyoto (WKY) rats used as the normotensive control.[31]

2.1 MasR signaling

MasR stimulation in response to Ang-(1-7) binding has been shown to be coupled to many downstream signaling pathways in different cell types, comprising phospholipase A2 (PLA2) activation, arachidonic acid liberation, nitric oxide (NO) production, activation of Akt, mitogen activated protein kinase (MAPKs, RhoA) and cyclic adenosine monophosphate (cAMP)/protein kinase A (PKA).[2] MasR is coupled to multiple signaling cascades elicited by Gαs, Gαq, Gαi, and Gα12/13 proteins in a constitutive manner.[19] Ang-(1-7) binding to MasR leads to the activation of different effectors like NO and cyclo-oxygenase-2 (COX-2).[8,33] In cardiomyocytes, Ang-(1-7) increases NO levels through the phosphatidylinositol 3-kinase (PI3K)/Akt signaling cascade and inhibits the MAPK/ERK1/2 pathway.[34,35] Most of the cardiac protective MasR-mediated signalings antagonize Ang II's actions, such as the activation of the PI3K/Akt/endothelial nitric oxide synthase (NOS) pathway and blockade of protein kinase C–p38 mitogen-activated protein kinase signaling cascades.[10,36,37] Ang-(1-7) prevents the translocation of NFAT (nuclear factor of activated T cells), which is a transcription factor involved in gene transcription that stimulate hypertrophy, through the inhibition of glycogen synthase kinase 3β.[36] Moreover, Ang-(1-7) reduced proinflammatory cytokines (tumor necrosis factor-α (TNF- α) and interleukin (IL) 6) levels in the myocardium, alleviating cardiac inflammation.[38]

Centrally, there is plenty of evidence showing that MasR activation induces NO generation.[39–41] MasR stimulation increases neuronal NOS in hypothalami from WKY and SHR rats, thus leading to NO generation.[42] The cerebral anti-inflammatory effect of Ang-(1-7) mediated by MasR is coupled to a decrease in inducible NOS expression, IL1α, IL6, IL8, TNFα, monocyte chemoattractant protein-1, nuclear factor kappa B (NF-κB), chemokine receptor type 4 and the marker of activation of macrophages/microglia CD11b demonstrating a reduction in microglia activation.[30,43,44] The MasR-mediated changes in norepinephrine transporter gene transcription in rat hypothalamic neurons is coupled to PI3K/Akt and MEK 1/2-ERK 1/2 pathways.[31] COX-2 has been shown to mediate the alterations in plasticity elicited by Ang-(1-7) in the lateral amygdala.[39] Stimulation of MasR in the paraventricular nucleus and rostral ventrolateral medulla leads to cAMP-protein kinase A activation, which acts as a mediator of the increased sympathetic outflow and cardiac sympathetic afferent reflex in response to Ang-(1-7).[45,46] Since MasR is a GPCR that is active in a constitutive manner, variations in protein amounts of the receptor can induce changes in constitutive signaling. Therefore, MasR can function without requiring a ligand.[19]

During the last years, it has been shown that GPCRs, including MasR, mediate distinct biological actions by stimulation of pleiotropic signaling pathways, in addition to their canonical mechanisms. The association of a GPCR with another receptor to form homomeric or heteromeric complexes is one of the available mechanisms leading to diversification of GPCRs signaling pathways.[47]

2.2 MasR heteromerization

GPCRs signaling can be diversified by receptor dimerization or oligomerization, to form homomers or heteromers.[47] Heteromerization of receptors affects important functions such as their biological synthesis, plasma membrane diffusion rate, binding to ligand, signaling and pharmacology.[48,49]

Heterodimeric interactions between MasR and AT1R have been described in transfected CHO-K1 cells. MasR-AT1R constitutive heterodimerization resulted in changes in AT1R trafficking and reduced production of inositol phosphate and intracellular Ca^{2+} mobilization induced by Ang II.[50] MasR-AT2R heteromer mediates the CX3C chemokine receptor-1 messenger RNA expression induced by Ang-(1-7) in cultured mouse astrocytes.[51] MasR-AT2 heteromer formation has also been described in obese Zucker rat kidney, in association with the induction of natriuresis, diuresis and NO generation.[52] The existence of heteromeric complexes involving MasR and the bradykinin type 2 receptor (B2R) has been reported in mesenteric vascular beds from Wistar rats and in glomerular endothelial cells from humans.[53] MasR-B2R heteromerization led to a delay in MasR sequestration from the cell membrane and an increment in the binding affinity to ligand of MasR in HEK293T cells transfected with both receptors.[53] The existence of heteromeric complexes between MasR and the dopamine D2R has been proved on the cell surface of transfected HEK-293 T cells. MasR-D2R heteromer formation was constitutive and was not modified by its ligands and was implicated in ERK1/2 and Akt activation elicited by Ang-(1-7) and dopamine.[54] MasR-D2R heteromers were also involved in the anti-inflammatory effects induced by Ang-(1-7) and the specific D2R agonist sumanirole in human macrophages.[54] Recently it has shown the existence of MasR-AT1R and MasR-AT2 in dopaminergic neurons and microglia, with a greater heteromer expression in microglia with respect to neurons.[55] Furthermore, the amount of heteromers between MasR and either AT1R or AT2R was higher in the striatum of the injured hemisphere in a model of Parkinson's Disease (PD) in rats.[55] Fig. 2 represents the biological responses coupled to MasR heteromerization with other receptors.

Altogether these results show that MasR interacts with each other, with another receptor, with G proteins, or with other cell constituents in a dynamic way, inducing different biological responses.

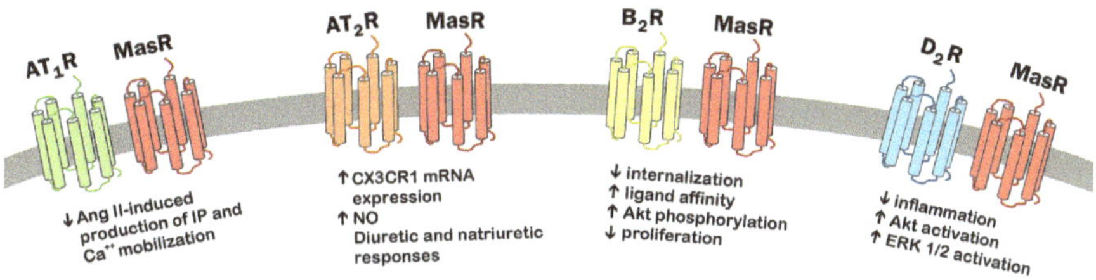

Fig. 2 Biological responses associated to MasR heteromerization with other receptors. Biological responses coupled to MasR heteromerization with AT1R, AT2R, B2R and D2R is depicted under each heteromer. Abbreviations: Ang II, Angiotensin II; AT1R, Ang II type 1 receptor; AT2R, Ang II type 2 receptor; B2R, Bradykinin type 2 receptor; CX3CR1, CX3C chemokine receptor; D2R, dopamine type 2 receptor; IP, inositol phosphate; NO, nitric oxide.

3. MasR trafficking

Endocytic trafficking affects the way that receptor signaling is organized and directed, influencing GRCPs signaling complexity and specificity.[48] Internalization of GRCPs can occur through caveolae, clathrin-coated vesicles, or non-coated vesicles.[56,57] Endocytosis by clathrin-coated vesicles is the best characterized mechanism.[57] These vesicles contain complexes of clathrin and adapter protein 2 (AP2). The β-arrestins facilitate the binding of GRCPs to proteins involved in the internalization and subsequent intracellular trafficking of the receptor, such as AP2 and clathrin.[57,58] β-arrestin1 (or arrestin 2) and β-arrestin2 (or arrestin 3) expression has been identified in several tissues and play key roles as adapters and scaffolds in this process. Phosphorylated GRCPs with β-arrestin attached to it is recruited into a clathrin shell. This coating invaginates and is released from the membrane by dynamin, a guanosine triphosphatase (GTPase), to form the early endosome. Some GRCPs are preferentially internalized through caveolae, invaginations of the plasma membrane rich in sphingolipids and cholesterol, covered with type 1, 2 or 3 caveolins, which are important for the formation and maintenance of caveolae.[59] GPCR trafficking offers different signaling platforms required for the establishment of specific receptor actions. Therefore, GPCRs can exhibit different signaling pathways regarding their localization in the cell.[60] After internalization, GPCRs may return back to the plasma membrane or be sorted to intracellular compartments such as lysosomes, endoplasmic reticulum, Golgi body, mitochondria, or, at last, be directed to the nuclear compartiment.[61]

After binding of its ligand, MasR is trafficked into early endosomes.[62] MasR is internalized via clathrin-coated pits and caveolae in a pathway dependent on dynamin.[63] After internalization, MasR activates Akt signaling pathway in a mechanism independent of β-arrestin2. On the contrary, β-arrestin2 participates in ERK1/2 activation elicited by MasR. Subsequently, MasR is returned back to the cell membrane via slow recycling vesicles in HEK293T cells transfected with MasR.[63] This mechanism of trafficking has also been observed in cultured brainstem neurons from Sprague-Dawley, WKY and SHR rats.[64] Nevertheless, in brainstem neuronal cells from SHRs, MasR exhibits different mechanisms of trafficking: the fraction of internalized MasRs into early endosomes is larger and the fraction of MasRs returned back to the cell membrane is smaller in brainstem neurons from SHRs in comparison to normotensive WKY rats.[64] Furthermore, after stimulation with its agonist Ang-(1-7), MasR is trafficked to the nuclear compartment only in hypothalamic neurons from SHRs and not from WKY[64] (Fig. 3).

The nucleus has all the machinery implicated in GPCRs signaling cascades including G-proteins, second messengers and ion channels.[61] MasR has been identified in the nucleus of dopaminergic neurons[32], brainstem neurons of normotensive and SHR rats[64] and in the sheep kidney[65], and to be functionally active. Ang-(1-7) promotes NO production via MasR activation present in nuclei obtained from young adult sheep kidneys[65] and substantia nigra of rodents[32]. The rise in NO induced by nuclear MasR activation is decreased in aged animals or those with fetal programmed hypertension[32,66,67]. Nuclear MasR activation by Ang-(1-7) also opposed to the rise in superoxide production elicited by Ang II in the nucleus and reduced the synthesis of AT2Rs mRNA. This action was reduced in aged animals.[32] All these biological responses mediated by nuclear MasR activation were not due to MasR trafficking to the nuclear compartment. Physiological consequences of agonist-dependent MasR nuclear translocation in SHR neurons[64] have not been determined to date.

4. MasR signaling and receptor trafficking in pathological situations

Given the physiological importance of GPCRs in various cellular responses, alterations in their signaling and trafficking have been linked to the development of several pathologies. Defects in receptor signaling and trafficking are implicated in the pathophysiology of several brain disorders,

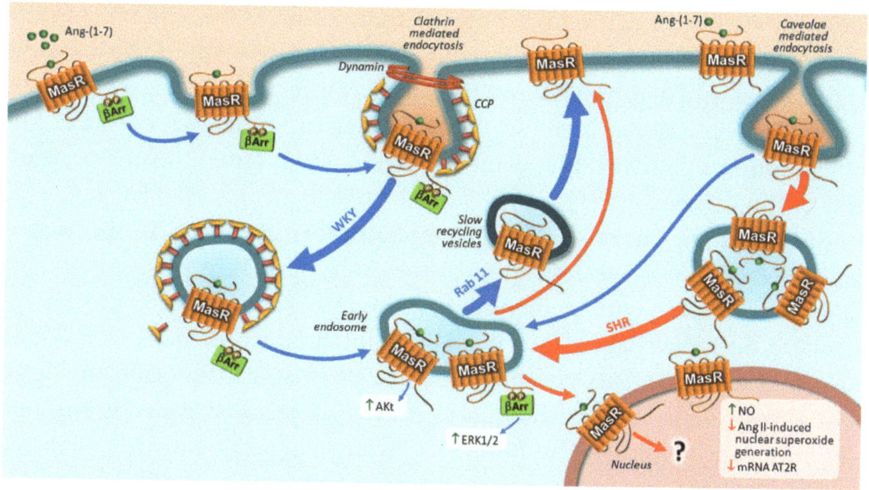

Fig. 3 Intracellular trafficking of MasR and main biological responses coupled to nuclear MasR stimulation. Upon Ang-(1-7) stimulation, MasR is endocyted into early endosome by clathrin-coated pits and caveolae in a dynamin-dependent pathway and then recycled back to the cell surface by slow recycling vesicles. Once in early endosome, MasR triggers Akt and ERK1/2 activation. MasR undergoes a unique trafficking in brainstem neurons from SHR: the number of MasRs internalized through caveolae was greater compared to that internalized by clathrin-coated pits and the number of receptors that were returned to the cell membrane was smaller, resulting in a lower amount of resensitized MasRs present at the plasma membrane. A fraction of MasRs is translocated to the nucleus bound to its ligand only in brainstem neurons obtained from SHRs and not from normotensive rats. Upon stimulation, constitutive nuclear MasR stimulates NO generation and opposes the increase in nuclear superoxide production and the decrease in AT2R mRNA expression induced by Ang II. Blue and red arrows represent MasR trafficking in brainstem neurons Wistar-Kyoto (WKY) and spontaneously hypertensive rats (SHR), respectively. Abbreviations: Ang-(1-7), Angiotensin-(1-7); Ang II, Angiotensin II; AT2R, Ang II receptor type 2; ERK1/2, extracellular signal-regulated kinase 1/2; NO, nitric oxide; SHR, spontaneously hypertensive rat; WKY, Wistar Kyoto; β-arr: β-arrestin. (For interpretation of the references to color in this figure legend, the reader is referred to the web version of this article.)

including Alzheimer's (AD)[68], Huntington's[69], and PD.[70] Changes in the cerebral Ang-(1-7)/MasR axis have been implicated in the pathogenesis of cardiovascular and neurodegenerative diseases.[15] The evidence regarding alterations in MasR trafficking in pathological states is still scarce.

4.1 Alzheimer's disease

AD is a neurodegenerative disease whose main features are loss of memory after progressive cognitive impairment and changes in personality.[71] One of the most important pathological mechanisms involves the accumulation of extracellular amyloid plaques formed by amyloid-β (Aβ).[72] Aβ peptides originate from the proteolysis of the amyloid precursor protein (APP) by β- and γ-secretases which act in a sequential order to conform the amyloidogenic pathway. Neuroinflammation is also crucial for the progression of the disease.[73]

Several studies have shown the implication of the RAS in the pathogenesis of AD via hyperactivation of the ACE/Ang II/AT1R axis, which has been correlated with an increased synthesis and accumulation of Aβ peptides, tau phosphorylation, increased inflammation and oxidative stress, and decreased release of acetylcholine.[74] It is known that AT1R stimulation leads to two main signaling pathways: a G protein-dependent pathway through Gaq stimulation followed by PLC-PKC activation and a pathway mediated by β-arrestin 2. It has been proven that only the pathway that depends on G protein mediates the Ang II actions on APP shedding and processing.[75]

On the other hand, stimulation of the brain ACE2/Ang-(1-7)/MasR axis counteracts the effects of the classic axis of the RAS. Ang-(1-7) decreases oxidative stress and exerts anti-inflammatory actions in AD.[32] Ang-(1-7) decreased the neuropathology that resembles the one that takes place in AD in the hippocampi of diabetic rats via MasR activation.[76] It has been recently shown that repeated central administration of Ang-(1-7) induced a reduction of the toxic effects at mitochondrial level in memory-sensitive brain areas, increased the function of the cholinergic system and decreased the accumulation of Aβ in a model of AD generated by Aβ in rats.[77] The MasR antagonist A779 abolished the behavioral, biochemical and molecular effects elicited by Ang-(1-7).[77] ACE2/Ang-(1-7)/MasR axis participation in the process of neuroinflammation has been recently proved by Dang et al., who recently showed that administration of either an ACE2 activator or a MasR agonist increased autophagy and suppressed the polarization of microglia induced by inflammation through the signaling of forkhead box class O1.[78]

The favorable actions of the ACE2/Ang-(1-7)/MasR axis on cognitive function including memory and learning could represent a novel therapeutic target for AD treatment. At clinical level, additional studies are needed to prove the contribution and importance of this system on AD.

4.2 Parkinson's disease (PD)

PD is a neurodegenerative disease whose main features are progressive loss of dopaminergic neurons in the substantia nigra pars compacta and the production of atypical α-synuclein aggregates (named Lewy bodies) in the surviving neurons.[70] Mitochondrial defective function, increased inflammation and oxidative stress, and activation of microglia that lastly cause neurodegeneration characterizes PD. AT1R overactivation induced by Ang II has been involved in PD pathogenesis and its complications, by exacerbating the cell death of dopaminergic neurons.[79] Ang II/AT1R signaling is associated to activation of the NADPH-oxidase by PKC stimulation, leading to augmented oxidative stress at the central level.[80] By reducing the levels of superoxide anion, the Ang-(1-7)/MasR axis opposes the prooxidative responses of the Ang II/AT1 axis in dopaminergic neurons.[32]

The beneficial effect of Ang-(1-7) in PD is linked to the activation of MasR and the stimulation of PI3K/Akt/ brain derived neurotrophic factor pathway to augment dopamine synthesis, apart from the blockade of AT1R/MAPK p38/NF-κB p65/NADPH oxidase signaling cascade.[81] Furthermore, the activation of signal transducer and activator of transcription (STAT) 3/ suppressor of cytokine signaling (SOCS) 3 cascade along with the inhibition of the high mobility group box (HMGB)-1/receptor for advanced glycation end products (RAGE)/NF-κB /TNF-α reduce symptoms of PD in association with increased dopamine production, actions correlated with MasR expression and its stimulation by Ang-(1-7).[82]

4.3 Anxiety disorders

Brain Ang-(1-7)/MasR axis activation is also related to anxiety modulation.[83] It has been shown that Ang-(1-7) can attenuate the normal responses to emotional stress that occur centrally.[84] Intracerebroventricular administration of Ang-(1-7) augmented the time expent and frequency of entrances in the open arms of the elevated plus maze, increased the head-dipping behavior in open arms and reduced stretching in closed arms. Altogether, these actions were related to an anxiolytic effect of Ang-(1-7).[85] Centrally administered Ang-(1-7) reduced the anxiety and depressive-like conduct in a model of anxiety-like behavior induced by low brain angiotensinogen.[86] Ang-(1-7) anxiolytic action was also observed in transgenic hypertensive (mRen2)27 rats. Pre-administration of the MasR antagonist A779 blocked the anxiolytic actions elicited by Ang-(1-7), demonstrating MasR involvement.[87] These findings are in agreement with a prior report demonstrating that mice with genetic deletion of MasR showed increased anxiety-like conduct.[88] Recently it has been proven that the MasR-mediated Ang-(1-7) anxiolytic effects are coupled to *N*-methyl-D-aspartate receptor/neuronal NOS/NO signaling cascade and serotonergic metabolism.[89] Overall, the

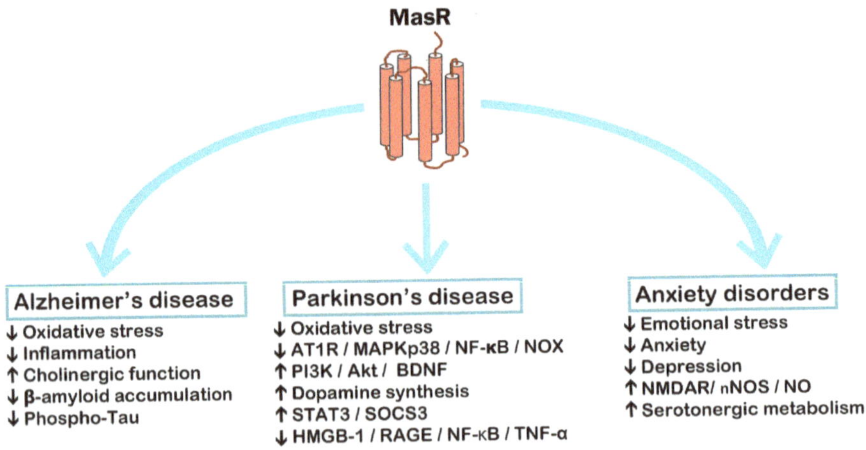

Fig. 4 Biological responses associated to MasR activation in several brain diseases. The main actions elicited by MasR stimulation in Alzheimer's disease, Parkinson's disease and anxiety disorders are depicted. Abbreviations: AT1R, Ang II type 1 receptor; BDNF, brain derived neurotrophic factor; HMGB-1, high mobility group box-1; MAPKp38, p38 mitogen activated protein kinase; NF-κB, nuclear factor kappa B; NOX, NADPH oxidase; PI3K, phosphatydilinositol-3-kinase; RAGE, receptor for advanced glycation end products; SOCS, suppressor of cytokine signaling; STAT, signal transducer and activator of transcription; TNF-α, tumoral necrosis factor alpha.

Ang-(1-7)/MasR axis may be further explored as a novel therapeutic target for anxiety-related conditions.

Fig. 4 represents biological responses coupled to MasR activation in several brain disorders.

5. Concluding remarks

Activation of receptors from the plasma membrane leads to specific signaling activation. However, receptors not only activate signaling pathways from plasma membrane but also from intracellular compartments. After stimulation, receptors are internalized and trafficked inside the cell, inducing the activation of different signaling from those activated from the cell membrane. Thus, the ultimate cellular response results from the integration of all the signaling activated from the plasma membrane and those resulting from receptor trafficking into cellular compartments. Regarding MasR, not only receptor trafficking but also the interaction with other receptors mediates MasR biological responses. Information regarding how trafficking influences MasR-mediated effects in diseases is scarce.

Further investigation should be performed. Expanding our understanding on MasR regulation would open novel therapeutic approaches.

References

1. Forrester SJ, Booz GW, Sigmund CD, et al. Angiotensin II signal transduction: An update on mechanisms of physiology and pathophysiology. *Physiol Rev*. 2018;98: 1627–1738.
2. Santos RAS, Oudit GY, Verano-Braga T, et al. The renin-angiotensin system: Going beyond the classical paradigms. *Am J Physiol Heart Circ Physiol*. 2019;316:H958–H970.
3. Lemos VS, Côrtes SF, Silva DM, Campagnole-Santos MJ, Santos RA. Angiotensin-(1-7) is involved in the endothelium-dependent modulation of phenylephrine-induced contraction in the aorta of mRen-2 transgenic rats. *Br J Pharmacol*. 2002;135:1743–1748.
4. Vickers C, Hales P, Kaushik V, et al. Hydrolysis of biological peptides by human angiotensin-converting enzyme-related carboxypeptidase. *J Biol Chem*. 2002;277: 14838–14843.
5. Karamyan VT, Speth RC. Enzymatic pathways of the brain renin-angiotensin system: Unsolved problems and continuing challenges. *Regul Pept*. 2007;143:15–27.
6. Pucci F, Annoni F, Dos Santos RAS, Taccone FS, Rooman M. Quantifying renin-angiotensin-system alterations in COVID-19. *Cell*. 2021;10:2755.
7. Nehme A, Zouein FA, Zayeri ZD, Zibara K. An update on the tissue renin angiotensin system and its role in physiology and pathology. *J Cardiovasc Dev Dis*. 2019;6:14.
8. Gironacci MM, Cerniello FM, Longo Carbajosa NA, Goldstein J, Cerrato BD. Protective axis of the renin-angiotensin system in the brain. *Clin Sci (Lond)*. 2014; 127:295–306.
9. Kangussu LM, Melo-Braga MN, de Souza Lima BS, et al. Angiotensin-(1-7) central mechanisms after ICV infusion in hypertensive transgenic (mRen2)27 rats. *Front Neurosci*. 2021;15, 624249.
10. Santos RAS, Sampaio WO, Alzamora AC, et al. The ACE2/angiotensin-(1-7)/MAS Axis of the renin-angiotensin system: Focus on angiotensin-(1-7). *Physiol Rev*. 2018; 98:505–553.
11. Jiang T, Gao L, Shi J, et al. Angiotensin-(1-7) modulates renin-angiotensin system associated with reducing oxidative stress and attenuating neuronal apoptosis in the brain of hypertensive rats. *Pharmacol Res*. 2013;67:84–93.
12. Bennion DM, Haltigan E, Regenhardt RW, Steckelings UM, Sumners C. Neuroprotective mechanisms of the ACE2-angiotensin-(1-7)-mas axis in stroke. *Curr Hypertens Rep*. 2015;17:3.
13. Goldstein J, Carden TR, Perez MJ, et al. Angiotensin-(1-7) protects from brain damage induced by Shiga toxin 2-producing enterohemorrhagic Escherichia coli. *Am J Physiol Regul Integr Comp Physiol*. 2016;311:R1173–R1185.
14. Gironacci MM, Vicario A, Cerezo G, Silva MG. The depressor axis of the renin-angiotensin system and brain disorders: A translational approach. *Clin Sci (Lond)*. 2018;132:1021–1038.
15. Abiodun OA, Ola MS. Role of brain renin angiotensin system in neurodegeneration: An update. *Saudi J Biol Sci*. 2020;27:905–912.
16. Jarajapu YPR. Targeting angiotensin-converting Enzyme-2/angiotensin-(1-7)/mas receptor Axis in the vascular progenitor cells for cardiovascular diseases. *Mol Pharmacol*. 2021;99:29–38.
17. Lissoni P, Messina G, Boniardi B, et al. Angiotensin 1-7 as the possible universal panacea of human systemic diseases and potentially the best anti-aging molecule of human body. *Int J Clin Stu Med Case*. 2021;9:1–5.

18. Young D, Waitches G, Birchmeier C, Fasano O, Wigler M. Isolation and characterization of a new cellular oncogene encoding a protein with multiple potential transmembrane domains. *Cell.* 1986;45:711–719.
19. Karnik SS, Singh KD, Tirupula K, Unal H. Significance of angiotensin 1-7 coupling with MAS1 receptor and other GPCRs to the renin-angiotensin system: IUPHAR review 22. *Br J Pharmacol.* 2017;174:737–753.
20. Bader M. Tissue renin-angiotensin-aldosterone systems: Targets for pharmacological therapy. *Annu Rev Pharmacol Toxicol.* 2010;50:439–465.
21. Santos RA, Simoes e Silva AC, Maric C, et al. Angiotensin-(1-7) is an endogenous ligand for the G protein-coupled receptor mas. *Proc Natl Acad Sci U S A.* 2003; 100:8258–8263.
22. Becker LK, Etelvino GM, Walther T, Santos RA, Campagnole-Santos MJ. Immunofluorescence localization of the receptor mas in cardiovascular-related areas of the rat brain. *Am J Physiol Heart Circ Physiol.* 2007;293:H1416–H1424.
23. Santos RA. Angiotensin-(1-7). *Hypertension.* 2014;63:1138–1147.
24. Alenina N, RAS dS. Angiotensin-(1–7) and Mas: A Brief History. *The Protective Arm of the Renin Angiotensin System (RAS).* 2015;155–159.
25. Young D, O'Neill K, Jessell T, Wigler M. Characterization of the rat mas oncogene and its high-level expression in the hippocampus and cerebral cortex of rat brain. *Proc Natl Acad Sci U S A.* 1988;85:5339–5342.
26. Metzger R, Bader M, Ludwig T, et al. Expression of the mouse and rat mas proto-oncogene in the brain and peripheral tissues. *FEBS Lett.* 1995;357:27–32.
27. Cebra-Thomas JA, Tsai JY, Pilder SH, et al. Localization of the mas proto-oncogene to a densely marked region of mouse chromosome 17 associated with genomic imprinting. *Genomics.* 1992;13:444–446.
28. Rabin M, Birnbaum D, Young D, et al. Human ros1 and mas1 oncogenes located in regions of chromosome 6 associated with tumor-specific rearrangements. *Oncogene Res.* 1987;1:169–178.
29. Freund M, Walther T, von Bohlen und Halbach O. Immunohistochemical localization of the angiotensin-(1-7) receptor mas in the murine forebrain. *Cell Tissue Res.* 2012;348:29–35.
30. Mecca AP, Regenhardt RW, O'Connor TE, et al. Cerebroprotection by angiotensin-(1-7) in endothelin-1-induced ischaemic stroke. *Exp Physiol.* 2011;96:1084–1096.
31. Lopez Verrilli MA, Rodriguez Fermepín M, Longo Carbajosa N, et al. Angiotensin-(1-7) through mas receptor up-regulates neuronal norepinephrine transporter via Akt and Erk1/2-dependent pathways. *J Neurochem.* 2012;120:46–55.
32. Costa-Besada MA, Valenzuela R, Garrido-Gil P, et al. Paracrine and Intracrine angiotensin 1-7/mas receptor Axis in the substantia Nigra of rodents, monkeys, and humans. *Mol Neurobiol.* 2018;55:5847–5867.
33. Passos-Silva DG, Verano-Braga T, Santos RA. Angiotensin-(1-7): Beyond the cardio-renal actions. *Clin Sci (Lond).* 2013;124:443–456.
34. Tallant EA, Ferrario CM, Gallagher PE. Angiotensin-(1-7) inhibits growth of cardiac myocytes through activation of the mas receptor. *Am J Physiol Heart Circ Physiol.* 2005;289:H1560–H1566.
35. Gomes ER, Santos RA, Guatimosim S. Angiotensin-(1-7)-mediated signaling in cardiomyocytes. *Int J Hypertens.* 2012;2012, 493129.
36. Gomes ERM, Lara AA, Almeida PWM, et al. Angiotensin-(1–7) prevents cardiomyocyte pathological remodeling through a nitric oxide/guanosine 3′,5′-cyclic monophosphate-dependent pathway. *Hypertension.* 2010;55:153–160.
37. Patel VB, Zhong JC, Grant MB, Oudit GY. Role of the ACE2/angiotensin 1-7 Axis of the renin-angiotensin system in heart failure. *Circ Res.* 2016;118:1313–1326.

38. Qi Y, Shenoy V, Wong F, et al. Lentivirus-mediated overexpression of angiotensin-(1–7) attenuated ischaemia-induced cardiac pathophysiology. *Exp Physiol.* 2011; 96:863–874.
39. Albrecht D. Angiotensin-(1-7)-induced plasticity changes in the lateral amygdala are mediated by COX-2 and NO. *Learn Mem.* 2007;14:177–184.
40. Xing J, Kong J, Lu J, Li J. Angiotensin-(1-7) inhibits neuronal activity of dorsolateral periaqueductal gray via a nitric oxide pathway. *Neurosci Lett.* 2012;522:156–161.
41. Gironacci MM, Valera MS, Yujnovsky I, Peña C. Angiotensin-(1-7) inhibitory mechanism of norepinephrine release in hypertensive rats. *Hypertension.* 2004;44: 783–787.
42. Cerrato BD, Frasch AP, Nakagawa P, et al. Angiotensin-(1-7) upregulates central nitric oxide synthase in spontaneously hypertensive rats. *Brain Res.* 2012;1453:1–7.
43. Regenhardt RW, Desland F, Mecca AP, et al. Anti-inflammatory effects of angiotensin-(1-7) in ischemic stroke. *Neuropharmacology.* 2013;71:154–163.
44. Regenhardt RW, Mecca AP, Desland F, et al. Centrally administered angiotensin-(1-7) increases the survival of stroke-prone spontaneously hypertensive rats. *Exp Physiol.* 2014;99:442–453.
45. Li P, Sun HJ, Cui BP, Zhou YB, Han Y. Angiotensin-(1-7) in the rostral ventrolateral medulla modulates enhanced cardiac sympathetic afferent reflex and sympathetic activation in renovascular hypertensive rats. *Hypertension.* 2013;61:820–827.
46. Han Y, Sun HJ, Li P, et al. Angiotensin-(1-7) in paraventricular nucleus modulates sympathetic activity and cardiac sympathetic afferent reflex in renovascular hypertensive rats. *PLoS One.* 2012;7:e48966.
47. Ferré S, Ciruela F, Casadó V, Pardo L. Oligomerization of G protein-coupled receptors: Still doubted? *Prog Mol Biol Transl Sci.* 2020;169:297–321.
48. Sposini S, Hanyaloglu AC. Evolving view of membrane trafficking and signaling Systems for G Protein-Coupled Receptors. *Prog Mol Subcell Biol.* 2018;57:273–299.
49. Bourque K, Jones-Tabah J, Devost D, Clarke PBS, Hébert TE. Exploring functional consequences of GPCR oligomerization requires a different lens. *Prog Mol Biol Transl Sci.* 2020;169:181 211.
50. Kostenis E, Milligan G, Christopoulos A, et al. G-protein-coupled receptor mas is a physiological antagonist of the angiotensin II type 1 receptor. *Circulation.* 2005;111:1806–1813.
51. Leonhardt J, Villela DC, Teichmann A, et al. Evidence for Heterodimerization and functional interaction of the angiotensin type 2 receptor and the receptor MAS. *Hypertension.* 2017;69:1128–1135.
52. Patel S, Rauf A, Khan H, Abu-Izneid T. Renin-angiotensin-aldosterone (RAAS): The ubiquitous system for homeostasis and pathologies. *Biomed Pharmacother.* 2017;94: 317–325.
53. Cerrato BD, Carretero OA, Janic B, Grecco HE, Gironacci MM. Heteromerization between the bradykinin B2 receptor and the angiotensin-(1-7) mas receptor: Functional consequences. *Hypertension.* 2016;68:1039–1048.
54. Rukavina Mikusic NL, Silva MG, Mazzitelli LR, et al. Interaction between the angiotensin-(1-7) mas receptor and the dopamine D2 receptor: Implications in inflammation. *Hypertension.* 2021;77:1659–1669.
55. Rivas-Santisteban R, Lillo J, Muñoz A, et al. Novel interactions involving the mas receptor show potential of the renin-angiotensin system in the regulation of microglia activation: Altered expression in parkinsonism and dyskinesia. *Neurotherapeutics.* 2021;18:998–1016.
56. Mayor S, Parton RG, Donaldson JG. Clathrin-independent pathways of endocytosis. *Cold Spring Harb Perspect Biol.* 2014;6:a016758.

57. West C, Hanyaloglu AC. Minireview: Spatial programming of G protein-coupled receptor activity: Decoding signaling in health and disease. *Mol Endocrinol.* 2015;29: 1095–1106.
58. Tian X, Kang DS, Benovic JL. β-Arrestins and G protein-coupled receptor trafficking. *Handb Exp Pharmacol.* 2014;219:173–186.
59. Busija AR, Patel HH, Insel PA. Caveolins and cavins in the trafficking, maturation, and degradation of caveolae: Implications for cell physiology. *Am J Physiol Cell Physiol.* 2017;312:C459–C477.
60. Weinberg ZY, Crilly SE, Puthenveedu MA. Spatial encoding of GPCR signaling in the nervous system. *Curr Opin Cell Biol.* 2019;57:83–89.
61. Ribeiro-Oliveira R, Vojtek M, Gonçalves-Monteiro S, et al. Nuclear G-protein-coupled receptors as putative novel pharmacological targets. *Drug Discov Today.* 2019;24:2192–2201.
62. Gironacci MM, Adamo HP, Corradi G, et al. Angiotensin (1-7) induces MAS receptor internalization. *Hypertension.* 2011;58:176–181.
63. Cerniello FM, Carretero OA, Longo Carbajosa NA, et al. MAS1 receptor trafficking involves ERK1/2 activation through a β-Arrestin2-dependent pathway. *Hypertension.* 2017;70:982–989.
64. Cerniello FM, Silva MG, Carretero OA, Gironacci MM. Mas receptor is translocated to the nucleus upon agonist stimulation in brainstem neurons from spontaneously hypertensive rats but not normotensive rats. *Cardiovasc Res.* 2020;116:1995–2008.
65. Gwathmey TM, Westwood BM, Pirro NT, et al. Nuclear angiotensin-(1-7) receptor is functionally coupled to the formation of nitric oxide. *Am J Physiol Renal Physiol.* 2010;299:F983–F990.
66. Gwathmey TM, Alzayadneh EM, Pendergrass KD, Chappell MC. Novel roles of nuclear angiotensin receptors and signaling mechanisms. *Am J Physiol Regul Integr Comp Physiol.* 2012;302:R518–R5130.
67. Chappell MC, Marshall AC, Alzayadneh EM, Shaltout HA, Diz DI. Update on the angiotensin converting enzyme 2-angiotensin (1-7)-MAS receptor axis: Fetal programing, sex differences, and intracellular pathways. *Front Endocrinol (Lausanne).* 2014;4:201.
68. Perdigão C, Barata MA, Araújo MN, et al. Intracellular trafficking mechanisms of synaptic dysfunction in Alzheimer's disease. *Front Cell Neurosci.* 2020;14:72.
69. Kang R, Wang L, Sanders SS, et al. Altered regulation of striatal neuronal N-methyl-D-aspartate receptor trafficking by Palmitoylation in Huntington disease mouse model. *Front Synaptic Neurosci.* 2019;11:3.
70. Singh PK, Muqit MMK. Parkinson's: A disease of aberrant vesicle trafficking. *Annu Rev Cell Dev Biol.* 2020;36:237–264.
71. DeTure MA, Dickson DW. The neuropathological diagnosis of Alzheimer's disease. *Mol Neurodegener.* 2019;14:32.
72. Chong FP, Ng KY, Koh RY, Chye SM. Tau proteins and Tauopathies in Alzheimer's disease. *Cell Mol Neurobiol.* 2018;38:965–980.
73. Ardura-Fabregat A, Boddeke EWGM, Boza-Serrano A, et al. Targeting Neuroinflammation to treat Alzheimer's disease. *CNS Drugs.* 2017;31:1057–1082.
74. Gebre AK, Altaye BM, Atey TM, Tuem KB, Berhe DF. Targeting Renin-Angiotensin System Against Alzheimer's Disease. *Front Pharmacol.* 2018;9:440.
75. Kanarek AM, Wagner A, Küppers J, et al. Crosstalk between angiotensin and the non-amyloidogenic pathway of Alzheimer's amyloid precursor protein. *FEBS J.* 2017;284: 742–753.
76. Chen JL, Zhang DL, Sun Y, et al. Angiotensin-(1-7) administration attenuates Alzheimer's disease-like neuropathology in rats with streptozotocin-induced diabetes via mas receptor activation. *Neuroscience.* 2017;346:267–277.

77. Varshney V, Garabadu D. Ang(1-7) exerts Nrf2-mediated neuroprotection against amyloid beta-induced cognitive deficits in rodents. *Mol Biol Rep.* 2021;48:4319–4331.
78. Dang R, Yang M, Cui C, et al. Activation of angiotensin-converting enzyme 2/angiotensin (1-7)/mas receptor axis triggers autophagy and suppresses microglia proinflammatory polarization via forkhead box class O1 signaling. *Aging Cell.* 2021;20:e13480.
79. Labandeira-García JL, Garrido-Gil P, Rodriguez-Pallares J, et al. Brain renin-angiotensin system and dopaminergic cell vulnerability. *Front Neuroanat.* 2014;8:67.
80. Saavedra JM. Beneficial effects of angiotensin II receptor blockers in brain disorders. *Pharmacol Res.* 2017;125:91–103.
81. Rabie MA, Abd El Fattah MA, Nassar NN, El-Abhar HS, Abdallah DM. Angiotensin 1-7 ameliorates 6-hydroxydopamine lesions in hemiparkinsonian rats through activation of MAS receptor/PI3K/Akt/BDNF pathway and inhibition of angiotensin II type-1 receptor/NF-κB axis. *Biochem Pharmacol.* 2018;151:126–134.
82. Rabie MA, Abd El Fattah MA, Nassar NN, Abdallah DM, El-Abhar HS. Correlation between angiotensin 1-7-mediated mas receptor expression with motor improvement, activated STAT3/SOCS3 cascade, and suppressed HMGB-1/RAGE/NF-κB signaling in 6-hydroxydopamine hemiparkinsonian rats. *Biochem Pharmacol.* 2020;171:113681.
83. Fontes MA, Martins Lima A, Santos RA. Brain angiotensin-(1-7)/mas axis: A new target to reduce the cardiovascular risk to emotional stress. *Neuropeptides.* 2016;56:9–17.
84. Bild W, Ciobica A. Angiotensin-(1-7) central administration induces anxiolytic-like effects in elevated plus maze and decreased oxidative stress in the amygdala. *J Affect Disord.* 2013;145:165–171.
85. Hoły Z, Braszko J, Kupryszewski G, Witczuk B, Wiśniewski K. Angiotensin II--derived peptides devoid of phenylalanine in position 8 have full psychotropic activity of the parent hormone. *J Physiol Pharmacol.* 1992;43:183–192.
86. Kangussu LM, Almeida-Santos AF, Bader M, et al. Angiotensin-(1-7) attenuates the anxiety and depression-like behaviors in transgenic rats with low brain angiotensinogen. *Behav Brain Res.* 2013;257:25–30.
87. Almeida-Santos AF, Kangussu LM, Moreira FA, et al. Anxiolytic- and antidepressant-like effects of angiotensin-(1-7) in hypertensive transgenic (mRen2)27 rats. *Clin Sci (Lond).* 2016;130:1247–1255.
88. Walther T, Balschun D, Voigt JP, et al. Sustained long term potentiation and anxiety in mice lacking the mas protooncogene. *J Biol Chem.* 1998;273:11867–11873.
89. Zhu D, Sun M, Liu Q, et al. Angiotensin (1-7) through modulation of the NMDAR-nNOS-NO pathway and serotonergic metabolism exerts an anxiolytic-like effect in rats. *Behav Brain Res.* 2020;390:112671.

CHAPTER FOUR

Regulation of transferrin receptor trafficking by optineurin and its disease-associated mutants

Shivranjani C. Moharir[†], **Kapil Sirohi**[‡], **and Ghanshyam Swarup***
Centre for Cellular and Molecular Biology, Council for Scientific and Industrial Research, Hyderabad, India
*Corresponding author: e-mail address: gshyam@ccmb.res.in

Contents

Abstract

Transferrin receptor (TFRC) is a transmembrane protein that plays a crucial role in mediating homeostasis of iron in the cell. The binding of transferrin (that is bound to iron) to TFRC at the cell membrane generally starts endocytosis of TFRC-transferrin complex, which leads to formation of vesicles that are positive for TFRC. These vesicles travel to the early endosomes and later to the endocytic recycling compartment. Release of iron occurs in the early endosomes because of acidic pH. Major fraction of the transferrin and TFRC is transported back to the cell membrane; however, a minor fraction of it is transported to lysosomes through the process of autophagy. Optineurin (OPTN) is a multi-functional adaptor protein that plays a pivotal role in the control of TFRC trafficking, recycling and autophagy dependent degradation. Optineurin also plays a role in cargo-selective and non-selective autophagy. Here, we review our

[†] Present address: Tata Institute for Genetics and Society, Bangalore, India. E-mail: shivranjani.c@tigs.res.in
[‡] Present address: Division of Allergy and Immunology, Department of Medicine, National Jewish Health, 1400 Jackson St., Denver, CO 80206. E-mail: sirohik@njhealth.org

Progress in Molecular Biology and Translational Science, Volume 194
ISSN 1877-1173
https://doi.org/10.1016/bs.pmbts.2022.06.019

understanding of the function of OPTN in regulating TFRC trafficking, recycling and autophagy dependent degradation. We also discuss the mechanisms by which certain disease-associated mutations of OPTN alter these processes.

1. Introduction

The adaptor protein Optineurin (OPTN) is present in the cytoplasm and mediates several functions in the cells such as signal transduction, protein transport by vesicles, autophagy, response to endoplasmic reticulum (ER) stress, etc.[1–8] Its expression is seen in most of the tissues and cells. It is localized in various vesicles and organelles involved in mediating transport like the endosomes, Golgi, autophagosomes and lysosomes.[7,9–15] It is also present in the cytosol. OPTN interacts with a wide array of proteins and often links two or more proteins for mediating its functions. It is involved in the vesicle mediated transport pathways between the plasma membrane and Golgi.[9–15] OPTN was identified as a protein that interacts with the activated form of Rab8 (that is bound to GTP) specifically, which does not interact with the inactive form of Rab8 (that is bound to GDP).[9] Due to this reason, OPTN is an effector of Rab8, a GTPase having role in several vesicle trafficking functions. OPTN performs an important function in regulating Rab8-dependent trafficking of transferrin receptor (TFRC, also termed as TFR1) through the endosomes, and its recycling.[10,14]

1.1 Trafficking of TFRC through early and recycling endosomes

Iron is taken up by the cells through iron-binding protein, transferrin, which binds to TFRC, a transmembrane protein present on the plasma membrane. Transferrin and TFRC have a role in homeostasis of iron. Interaction of transferrin (which is bound to iron) with TFRC at the cell membrane generally initiates endocytosis, which leads to formation of vesicles that are positive for TFRC. The endocytosis of TFRC is mediated by a clathrin-dependent process. The vesicles positive for TFRC travel to the early endosomes and later to the recycling endosomes.[16,17] Due to the acidic pH in the endosomes, iron is released. Most of the transferrin and TFRC is then transported back to the cell membrane either from early endosomes through the Rab4-dependent fast recycling pathway, or from recycling endosomes through Rab11-dependent slow recycling pathway. But a minor fraction of TFRC is delivered through the autophagosomes to the lysosomes for degradation.[18,19] Several Rab GTPases such as Rab4, Rab5, Rab8 and

Rab11, and their effector proteins are involved in regulating various steps involved in TFRC trafficking and recycling. TFRC trafficking and recycling plays an essential role in the survival of cells, and impaired TFRC trafficking or recycling can contribute to the death of retinal cells.

2. Role of OPTN in TFRC trafficking and recycling

The first indication that OPTN may be involved in TFRC trafficking was provided by the observation that a glaucoma-associated mutant of OPTN, E50K, forms large vesicles. These vesicles were characterized using various markers. It was observed that most of the vesicles formed by E50K-OPTN were positive for TFRC.[12,13] Normal OPTN also forms vesicles, which are positive for TFRC, but the mutant OPTN formed more number of vesicles with larger size.[12,13] OPTN complexes with TFRC, as determined by immunoprecipitation. Presence of OPTN in TFRC-positive vesicles indicated that OPTN may be involved in TFRC trafficking. Knockdown of OPTN results in lower transferrin uptake, and reduces trafficking of TFRC to the juxtanuclear region from the periphery of the cell.[12,13] These observations suggested that OPTN regulates TFRC trafficking from early endosomes to recycling endosomes. Reduced uptake of transferrin upon OPTN knockdown is possibly due to impaired transport of TFRC to cell membrane.

The mechanisms contributing to the control of trafficking of TFRC and its recycling, and Rab8 function by OPTN have been explored. TBC1D17, a member of the Rab GTPase activating protein (GAP) family interacts with OPTN as revealed by a screen for identifying OPTN-interacting proteins.[14,20] Further functional analysis revealed that TBC1D17 is a GAP for Rab8 that inactivates activated form of Rab8, and this is facilitated by OPTN.[14] The interaction of TBC1D17 with Rab8 is mediated by OPTN.[14] The binding sites for Rab8 and TBC1D17 on OPTN are adjacent but non-overlapping. The N-terminal region of OPTN binds to activated Rab8, whereas central region of OPTN is involved in interaction with TBC1D17 (Fig. 1). Rab8 plays a very significant role in controlling TFRC transport and recycling. OPTN, as an effector of Rab8, mediates endocytic trafficking and recycling of TFRC from recycling endosomes to plasma membrane. Then OPTN recruits TBC1D17 to inactivate Rab8. This is shown schematically in Fig. 2. Thus, by regulating the function of Rab8, OPTN controls TFRC trafficking and recycling.[14] OPTN

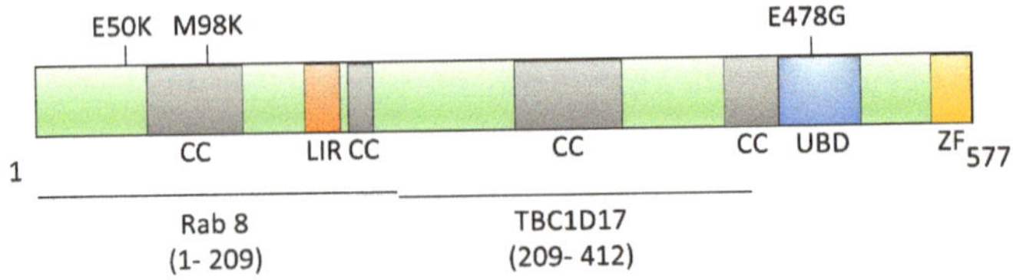

Fig. 1 Domain organization and disease-associated mutations of OPTN. The schematic shows various domains of human OPTN, and two important glaucoma-associated mutations, E50K and M98K, and an ALS-associated mutation, E478G. These mutations affect TFRC trafficking and/or recycling. Regions of OPTN involved in interaction with Rab8 GTPase and TBC1D17, a Rab GAP protein, are also shown. These proteins (Rab8 and TBC1D17) are involved in controlling trafficking and recycling of TFRC. CC, coiled coil domain; LIR, LC3-interacting region; UBD, ubiquitin-binding domain; ZF, zinc finger.

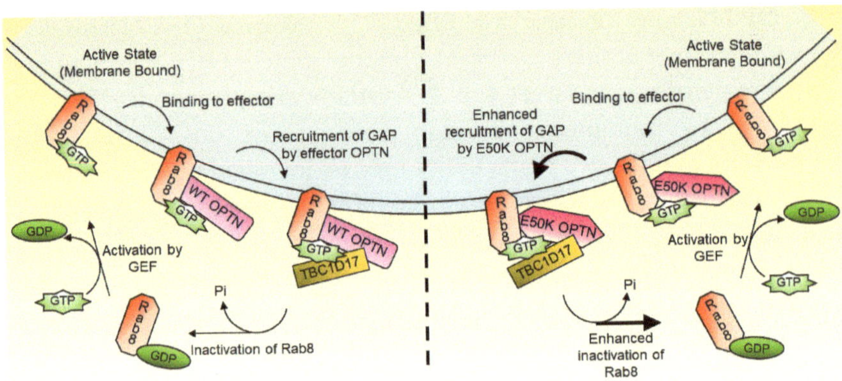

Fig. 2 Model depicting regulation of Rab8 by OPTN and its impaired regulation by a glaucoma-causing mutant, E50K. Active GTP-bound form of Rab8, which is associated with membrane, performs its vesicle trafficking function of TFRC trafficking and recycling. Upon binding to the activated form of Rab8, OPTN is able to recruit TBC1D17, a GAP protein, leading to inactivation of Rab8. The OPTN mutant E50K is much more efficient in recruiting TBC1D17 to the multi-molecular complex, which results in enhanced inactivation of Rab8. GEF, guanine nucleotide exchange factor. GAP, GTPase activating protein.

functions as a molecular switch at recycling endosome compartment to maintain homeostasis of TFRC recycling to plasma membrane.

3. Glaucoma-associated mutations of OPTN impair TFRC trafficking and recycling

OPTN mutations are associated with amyotrophic lateral sclerosis (ALS) and primary open angle glaucoma in adults.[21,22] Both of these are neurodegenerative disorders. Glaucoma is an eye disease that leads to irreversible loss of vision because of death of ganglion cells in the retina and their axons in the optic nerve head, whereas ALS is a motor neuron disease that causes progressive paralysis due to the degeneration of motor neurons, respiratory depression and death. OPTN mutations, associated with glaucoma are mostly single copy point mutations.[3,6,21] Therefore, these are likely to be dominant. ALS-causing mutations of OPTN include missense, nonsense and deletion mutations.[22,23] The E50K mutation of OPTN that is associated with normal tension glaucoma, has been studied extensively. This mutation segregates with disease phenotype in a very large family, which suggests that this mutation causes the disease.[21] This mutant forms large vesicles that are positive for TFRC.[12,13,24] Expression of E50K-OPTN causes depletion of TFRC from the plasma membrane and that results in lowering of the uptake of transferrin by the cells, which shows that this mutant impairs TFRC trafficking and recycling.[12,13] The E50K mutant shows enhanced interaction and colocalization with TFRC, as seen by immunoprecipitation and confocal microscopy experiments.[12,13] This mutant causes increased inactivation of Rab8 by the GAP protein TBC1D17, which leads to impaired TFRC recycling.[14] This is supported by the finding that endogenous TBC1D17 shows enhanced colocalization with E50K-OPTN in the vesicles.[25] The inhibition of TFRC trafficking/recycling by E50K mutant could be partially abrogated by a catalytic mutant of TBC1D17, or by knockdown of TBC1D17.[14] OPTN, Rab8, TFRC and TBC1D17 form a multimolecular complex. In this complex, direct interaction of E50K mutant with Rab8 is abrogated but indirect interaction is enhanced.[14] Impaired regulation of Rab8 by the GAP protein TBC1D17 in E50K-OPTN expressing cells is depicted in Fig. 2.

Overexpression of E50K-OPTN in a retinal cell line causes cell death which could be inhibited by TBC1D17 knockdown, and also by a catalytic mutant of TBC1D17.[25–27] Overexpression of TFRC also partially protected these cells from apoptosis induced by E50K-OPTN. This suggests that impaired TFRC trafficking/recycling by E50K-OPTN contributes to cytotoxicity in retinal cells.[26] It is interesting to note that this cytotoxic effect of E50K-OPTN is seen in RGC-5 and 661 W cells (which are retinal ganglion precursor-like cells) but not in many other types of cells that have been examined.[25,27]

M98K is an interesting polymorphism of OPTN that shows association with normal tension glaucoma in several populations in Asia.[28–33] Expression of M98K-OPTN leads to impaired TFRC trafficking/recycling as seen by reduced uptake of transferrin by retinal cells.[18] M98K-OPTN forms bigger and more vesicles in the cells in comparison with normal OPTN. It shows enhanced colocalization and interaction with TFRC. In M98K-OPTN expressing cells enhanced delivery of TFRC to autophagosomes was observed as seen by increased colocalization of TFRC and LC3 in vesicles (Fig. 3).[18] This leads to enhanced TFRC degradation by autophagy, resulting in reduced TFRC levels and enhanced cell death.[18] Like E50K-OPTN, M98K-OPTN also induces apoptosis in retinal cells selectively, which can be inhibited by overexpression of TFRC, or supplementation of iron in the media.[18,26,27] Thus, it appears that TFRC trafficking/recycling plays a particularly vital role in retinal cells in comparison with many other types of cells. Unlike E50K-OPTN, the M98K-OPTN did not show altered interaction with Rab8 in yeast two-hybrid assay.[18] The impaired TFRC trafficking/recycling caused by expression of M98K-OPTN is likely to be due to enhanced autophagic degradation of TFRC in retinal cells induced by this mutant.[18] However, it is also possible that lower recycling of TFRC by M98K-OPTN results in enhanced delivery of TFRC to autophagosomes for degradation. Thus, it seems that OPTN functions as a molecular switch, which regulates TFRC trafficking, recycling and autophagic degradation. This OPTN switch is altered by M98K mutation.[18] This is shown schematically in Fig. 4. Rab12 GTPase carries out an essential function in the TFRC transport to autophagosomes

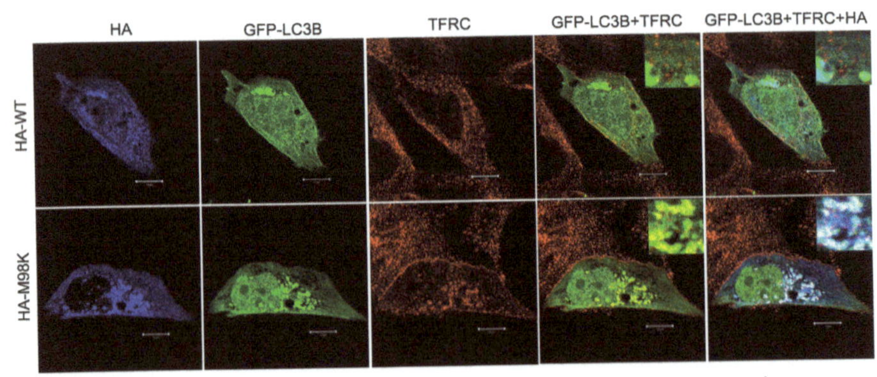

Fig. 3 Expression of M98K-OPTN increases localization of TFRC in autophagosomes. Confocal microscopy images show colocalization of TFRC with LC3 in vesicles. RGC-5 cells were transfected with GFP-LC3 and hemagglutinin (HA) tagged WT-OPTN (HA-WT) or M98K-OPTN (HA-M98K) and stained with TFRC antibody (shown in red) and HA antibody (shown in blue) as described in Ref. [18]. Scale bar: 10 μm. Insets show magnified areas.

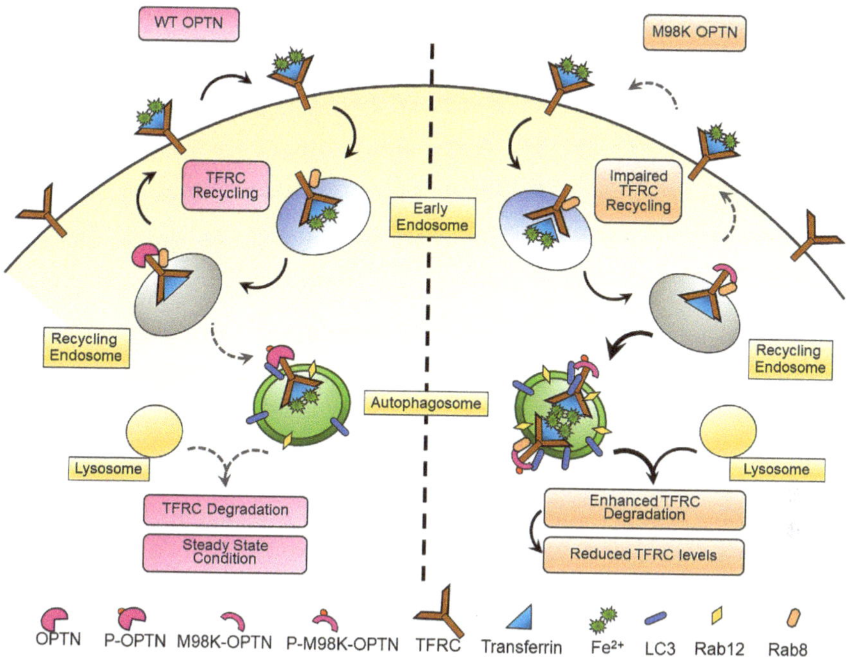

Fig. 4 Model depicting the role of OPTN in controlling TFRC trafficking and recycling. Iron-bound transferrin binds to TFRC, and transferrin-TFRC complex is endocytosed into vesicles that travel to early endosome and subsequently to the recycling endosome. Release of iron occurs in the endosome and major fraction of TFRC-transferrin is transported back to the cell membrane in normal cells expressing wild-type (WT) OPTN. OPTN along with Rab8 plays an essential role in controlling TFRC trafficking to recycling endosome, and its transport back to the cell membrane. Minor fraction of TFRC is not recycled back to plasma membrane and is delivered to autophagosome for degradation. In retinal cells expressing M98K-OPTN, the TFRC recycling is reduced and much more TFRC is delivered to autophagosomes for degradation. Thus, it appears that OPTN functions like a molecular switch that not only regulates TFRC trafficking and recycling but also regulates its delivery to autophagosomes leading to its autophagic degradation.

for degradation by autophagy in M98K-OPTN expressing cells.[18] It has been shown that Rab12 is localized in recycling endosomes and lysosomes but not in early endosomes, and it mediates transport of TFRC from recycling endosomes to lysosomes for degradation.[19,34] M98K-OPTN shows increased colocalization and interaction with Rab12 compared to WT-OPTN and this enhanced complex formation of M98K with Rab12 could be the possible molecular mechanism of increased TFRC degradation in M98K expressing retinal cells.[18]

4. Function of OPTN in trafficking of TFRC and associated membrane to autophagosomes

Autophagy is one of two major degradative mechanisms involved in maintaining homeostasis in the cells. It is a waste disposal mechanism deployed by various cells to get rid of aggregated proteins, invading pathogens and damaged organelles, by degradation in lysosomes.[35–37] Macroautophagy, also simply termed as autophagy, is initiated through the generation of a cup-shaped double-membrane structure known as phagophore.[35,36] The phagophore grows by the addition of membrane from several sources and matures into autophagosome that are closed organelles. Autophagosome then fuses with lysosomes, which results in the formation of autolysosome where the cargo is degraded. In addition to various autophagy related (Atg) proteins, several other proteins also control maturation of phagophore into autophagosome.[35,36,38,39] The autophagosomal protein microtubule-associated protein 1 light chain protein 3 (LC3) performs several functions such as closure of autophagosome, and recruitment of the cargo to the autophagosome. Membrane is required for biogenesis of autophagosomes that is delivered from several organelles such as mitochondria, ER, Golgi, early and recycling endosomes, and plasma membrane.[40–46] However, the mechanisms involved and proteins that mediate the transport of the membrane to autophagosome during their biogenesis are not fully clear.

OPTN has been shown to control several vesicle trafficking pathways; therefore, it is one of the candidates for mediating delivery of membrane to autophagosomes during their biogenesis. It was reported that vesicles, which are derived from recycling endosomes, and are positive for TFRC, are delivered to autophagosomes during their biogenesis.[40,41] OPTN has been shown to promote the production of LC3-II, the autophagosomal membrane protein, which enhances maturation of phagophore (pre-autophagosome) into autophagosome.[47] Recently, it has been shown that OPTN has a role in the transport of membrane from recycling endosomes to autophagosomes.[48] In mouse embryonic fibroblasts that are deficient in OPTN, the percentage of autophagosomes positive for TFRC was reduced compared to normal cells, and overexpression of OPTN enhanced the percentage of autophagosomes positive for TFRC. This suggests that OPTN facilitates the transport of TFRC and the associated membrane from the endosomes to the autophagosomes. A fraction of autophagosomes were found to be positive for TFRC, which suggests that endosomes (or recycling endosomes) that contain TFRC provide membrane to only a fraction of the autophagosomes.[48] Optineurin deficient cells also showed reduced degradation of TFRC through autophagy. This could be caused by the impaired transport of TFRC-positive membrane to the autophagosomes in optineurin deficient cells.[48] An alternate possibility is that TFRC degradation by autophagosomes may require OPTN independently of its role in membrane delivery. E478G-OPTN, that is associated with ALS, and impaired in ubiquitin-binding and autophagy, is also impaired in increasing the fraction of autophagosomes that contain TFRC. This shows that interaction of OPTN with ubiquitin is required for the transport of TFRC and associated membrane from the endosomes to autophagosomes.[48] By enhancing the production of LC3-II, OPTN promotes maturation of phagophores into autophagosomes.[47] Formation of autophagosomes is promoted by the delivery of membrane from different cellular organelles. Thus, it seems likely that OPTN promotes autophagosome formation in two different ways: by enhancing LC3-II production and also by facilitating the transport of membrane from endosomes (that are TFRC-positive) to forming autophagosomes.

5. Unconventional endocytic trafficking of TFRC

Recently, it has been shown that under conditions of iron overload, degradation of TFRC takes place in lysosomes.[49] This results in reduced level of TFRC at the cell surface. The level of TFRC mRNA is also reduced under condition of iron overload. This iron-induced degradation of TFRC is independent of transferrin, and clathrin adaptor protein AP2, which is required for clathrin-dependent endocytosis. Thus, it appears that iron-induced endocytosis of TFRC that leads to its lysosomal degradation, occurs through an unconventional endocytic pathway.[49] However, the role of Rab proteins, autophagy proteins and OPTN, if any, in this iron-induced unconventional endocytic trafficking pathway of TFRC leading to its degradation in lysosomes is yet to be investigated.

6. Concluding remarks

OPTN, as an adaptor protein carries out crucial functions in the control of TFRC trafficking and recycling, and its degradation in the lysosomes through autophagy. By mediating the negative regulation of Rab8 through TBC1D17, OPTN carries out an important function in controlling TFRC transport and recycling. OPTN also mediates the transport of TFRC to the autophagosomes. It appears that OPTN functions as a molecular switch that controls TFRC recycling and autophagy-dependent degradation, and M98K polymorphism of OPTN modifies this switch. However, some details of the mechanisms involved are yet to be understood. For example, it is not clear what role OPTN plays in autophagic degradation of TFRC. Is the function of OPTN limited to the transport of TFRC to autophagosomes? Does OPTN play an additional role in autophagic degradation of TFRC? Does OPTN play any role in iron-induced TFRC degradation in lysosomes? Disease-associated mutants of OPTN such as E50K, M98K and E478G have provided useful tools for exploring the OPTN function in TFRC trafficking and recycling, and its autophagy mediated degradation.

Acknowledgment

G.S. thanks the Indian National Science Academy, New Delhi, for Senior Scientist position.

Author contributions

S.C.M. and K.S. prepared figures/illustrations, and helped in drafting and editing the manuscript; G.S. wrote the manuscript.

References

1. Ying H, Yue BY. Cellular and molecular biology of optineurin. *Int Rev Cell Mol Biol.* 2012;294:223–258.
2. Kachaner D, Genin P, Laplantine E, Weil R. Toward an integrative view of optineurin functions in *cell. Cycle.* 2012;11(15):2808–2818.
3. Minegishi Y, Nakayama M, Iejima D, Kawase K, Iwata T. Significance of optineurin mutations in glaucoma and other diseases. *Prog Retin Eye Res.* 2016;55:149–181.
4. Markovinovic A, Cimbro R, Ljutic T, Kriz J, Rogelj B, Munitic I. Optineurin in amyotrophic lateral sclerosis: Multifunctional adaptor protein at the crossroads of different neuroprotective mechanisms. *Prog Neurobiol.* 2017;154:1–20.
5. Toth RP, Atkin JD. Dysfunction of optineurin in amyotrophic lateral sclerosis and glaucoma. *Front Immunol.* 2018;9:1017.
6. Swarup G, Sayyad Z. Altered functions and interactions of glaucoma-associated mutants of optineurin. *Front Immunol.* 2018;9:1287.
7. Wild P, Farhan H, McEwan DG, et al. Phosphorylation of the autophagy receptor optineurin restricts salmonella growth. *Science.* 2011;333:228–233.
8. Ramachandran G, Moharir SC, Raghunand TR, Swarup G. Optineurin modulates ER stress-induced signaling pathways and cell death. *Biochem Biophys Res Commun.* 2021;534:297–302.
9. Hattula K, Peranen J. FIP-2, a coiled-coil protein, links huntingtin to Rab8 and modulates cellular morphogenesis. *Curr Biol.* 2000;10:1603–1606.
10. Hattula K, Furuhjelm J, Tikkanen J, Tanhuanpää K, Laakkonen P, Peränen J. Characterization of the Rab8-specific membrane traffic route linked to protrusion formation. *J Cell Sci.* 2006;119:4866–4877.

11. Sahlender DA, Roberts RC, Arden SD, et al. Optineurin links myosin VI to the Golgi complex and is involved in Golgi organization and exocytosis. *J Cell Biol.* 2005; 169:285–295.
12. Nagabhushana A, Chalasani ML, Jain N, et al. Regulation of endocytic trafficking of transferrin receptor by optineurin and its impairment by a glaucoma-associated mutant. *BMC Cell Biol.* 2010;11:4.
13. Park B, Ying H, Shen X, et al. Impairment of protein trafficking upon overexpression and mutation of optineurin. *PLoS One.* 2010;5(7):e11547.
14. Vaibhava V, Nagabhushana A, Chalasani ML, Sudhakar C, Kumari A, Swarup G. Optineurin mediates a negative regulation of Rab8 by the GTPase-activating protein TBC1D17. *J Cell Sci.* 2012;125:5026–5039.
15. Ryan TA, Tumbarello DA. Optineurin: A coordinator of membrane-associated cargo trafficking and autophagy. *Front Immunol.* 2018;9:1024.
16. MacKenzie EL, Iwasaki K, Tsuji Y. Intracellular iron transport and storage: From molecular mechanisms to health implications. *Antioxid Redox Signal.* 2008;10:997–1030.
17. Lee DA, Goodfellow JM. The pH-induced release of iron from transferrin investigated with a continuum electrostatic model. *Biophys J.* 1998;74:2747–2759. https://doi.org/10.1016/S0006-3495(98)77983-4.
18. Sirohi K, Chalasani ML, Sudhakar C, Kumari A, Radha V, Swarup G. M98K-OPTN induces transferrin receptor degradation and RAB12-mediated autophagic death in retinal ganglion cells. *Autophagy.* 2013;9(4):510–527.
19. Matsui T, Itoh T, Fukuda M. Small GTPase Rab12 regulates constitutive degradation of transferrin receptor. *Traffic.* 2011;12:1432–1443.
20. Chalasani ML, Swarup G, Balasubramanian D. Optineurin and its mutants: molecules associated with some forms of glaucoma. *Ophthalmic Res.* 2009;42(4):176–184.
21. Rezaie T, Child A, Hitchings R, et al. Adult-onset primary open-angle glaucoma caused by mutations in optineurin. *Science.* 2002;295:1077–1079.
22. Maruyama H, Morino H, Ito H, et al. Mutations of optineurin in amyotrophic lateral sclerosis. *Nature.* 2010;465:223–226.
23. Ito H, Nakamura M, Komure O, et al. Clinicopathologic study on an ALS family with a heterozygous E478G optineurin mutation. *Acta Neuropathol.* 2011;122:223–229.
24. Park BC, Shen X, Samaraweera M, Yue BY. Studies of optineurin, a glaucoma gene: Golgi fragmentation and cell death from overexpression of wild-type and mutant optineurin in two ocular cell types. *Am J Pathol.* 2006;169(6):1976–1989. https://doi.org/10.2353/ajpath.2006.060400.
25. Chalasani ML, Radha V, Gupta V, Agarwal N, Balasubramanian D, Swarup G. A glaucoma-associated mutant of optineurin selectively induces death of retinal ganglion cells which is inhibited by antioxidants. *Invest Ophthalmol Vis Sci.* 2007;48(4):1607–1614.
26. Chalasani ML, Kumari A, Radha V, Swarup G. E50K-OPTN-induced retinal cell death involves the Rab GTPase-activating protein, TBC1D17 mediated block in autophagy. *PLoS One.* 2014;9:e95758.
27. Sayyad Z, Sirohi K, Radha V, Swarup G. 661W is a retinal ganglion precursor-like cell line in which glaucoma-associated optineurin mutants induce cell death selectively. *Sci Rep.* 2017;7:16855.
28. Alward WL, et al. Evaluation of optineurin sequence variations in 1,048 patients with open-angle glaucoma. *Am J Ophthalmol.* 2003;136:904–910.
29. Fuse N, et al. Molecular genetic analysis of optineurin gene for primary open-angle and normal tension glaucoma in the Japanese population. *J Glaucoma.* 2004;13:299–303.
30. Ayala-Lugo RM, et al. Variation in optineurin (OPTN) allele frequencies between and within populations. *Mol Vis.* 2007;13:151–163.
31. Mukhopadhyay A, et al. Evaluation of optineurin as a candidate gene in Indian patients with primary open angle glaucoma. *Mol Vis.* 2005;11:792–797.

32. Sripriya S, et al. OPTN gene: Profile of patients with glaucoma from India. *Mol Vis.* 2006;12:816–820.
33. Chen M, Yu X, Xu J, et al. Association of Gene Polymorphisms with Primary Open Angle Glaucoma: A systematic review and meta-analysis. *Invest Ophthalmol Vis Sci.* 2019;60(4):1105–1121. https://doi.org/10.1167/iovs.18-25922.
34. Matsui T, Fukuda M. Small GTPase Rab12 regulates transferrin receptor degradation: Implications for a novel membrane trafficking pathway from recycling endosomes to lysosomes. *Cell Logist.* 2011;1(4):155–158.
35. Yang Z, Klionsky DJ. An overview of the molecular mechanism of autophagy. *Curr Top Microbiol Immunol.* 2009;335:1–32.
36. Stolz A, Ernst A, Dikic I. Cargo recognition and trafficking in selective autophagy. *Nat Cell Biol.* 2014;16(6):495–501.
37. Levine B, Packer M, Codogno P. Development of autophagy inducers in clinical medicine. *J Clin Invest.* 2015;125(1):14–24.
38. Tsukada M, Ohsumi Y. Isolation and characterization of autophagy-defective mutants of Saccharomyces cerevisiae. *FEBS Lett.* 1993;333(1–2):169–174.
39. Mizushima N. The role of the Atg1/ULK1 complex in autophagy regulation. *Curr Opin Cell Biol.* 2010;22(2):132–139.
40. Tooze SA, Abada A, Elazar Z. Endocytosis and autophagy: Exploitation or cooperation? *Cold Spring Harb Perspect Biol.* 2014;6:a018358.
41. Longatti A, Lamb CA, Razi M, Yoshimura S, Barr FA, Tooze SA. TBC1D14 regulates autophagosome formation via Rab11- and ULK1-positive recycling endosomes. *J Cell Biol.* 2012;197:659–675.
42. Razi M, Chan EY, Tooze SA. Early endosomes and endosomal coatomer are required for autophagy. *J Cell Biol.* 2009;185:305–321.
43. Ravikumar B, Moreau K, Jahreiss L, Puri C, Rubinsztein DC. Plasma membrane contributes to the formation of pre-autophagosomal structures. *Nat Cell Biol.* 2010; 12:747–757.
44. Hailey DW, Rambold AS, Satpute-Krishnan P, et al. Mitochondria supply membranes for autophagosome biogenesis during starvation. *Cell.* 2010;141:656–667.
45. Axe EL, Walker SA, Manifava M, et al. Autophagosome formation from membrane compartments enriched in phosphatidylinositol 3-phosphate and dynamically connected to the endoplasmic reticulum. *J Cell Biol.* 2008;182:685–701.
46. Geng J, Nair U, Yasumura-Yorimitsu K, Klionsky DJ. Post-Golgi sec proteins are required for autophagy in Saccharomyces cerevisiae. *Mol Biol Cell.* 2010;21:2257–2269.
47. Bansal M, Moharir SC, Sailasree SP, et al. Optineurin promotes autophagosome formation by recruiting the autophagy-related Atg12-5-16L1 complex to phagophores containing the Wipi2 protein. *J Biol Chem.* 2018;293:132–147.
48. Bansal, M., Sirohi, K., Moharir, S. C. and Swarup, G. Optineurin mediates delivery of membrane from transferrin receptor-positive endosomes to autophagosomes. bioRxiv https://doi.org/10.1101/2022.01.12.476135.
49. Gammella E, Lomoriello IS, Conte A, et al. (2021) unconventional endocytosis and trafficking of transferrin receptor induced by iron. *Mol Biol Cell.* 2021;32(2):98–108.

CHAPTER FIVE

The insulin receptor endocytosis

Jiayi Wu†, Stella Hayoung Park†, and Eunhee Choi*
Department of Pathology and Cell Biology, Vagelos College of Physicians and Surgeons, Columbia University, New York, NY, United States
*Corresponding author: e-mail address: ec3477@cumc.columbia.edu

Contents

Abstract

Insulin signaling controls multiple aspects of animal physiology. At the cell surface, insulin binds and activates the insulin receptor (IR), a receptor tyrosine kinase. Insulin promotes a large conformational change of IR and stabilizes the active conformation. The insulin-activated IR triggers signaling cascades, thus controlling metabolism, growth, and proliferation. The activated IR undergoes internalization by clathrin- or caveolae-mediated endocytosis. The IR endocytosis plays important roles in insulin clearance from blood, and distribution and termination of the insulin signaling. Despite decades of extensive studies, the mechanism and regulation of IR endocytosis and its contribution to pathophysiology remain incompletely understood. Here we discuss recent findings that provide insights into the molecular mechanisms and regulatory pathways that mediate the IR endocytosis.

† These authors contributed equally.

Progress in Molecular Biology and Translational Science, Volume 194
ISSN 1877-1173
https://doi.org/10.1016/bs.pmbts.2022.06.020

1. Introduction

Insulin is a hormone secreted from the pancreas that controls multiple aspects of the human body.[1,2] Insulin promotes glucose uptake into cells to lower blood glucose levels, and regulates carbohydrate, lipid, and protein metabolism. As a growth factor, insulin also promotes cell growth and proliferation. At the plasma membrane, insulin binds and activates the insulin receptor. The insulin receptor (IR) forms a stable homodimer, of which the protomers are covalently linked by multiple disulfide bonds (Fig. 1).[3–5] Each protomer consists of an α-subunit and a β-subunit generated by

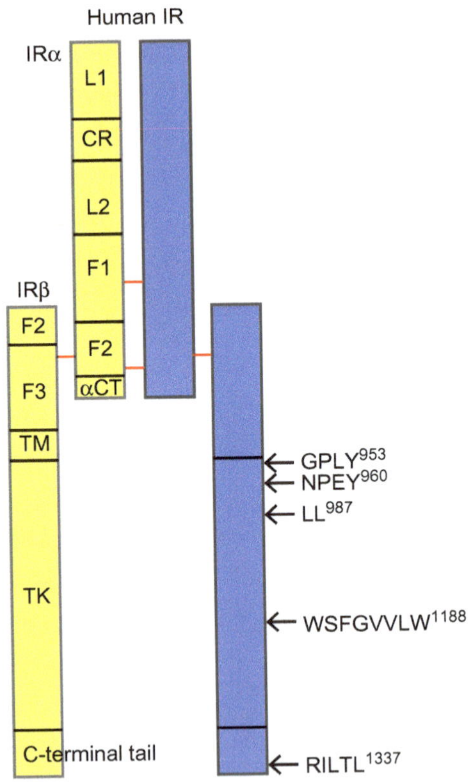

Fig. 1 Schematic illustration of domains and sequence motifs of insulin receptor (IR) endocytosis. L1 and L2, leucine rich domains 1 and 2. CR, cysteine rich domain. F1, F2, and F3, fibronectin III (FnIII) domains. αCT, C-terminal helix of the a-chain. TM, transmembrane domain. TK, tyrosine kinase domain. Disulfide bonds of IR are marked as red. Sequence motifs for IR trafficking are marked.

proteolytic cleavage. The extracellular sequences of IR include the entire α-subunit and about 190 residues at the N-terminus of the β-subunit, comprising leucine rich repeat 1 (L1), cysteine rich (CR), L2, and fibronectin type 3 (FnIII)-1, -2, and -3 (F1, F2, and F3) domains. The rest of β-subunit sequence contains a transmembrane domain, a tyrosine kinase domain capable of trans-autophosphorylation for IR activation, and a C-terminal tail.

Each IR dimer can maximally bind four insulin molecules at four binding sites (named site 1, 1′, 2 and 2′) related by twofold symmetry.[6–8] The insulin-free IR dimer (apo-IR) forms an inverted "V"-shaped architecture.[9] The large distance between the two membrane-proximal F3 domains connected to tyrosine kinase domains prevents the trans-autophosphorylation of IR, indicating that the apo-IR is in an inhibitory conformation.[9] Binding of multiple insulin molecules to the IR dimer breaks the auto-inhibitory conformation, induces a large conformational change, and stabilizes a "T"-shaped active IR dimer.[6–8] In this "T"-shape architecture, tyrosine kinase domains are in close proximity, thus enabling trans-autophosphorylation and activation of IR.

IR has two isoforms, A and B, which result from alternative splicing of the primary transcript.[4,5] The only difference between the isoforms is the insertion of 12 amino acids in exon 11 of IR-B at the C-terminus of the α-subunit.[10] While IR-A is expressed predominantly during fetal life, IR-B is the main isoform in adult tissues, such as liver and adipose tissue.[10] These results suggest that IR-A primarily controls fetal growth and development, whereas IR-B controls metabolic action of insulin. The two IR isoforms have similar binding affinity for insulin and insulin-like growth factor 1 (IGF1), but the binding affinity of IGF2 to IR-A is about 10-fold higher than IR-B.[10–12] Whether the two IR isoforms differ in trafficking and the function of their ligands in this process are controversial and still poorly characterized.[13]

The autophosphorylation of IR is essential for the receptor endocytosis.[14–17] The activation of the intrinsic tyrosine kinase releases a constraint that keeps unbounded apo-IR on cytoskeleton-rich microvilli. The insulin-activated IR leaves microvilli and concentrates on non-villous region of the plasma membrane, accelerating clathrin-mediated IR endocytosis. However, further observations suggest that different routes of endocytosis (i.e., caveolae-mediated endocytosis) are used depending on the cell types and insulin concentrations.[18–20] In addition, recent studies show that a regulatory feedback mechanism of IR endocytosis by the mitogen-activated protein kinase (MAPK)-Src homology phosphatase 2 (SHP2) pathway ensures the insulin-activated IR endocytosis.[21] Inhibition of this feedback loop delays insulin-mediated IR endocytosis and improves insulin sensitivity in mice, suggesting that targeting the IR endocytosis might be beneficial for insulin resistance treatment. In this book chapter, we summarize the current knowledge on the molecular mechanisms and regulation of IR endocytosis. We further compare mouse phenotypes targeting key regulators in IR endocytosis and introduce the impact of IR endocytosis on the pathophysiology of insulin resistance.

2. Clathrin-mediated endocytosis of the insulin receptor

Clathrin-mediated endocytosis is important for a variety of physiological processes in eukaryotic cells, including maintaining membrane homeostasis and regulating intracellular signaling.[22] At the plasma membrane, insulin binds to IR and triggers two major signaling cascades to control metabolism and growth. The activated IR can be internalized by clathrin-mediated endocytosis. The assembly polypeptide 2 (AP2) complex is a key clathrin-mediated endocytosis adaptor that links the active IR to both clathrin and lipids on the plasma membrane, thus accelerating IR endocytosis.[17,23] The AP2 is composed of AP2A, AP2B1, AP2M1, and AP2S1 subunits.[24,25] The large globular core (AP2M1, AP2S1 and the N-terminal trunk domains of AP2A and AP2B1) of AP2 complex recognizes sorting signals such as YXXΦ (X, any amino acid; Φ, hydrophobic residues, Val, Leu, Ile, Phe, or Met) and dileucine motifs. The appendages of the AP2A and AP2B1 bind to clathrin and other accessory proteins of clathrin-mediated endocytosis, thereby promoting clathrin pit formation.

2.1 Endocytosis motifs in insulin receptor

1. *NPXY motif*: The NPXY (X, any amino acid) plays important roles in ligand-receptor endocytosis such as low-density lipoprotein receptor and integrin receptor.[26] IR has the NPXY motif at residues 957–960 (numbering of mature, short isoform of human IR) in the juxtamembrane region of the β-subunit, and it is required for the insulin-activated IR endocytosis (Fig. 1).[27,28] In Chinese hamster ovary (CHO) cells, around 40–50% of the ^{125}I-insulin associated with IR wild-type (WT) is internalized, whereas only 10% of the ^{125}I-insulin associated with the IR-Δ960 (deleted the NPEY^{960} motif) undergoes internalization. In addition to the indirect method used by Backer et al.[27] that examines the IR internalization using the isotope-labeled

insulin bound to the IR, recent studies have shown the importance of the NPXY motif on IR endocytosis by directly monitoring the cellular localization of IR in HepG2 cells. In the absence of insulin stimulation, IR-WT and IR-Y960F (replaced Tyr^{960} with phenylalanine) localize to the plasma membrane. In insulin-stimulated HepG2 cells, the level of IR-WT at the plasma membrane is significantly decreased, and the IR-WT is enriched in the intracellular compartment, indicating insulin-induced IR internalization. Consistent with the previous data,[27,28] IR-Y960F is less efficiently internalized after insulin stimulation,[21] although IR-Y960F exhibits normal kinase activation. These data suggest that the NPXY motif of IR is required for insulin-induced IR endocytosis. We will further discuss how the NPXY motif ensures the insulin-activated IR endocytosis in the next section.

2. *GPLY motif:* The GPLY motif at residues 950–953 (numbering of mature, short isoform of human IR) is in the juxtamembrane region of the β-subunit (Fig. 1). Due to the sequence similarity and proximity between the GPLY and NPXY motifs, the GPLY motif was also considered to play a role in IR internalization.[29] The IR GPLY/APLA mutant, where $GPLY^{953}$ is replaced with APLA, underwent normal insulin-mediated autophosphorylation. The amounts of internalized ^{125}I-insulin and surface-bound ^{125}I-insulin at different times were measured to calculate the specific internalization rate. The IR GPLY/APLA mutant exhibited much slower internalization compared to IR-WT. These data suggest that the GPLY motif also contributes to the IR endocytosis.
3. *Di-leucine motif:* A dileucine motif, Leu^{986} and Leu^{987}, is present within exon 17 in the juxtamembrane domain of the β-subunit in the context of $EKITL^{986}L^{987}$ (numbering of mature, short isoform of human IR) (Fig. 1). The dileucine motif had been known as an important motif in the regulation of IR internalization and sorting inside the cell.[30,31] Using insulin binding assay with ^{125}I-insulin in NIH-3T3 cells, previous studies showed that cells expressing IR-AA (replacement of the two leucine residues with alanine, IR-AA; $EKITA^{986}A^{987}$) exhibited a five-fold decrease of insulin uptake. Therefore, the dileucine motif of IR was thought to be essential for insulin uptake and the insulin-bound IR internalization.

However, a recent study has found evidence that disagrees with this conclusion.[21] Without insulin stimulation, IR-AA was not expressed on the plasma membrane and was enriched in intracellular compartment in

HepG2 cells. A major fraction of IR-AA was co-stained with the late endosome marker ras-associated protein 7 (RAB7), suggesting that IR-AA constitutively undergoes internalization and accumulates in the late endosomes, or that IR-AA has a processing defect. Thus, the lower rate of insulin uptake in the cells expressing IR-AA found in the previous studies[30,31] could be due to decreased levels of IR-AA at the plasma membrane rather than a defect of insulin-mediated IR endocytosis. Because the dileucine motif locates in the β strand of the N-terminal lobe of the IR kinase domain,[32] the mutation of the dileucine motif might alter the structural integrity of IR. Future studies are required to test this possibility.

4. *MAD2-interacting motif (MIM)*: Recent studies demonstrated that the spindle checkpoint regulators play an important role in insulin signaling and metabolic homeostasis by governing IR endocytosis.[16,17,21,33–35] IR has a well-conserved mitotic arrest deficient 2 (MAD2)-interacting motif (MIM: K/RΦΦXΦ$X_{3\text{-}4}$P; Φ, a hydrophobic residue; X, any residue) at the C-terminal tail (Fig. 1). Mutation of the MIM (IR-4A, replacement of four critical residues with four alanine) did not alter the basal IR levels at the plasma membrane. However, IR-4A and IR-Y960F/4A mutants displayed significantly delayed insulin-induced receptor endocytosis in HepG2 cells without affecting the autophosphorylation of their kinase. These data suggest that the MIM of IR is required for insulin-induced IR endocytosis. The function and regulatory mechanism of the MIM on IR endocytosis will be discussed in the next section.

2.2 Spindle checkpoint regulators in insulin receptor endocytosis

The spindle checkpoint ensures the accurate chromosome segregation and genome stability.[33,36–52] In mitosis, when kinetochores are not properly attached by bipolar microtubules, the spindle checkpoint inhibits a multi-subunit ubiquitin ligase called the anaphase promoting complex/cyclosome (APC/C) in complex with its mitotic activator CDC20 (cell division cycle 20). The spindle checkpoint promotes the formation of the mitotic checkpoint complex (MCC) consisting of mitotic arrest deficient 2 (MAD2), BUB1 mitotic checkpoint serine/threonine kinase B (BUB1B, also known as BUBR1), BUB3 mitotic checkpoint protein (BUB3) and CDC20, which directly inhibits the activity of APC/C-CDC20. Therefore, the level of MCC indicates the strength of the spindle checkpoint.

MAD2 exists in at least two conformations: the active conformation, C-MAD2, and the inactive conformation, O-MAD2. MAD2 binds to its activator mitotic arrest deficient 1 (MAD1) and to its effector CDC20 via a short hydrophobic residue called MAD2-interacting motif (MIM) and adopts C-MAD2.[33,36–39,53] The MAD1-C-MAD2 complex at unattached kinetochores binds an extra copy of the inactive MAD2 (O-MAD2) and facilitates the binding to MIM of CDC20. The stable CDC20-C-MAD2 complex further binds with the BUBR1-BUB3 complex, thus promoting new MCC formation. When all kinetochores are correctly attached by bi-polar spindles, the MAD2 inhibitor p31^{comet} (MAD2L1 binding protein) binds to MAD1-bound C-MAD2 and prevents further MCC formation. In addition, p31^{comet} competes with BUBR1 for the same binding surface of MAD2 and accelerates the MCC disassembly, which promotes anaphase onset.

It is notable that MCC plays an important role not only during cell division but also in the internalization of IR (Fig. 2A).[16,21,33] The spindle checkpoint proteins, MAD2-BUBR1-CDC20-p31^{comet}, control IR endocytosis through a similar mechanism to that in mitosis. MAD2 interacts with IR via a well-conserved MAD2-interacting motif (MIM) in the extreme C-terminal region of IR (Fig. 1). The IR-bound MAD2 adopts the active C-MAD2 conformation like in MAD1-C-MAD2 and CDC20-C-MAD2 complexes in mitosis. The IR-MAD2 interaction is not regulated by insulin stimulation. The IR-bound C-MAD2 binds BUBR1-CDC20 and recruits AP2B1, thereby promoting clathrin-mediated IR endocytosis. Here, as in mitosis, p31^{comet} plays an inhibitory role in IR endocytosis. p31^{comet} binds to IR-bound MAD2 in the insulin-free, apo-IR and interrupts the BUBR1-AP2 association with IR-bound MAD2, thereby preventing precocious IR endocytosis (Fig. 2A).

2.3 Insulin receptor signaling pathways in insulin receptor endocytosis

Insulin-activated IR triggers trans-autophosphorylation of its kinase domains and recruits IR substrate (IRS) and/or SRC homology 2 domain-containing (SHC) proteins.[54] The activated IR phosphorylates several tyrosine residues on IRS and SHC, and further recruits additional effectors and adaptors, leading to the activation of two major signaling pathways: the mitogen-activated protein kinase (MAPK) pathway and the phosphatidylinositol 3-kinase (PI3K)-protein kinase B/AKT (PI3K-AKT) pathway (Fig. 2).

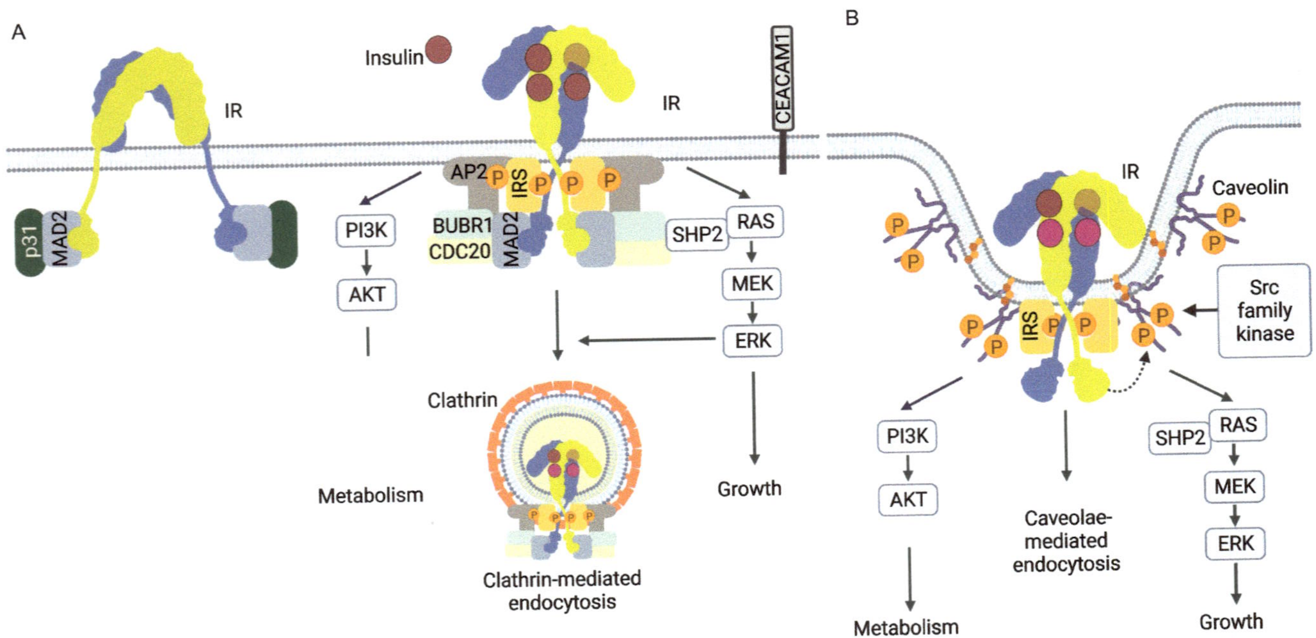

Fig. 2 See figure legend on opposite page.

1. *MAPK Pathway in IR endocytosis*: The insulin-activated IR phosphorylates the tyrosine residue on the NPEY motif and recruits IRS1/2 (Fig. 2A). The active IR then phosphorylates the tyrosine residues in the YXXΦ motifs of IRS1/2. Src homology phosphatase 2 (SHP2), a tyrosine phosphatase plays a counteracting role by dephosphorylating these tyrosine residues.[55,56] In the MAPK pathway, the adaptor/guanine nucleotide exchange factor complex GRB2/SOS is recruited to phosphorylated SHC and/or IRS1/2.[57,58] SHP2 also binds to the IRS1/2 at their C-terminal phosphotyrosine sites and promotes the activation of the MAPK pathway.[59–61]

   IRS1/2 directly bind to another clathrin adaptor AP2M1 via the YXXΦ motifs and promote insulin-activated IR endocytosis (Fig. 2A).[16,17,21,33] Insulin signaling also increases the phosphorylation of Ser/Thr residues immediately following the YXXΦ motifs on IRS1/2. For example, human IRS1 has three YXXΦ motifs, $Y^{612}MPMS^{616}P$, $Y^{632}MPMS^{636}P$, and $Y^{662}MMMS^{666}P$.[17,21] The Ser/Thr residues following YXXΦ motifs are in the context of PXS/

Fig. 2 Models of the regulation of insulin-activated insulin receptor (IR) endocytosis. (A) Insulin-free IR exhibits a Λ-shaped conformation (left). The large distance between two kinase domains might prevent the trans-autophosphorylation. IR binds to MAD2 through a well conserved MAD2-interacting motif (MIM). p31^comet^ binds to IR-bound MAD2 in this auto-inhibitory conformation, thus preventing IR endocytosis. Multiple insulin binding to two distinct sites induces a large conformational change of IR and activates the IR kinase. The insulin-bound IR undergoes trans-autophosphorylation and phosphorylates IRS1/2, thereby activating the PI3K-AKT and MAPK pathways. The upstream regulator SHP2 promotes the activation of MAPK pathway. The activated ERK and SHP2 change the phosphorylation status of IRS1/2 and facilitates the IRS1/2-AP2 interaction. p31^comet^ is released from IR-bound MAD2 by unknown mechanisms upon insulin stimulation. The IR-bound MAD2 binds to BUBR1 and CDC20 and forms a mitotic checkpoint complex-like complex on the IR. The BUBR1 binds to AP2 directly. The MAD2-BUBR1 and IRS1/2-dependent AP2 recruitment promotes clathrin-mediated IR endocytosis. (B) IR is internalized via caveolae-mediated endocytosis. IR directly binds to caveolin through a caveolin scaffolding domain. Insulin stimulation increases tyrosine phosphorylation of caveolin by Src family kinases and potentially by IR. The insulin-specific caveolin phosphorylation facilitates IR signaling in differentiated adipocytes and pancreatic beta cells. IR, insulin receptor; IRS, insulin receptor substrate; PI3K, phosphatidylinositol 3-kinase; AKT, protein kinase B; MEK, mitogen-activated protein kinase kinase; ERK, extracellular signal-regulated kinase; RAS, rat sarcoma virus oncogene; SHP2, Src homology phosphatase 2; AP2, assembly polypeptide-2; BUBR1, BUB1 mitotic checkpoint serine/threonine kinase B; CDC20, cell division cycle 20; MAD2, mitotic arrest deficient 2; p31^comet^, MAD2L1 binding protein; CEACAM1, CEA cell adhesion molecule 1.

TP, a well-known MAPK substrate motif. Extracellular signal-regulated kinase 1 and 2 (ERK1/2) phosphorylate the Ser/Thr residues following YXXΦ motifs.[62–64] The increased Ser/Thr phosphorylation of IRS proteins is positively related with insulin resistance in human patients.[65–67] Consistent with this, the ERK1/2-dependent IRS phosphorylation negatively controls insulin signaling. Structural study shows that tyrosine and methionine residues on YXXΦ motifs of IRS1/2 form extensive hydrophobic interactions with AP2M1. Due to tyrosine phosphorylation on YXXΦ motifs of IRS1/2 will cause static hindrance as well as unfavorable electrostatic interactions, it is reasonable to anticipate that IRS1/2-AP2M1 interaction would be prevented by IR-dependent tyrosine phosphorylation on the YXXΦ motifs of IRS1/2. Meanwhile, SHP2-dependent tyrosine dephosphorylation on the YXXΦ motifs of IRS1 is expected to stabilize the IRS1/2-AP2M1 interaction and promote timely IR endocytosis. Strikingly, ERK-mediated Ser/Thr phosphorylation accelerates SHP2-mediated tyrosine dephosphorylation on the YXXΦ motifs of IRS1. Given the fact that SHP2 is the upstream activator of the MAPK pathway, SHP2 promotes the conversion of IRS1/2 from phosphor-Tyr form to phosphor-Ser/Thr form, thereby accelerating IRS1/2-AP2M1 interaction and IR endocytosis. Consistent with the structural study, chemical inhibitors of MEK (U0126 and PD0325901) and SHP2 (SHP099) significantly delay insulin-dependent IR endocytosis.[21] The requirement of the phosphorylation conversion of IRS1/2 by SHP2 and ERK for IRS1/2-AP2 interaction explains why the activated IR is preferentially internalized.

2. *PI3K-AKT pathway in IR endocytosis*: The tyrosine phosphorylated IRS1/2 recruit and activate PI3K.[54,57] On the plasma membrane, the activated PI3K converts phosphatidylinositol 4,5-bisphosphate (PIP2) to phosphatidylinositol(3,4,5)-triphosphate (PIP3), which in turn co-recruits the serine/threonine kinases, phosphoinositide-dependent protein kinases (PDKs), and their substrate kinase AKT to the membrane through their pleckstrin homology (pH) domains. The PDK then phosphorylates AKT at serine and threonine residues and activates the metabolic signaling (Fig. 2A).

PIP2 serves an important role in clathrin-mediated endocytosis.[22,68] The synthesis and accumulation of PIP2 are required during the assembly of the clathrin-coated pit, whereas its dephosphorylation triggers coat disassembly. To initiate an endocytic event, early endocytic proteins need to be clustered at the plasma membrane by binding to PIP2, thus assembling the vesicle coat. A pioneer module is then formed by several endocytic proteins, including the adaptor proteins FCHO1/2 (F-BAR domain only protein 1/2), AP2, and the scaffold proteins EPS15 (epidermal growth factor receptor pathway substrate 15), EPS15L1 (epidermal growth factor receptor pathway substrate 15 like 1, also known as EPS15R) and intersectins 1 and 2. When AP2 binds to PIP2 and cargo, a large conformational change in AP2 is induced, exposing the clathrin-binding site of AP2. As a result, the AP2 complex is able to recruit clathrin to the assembling coat and coordinates cargo recruitment. Epsins and Huntingtin-interacting protein 1-related protein (HIP1R) bridge the actin filaments to PIP2 and clathrin during membrane bending, and facilitate in membrane shaping and coat formation. Besides, dephosphorylation of PIP2 and its conversion to phosphoinositol 4-phosphate control the scission and uncoating stages of clathrin-mediated endocytosis.

Because PI3K catalyzes PIP2 into PIP3, and the PI3K-AKT pathway is one of the major downstream signaling cascades of IR, the effect of PI3K-AKT pathway on the IR endocytosis was examined. Early study showed that insulin induces a rapid reorganization of the actin filament that depends on PI3K activity in L6 myotubes.[69] This actin cytoskeleton remodeling facilitates the insulin signaling by modulating subcellular localization of IR signaling molecules. Interestingly, in the presence of inhibitors of AKT (AKT inhibitor VIII) or PI3K (GDC-0941), insulin-dependent IR endocytosis was not altered in HepG2 cells.[21] These results suggest that the PI3K-AKT pathways may not control the insulin-activated IR endocytosis.

3. *Negative feedback loop in IR endocytosis*: IR kinase activation and tyrosine phosphorylation of IRS1/2 are essential for IR signaling. By contrast, serine/threonine phosphorylation of IRS1/2 primarily produces a negative effect on IR signaling.[62] For example, Ser/Thr phosphorylation of IRS1 by mammalian target of rapamycin complex 1/ribosomal protein S6 kinase B1 (mTORC1/S6K1) and/or MAPK inhibits IR signaling and promotes insulin resistance.[70–72] The MAPK pathway promotes the formation of IR-IRS1/2-AP2 complex and accelerates the activated IR endocytosis. Future studies using specific inhibitors of S6K1 and mTORC are required for determining the effect of this negative feedback loop on IR endocytosis.

The IR kinase activity is also regulated by protein kinase c-ε (PKCε).[73,74] Diacylglycerol (DAG)-activated PKCε phosphorylates at

Thr^{1160} (numbering of mature, long isoform of human IR) in the activation loop of kinase domain of IR. The phosphorylation of T1160 of IR is expected to destabilize the active conformation of the activation loop. Consistent with this structural prediction, mice expressing IR T1150A (PKCε-mediated phosphorylation defective mutant, numbering of mature, long isoform of mouse IR) exhibit enhanced autophosphorylation of IR and active phosphorylation of AKT upon insulin stimulation. In addition, IR T1150A mice are resistant to diet-induced hepatic insulin resistance, indicating that phosphorylation of IR T1150 contributes to optimal IR kinase activation. It will be interesting to examine whether DAG and PKCε activity also regulate the distribution of IR to the cell surface.

3. Caveolae-mediated IR endocytosis

Caveolae are invaginations of the plasma membrane that are formed by assembly of caveolins.[75,76] Caveolins are integral membrane proteins that bind directly to membrane cholesterol and exist in three major isoforms, caveolin-1, -2 and -3. In addition to clathrin-mediated endocytosis, IR can also be internalized via caveolae-mediated endocytosis upon insulin binding (Fig. 2B).[77–84]

3.1 Caveolin and insulin signaling

Caveolae contain high concentrations of cholesterol, and their function depends on the level of cholesterol in the plasma membrane.[79] Although caveolae are found in many cell types, they are particularly abundant in the plasma membrane of adipocytes.[83,85] In adipocytes, the insulin-triggered IR has been shown to be localized to caveolae in the plasma membrane.[82] Tyrosine residue 14 of caveolin-1 and tyrosine residue 19 of caveolin-2 are phosphorylated by Src family kinase, which induces caveolar internalization.[75] Saltiel and colleagues showed that insulin stimulation, but not epidermal growth factor (EGF) or platelet-derived growth factor (PDGF), also increases tyrosine phosphorylation of caveolin.[86–89] In 3T3-L1 adipocytes, the phosphorylation of caveolin is greatly increased within 5 min after insulin stimulation and lasts for at least 30 min. In addition, inhibitors of the MAPK pathway or the PI3K pathways do not change the insulin-mediated tyrosine phosphorylation of caveolin.

The insulin-specific tyrosine phosphorylation of caveolin is observed in differentiated adipocytes, but not in preadipocytes. The phosphorylation of caveolin-1 and its colocalization with IR are also observed in pancreatic beta cells.[90] Johnson and colleagues showed that the caveolin controls the insulin-activated IR internalization and promotes the MAPK activation. In contrast, Ikonen and colleagues showed that caveolae are not essential for IR signaling in cells derived from liver.[91] The insulin-specific caveolin phosphorylation may facilitate IR signaling by recruiting SH2-domain-containing proteins or by modulating caveolae-mediated IR endocytosis in different cell types and metabolic conditions.

Caveolin-1 or caveolin-3 knockout mouse models demonstrate the important roles of caveolin in IR signaling. Caveolin-1 knockout mice do not develop diabetic phenotypes in normal chow conditions. However, caveolin-1 knockout mice are resistant to diet-induced obesity and display reduced insulin sensitivity and abnormalities of lipid homeostasis.[83,92] Strikingly, the expression levels of IR in adipose tissues are greatly reduced in the caveolin-1 knockout mice, without alteration of mRNA abundance. These results suggest that caveolin-1 controls the stability of IR protein, most likely during IR processing, thus enhancing IR signaling in adipose tissue. Unlike caveolin-1 and -2, which are expressed in various differentiated cell types, caveolin-3 is mainly expressed in myoblast lineage, such as skeletal muscle.[93] Caveolin-3 knockout mice display reduced glucose and insulin sensitivity, and postprandial hyperinsulinemia.[94] In fasting condition, caveolin-3 knockout mice express normal levels of IR in the skeletal muscle, suggesting that caveolin-3 does not control the stability of IR protein and its trafficking during processing. However, insulin stimulation greatly reduces the IR levels in the skeletal muscle from caveolin-3 knockout mice, compared to those in wild-type mice. It suggests that caveolin-3 might prevent a rapid degradation of IR after insulin-dependent internalization in the skeletal muscle.

3.2 Caveolin-binding motifs in insulin receptor

There are two caveolin-binding motifs that interact with the caveolin scaffolding domain: (1) ΦXXXXΦXXΦ and (2) ΦXΦXXXXΦ (Φ represents an aromatic amino acid).[95,96] IR interacts with the caveolin through the characteristic caveolin-binding motif, ^{1181}WSFGVVLW^{1188}, in its cytoplasmic tyrosine kinase domain (numbering of mature, short isoform of human IR) (Fig. 1). Yamamoto et al. found that overexpression of caveolin-3 enhances IRS1 phosphorylation in 293T cells upon insulin stimulation without affecting IR autophosphorylation.[97] Although the molecular mechanism by which caveolin-3 stimulates the IR kinase activity remains unclear, a direct interaction between caveolin and the caveolin-binding motif of IR was observed. Because the caveolin-binding motif of IR locates close to its activation loop in the kinase domain, it is reasonable to imagine that (1) caveolin might stabilize the active conformation of the activation loop of IR kinase and/or (2) caveolin accumulates IR and its downstream substrates (e.g., IRS1/2) into caveolae and facilitates IR kinase activation and IR signaling (Fig. 2B).

Mutations in the caveolin-binding motif of IR causes the failure of IR trafficking and insulin-dependent IR activation.[17,98] The genetic mutation of the caveolin-binding motif, W1181L, was found in a patient with severe insulin resistance syndrome.[99] Without insulin stimulation, IR W1181L remains in the endoplasmic reticulum and the Golgi apparatus.[17] Inhibitors of IR kinase, SHP2 and MAPK pathway do not increase the cell surface levels of IR W1181L, indicating that mutation of caveolin-binding motif of IR causes defects of IR processing and trafficking. These findings demonstrate that caveolin plays a role in both IR trafficking and signaling *in vivo*.

4. Dysfunction of insulin receptor endocytosis

4.1 Animal model

1. *p31^{comet} knockout mice*: IR and its endocytosis play important roles in regulation of blood glucose levels and diverse related biochemical mechanisms in the animal body. The clathrin-mediated endocytosis of IR is promoted by spindle checkpoint proteins MAD2, CDC20 and BUBR1.[16,17,21,33] In cultured cells and mouse liver, ablation of p31^{comet} reduces IR levels on the plasma membrane. Whole-body p31^{comet} knockout mice display neonatal lethality with liver glycogen shortage. Glycogen storage in the liver largely impacts energy homeostasis in newborn mice, and hypoglycemia is a critical factor in early deaths. Due to the crucial role of p31^{comet} in a steady state of IR on the cell surface, p31^{comet} knockout mice lack the ability to prevent unscheduled endocytosis of non-stimulated IR, resulting in low levels of expression of IR at the cell surface and unsuccessful activation of IR. This leads to the inability to synthesize glycogen in liver and causes neonatal lethality in the mouse model.

   In liver-specific p31^{comet} knockout mice newborns, the hepatic glycogen level is reduced but not as severely as whole-body p31^{comet}

knockout mice, allowing them to survive.[16,33] Two-month-old liver-specific p31^{comet} knockout mice display hyperinsulinemia and hyperglycemia after feeding due to excess IR endocytosis prior to the insulin binding. The liver-specific p31^{comet} knockout mice also exhibit glucose and insulin intolerance. These phenotypes are very similar to the characteristics of IR knockout mice[100] but to a lesser degree. It is unclear whether the premature IR endocytosis in p31^{comet} knockout mice is caused by basal insulin or insulin-independent way (or both). Uncontrolled IR endocytosis in p31^{comet} knockout mice results in failure in normal IR signaling, further leading up to hyperinsulinemia and hyperglycemia.

2. *Bubr1 insufficient mice*: Whereas p31^{comet} knockout mice exhibit insulin resistance and glucose intolerance, it is interesting yet understandable that BUBR1 insufficient mice exhibit opposite phenotypes. BUBR1 insufficient mice display improved insulin and glucose sensitivity with reduced blood glucose levels.[16,101] Because BUBR1-dependent AP2 recruitment to IR is required for IR endocytosis, BUBR1 insufficiency delays insulin-mediated internalization of IR and prolongs IR signaling on the plasma membrane, thus improving insulin sensitivity. BUBR1 deficiency mice along with p31^{comet} knockout mice show that IR endocytosis is closely related with insulin resistance and sensitivity.
3. *Lmbd1 knockout mice: LMBRD1* gene is in the limb region 1 (LMBR1) domain which is responsible for encoding proteins belonging to the membrane protein family of lipocalin-1-interacting membrane receptors.[102] The *LMBRD1* gene encodes two major proteins: nuclear export signal-interacting protein (NESI, 467 amino acids) and LMBD1 (73 additional residues at the N-terminus of the NESI, 540 amino acids). The NESI plays an important role in the assembly of the hepatitis δ virus. The LMBD1 is a lysosomal membrane protein with homology to lipocalin-interacting membrane receptor (LIMR). Mutations in LMBRD1 resulting in loss of LMBD1 function were found in patients with an inborn error of Vitamin B-12 metabolism, suggesting that LMBD1 may control the lysosomal Vitamin B-12 (cobalamin) export.[103] Although the exact mechanism remains unclear, it is known that LMBD1 protein participates in clathrin-mediated endocytosis of IR by interacting with AP2. The evidence includes the retention of IR on the plasma membrane in LMBD1 deficient cells, increased IR autophosphorylation, and the existence of LMBD1 in clathrin-coated vesicles. *Lmbrd1* heterozygote mice exhibit increased glucose uptake and

IR signaling. LMBD1 directly binds to AP2 complex through two YXXΦ motifs and a WXX(F/W) motif. It is unclear whether and how LMBD1 directly interacts with IR, however, these data suggest that LMBD1 may control IR endocytosis and IR signaling in mice.

4. *Inceptor knockout mice*: The estrogen-induced gene 121 protein (EIG121), which has been mainly considered in the field of cancer, has adopted a new name, the insulin inhibitory receptor (INCEPTOR), due to the recent study revealing its newly understood characteristics of inhibiting IR.[104–106] Despite the fact that the INCEPTOR lacks a kinase domain like IR, the researchers have found out many similarities between the INCEPTOR and IR. INCEPTOR is internalized via clathrin-mediated endocytosis, and it directly recruits AP2M1 and facilitates endocytosis of IR in β-cells. Whole-body *Inceptor* knockout mice are born in an expected Mendelian ratio but show neonatal lethality. The *Inceptor* knockout mice exhibit decreased IR endocytosis, hyperinsulinemia, and hypoglycemia with improved glucose tolerance. Future studies should aim to determine the molecular mechanism of INCEPTOR-mediated IR endocytosis in β-cells.

5. *Ceacam1 mutant mice:* CEA cell adhesion molecule 1 (CEACAM1) is a transmembrane glycoprotein that is phosphorylated by IR in hepatocytes. Insulin-activated IR phosphorylates CEACAM1 at Tyr^{488}, which is in its intracellular domain. Phosphorylation at Ser^{503} occurs in the absence of insulin and is required for the phosphorylation at Tyr^{488} by IR.[107,108] The serine/threonine kinase for Ser^{503} is unknown yet. The studies from Najjar and colleagues reveal that CEACAM1 is involved in IR endocytosis.[109] CEACAM1 accelerates insulin-activated, clathrin-mediated IR endocytosis and insulin clearance. In a CEACAM1 mutant mouse model where Ser^{503} is substituted to alanine in the liver, hyperinsulinemia with secondary hepatic insulin resistance and impaired glucose tolerance are observed.[107]

   The CEACAM1 does not regulate insulin-like growth factor 1 (IGF1) clearance[110], suggesting the functional selectivity of CEACAM1 on IR endocytosis. Although the binding region of CEACAM1 in IR is unclear, it is proposed that at least two separate domains (juxtamembrane and C-terminal tail) are involved. Identification of binding sites and regulatory modules of the IR-CEACAM1 interaction will provide insight into IR endocytosis and insulin clearance.

6. *March1 knockout mice*: E3 ubiquitin ligases regulate both IR and its substrates. A recent large-scale RNAi screening identified MARCH1 as a negative regulator of IR signaling. The E3 ubiquitin ligase

MARCH1 (membrane-associated ring-CH-type finger protein 1) down regulates IR at the cell surface by facilitating IR endocytosis.[111] Interestingly, unlike other IR ubiquitin ligases, MARCH1 removes IR from the membrane prior to the insulin stimulation, which causes attenuated insulin response due to insufficient levels of IR at the cell surface. The MARCH1 knockdown mice exhibit improved glucose tolerance and insulin sensitivity without a change in insulin levels. Meanwhile, overexpression of MARCH1 impairs hepatic insulin sensitivity in the mouse. It is notable that in white adipose tissue of obese patients, increased MARCH1 expression has been reported. Further studies revealing the physiological mechanism of MARCH1, and IR endocytosis can be a promising flag for diabetes treatment.

7. *MG53 knockout mice*: Skeletal muscle cells account for most of the insulin-stimulated glucose metabolism.[112] MG53 (also called TRIM72), specifically expressed in heart and skeletal muscle, is a tripartite motif-containing (TRIM) family containing E3 ligase activity.[112,113] It is also identified as a major regulator for membrane repair. The E3 ligase activity of MG53 targets IR and IRS1 for ubiquitin-dependent degradation, thus regulating IR abundance on the plasma membrane. MG53 binds to IR and IRS1 in skeletal muscle regardless of the presence of insulin. Overexpression of MG53 in mice significantly increases the ubiquitination of both IR and IRS1 in the skeletal muscle. The ubiquitination of IR and IRS1 is largely increased in wild-type mice fed a high-fat diet, but not in $MG53^{-/-}$ mice, confirming that MG53 is the E3 ligase for IR and IRS1 *in vivo*. $MG53^{-/-}$ mice exhibit increased levels of IRS1, and active phosphorylations of AKT and ERK1/2.[112] Moreover, $MG53^{-/-}$ mice are resistant to the diet-induced metabolic syndrome or insulin resistance.[112,113] In contrast, MG53 overexpression in mice promotes insulin resistance along with obesity, hypertension, and dyslipidaemia.[113]

Strikingly, a recent study showed that MG53 acts as a glucose-sensitive myokine/cardiokine by allosterically inhibiting IR. Xiao and colleagues found that hyperglycemia and hyperinsulinemia induce MG53 secretion from skeletal muscle, and that the circulating MG53 directly binds to the ectodomain of IR.[114] Insulin does not affect the MG53-IR interaction, indicating that MG53 might prevent the insulin-dependent IR conformational transition, thereby suppressing normal IR activation. These studies suggest that MG53 could be a potential therapeutic target to treat metabolic diseases associated with insulin resistance.

8. *Apolipoprotein E4-targeted replaced mice*: Impaired insulin signaling in the brain is closely related to the pathogenesis of Alzheimer's disease (AD). It is previously studied that low levels of insulin, IR, and insulin signaling are observed in brains from AD patients, and patients with diabetes are at higher risk for AD.[115] The apolipoprotein E (APOE) is a cholesterol carrier that supports lipid transport and injury repair in the brain. Individuals carrying polymorphic alleles of APOE, APOE4, exhibit increased risk of AD and impaired IR signaling. Clinical studies show that the positive impacts of intranasal insulin treatment on cognitive improvement in AD patients are dependent on their APOE genotype status. In the mouse brain where murine *Apoe* is replaced by human *APOE4*, impaired IR signaling and insulin resistance are observed in an age-dependent manner. Notably, IR directly binds to APOE4 with high binding affinity (Kd = ~200 nM). It further suggests that APOE4 suppresses the level of IR on the cell surface by trapping IR in the endosome, thus weakening IR signaling in the brain. In aged brain, APOE4 aggregation and dysfunctional endosome further accelerate APOE4-dependent inhibition of IR signaling. This study suggests the pathogenic mechanism of APOE4-mediated insulin resistance in AD and the importance of IR trafficking in the brain.

4.2 Obesity, diabetes, and insulin receptor endocytosis

Diabetes is an abnormal state of high levels of blood glucose that often results from malfunctions in insulin synthesis or IR regulation.[116,117] Diabetic patients with high levels of blood glucose can develop serious health problems such as heart attack, stroke, nerve damage due to the weakened walls of capillaries, kidney damage, and malfunctions in filtering system. The most common types of diabetes are type 1 and type 2 diabetes. Type 1 diabetes is where little or no insulin is produced in the pancreas, specifically by the islets of Langerhans. Genetic and environmental factors may contribute to development of type 1 diabetes.[117] Type 2 diabetes is the result of insulin resistance, often a characteristic of malfunctioning IR or decreased surface IR contents potentially caused by impairment of IR trafficking.[116] The availability of IR on the plasma membrane and its trafficking process can largely affect the response to insulin.[118] The trafficking process including endocytosis is critical for steady state of IR on the cell surface. Dysregulation of IR endocytosis can lead to insufficient IR on the plasma membrane, resulting in unbalanced insulin response and insulin resistance. In response to the high glucose level, the pancreas may produce more insulin resulting in hyperinsulinemia, and it may contribute to severe insulin resistance by accelerating IR endocytosis.

Previous studies show that insulin binding to liver plasma membrane is decreased in obesity and diabetic patients.[119] Observing the liver sections from type 2 diabetes patients, the level of IR at the plasma membrane is significantly lower than that of the non-diabetic patients.[21] This suggests that reduced IR at the plasma membrane might be (1) a contributing factor to the insulin resistance and the development of type 2 diabetes in human patients or (2) a result from dysfunction of IR endocytosis, where excessive endocytosis occurs and there are insufficient IR at the plasma membrane for the normal insulin response. In addition, it is well known that the IR down-regulation is commonly observed in patients with hyperinsulinemic state, regardless of the presence of diabetes; the dysfunction in IR endocytosis brings hyperglycemic diabetic state.[120]

The levels of IR at the plasma membrane and the insulin-induced IR endocytosis are also significantly decreased in the monocytes from obese type 2 diabetic patients compared to healthy individuals.[121,122] Although IR internalization in adipocytes of type 2 diabetes patients does not show significant difference from that of non-diabetes patients, the study suggests that intracellular processing of insulin related to chloroquine-sensitive pathways may be impaired in the adipocytes of type 2 diabetes patients.[123] In rat hepatocytes and human hepatoma cell lines, it has been shown that there are two independent, distinct pathways for IR endocytosis depending on the concentration of insulin[20]: at a low concentration, IR internalization involves a pathway that includes receptor segregation, degradation and recycling, which are blocked by the inhibition of coated pit formation; whereas at a high concentration, IR enters the cell through a more rapidly recycling pathway that is nondegradative and independent of coated pits and endosomal acidification. In CHO cells expressing human IR, it has been shown that IR can be internalized in an insulin-independent manner.[124] These results from different tissues and cultured cells suggest that IR endocytosis mechanisms vary in different types of tissues and different metabolic conditions. Further studies on tissue-specific IR endocytosis mechanisms are required.

4.3 Insulin receptor endocytosis and gene expression

IR signaling regulates insulin-dependent gene expression that controls metabolism and cell growth. One model is that insulin-bound IR activates the PI3K-AKT pathway and the forkhead family box O (FOXO) transcription factors.[125] The phosphorylated FOXOs then bind to 14-3-3 proteins, which is followed by their exclusion from the nucleus, resulting in inhibition of gene expression. Another model is that activated IR phosphorylates forkhead transcription factors (FOXK)[126] and promotes sterol regulatory element binding protein-1 (SREBP-1c) processing,[127] which result in migration of FOXKs and SREBP-1 into the nucleus and regulation of transcription.

A recent study proposed that insulin-activated IR itself can relocate into the nucleus and directly regulate gene expression.[128] The majority of the genomic regions at which IR is detected overlapped with the region occupied by RNA polymerase II. These regions are actively transcribed promoter regions. It further shows that IR forms a transcriptional complex with RNA polymerase II, host cell factor-1 (HCF-1), and transcription factors such as THAP11 (THAP domain-containing protein 11) to regulate gene expression. Interestingly, the IR-bound promoters are enriched with genes previously known to be involved in insulin-related functions such as lipid metabolism and protein synthesis. The target genes also include those involved in disease pathways for diabetes, neurodegeneration, and cancer. It is notable that these regions do not include genes involved in glucose metabolism pathways, in which insulin plays the major regulatory role in liver. Although the specific mechanism under this new finding remains unclear, it is reasonable to draw associations between IR endocytosis and IR found in nucleus for regulation of gene expression as the gene expression regulatory function of IR will be impaired by dysfunction of IR endocytosis.

5. Discussion

1. *SHP2 inhibitors and the treatments of cancer and diabetes:* SHP2 plays an important role as a regulator of cell growth and proliferation in multiple cell types. The mutation in PTPN11 encoding SHP2 may induce Noonan syndrome,[129] which possibly leads to development of childhood leukemia, including juvenile myelomonocytic leukemia. Previous studies have reported the association between SHP2 mutation and solid tumors; in particular, mutations in PTPN11 unblock PTP activity through disrupting the auto-inhibition status, leading to disease progression.[130,131] SHP2 has emerged as a promising target for cancer treatment due to its association with tumor proliferation, metastasis, and chemotherapeutic resistance.[132]

SHP2 is autoinhibited in the absence of insulin-stimulation. SHP099 is an allosteric inhibitor that maintains the autoinhibited, inactive form of SHP2.[17,60,133] It is known that SHP099 binds specifically to SHP2, but not SHP1, which is the closest homolog of SHP2, confirming its specificity. SHP099 was developed for cancer therapy, especially receptor tyrosine kinase mutated cancers. With MEK inhibitors, SHP099 has successfully decreased cancer cell proliferation in diverse cancer types. Using SHP099 along with MEK inhibitors has an advantage of avoiding adaptive resistance to MEK inhibition that might develop when MEK inhibitors are used alone.[134] Also, in mouse models with grafted colon cancer cells, it is reported that SHP099 successfully decreases tumor burden and enhances anti-tumor immunity.[135] Such findings suggest SHP099 as a potential cancer therapy agent either by itself or with other treatments.

Given the function of SHP2-MAPK pathway on IR endocytosis, recent studies examined the effect of SHP099 on IR signaling and showed that the allosteric SHP2 inhibitor could improve insulin sensitivity in mice.[21] In addition, the fact that the patients with LEOPARD syndrome-related SHP2 mutations are resistant to diet-induced obesity and show insulin hypersensitivity[136] suggests that SHP2 is a potential therapeutic target for insulin resistance treatment and type 2 diabetes. There are ongoing clinical trials on dose-escalation of SHPs inhibitors (e.g., JAB-3068, BBP-398) in adult patients with advanced solid tumors (clinical ID: NCT03518554 and NCT04528836). Since cancer and diabetes are closely related with sharing risk factors including obesity, hyperinsulinemia and aging, inhibition of SHP2 is a potential therapeutic method for both cancer and diabetes, and it could be particularly advantageous for patients who have both.

2. *Targeting IR endocytosis for insulin resistance treatment:* IR mutations cause severe insulin resistance syndromes.[17,137–144] Based on subcellular localization of IR in the unstimulated state, three classes of IR mutants are identified (1): Class I, IR mutants that localize to the plasma membrane, (2) Class II, IR mutants that localize in intracellular compartments such as RAB7-positive late endosome, and (3) Class III, IR mutants that remain in the endoplasmic reticulum and Golgi apparatus.[17] When a chemical inhibitor of dynamin (dynasore) is added to the Class II of IR mutants, the IR level is increased at the plasma membrane, indicating that these mutants undergo uncontrolled clathrin-mediated endocytosis. When the Class II mutants is treated with SHP099 (SHP2 inhibitor) and U0125 (MEK inhibitor), but not other classes of IR mutants, the levels

of IR at the plasma membrane are dramatically increased. These results suggest that targeting IR endocytosis may improve insulin sensitivity in patients with these Class II IR mutants.

6. Conclusion

Increasing interest has grown in understanding the molecular basis and regulation of IR endocytosis, as well as its applications in treating diseases. Hyperinsulinemia is associated with type 2 diabetes and a high risk of cancer death. Because insulin-induced IR activation is essential for the IR endocytosis, it is reasonable to expect that hyperinsulinemia reduces the functional IR level on the cell surface by promoting IR endocytosis. Studying how relevant proteins involved in IR endocytosis impact a variety of disease pathogenesis is an important next step. Additionally, understanding how endocytosed IR can activate specific signaling pathways in subcellular compartments, as well as how dysregulation of IR endocytosis contributes to human diseases will provide insight into the function, regulation, and mechanism of IR endocytosis.

References

1. Saltiel AR, Kahn CR. Insulin signalling and the regulation of glucose and lipid metabolism. *Nature*. 2001;414(6865):799–806.
2. White MF, Kahn CR. Insulin action at a molecular level—100 years of progress. *Mol Metab*. 2021;52:101304.
3. Goh LK, Sorkin A. Endocytosis of receptor tyrosine kinases. *Cold Spring Harb Perspect Biol*. 2013;5(5):a017459.
4. Haeusler RA, McGraw TE, Accili D. Biochemical and cellular properties of insulin receptor signalling. *Nat Rev Mol Cell Biol*. 2018;19(1):31–44.
5. Lawrence MC. Understanding insulin and its receptor from their three-dimensional structures. *Mol Metab*. 2021;52:101255.
6. Uchikawa E, Choi E, Shang G, Yu H, Bai XC. Activation mechanism of the insulin receptor revealed by cryo-EM structure of the fully liganded receptor-ligand complex. *eLife*. 2019;8:e48630.
7. Gutmann T, Schafer IB, Poojari C, et al. Cryo-EM structure of the complete and ligand-saturated insulin receptor ectodomain. *J Cell Biol*. 2020;219(1):e201907210.
8. Li J, Park J, Mayer JP, et al. Synergistic activation of the insulin receptor via two distinct sites. *Nat Struct Mol Biol*. 2022;29(4):357–368.
9. McKern NM, Lawrence MC, Streltsov VA, et al. Structure of the insulin receptor ectodomain reveals a folded-over conformation. *Nature*. 2006;443(7108):218–221.
10. Frasca F, Pandini G, Scalia P, et al. Insulin receptor isoform a, a newly recognized, high-affinity insulin-like growth factor II receptor in fetal and cancer cells. *Mol Cell Biol*. 1999;19(5):3278–3288.
11. Sciacca L, Cassarino MF, Genua M, et al. Insulin analogues differently activate insulin receptor isoforms and post-receptor signalling. *Diabetologia*. 2010;53(8):1743–1753.

12. Pierre-Eugene C, Pagesy P, Nguyen TT, et al. Effect of insulin analogues on insulin/IGF1 hybrid receptors: increased activation by glargine but not by its metabolites M1 and M2. *PLoS One*. 2012;7(7):e41992.
13. Belfiore A, Malaguarnera R, Vella V, et al. Insulin receptor isoforms in physiology and disease: an updated view. *Endocr Rev*. 2017;38(5):379–431.
14. Carpentier JL, Paccaud JP, Gorden P, Rutter WJ, Orci L. Insulin-induced surface redistribution regulates internalization of the insulin receptor and requires its autophosphorylation. *Proc Natl Acad Sci U S A*. 1992;89(1):162–166.
15. Carpentier JL, Paccaud JP, Backer J, et al. Two steps of insulin receptor internalization depend on different domains of the beta-subunit. *J Cell Biol*. 1993;122 (6):1243–1252.
16. Choi E, Zhang X, Xing C, Yu H. Mitotic checkpoint regulators control insulin signaling and metabolic homeostasis. *Cell*. 2016;166(3):567–581.
17. Hall C, Yu H, Choi E. Insulin receptor endocytosis in the pathophysiology of insulin resistance. *Exp Mol Med*. 2020;52:911–920.
18. Smith RM, Jarett L. Receptor-mediated endocytosis and intracellular processing of insulin: ultrastructural and biochemical evidence for cell-specific heterogeneity and distinction from nonhormonal ligands. *Lab Invest*. 1988;58(6):613–629.
19. McClain DA. Mechanism and role of insulin receptor endocytosis. *Am J Med Sci*. 1992;304(3):192–201.
20. McClain DA, Olefsky JM. Evidence for two independent pathways of insulin-receptor internalization in hepatocytes and hepatoma cells. *Diabetes*. 1988;37(6):806–815.
21. Choi E, Kikuchi S, Gao H, et al. Mitotic regulators and the SHP2-MAPK pathway promote IR endocytosis and feedback regulation of insulin signaling. *Nat Commun*. 2019;10(1):1473.
22. Kaksonen M, Roux A. Mechanisms of clathrin-mediated endocytosis. *Nat Rev Mol Cell Biol*. 2018;19(5):313–326.
23. Najjar SM, Perdomo G. Hepatic insulin clearance: mechanism and physiology. *Physiology (Bethesda)*. 2019;34(3):198–215.
24. Collins BM, McCoy AJ, Kent HM, Evans PR, Owen DJ. Molecular architecture and functional model of the endocytic AP2 complex. *Cell*. 2002;109(4):523–535.
25. Owen DJ, Evans PR. A structural explanation for the recognition of tyrosine-based endocytotic signals. *Science*. 1998;282(5392):1327–1332.
26. Chen WJ, Goldstein JL, Brown MS. NPXY, a sequence often found in cytoplasmic tails, is required for coated pit-mediated internalization of the low density lipoprotein receptor. *J Biol Chem*. 1990;265(6):3116–3123.
27. Backer JM, Kahn CR, Cahill DA, Ullrich A, White MF. Receptor-mediated internalization of insulin requires a 12-amino acid sequence in the juxtamembrane region of the insulin receptor beta-subunit. *J Biol Chem*. 1990;265(27):16450–16454.
28. Backer JM, Shoelson SE, Haring E, White MF. Insulin receptors internalize by a rapid, saturable pathway requiring receptor autophosphorylation and an intact juxtamembrane region. *J Cell Biol*. 1991;115(6):1535–1545.
29. Rajagopalan M, Neidigh JL, McClain DA. Amino acid sequences Gly-pro-Leu-Tyr and Asn-pro-Glu-Tyr in the submembranous domain of the insulin receptor are required for normal endocytosis. *J Biol Chem*. 1991;266(34):23068–23073.
30. Hamer I, Haft CR, Paccaud JP, Maeder C, Taylor S, Carpentier JL. Dual role of a dileucine motif in insulin receptor endocytosis. *J Biol Chem*. 1997;272 (35):21685–21691.
31. Haft CR, Klausner RD, Taylor SI. Involvement of dileucine motifs in the internalization and degradation of the insulin receptor. *J Biol Chem*. 1994;269 (42):26286–26294.
32. Hubbard SR, Wei L, Ellis L, Hendrickson WA. Crystal structure of the tyrosine kinase domain of the human insulin receptor. *Nature*. 1994;372(6508):746–754.

33. Choi E, Yu H. Spindle checkpoint regulators in insulin signaling. *Front Cell Dev Biol.* 2018;6:161.
34. Tang R, Jiang Z, Chen F, et al. The kinase activity of Drosophila BubR1 is required for insulin signaling-dependent stem cell maintenance. *Cell Rep.* 2020;31(12):107794.
35. Yu H, Li Y, Li L, et al. Functional reciprocity of proteins involved in mitosis and endocytosis. *FEBS J.* 2021;288(20):5850–5866.
36. Luo X, Tang Z, Rizo J, Yu H. The Mad2 spindle checkpoint protein undergoes similar major conformational changes upon binding to either Mad1 or Cdc20. *Mol Cell.* 2002;9(1):59–71.
37. Yu H. Regulation of APC-Cdc20 by the spindle checkpoint. *Curr Opin Cell Biol.* 2002;14(6):706–714.
38. Luo X, Tang Z, Xia G, et al. The Mad2 spindle checkpoint protein has two distinct natively folded states. *Nat Struct Mol Biol.* 2004;11(4):338–345.
39. Yu H. Cdc20: a WD40 activator for a cell cycle degradation machine. *Mol Cell.* 2007;27(1):3–16.
40. Luo X, Yu H. Protein metamorphosis: the two-state behavior of Mad2. *Structure.* 2008;16(11):1616–1625.
41. Yang M, Li B, Liu CJ, et al. Insights into mad2 regulation in the spindle checkpoint revealed by the crystal structure of the symmetric mad2 dimer. *PLoS Biol.* 2008;6 (3):e50.
42. Jia L, Kim S, Yu H. Tracking spindle checkpoint signals from kinetochores to APC/C. *Trends Biochem Sci.* 2013;38(6):302–311.
43. Nilsson J, Yekezare M, Minshull J, Pines J. The APC/C maintains the spindle assembly checkpoint by targeting Cdc20 for destruction. *Nat Cell Biol.* 2008;10(12):1411–1420.
44. Izawa D, Pines J. The mitotic checkpoint complex binds a second CDC20 to inhibit active APC/C. *Nature.* 2015;517(7536):631–634.
45. Weaver BA, Silk AD, Montagna C, Verdier-Pinard P, Cleveland DW. Aneuploidy acts both oncogenically and as a tumor suppressor. *Cancer Cell.* 2007;11(1):25–36.
46. Holland AJ, Cleveland DW. Boveri revisited: chromosomal instability, aneuploidy and tumorigenesis. *Nat Rev Mol Cell Biol.* 2009;10(7):478–487.
47. Holland AJ, Cleveland DW. Losing balance: the origin and impact of aneuploidy in cancer. *EMBO Rep.* 2012;13(6):501–514.
48. Choi E, Lee H. Chromosome damage in mitosis induces BubR1 activation and prometaphase arrest. *FEBS Lett.* 2008;582(12):1700–1706.
49. Choi E, Choe H, Min J, Choi JY, Kim J, Lee H. BubR1 acetylation at prometaphase is required for modulating APC/C activity and timing of mitosis. *EMBO J.* 2009;28 (14):2077–2089.
50. Lee YK, Choi E, Kim MA, Park PG, Park NH, Lee H. BubR1 as a prognostic marker for recurrence-free survival rates in epithelial ovarian cancers. *Br J Cancer.* 2009;101 (3):504–510.
51. Choi E, Park PG, Lee HO, et al. BRCA2 fine-tunes the spindle assembly checkpoint through reinforcement of BubR1 acetylation. *Dev Cell.* 2012;22(2):295–308.
52. Park I, Lee HO, Choi E, et al. Loss of BubR1 acetylation causes defects in spindle assembly checkpoint signaling and promotes tumor formation. *J Cell Biol.* 2013;202 (2):295–309.
53. Sironi L, Mapelli M, Knapp S, De Antoni A, Jeang KT, Musacchio A. Crystal structure of the tetrameric Mad1-Mad2 core complex: implications of a 'safety belt' binding mechanism for the spindle checkpoint. *EMBO J.* 2002;21(10):2496–2506.
54. Guo S. Insulin signaling, resistance, and the metabolic syndrome: insights from mouse models into disease mechanisms. *J Endocrinol.* 2014;220(2):T1–T23.
55. White MF. Insulin signaling in health and disease. *Science.* 2003;302(5651):1710–1711.

56. Myers Jr MG, Mendez R, Shi P, Pierce JH, Rhoads R, White MF. The COOH-terminal tyrosine phosphorylation sites on IRS-1 bind SHP-2 and negatively regulate insulin signaling. *J Biol Chem*. 1998;273(41):26908–26914.
57. Siddle K. Signalling by insulin and IGF receptors: supporting acts and new players. *J Mol Endocrinol*. 2011;47(1):R1–10.
58. Gehart H, Kumpf S, Ittner A, Ricci R. MAPK signalling in cellular metabolism: stress or wellness? *EMBO Rep*. 2010;11(11):834–840.
59. Nichols RJ, Haderk F, Stahlhut C, et al. RAS nucleotide cycling underlies the SHP2 phosphatase dependence of mutant BRAF-, NF1- and RAS-driven cancers. *Nat Cell Biol*. 2018;20(9):1064–1073.
60. Chen YN, LaMarche MJ, Chan HM, et al. Allosteric inhibition of SHP2 phosphatase inhibits cancers driven by receptor tyrosine kinases. *Nature*. 2016;535(7610):148–152.
61. Hof P, Pluskey S, Dhe-Paganon S, Eck MJ, Shoelson SE. Crystal structure of the tyrosine phosphatase SHP-2. *Cell*. 1998;92(4):441–450.
62. Copps KD, White MF. Regulation of insulin sensitivity by serine/threonine phosphorylation of insulin receptor substrate proteins IRS1 and IRS2. *Diabetologia*. 2012;55 (10):2565–2582.
63. De Fea K, Roth RA. Modulation of insulin receptor substrate-1 tyrosine phosphorylation and function by mitogen-activated protein kinase. *J Biol Chem*. 1997;272 (50):31400–31406.
64. Hers I, Tavare JM. Mechanism of feedback regulation of insulin receptor substrate-1 phosphorylation in primary adipocytes. *Biochem J*. 2005;388(Pt. 2):713–720.
65. Bouzakri K, Roques M, Gual P, et al. Reduced activation of phosphatidylinositol-3 kinase and increased serine 636 phosphorylation of insulin receptor substrate-1 in primary culture of skeletal muscle cells from patients with type 2 diabetes. *Diabetes*. 2003;52(6):1319–1325.
66. Carlson CJ, Koterski S, Sciotti RJ, Poccard GB, Rondinone CM. Enhanced basal activation of mitogen-activated protein kinases in adipocytes from type 2 diabetes: potential role of p38 in the downregulation of GLUT4 expression. *Diabetes*. 2003;52(3):634–641.
67. Ozaki KI, Awazu M, Tamiya M, et al. Targeting the ERK signaling pathway as a potential treatment for insulin resistance and type 2 diabetes. *Am J Physiol Endocrinol Metab*. 2016;310(8):E643–E651.
68. Czech MP. PIP2 and PIP3: complex roles at the cell surface. *Cell*. 2000;100(6):603–606.
69. Tsakiridis T, Tong P, Matthews B, et al. Role of the actin cytoskeleton in insulin action. *Microsc Res Tech*. 1999;47(2):79–92.
70. Zhang J, Gao Z, Yin J, Quon MJ, Ye J. S6K directly phosphorylates IRS-1 on Ser-270 to promote insulin resistance in response to TNF-(alpha) signaling through IKK2. *J Biol Chem*. 2008;283(51):35375–35382.
71. Shah OJ, Wang Z, Hunter T. Inappropriate activation of the TSC/Rheb/mTOR/S6K cassette induces IRS1/2 depletion, insulin resistance, and cell survival deficiencies. *Curr Biol*. 2004;14(18):1650–1656.
72. Harrington LS, Findlay GM, Gray A, et al. The TSC1-2 tumor suppressor controls insulin-PI3K signaling via regulation of IRS proteins. *J Cell Biol*. 2004;166(2):213–223.
73. Lyu K, Zhang Y, Zhang D, et al. A membrane-bound diacylglycerol species induces PKC-mediated hepatic insulin resistance. *Cell Metab*. 2020;32(4):654–664.e655.
74. Petersen MC, Madiraju AK, Gassaway BM, et al. Insulin receptor Thr1160 phosphorylation mediates lipid-induced hepatic insulin resistance. *J Clin Invest*. 2016;126 (11):4361–4371.
75. Kiss AL, Botos E. Endocytosis via caveolae: alternative pathway with distinct cellular compartments to avoid lysosomal degradation? *J Cell Mol Med*. 2009;13(7):1228–1237.

76. Kumari S, Mg S, Mayor S. Endocytosis unplugged: multiple ways to enter the cell. *Cell Res*. 2010;20(3):256–275.
77. Kabayama K, Sato T, Saito K, et al. Dissociation of the insulin receptor and caveolin-1 complex by ganglioside GM3 in the state of insulin resistance. *Proc Natl Acad Sci U S A*. 2007;104(34):13678–13683.
78. Brannmark C, Palmer R, Glad ST, Cedersund G, Stralfors P. Mass and information feedbacks through receptor endocytosis govern insulin signaling as revealed using a parameter-free modeling framework. *J Biol Chem*. 2010;285(26):20171–20179.
79. Gustavsson J, Parpal S, Karlsson M, et al. Localization of the insulin receptor in caveolae of adipocyte plasma membrane. *FASEB J*. 1999;13(14):1961–1971.
80. Karlsson M, Thorn H, Danielsson A, et al. Colocalization of insulin receptor and insulin receptor substrate-1 to caveolae in primary human adipocytes. Cholesterol depletion blocks insulin signalling for metabolic and mitogenic control. *Eur J Biochem*. 2004;271(12):2471–2479.
81. Foti M, Porcheron G, Fournier M, Maeder C, Carpentier JL. The neck of caveolae is a distinct plasma membrane subdomain that concentrates insulin receptors in 3T3-L1 adipocytes. *Proc Natl Acad Sci U S A*. 2007;104(4):1242–1247.
82. Fagerholm S, Ortegren U, Karlsson M, Ruishalme I, Stralfors P. Rapid insulin-dependent endocytosis of the insulin receptor by caveolae in primary adipocytes. *PLoS One*. 2009;4(6):e5985.
83. Cohen AW, Combs TP, Scherer PE, Lisanti MP. Role of caveolin and caveolae in insulin signaling and diabetes. *Am J Physiol Endocrinol Metab*. 2003;285(6): E1151–E1160.
84. Kabayama K, Sato T, Kitamura F, et al. TNF alpha-induced insulin resistance in adipocytes as a membrane microdomain disorder: involvement of ganglioside GM3. *Glycobiology*. 2005;15(1):21–29.
85. Razani B, Woodman SE, Lisanti MP. Caveolae: from cell biology to animal physiology. *Pharmacol Rev*. 2002;54(3):431–467.
86. Mastick CC, Brady MJ, Saltiel AR. Insulin stimulates the tyrosine phosphorylation of caveolin. *J Cell Biol*. 1995;129(6):1523–1531.
87. Mastick CC, Saltiel AR. Insulin-stimulated tyrosine phosphorylation of caveolin is specific for the differentiated adipocyte phenotype in 3T3-L1 cells. *J Biol Chem*. 1997;272 (33):20706–20714.
88. Baumann CA, Ribon V, Kanzaki M, et al. CAP defines a second signalling pathway required for insulin-stimulated glucose transport. *Nature*. 2000;407(6801):202–207.
89. Kimura A, Mora S, Shigematsu S, Pessin JE, Saltiel AR. The insulin receptor catalyzes the tyrosine phosphorylation of caveolin-1. *J Biol Chem*. 2002;277(33):30153–30158.
90. Boothe T, Lim GE, Cen H, et al. Inter-domain tagging implicates caveolin-1 in insulin receptor trafficking and Erk signaling bias in pancreatic beta-cells. *Mol Metab*. 2016;5 (5):366–378.
91. Vainio S, Heino S, Mansson JE, et al. Dynamic association of human insulin receptor with lipid rafts in cells lacking caveolae. *EMBO Rep*. 2002;3(1):95–100.
92. Cohen AW, Razani B, Wang XB, et al. Caveolin-1-deficient mice show insulin resistance and defective insulin receptor protein expression in adipose tissue. *Am J Physiol Cell Physiol*. 2003;285(1):C222–C235.
93. Song KS, Scherer PE, Tang Z, et al. Expression of caveolin-3 in skeletal, cardiac, and smooth muscle cells. Caveolin-3 is a component of the sarcolemma and co-fractionates with dystrophin and dystrophin-associated glycoproteins. *J Biol Chem*. 1996;271 (25):15160–15165.
94. Capozza F, Combs TP, Cohen AW, et al. Caveolin-3 knockout mice show increased adiposity and whole body insulin resistance, with ligand-induced insulin receptor instability in skeletal muscle. *Am J Physiol Cell Physiol*. 2005;288(6):C1317–C1331.

95. Couet J, Li S, Okamoto T, Ikezu T, Lisanti MP. Identification of peptide and protein ligands for the caveolin-scaffolding domain. Implications for the interaction of caveolin with caveolae-associated proteins. *J Biol Chem.* 1997;272(10):6525–6533.
96. Couet J, Sargiacomo M, Lisanti MP. Interaction of a receptor tyrosine kinase, EGF-R, with caveolins. Caveolin binding negatively regulates tyrosine and serine/threonine kinase activities. *J Biol Chem.* 1997;272(48):30429–30438.
97. Yamamoto M, Toya Y, Schwencke C, Lisanti MP, Myers Jr MG, Ishikawa Y. Caveolin is an activator of insulin receptor signaling. *J Biol Chem.* 1998;273(41):26962–26968.
98. Nystrom FH, Chen H, Cong LN, Li Y, Quon MJ. Caveolin-1 interacts with the insulin receptor and can differentially modulate insulin signaling in transfected Cos-7 cells and rat adipose cells. *Mol Endocrinol.* 1999;13(12):2013–2024.
99. Imamura T, Takata Y, Sasaoka T, et al. Two naturally occurring mutations in the kinase domain of insulin receptor accelerate degradation of the insulin receptor and impair the kinase activity. *J Biol Chem.* 1994;269(49):31019–31027.
100. Michael MD, Kulkarni RN, Postic C, et al. Loss of insulin signaling in hepatocytes leads to severe insulin resistance and progressive hepatic dysfunction. *Mol Cell.* 2000;6(1):87–97.
101. Baker DJ, Perez-Terzic C, Jin F, et al. Opposing roles for p16Ink4a and p19Arf in senescence and ageing caused by BubR1 insufficiency. *Nat Cell Biol.* 2008;10(7):825–836.
102. Tseng LT, Lin CL, Tzen KY, Chang SC, Chang MF. LMBD1 protein serves as a specific adaptor for insulin receptor internalization. *J Biol Chem.* 2013;288(45):32424–32432.
103. Rutsch F, Gailus S, Miousse IR, et al. Identification of a putative lysosomal cobalamin exporter altered in the cblF defect of vitamin B12 metabolism. *Nat Genet.* 2009;41(2):234–239.
104. Ansarullah JC, Far FF, et al. Inceptor counteracts insulin signalling in beta-cells to control glycaemia. *Nature.* 2021;590(7845):326–331.
105. Hall C, Choi E. New scavenger to fine-tune insulin action in beta cells. *Cell Metab.* 2021;33(4):707–708.
106. Kulkarni RN. New-found brake calibrates insulin action in beta-cells. *Nature.* 2021;590(7845):221–223.
107. Poy MN, Yang Y, Rezaei K, et al. CEACAM1 regulates insulin clearance in liver. *Nat Genet.* 2002;30(3):270–276.
108. Najjar SM, Philippe N, Suzuki Y, et al. Insulin-stimulated phosphorylation of recombinant pp120/HA4, an endogenous substrate of the insulin receptor tyrosine kinase. *Biochemistry.* 1995;34(29):9341–9349.
109. Najjar SM. Regulation of insulin action by CEACAM1. *Trends Endocrinol Metab.* 2002;13(6):240–245.
110. Najjar SM, Choice CV, Soni P, Whitman CM, Poy MN. Effect of pp120 on receptor-mediated insulin endocytosis is regulated by the juxtamembrane domain of the insulin receptor. *J Biol Chem.* 1998;273(21):12923–12928.
111. Nagarajan A, Petersen MC, Nasiri AR, et al. MARCH1 regulates insulin sensitivity by controlling cell surface insulin receptor levels. *Nat Commun.* 2016;7:12639.
112. Yi JS, Park JS, Ham YM, et al. MG53-induced IRS-1 ubiquitination negatively regulates skeletal myogenesis and insulin signalling. *Nat Commun.* 2013;4:2354.
113. Song R, Peng W, Zhang Y, et al. Central role of E3 ubiquitin ligase MG53 in insulin resistance and metabolic disorders. *Nature.* 2013;494(7437):375–379.
114. Wu HK, Zhang Y, Cao CM, et al. Glucose-sensitive Myokine/Cardiokine MG53 regulates systemic insulin response and metabolic homeostasis. *Circulation.* 2019;139(7):901–914.

115. Zhao N, Liu CC, Van Ingelgom AJ, et al. Apolipoprotein E4 impairs neuronal insulin signaling by trapping insulin receptor in the endosomes. *Neuron*. 2017;96(1):115–129. e115.
116. Chatterjee S, Khunti K, Davies MJ. Type 2 diabetes. *Lancet*. 2017;389(10085): 2239–2251.
117. DiMeglio LA, Evans-Molina C, Oram RA. Type 1 diabetes. *Lancet*. 2018;391 (10138):2449–2462.
118. Chen Y, Huang L, Qi X, Chen C. Insulin receptor trafficking: consequences for insulin sensitivity and diabetes. *Int J Mol Sci*. 2019;20(20).
119. Caro JF, Ittoop O, Pories WJ, et al. Studies on the mechanism of insulin resistance in the liver from humans with noninsulin-dependent diabetes. Insulin action and binding in isolated hepatocytes, insulin receptor structure, and kinase activity. *J Clin Invest*. 1986;78(1):249–258.
120. Gorden P, Arakaki R, Collier E, Carpentier JL. Biosynthesis and regulation of the insulin receptor. *Yale J Biol Med*. 1989;62(5):521–531.
121. Trischitta V, Gullo D, Squatrito S, Pezzino V, Goldfine ID, Vigneri R. Insulin internalization into monocytes is decreased in patients with type II diabetes mellitus. *J Clin Endocrinol Metab*. 1986;62(3):522–528.
122. Trischitta V, Brunetti A, Chiavetta A, Benzi L, Papa V, Vigneri R. Defects in insulin-receptor internalization and processing in monocytes of obese subjects and obese NIDDM patients. *Diabetes*. 1989;38(12):1579–1584.
123. Jochen AL, Berhanu P, Olefsky JM. Insulin internalization and degradation in adipocytes from normal and type II diabetic subjects. *J Clin Endocrinol Metab*. 1986;62 (2):268–274.
124. Paccaud JP, Siddle K, Carpentier JL. Internalization of the human insulin receptor. The insulin-independent pathway. *J Biol Chem*. 1992;267(18):13101–13106.
125. Nakae J, Barr V, Accili D. Differential regulation of gene expression by insulin and IGF-1 receptors correlates with phosphorylation of a single amino acid residue in the forkhead transcription factor FKHR. *EMBO J*. 2000;19(5):989–996.
126. Sakaguchi M, Cai W, Wang CH, et al. FoxK1 and FoxK2 in insulin regulation of cellular and mitochondrial metabolism. *Nat Commun*. 2019;10(1):1582.
127. Shimano H, Sato R. SREBP-regulated lipid metabolism: convergent physiology—divergent pathophysiology. *Nat Rev Endocrinol*. 2017;13(12):710–730.
128. Hancock ML, Meyer RC, Mistry M, et al. Insulin receptor associates with promoters genome-wide and regulates gene expression. *Cell*. 2019;177(3):722–736 e722.
129. Pierpont ME, Digilio MC. Cardiovascular disease in Noonan syndrome. *Curr Opin Pediatr*. 2018;30(5):601–608.
130. Guo W, Liu W, Chen Z, et al. Tyrosine phosphatase SHP2 negatively regulates NLRP3 inflammasome activation via ANT1-dependent mitochondrial homeostasis. *Nat Commun*. 2017;8(1):2168.
131. Dong L, Han D, Meng X, Xu M, Zheng C, Xia Q. Activating mutation of SHP2 establishes a tumorigenic phonotype through cell-autonomous and non-cell-autonomous mechanisms. *Front Cell Dev Biol*. 2021;9:630712.
132. Zhang J, Zhang F, Niu R. Functions of Shp2 in cancer. *J Cell Mol Med*. 2015;19 (9):2075–2083.
133. Zhang RY, Yu ZH, Zeng L, et al. SHP2 phosphatase as a novel therapeutic target for melanoma treatment. *Oncotarget*. 2016;7(45):73817–73829.
134. Fedele C, Ran H, Diskin B, et al. SHP2 inhibition prevents adaptive resistance to MEK inhibitors in multiple cancer models. *Cancer Discov*. 2018;8(10):1237–1249.
135. Zhao M, Guo W, Wu Y, et al. SHP2 inhibition triggers anti-tumor immunity and synergizes with PD-1 blockade. *Acta Pharm Sin B*. 2019;9(2):304–315.

136. Tajan M, Batut A, Cadoudal T, et al. LEOPARD syndrome-associated SHP2 mutation confers leanness and protection from diet-induced obesity. *Proc Natl Acad Sci U S A.* 2014;111(42):E4494–E4503.
137. Ardon O, Procter M, Tvrdik T, Longo N, Mao R. Sequencing analysis of insulin receptor defects and detection of two novel mutations in INSR gene. *Mol Genet Metab Rep.* 2014;1:71–84.
138. Donohue WL, Uchida I. Leprechaunism: a euphemism for a rare familial disorder. *J Pediatr.* 1954;45(5):505–519.
139. Rogers DR. Leprechaunism (Donohue's syndrome). A possible case, with emphasis on changes in the adenohypophysis. *Am J Clin Pathol.* 1966;45(5):614–619.
140. Desbois-Mouthon C, Danan C, Amselem S, et al. Severe resistance to insulin and insulin-like growth factor-I in cells from a patient with leprechaunism as a result of two mutations in the tyrosine kinase domain of the insulin receptor. *Metabolism.* 1996;45(12):1493–1500.
141. Hart LM, Lindhout D, Van der Zon GC, et al. An insulin receptor mutant (Asp707 - -> Ala), involved in leprechaunism, is processed and transported to the cell surface but unable to bind insulin. *J Biol Chem.* 1996;271(31):18719–18724.
142. Takahashi Y, Kadowaki H, Momomura K, et al. A homozygous kinase-defective mutation in the insulin receptor gene in a patient with leprechaunism. *Diabetologia.* 1997;40(4):412–420.
143. Longo N, Singh R, Griffin LD, Langley SD, Parks JS, Elsas LJ. Impaired growth in Rabson-Mendenhall syndrome: lack of effect of growth hormone and insulin-like growth factor-I. *J Clin Endocrinol Metab.* 1994;79(3):799–805.
144. Dutta D, Maisnam I, Ghosh S, Mukhopadhyay S, Chowdhury S. Syndrome of extreme insulin resistance (Rabson-Mendenhall phenotype) with atrial septal defect: clinical presentation and treatment outcomes. *J Clin Res Pediatr Endocrinol.* 2013;5(1):58–61.

CHAPTER SIX

VEGFR endocytosis: Implications for angiogenesis

Queen Saikia[a], Hannah Reeve[a], Areej Alzahrani[a], William R. Critchley[a], Elton Zeqiraj[a], Aysha Divan[a], Michael A. Harrison[b], and Sreenivasan Ponnambalam[a,*]

[a]School of Molecular & Cellular Biology, University of Leeds, Leeds, United Kingdom
[b]School of Biomedical Sciences, University of Leeds, Leeds, United Kingdom
*Corresponding author: e-mail address: s.ponnambalam@leeds.ac.uk

Contents

Abstract

The binding of vascular endothelial growth factor (VEGF) superfamily to VEGF receptor tyrosine kinases (VEGFRs) and co-receptors regulates vasculogenesis, angiogenesis and lymphangiogenesis. A recurring theme is that dysfunction in VEGF signaling promotes pathological angiogenesis, an important feature of cancer and pro-inflammatory disease states. Endocytosis of basal (resting) or activated VEGFRs facilitates signal attenuation and endothelial quiescence. However, increasing evidence suggest that activated VEGFRs can continue to signal from intracellular compartments such as endosomes. In this chapter, we focus on the evolving link between VEGFR endocytosis,

Progress in Molecular Biology and Translational Science, Volume 194
ISSN 1877-1173
https://doi.org/10.1016/bs.pmbts.2022.06.021

signaling and turnover and the implications for angiogenesis. There is much interest in how such understanding of VEGFR dynamics can be harnessed therapeutically for a wide range of human disease states.

Abbreviations

CDE clathrin-dependent endocytosis
CIE clathrin-independent endocytosis
NRP neuropilin
PTM post-translational modification
RTK receptor tyrosine kinase
VEGF vascular endothelial growth factor
VEGFR vascular endothelial growth factor receptor

1. Introduction

The *de novo* development of a mature adult vasculature during embryogenesis involves endothelial cell precursor (angioblast) differentiation into endothelial cells which form immature vascular tubes (blood vessels), as a precursor to a fully mature vascular network i.e. vasculogenesis.[1] Simultaneously, the vascular network is progressively remodeled by new blood vessel sprouting from pre-existing ones i.e. angiogenesis. In adults, angiogenesis can be harnessed to promote normal physiological processes: growth and development, menstruation and wound healing.[2] However, dysregulated angiogenesis contributes to pathological conditions including tumor growth, age-related macular degeneration and inflammatory diseases.[3–5]

A family of pro-angiogenic molecules termed vascular endothelial growth factors (VEGFs) bind to receptor tyrosine kinase receptors (VEGFRs) and co-receptors (neuropilins, NRPs), regulating diverse cellular responses including angiogenesis.[6] Interestingly, it is becoming clear that VEGF-dependent responses can be elicited from both endothelial and non-endothelial cells. The modulation of VEGFR and VEGF functionality in pathological disease states is an intensive area of study. One long-standing view, which is increasingly challenged, is that VEGFR endocytosis causes the cessation of signal transduction due to lysosomal proteolysis. This is because there is increasing evidence for membrane receptor signaling from intracellular compartments e.g. endosomes. In this chapter, we first describe the key signaling players i.e. VEGF ligands and membrane receptors (Section 2) and then discuss VEGFR endocytosis (Sections 3 and 4). Finally, we integrate VEGFR trafficking and signaling with post-translational modifications (PTMs), protein turnover and disease outcomes (Sections 5–7).

2. Vascular endothelial growth factors and membrane receptors

2.1 Vascular endothelial growth factors (VEGFs)

The VEGF family in mammals comprises VEGF-A, B, C, D and placental growth factor (PlGF). These regulate normal physiological processes—angiogenesis, lymphangiogenesis and vasculogenesis. Non-human VEGFs such as cytomegalovirus VEGF-E and snake venom VEGF-F mediate pathological effects by subverting physiological response(s). VEGFs are characterized by a cysteine-knot motif which facilitates intertwining of a VEGF dimer via three disulfide bridges formed by six cysteine residues. The VEGF cysteine-knot motif is entropically favorable, provides thermal stability and proteolytic protection.[7–9]

VEGF gene transcription is regulated by multiple mechanisms; however, hypoxic (<1% oxygen) regulation of transcription of the *VEGFA* gene locus is linked to tumor development, progression and spread i.e. tumor angiogenesis.[9] In solid tumors ~1–2 mm in diameter, oxygen levels drop (<1%) due to insufficient vascular delivery of oxygenated blood. Under these conditions, *VEGFA* transcription is elevated via hypoxia-inducible factor-1 (HIF-1). Under normoxic conditions, HIF-1α is hydroxylated on proline residues by prolyl hydroxylases (PHDs), leading to recognition of these sites by the E3 ubiquitin ligase, Von Hippel Lindau (VHL) factor. HIF-1α ubiquitination by VHL programs its degradation by the 26S proteasome. However, under hypoxic conditions, PHD activity is decreased due to the lack of molecular oxygen, leading to stabilization of HIF-1α; this polypeptide then translocates into the nucleus, binding HIF-1β and forming a HIF-1 heterodimer which recognizes short DNA motifs within gene promoters termed hypoxia-responsive enhancer elements (HRE). The presence of HREs within the *VEGFA* promoter region enables HIF-1 binding, and co-activation alongside transcription factor (CBP/p300), promotes transcription of the *VEGFA* locus by RNA polymerase II. Other transcription factors are also implicated in angiogenesis, including the p53 tumor suppressor which represses VEGF-A expression in a p21/pRb-dependent manner.[8] Other factors such as TGF-α, TGF-β, IGF-I, FGF, PlGF and keratinocyte growth factor,[10–12] inflammatory cytokines, and hormones (e.g. Thyroid stimulating hormone, Adrenocorticotrophin hormone and gonadotropins)[13–15] play roles in upregulating VEGF-A expression. Notably, hormones such as progestins and estradiol can elevate VEGF-A levels; these factors are strongly linked to diseases such as endometrial cancer.[16,17]

VEGF-A (originally termed vascular permeability factor) is the founding member of the VEGF family and is broadly expressed across numerous cell types. However, VEGF-A levels are consistently upregulated in many solid tumors and implicated in disease development and progression. In *H. sapiens,* the *VEGFA* locus located at 6p21.3 consists of at least eight exons separated by seven introns. The first five *VEGFA* exons are present in all isoforms and encode a cleavable signal sequence (exon 1), N-terminal dimerization domain (exon 2), N-glycosylation site and VEGFR1 binding site (exon 3), VEGFR2 binding site (exon 4) and a plasmin cleavage site (exon 5).[6] Exons 6 and 7 encode motifs that interact with heparan sulfate proteoglycans in the extracellular matrix; alternative splicing events in exons 6 and 7 produce a number of VEGF-A isoforms, which differ in their biochemical properties e.g. solubility and extracellular matrix binding. Pro-angiogenic VEGF-A_{xxx} isoforms are generated by proximal splicing of exon 8, whereas anti-angiogenic isoforms ($VEGFA_{xxxb}$) arise from distal splicing of exon 8. There are at least seven pro-angiogenic and five anti-angiogenic VEGF-A isoforms. VEGF-A_{165} is the most widely expressed and extensively studied isoform with powerful pro-angiogenic properties.[18,19] Alternative splicing of *VEGFA* pre-mRNA also generates VEGF-A_{121}, VEGF-A_{145}, VEGF-A_{189}, and VEGF-A_{206}, which promote distinct endothelial responses.[18,20]

Although different VEGFs display distinct roles, these genes are essential as *VEGF* knockouts are invariably embryonically lethal. Deletion of even one mouse *VEGFA* allele causes embryonic lethality with disrupted angiogenesis, insufficient vascular and cardiac development.[21,22] In contrast, VEGF-B and PlGF play minor role(s) in vascular physiology but regulate cardiac development and function.[23,24] VEGF-C and VEGF-D regulate lymphangiogenesis i.e. the sprouting of lymphatic vessels from pre-existing ones.[25]

2.2 VEGF receptors

VEGFs bind to membrane proteins termed VEGFRs or neuropilins (NRPs). These are synthesized, co-translationally inserted into the endoplasmic reticulum (ER) and transported via the Golgi complex to the plasma membrane and the endosome-lysosome network. Cell surface VEGFR bioavailability relies on the dynamic balance of VEGFR and NRP pools within endocytic and secretory compartments linked to endocytosis, recycling and proteolysis. Both VEGFRs and NRPs are Type I membrane proteins and display multiple PTMs in basal and activated states. Tripartite VEGFR-VEGF-NRP complexes are implicated in regulating endothelial cell proliferation, migration and vasopermeability linked to vasculogenesis, angiogenesis and lymphangiogenesis.

VEGFRs are encoded by three separate but related gene loci, comprising seven extracellular immunoglobulin (Ig)-like domains, a single α-helical transmembrane domain, a split intracellular tyrosine kinase domain and a flexible carboxy-terminal tail. VEGF binding to the extracellular domain promotes VEGFR homo- or heterodimerization, causing conformational changes transmitted through to the cytoplasmic domain, causing rapid elevation in tyrosine kinase activity. It is thought that such conformational changes increase ADP/ATP exchange within the tyrosine kinase cleft, also triggering trans-autophosphorylation of specific tyrosine residues within the VEGFR dimer. VEGFR phosphotyrosine epitopes can then serve as docking sites for cytoplasmic adaptors or effectors which activate or modulate multiple signal transduction pathways. In this way, VEGFR activation promotes distinct cellular responses which impact on animal physiology and development (Fig. 1).

2.2.1 VEGFR1

This receptor tyrosine kinase (RTK) termed Flt-1, Flt1 or VEGFR1 is an ~180 kDa Type 1 membrane glycoprotein with high affinity (K_d ~10 pM) for VEGF-A, VEGF-B and PlGF (Fig. 1). However, VEGF-A binding to the second Ig-like domain of VEGFR1[20,26] only causes a small increase in tyrosine kinase activity compared to VEGFR2; thus it is postulated that VEGFR1 is not a major signaling node.[27] Nonetheless, the 6–8 tyrosine phosphorylation epitopes present in VEGFR1 (Table 1), raises questions as to the exact functional role of VEGFR1 in different tissues.[42,43] VEGFR1 tyrosine phosphorylation and downstream signaling has so far been observed using transfected cells overexpressing VEGFR1.[44,45]

VEGFR1 is widely expressed in endothelium, immune cells, brain, epithelial tissues and tumors.[7] *VEGFR1* pre-mRNA undergoes alternative splicing; further proteolysis can generate four smaller soluble VEGFR1 isoforms (sVEGFR1).[46] Most sVEGFR1 isoforms retain the first six Ig-like domain essential for binding VEGFs and heparin sulfate proteoglycans.

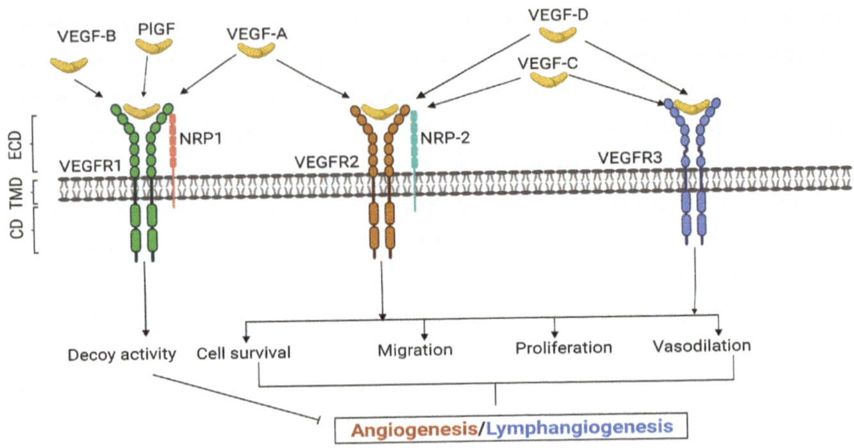

Fig. 1 VEGF and receptor complexity. VEGF family ligands (including PlGF) act as soluble dimers which bind membrane receptors. VEGFR-VEGF-NRP signaling impacts on cell survival, migration, proliferation and vasodilation regulating angiogenesis and lymphangiogenesis. *Abbreviations: extracellular domain, ECD; transmembrane domain, TMD; cytoplasmic domain, CD; vascular endothelial growth factor, VEGF; placental growth factor, PlGF; vascular endothelial growth factor receptor, VEGFR; neuropilin, NRP.*

However, the seventh Ig-like domain, membrane-anchoring region, juxtamembrane domain and cytoplasmic domains are absent from sVEGFR1 isoforms.[46] Hypoxia stimulates expression of both full-length VEGFR1 and sVEGFR1 isoforms in a HIF-1-dependent manner.[47]

The biological role(s) of VEGFR1 remain obscure despite being cloned more than 25 years ago. Knockout of mouse *VEGFR1* results in embryonic lethality caused by an excessive and disorganized vasculature, suggesting a negative regulatory role for VEGFR1 during embryonic angiogenesis.[48] Such a role could be attributed to the high affinity of the VEGFR1 extracellular domain for VEGF-A. It is postulated that VEGFR1 is a "decoy receptor" which does not function as a "signaling switch" but instead traps VEGF-A to sequester it away from the pro-angiogenic receptor, VEGFR2.[49] In pre-eclampsia, elevated sVEGFR1 levels are implicated in VEGF-A sequestration leading to hypertension, suggesting that this may be true of the soluble form of the receptor.[50,51] However, the conservation of a functional tyrosine kinase domain within the different VEGFR1 orthologs in metazoan organisms implies important signaling role(s) for this molecule that are yet to be discovered.

Table 1 Post-translational modifications in angiogenic regulators and their function.

	PTMs	Angiogenic modulator	Residues	Function	References
1.	N-glycosylation	VEGFR1	N	Ligand binding	28
		VEGFR2	N145, N160 N247	Ligand binding	29
2.	Phosphorylation	VEGFR1	Y794, Y1169, Y1213, Y1242, Y1309, Y1333	Receptor activation and signaling	30,31
		VEGFR2	Y1054,Y1059, Y951, Y1175, Y1214	Receptor activation and signaling	32,33
		VEGFR3	Y1063, 1068, 1230, 1231, 1265, 1337 and 1363	Receptor activation and signaling	34
3.	Ubiquitination	VEGFR2	K (unidentified)	Receptor degradation	35
4.	Palmitoylation	VEGFR1	C (unidentified)	Prevents VEGFR1 degradation	36
5.	SUMOylation	VEGFR2	K1270	Attenuates signaling	37
		Prox1	K556	Enhances VEGFR3 mediated lymphangiogenesis	38
6.	Methylation	HIF-1α	K32	Transcriptional regulation of VEGFA	39
7.	Acetylation	VEGFR2	K929, K937, K939, K937, K1053	Receptor activation and signaling	40
8.	Hydroxylation	HIF-1α		Prevents angiogenesis	41

Abbreviations: SUMO, small ubiquitin-like modifier; HIF, hypoxia-inducible factor; Prox1, prospero homeobox 1.

2.2.2 VEGFR2

This membrane receptor is alternatively termed KDR (*H. sapiens*) and Flk-1 (*Mus musculus*) and represents a glycoprotein of ~230 kDa which binds VEGF-A, VEGF-C, VEGF-D, and viral VEGF-E (Fig. 1).[51] VEGFR2 is expressed on vascular and lymphatic endothelial cells, megakaryocytes, platelets and tumor cells.[52] It is postulated that phosphorylation of the VEGFR2-Y951 epitope promotes nuclear localization, binding proximal to the Sp1 enhancer within the *VEGFR2* promoter to regulate gene transcription.[53,54] However, translocation of transmembrane regions from lipid bilayers into a hydrophilic environment is energetically unfavorable, suggesting it is likely that a soluble VEGFR2 polypeptide is functional in this context. The *VEGFR2* pre-mRNA can be alternatively spliced into a soluble, secreted isoform (sVEGFR2) comprising the first six Ig-like domains fused to a unique 16-amino-acid region at the C-terminus encoded within intron 13 in the *VEGFR2* locus.[54] Soluble VEGFR2 binding to VEGF-C sequesters it away from VEGFR3, thereby inhibiting cellular responses linked to lymphangiogenesis.[54,55]

VEGFR2 is the most potent pro-angiogenic receptor as *VEGFR2*-null mice die early in embryogenesis due to decreased vasculogenesis and angiogenesis and thus lack of vascular formation.[55] VEGF-A binding to VEGFR2 is primarily dependent on Ig-like domains 2 and 3 and further stabilized by low-affinity homotypic contacts between Ig-like domains 4, 5 and 7.[56,57] VEGF-A binding to VEGFR2 (K_d ~75–125 pM) has ~10-fold lower affinity than for VEGFR1.[20] Nonetheless, the role of VEGF-A as a pro-angiogenic molecule is largely attributed to its interactions with VEGFR2 and subsequent downstream signaling events. VEGFR2 is often overexpressed in endothelial cells in close proximity to rapidly growing solid tumors.[58] VEGFR2 levels are regulated by PTMs such as tyrosine phosphorylation and ubiquitination which impact on pro-angiogenic cellular responses. VEGFR2 is N-glycosylated upon constitutive secretion.[51] VEGFR2 undergoes ligand-dependent acetylation at specific lysine residues K929, K937, K939, K947 and K1053 within the tyrosine kinase module, which elevates kinase activity.[40] Similarly, specific arginine and lysine methylation (e.g. K1014) may regulate tyrosine phosphorylation and VEGFR2 kinase activity.[59]

2.2.3 VEGFR3

This ~150 kDa membrane glycoprotein is expressed on the embryonic vascular endothelium but restricted to the adult lymphatic endothelium.

VEGFR3 is initially synthesized as a ~195 kDa precursor in the ER which undergoes proteolytic cleavage and N-glycosylation to give rise to a mature ~150 kDa heterodimer comprising a membrane-bound subunit linked to a soluble fragment by a single disulfide bond within Ig-like domain 5 (Fig. 1). The *H. sapiens VEGFR3* pre-mRNA can undergo alternative splicing which generates two isoforms with differing functionality.[60]

VEGF-C and VEGF-D can bind VEGFR3 on endothelial, lympho-endothelial and immune cells and promote cellular responses.[61–63] VEGFR3 expression is also documented in other cell types including osteoblasts and neuronal progenitor cells.[64] VEGFR3 essentiality is highlighted in *VEGFR3*-null mice which die during embryogenesis at day E10.5, corresponding to an embryonic stage important for development of lymphatic structures.[65,66] However, profiling studies shows increased expression of VEGFR3 compared to VEGFR1 and VEGFR2 in endometrial adenocarcinoma.[67] The essential role of VEGFR3 in physiology and development is well-established but the disease implications require further study.

2.2.4 VEGF co-receptors

Neuropilins (NRPs) are highly conserved Type I membrane glycoproteins that act as co-receptors for VEGFs alongside VEGFRs.[68] NRP ligands include Class 3 semaphorins and VEGFs. The neuropilin gene loci comprise *NRP1* and *NRP2* which encode proteins of ~920 residues which undergo both N- and O-glycosylation producing mature proteins of 120–140 kDa.[69] Both NRP homologs are ubiquitously expressed in the heart, glomeruli and osteoblasts but exhibit differential expression in the adult vasculature. High levels of NRP1 is detected on the arterial endothelium, whereas NRP2 expression is restricted to lymphatic and venous endothelial cells.[70,71] Both NRP homologs share 44% sequence similarity, domain organization and structural features.[72]

NRP domains are organized into a1-a2-b1-b2-c configuration in the extracellular domain with a single transmembrane region followed by a 40-residue cytoplasmic domain (Fig. 1). VEGF-A binding to the b1-b2 domain is thought to promote G-protein activation leading to increased angiogenesis.[73,74] VEGF-A binding to NRP1 is mediated through exon 7-encoded carboxy-proximal region within the VEGF-A_{165} isoform, distinct from VEGF-A exon 3 and 4 peptide sequences which enable VEGFR1 and VEGFR2 recognition respectively.[75,76] The different binding activities encoded by the VEGF-A_{165} isoform facilitates assembly of a tripartite VEGFR-VEGF-NRP complex. NRP1 recruitment by VEGF-A promotes an additional fourfold synergistic increase in VEGFR2 affinity for VEGF-A.[76] In a similar manner, VEGF-C binding simultaneously to NRP2 and VEGFR3 regulates signaling and lymphangiogenesis.[71] Deregulation of VEGF-C and NRP2 interactions is implicated in tumor metastasis; mutations in NRP2 can cause abnormal lymph development.[77,78] Thus VEGF co-receptors NRP1 and NRP2 provide vital accessory roles in modulating endothelial and lymphatic development and animal physiology.

3. VEGFR endocytosis

VEGF receptor internalization occurs via clathrin-dependent endocytosis (CDE) or clathrin-independent endocytosis (CIE); the latter includes caveolin-dependent endocytosis and micropinocytosis.[19,79] The endocytic route taken is likely to depend on membrane receptor status and ligand concentration. VEGF-A binding to VEGFR2 triggers tyrosine phosphorylation and trafficking to endocytic, biosynthetic and degradative compartments, which in turn influence signaling magnitude and duration. In the absence of ligand, endocytosed VEGFR2 undergoes rapid recycling: such trafficking protects against ectodomain shedding.[79] The major VEGFR2 pool (~60%) is in a resting or basal state located at the plasma membrane whereas ~40% is located within internal pools in endosomes and the Golgi.[9] VEGF-A binding to VEGFR2 triggers increased endocytosis and delivery to the endosome-lysosome network.[80] Endosomes are increasingly viewed as convenient locations for protein-protein interactions between membrane receptors and signaling machinery. Two further types of VEGFR endocytosis have been observed i.e. slow constitutive recycling by basal or resting VEGFRs and shear stress-induced endocytosis.[81] Endothelial shear stress promotes CD31 phosphorylation, a membrane receptor and shear stress sensor; this in turn promotes VEGFR2 endocytosis. This mode of endocytosis is caveolae-dependent (discussed below).[81]

In contrast, plasma membrane VEGFR1 bioavailability is low, with ~80% of VEGFR1 located within a Golgi-like pool.[82] VEGF-A stimulation prompts VEGFR1 translocation from Golgi-to-plasma membrane which is dependent on cytosolic calcium ions. A positive feedback model could explain how cytosolic calcium ions could regulate VEGFR1 trafficking and modulate VEGFR2 signaling.[3]

3.1 Clathrin-dependent endocytosis (CDE)

Cell surface endocytosis of soluble and membrane proteins can occur through clathrin-coated pits and vesicles, influencing different signal transduction pathways.[83] Plasma membrane VEGFR2 can form a complex with many transmembrane proteins, including vascular endothelial cadherin (VE-cadherin) which can indirectly promote VEGFR2 dephosphorylation and also inhibits endocytosis (Fig. 2).[84] VEGF-A binding and VEGFR2 activation can destabilize the VEGFR2-VE-cadherin complex releasing VEGFR2 to undergo CDE. Plasma membrane VEGFR2 activation enables PLCγ1(Phospholipase C γ1) recruitment and phosphorylation, promoting breakdown of phosphatidylinositol-4,5-bisphosphate (PIP2) to inositol-1,4,5-triphosphate (IP3) and diacylglycerol (DAG).[7,85–87] As PIP2 binding to the endocytic AP2 adaptor complex promotes CDE, VEGFR2 activation thus modulates trafficking through CDE and CIE.

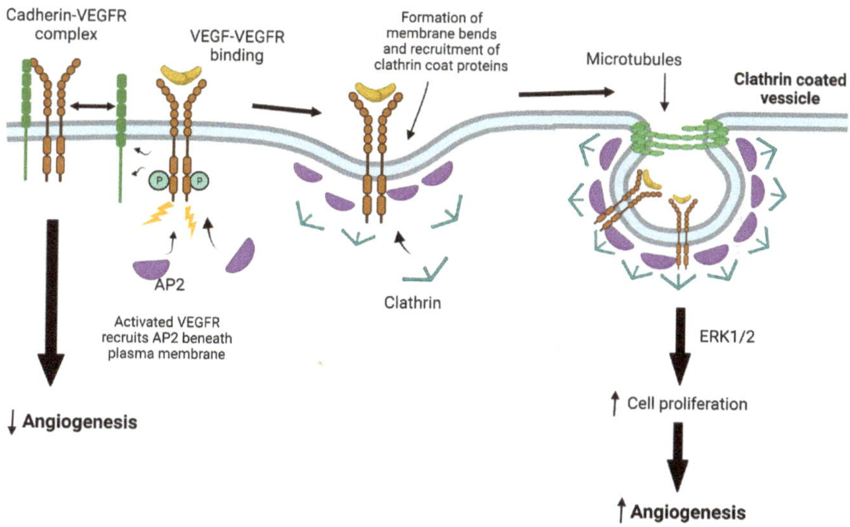

Fig. 2 Clathrin-dependent endocytosis (CDE) of VEF receptors. Association of VE-cadherin (green membrane protein) to VEGFR prevents its activation and signaling and ligand binding destabilises VEGFR-VE-cadherin complexes. VEGFR activation may increase recruitment to clathrin-coated pits and vesicles. Dynamin-2 GTPase (green oligomers) and actin microfilaments promote CDE and sorting to endosomes. VEGFR signaling from the plasma membrane and endosomes could result in different cellular responses and physiological outcomes. *Abbreviations: adaptor protein complex 2, AP2; extracellular-regulated kinase 1 and 2, ERK1/2.*

Accumulation of endocytic coat proteins facilitate plasma membrane curvature leading to clathrin-coated pits followed by fission into 80–120 nm diameter clathrin-coated vesicles. A classical view is that membrane protein endocytosis reduces receptor bioavailability for exogenous ligand, thus reducing receptor activation and signal transduction. However, increasing evidence indicates that endosomal VEGFR signaling promotes angiogenesis and vasopermeability.[35,88,89] Endosomal VEGFR2 pools are associated with signaling between different intracellular locations i.e. intracrine signaling.[9] For example, activated tyrosine phosphorylated VEGFR2 present within early endosomes co-distributes with PLCγ1; this is implicated in increased activation of the canonical MAPK pathway leading to cell proliferation.[84] Furthermore, peak activity in VEGF-A-stimulated ERK1/2 and AKT protein kinases correlates with an increase in the VEGFR2 endosomal pool (Fig. 2).[90,91] On the other hand, cell surface VEGFR2 is closely linked to the p38 MAPK signal transduction pathway.[19] Under resting conditions, basal or non-activated VEGFR2 at the plasma membrane can undergo specific proteolysis resulting in extracellular domain shedding; CDE protects against such cleavage.[92] Interestingly, VEGFR1 activation is linked to binding to an E3 ubiquitin ligase, Cbl (Casitas B-lineage Lymphoma) and CD2-asssociated protein (CD2AP), regulators of CDE.[93,94] VEGF-A also protects VEGFR1 from proteolytic cleavage and ectodomain shedding.[95] VEGFR1-VEGFR2 heterodimers have been detected[96–98] but the endocytic properties have not been explored.

Interestingly, VEGFR3 endocytosis regulates lymphangiogenesis in mammals and zebrafish.[99] This is linked to CDE, as deficiency in epsin, a key endocytic regulator, causes defects in lymphatic valve formation.[100] However, it is unclear to the extent by which CDE controls VEGFR3 internalization, recycling and degradation. Complexes containing both VEGFR3 homodimers and heterodimers (VEGFR2/VEGFR3) have been detected, with differences in signaling properties.[101]

Neuropilins act as co-receptors which bind to VEGFRs and modulate trafficking, signaling and cellular responses. NRP1 binding to VEGF-A promotes CDE and exhibits dependence on myosin VI[102] and synectin.[103] As NRP1 forms a complex with VEGFR2 (Fig. 1), this could be one mechanism to specifically recruit activated VEGFR2 to the CDE portal for subsequent sequestration into signaling and degradative outcomes (Fig. 2A). Both NRP1[104] and NRP2[105] are found to also regulate endosomal trafficking of another RTK, epidermal growth factor receptor (EGFR/ErbB1). Recently, NRP1 was identified as a SARS-COV-2 receptor and blocking such interactions reduced viral entry and infection.[106,107] NRP1 is also a receptor for Epstein-Barr Virus (EBV),[108] indicating that it may be a frequent pathogen receptor. It remains to be determined what different endocytic routes play in SARS-CoV-2 recognition of NRP1, and the effects on viral entry and infectivity.

3.2 Clathrin-independent endocytosis (CIE)

3.2.1 Macropinocytosis

Although the canonical CDE has long been considered the primary route for VEGFR2 endocytosis, blocking this pathway using inhibitors[109] or clathrin depletion[110] does not prevent VEGFR2 endocytosis and degradation. The endocytosis of activated VEGFR2 is thought to occur via micropinocytosis (Fig. 3), involving vesicular profiles with diameter >0.2 μm, whereas basal or non-activated VEGFR2 is postulated to undergo CDE.[79] Such trafficking is postulated to require actin-dependent plasma membrane ruffles, with subsequent delivery of membrane cargo to macropinosomes. Such VEGFR2 trafficking is linked to ERK1/2 and Akt signaling which are key features of the endothelial response to VEGF-A including angiogenesis (Fig. 3A). The CDC42 GTPase, rabankyrin-5 and EEA1 co-distribute to varying degrees with VEGFR2 within macropinosomes[79] raising questions as to the signaling properties of VEGFR2 pools associated with different endosome populations (Fig. 3).

3.2.2 Caveolae, lipid rafts and endocytosis

Nearly 50% of the endothelial plasma membrane is studded with ~50 nm diameter flask-shaped invaginations called caveolae. A main component of these structures is caveolin-1, a protein which forms oligomers alongside accessory factors e.g. cholesterol and sphingolipids (Fig. 3B). Caveolae are a subset of plasma membrane lipid rafts which mediate specific trafficking and signaling outcomes distinct from other endocytic pathways. One feature of lipid rafts is that the main lipid components are sphingolipids and cholesterol. Cav-1 and cholesterol association with VEGFR2 regulates tyrosine kinase activity, endothelial permeability and angiogenesis.[90,93] Endothelial ARF6 and Rac1 GTPases can be detected within a pool of VEGFR2 and Cav-1 linked to VEGF-A-regulated signaling, endothelial cell migration, proliferation and angiogenesis.[111] Interaction between activated VEGFR2 and Cav-1, leading to phosphorylated Cav-1-pY14, is implicated in signal transduction leading to activation of the RhoA GTPase, altered actin dynamics and increased endothelial cell migration.[112]

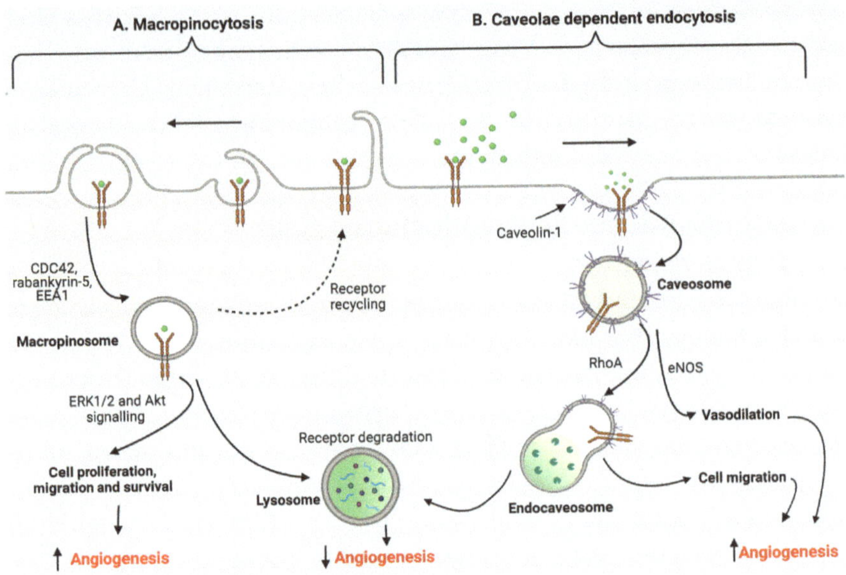

Fig. 3 Clathrin-independent endocytosis (CIE) of VEGF membrane receptors. (A) Macropinocytosis is depicted with membrane ruffles projecting into the extracellular spaces enabling capture of VEGFR2 into macropinosomes. This pathway facilitates VEGFR2 signaling to MAPK and Akt pathways. (B) Caveolae-regulated endocytosis or internalization is depicted as caveolin-rich structures which are a subset of lipid rafts present at the plasma membrane. VEGFRs sequestered into caveolae may be in close proximity to eNOS, facilitating rapid nitric oxide (NO) production leading to vasodilation. Both CIE pathways may also facilitate VEGF-regulated recruitment of Rho GTPases such as CDC42 and RhoA to promote endothelial cell migration during angiogenesis. *Abbreviations: mitogen-activated protein kinase, MAPK; early endosomal regulator, EEA1; small GTPases, RhoA and CDC42; endothelial nitric oxide synthase, eNOS.*

VEGF-A-regulated signaling in endothelial cells is characterized by phosphorylation and activation of AKT, a serine/threonine protein kinase which phosphorylates and activates endothelial nitric oxide synthase (eNOS) (Fig. 3B). The close proximity of VEGFR2 and eNOS within caveolae impacts on outcomes such as angiogenesis.[113] VEGFR3 association with caveolae is linked to Cav-1 tyrosine phosphorylation by soluble tyrosine kinases such as Src and Abl.[114]

4. Integration of VEGFR signaling, trafficking and proteolysis

Studies on angiogenesis in the mouse retina shows high rates of VEGFR endocytosis and turnover.[115] Such VEGFR kinetics determines duration and magnitude of VEGF-A-regulated signaling. Regardless of the route of entry, the ligand-receptor complex enters early endosomes enriched for the Rab5 GTPase.[7] VEGFR2-containing endosomes are trafficked away from the cell surface by a protein complex containing synectin (GPCI) and myosinVI (myosin6).[116] Although the molecular mechanism by which VEGFR2 interacts with the synectin/myosinVI complex is not established, several lines of evidence point towards NRP1-regulated VEGFR2 trafficking. VEGFR2 association with NRP1 via a carboxy-proximal serine-glutamic acid-alanine motif (SEA). NRP1 also interacts with synectin via a PDZ protein-protein interaction motif, facilitating myosinVI-dependent trafficking towards the minus ends of actin filaments.[116,117] Endosomal activated VEGFR2 undergoes dephosphorylation by the protein tyrosine phosphatase 1b (PTP1b), highlighting the dynamic nature of signals emanating from different VEGFR2 pools.[116]

Balancing VEGFR degradation vs. recycling within the endosome-lysosome network is important for physiological responses including angiogenesis. VEGFR2 undergoes endocytosis and trafficking through endosomes enriched for either Rab5 or Rab7 GTPases regulates degradation in late endosomes and lysosomes. However, VEGFR2 can undergo sorting and recycling back to the plasma membrane via recycling endosomes enriched for either Rab4 or Rab11 GTPases. VEGFR2 recycling occurs via the rapid short-loop Rab4-dependent pathway, or more slowly via the long-loop Rab11 and NRP1-dependent pathway. Furthermore, the transport of biosynthetic VEGFR2 through the secretory pathway is regulated by the Golgi SNARE, syntaxin 6.[118] Thus, the anterograde (forward) flow of newly synthesized VEGFRs is essential for plasma membrane bioavailability, and regulating the endothelial response to different exogenous VEGFs. How activated VEGFRs communicate to downstream signal transduction pathways is dependent on both positive and negative feedback loops which control VEGFR bioavailability, thus determining the extent of pro- or anti-angiogenic responses.

5. VEGFR post-translational modifications

It is increasingly clear that a variety of post-translational modifications (PTMs) influence VEGFR signaling, trafficking, turnover and cellular responses (Table 1). Such PTMs are largely documented to occur within the VEGFR cytoplasmic domains, but the physiological role of PTMs in the VEGFR extracellular domains remains relatively unexplored. The cell may use PTMs to discriminate between different intracellular VEGFR pools to control signaling outcomes and cellular responses.

5.1 Phosphorylation

All VEGFRs contain a ~330 residue tyrosine kinase (TK) module embedded within a cytoplasmic domain (550–600 residues), whereas NRPs contain a short ~40 residue cytoplasmic domain (Fig. 1). VEGF binding to the extracellular domain causes conformational changes leading to activation of the TK module, stimulating ADP/ATP exchange with subsequent transfer the γ-phosphate in ATP to specific tyrosine side chains on substrates including VEGFRs. The generation of a phosphotyrosine epitope embedded within a unique amino acid sequence forms new binding or docking sites for signaling factors which contain Src-homology 2 (SH2) or phosphotyrosine-binding (PTB) modules. Activated VEGFR phosphorylation is associated with increased CDE (Fig. 2); however, CIE is also implicated in VEGFR activation and signaling (Fig. 3).

Activated VEGFR2 undergoes tyrosine autophosphorylation at major (Y801, Y951, Y996, Y1054, Y1059, Y1175, Y1214) and minor (Y1223, Y1305, Y1309, Y1319) sites. The most well-characterized VEGFR2-regulated signal transduction pathway involves PLCγ1 recruitment, tyrosine phosphorylation and activation, thus generating signals which activate canonical MAPK pathway.[119] VEGF-A binding to VEGFR2 stimulates production of the VEGFR2-pY1175 epitope within the flexible carboxy-terminal tail[120]; this now creates a binding site for PLCγ1 via its SH2 domains.[86] Furthermore, activated VEGFR2 phosphorylates PLCγ1 on residue Y783.[110,121] PLCγ1-mediated cleavage of plasma membrane PIP2 leads to the production of two different second messengers: DAG (a protein kinase C activator) and IP3, a potent stimulator of calcium ion efflux from the ER to the cytosol. VEGFR2 endocytosis and delivery to endosomes thus inhibits PLCγ1-mediated breakdown of plasma membrane PIP2 and downstream signaling responses.

The mechanism by which VEGFR2 activates the PI3K-AKT or p38 MAPK signal transduction pathways is less well-understood. VEGFR2 phosphorylation at Y1214 promotes p38 MAPK-mediated actin remodeling and endothelial cell migration.[122] However, the VEGFR2-pY801 epitope within the juxtamembrane region is linked to signaling to the PI3K-AKT pathway which is critical for eNOS phosphorylation and activation.[123] Appearance of the VEGFR2-pY951 epitope within the N-terminal lobe of the TK module triggers AKT activation and endothelial cell migration.[32] The VEGFR2-pY1059 epitope lies within the activation loop in the C-terminal lobe of the TK module and is linked to increased MAPK pathway activation, cytosolic calcium ions and endothelial cell proliferation.[32] However, in murine VEGFR2, whereas Y1052F or 1057F single point mutations (equivalent to human VEGFR2 residues Y1054 and Y1059) had no effect, the Y1052F/Y1057F double mutation inhibits TK activity and endothelial function.[124] Casein kinase I activity leading to VEGFR2 phosphorylation on Ser/Thr residues is linked to increased proteolysis.[125] However, there are a number of VEGFR2 phosphoserine and phosphothreonine epitopes documented within the PhosphositePlus database,[126] but the regulatory effects are unknown. Of note, PKC-mediated phosphorylation of a serine-based motif within the VEGFR2 cytoplasmic domain regulates VEGFR2 turnover.[125,127]

VEGFR1 is associated with low TK activity but phosphorylation sites in the cytoplasmic domain have been identified (Y794, Y1169, Y1213, Y1242, Y1309, Y1327, Y1333). Phosphorylation of VEGFR2-Y1169 (equivalent to VEGFR2-Y1175) mediates recruitment of phospholipase PLCγ1; also other adaptor proteins can bind to VEGFR1.[45,128–130] Deletion of the murine VEGFR1 cytoplasmic domain in transgenic mice is not lethal, suggesting that VEGFR1 TK activity is not essential for gene function.[48] However, studies suggest a functional role for VEGFR1 in different contexts such as epithelial cancer cell lines.[131–133] Cytoplasmic VEGFR1 serine phosphorylation has been found; however, both serine and threonine phosphorylation of extracellular domain are also documented.[126] The functional significance of these modifications is unclear.

VEGFR3 activation and increased TK activity triggers cytoplasmic phosphorylation (Y830, Y833, Y853, Y1063, Y1068, Y1230, Y1231, Y1333, Y1337). Phosphorylation of VEGFR3 on residues within the TK N-lobe (Y830, Y831, Y853) are associated with Src recruitment and endothelial cell migration.[134–136] VEGFR3 tyrosine phosphorylation is also associated with recruitment of signaling factors including PLCγ1, Shc, Grb2, SOS, PI3K regulatory subunit p85 and protein tyrosine phosphatase SHP2.[134,137] In endothelial and lymphatic cells, VEGFR3 activation is associated with signal transduction through MAPK, PI3K-AKT, p38 MAPK and JNK pathways.[134,137] Again, both serine and threonine phosphorylation of VEGFR3 cytoplasmic and extracellular domains is also documented[126] but the functional significance is unclear.

NRP1 displays phosphorylation sites for serine, threonine and tyrosine within both the extracellular and cytoplasmic domains. Notably, the NRP1-pY920 epitope is conserved in human, mouse and rat[126] but how this modulates signaling, trafficking and turnover of the VEGFR-NRP1 complexes is unclear. In contrast, NRP2 displays phosphorylation sites for serine and threonine within the extracellular domain only.[126]

5.2 Ubiquitination

Ubiquitin-like modification of VEGF membrane receptors is an increasingly important PTM which regulates trafficking and turnover and impacts on signal transduction and cellular responses. VEGFR1 and VEGFR2 both undergo ubiquitination linked to endocytosis and degradation in the endosome-lysosome system.[138–140] The extent of VEGFR3 and NRP ubiquitination is unclear but likely to occur,[126] depending on cell type and context. The exact nature of ubiquitination-like modifications i.e. monomer, polymer, chain linkage and site of VEGF receptor attachment is unknown at present. It is also likely that a range of ubiquitin-like PTMs will regulate the dynamics of VEGF receptor endocytosis, sorting, recycling and proteolysis.

VEGFR2 ubiquitin-like PTMs are implicated in controlling both biosynthetic secretion, plasma membrane trafficking and turnover. Both basal (resting) VEGFR2[35,141] and activated VEGFR2[139,142,143] display ubiquitination linked to endocytosis and degradation in the endosome-lysosome system. Although one study has reported that VEGFR2 is largely monoubiquitinated, other work suggests both monoubiquitin and polyubiquitin modification of VEGFR2.[125,139,144] Depletion of the ubiquitin ligase E1 enzyme (UBA1) elevates basal plasma membrane VEGFR2 with increased endosome-to-plasma membrane recycling but similar kinetics in VEGFR2 endocytosis and proteolysis.[35] Such work implies a close link between VEGFR2 ubiquitination, endocytosis, endosomal sorting and degradation impacting on cellular responses including angiogenesis.

Earlier studies suggested that the proto-oncoprotein and E3 ubiquitin ligase CBL ubiquitinates activated VEGFR2.[138] However, subsequent work provided evidence for such PTM undertaken by another E3 ubiquitin ligase, βTrcP2.[145,146] There is also other evidence for regulatory roles played by other E3 enzymes such as βTrcP1,[125] Nedd4[147] and Cullin-3.[148] Interestingly, the E3 ubiquitin ligase RNF121 is implicated in controlling biosynthetic VEGFR2 turnover within the secretory pathway.[149]

Multiple lines of evidence suggest a close link between VEGFR2 phosphorylation and ubiquitination.[139] Blocking VEGFR1 or VEGFR2 tyrosine kinase activity is associated with decreased tyrosine phosphorylation, endocytosis and proteolysis.[139,140] VEGF-A-stimulated phosphorylation of VEGFR1-Y1333 is linked to increased association with CBL, endosomal localization and VEGFR1 ubiquitination.[140] In contrast, depletion of the UBA1 ubiquitin ligase caused increased plasma membrane VEGFR2 levels consistent with reduced ubiquitination, decreased endocytosis and proteolysis in the endosome-lysosome system.[35]

SUMOylation is the attachment of SUMO (small ubiquitin-like modifier) to target proteins and can modulate substrate protein localization, function and turnover. VEGFR2 SUMOylation influences biosynthetic VEGFR2 trafficking.[37] SUMOylation of residue K1270 influences VEGFR2 Golgi targeting; this aspect is regulated by the SENP1 deSUMOylating enzyme.[37] This raises the likelihood that a SUMOylation-deSUMOylation cycle controls VEGFR2 delivery to the plasma membrane to influence VEGF receptor bioavailability for signal transduction and cellular responses.

Interestingly, epsins are endocytic adaptors which contain ubiquitin interaction motifs (UIMs) which mediate binding to ubiquitinated cargo. Interaction between VEGFR3 and epsins is implicated in regulating lymphangiogenesis, thus suggesting that VEGFR3 also undergoes ubiquitination, trafficking and turnover in the endosome-lysosome network.[100] There are multiple ubiquitination sites within the VEGFR3 cytoplasmic domain.[126] Although VEGFR1 undergoes ubiquitination[144] with a potential role for CBL and associated factors, this mechanism and functional implications for trafficking and turnover remain ill-defined. There are multiple potential ubiquitination sites in the VEGFR1 cytoplasmic domain and NRP1 has potential ubiquitination sites at cytoplasmic residues K908 and K913.[126]

5.3 Other PTMs

The proteolysis of VEGFRs is complex and has substantial implications for functional outcomes including angiogenesis. Regulation of NRP proteolysis and functional implications have not been explored at present but membrane trafficking is fundamental in such regulation. It is clear that both biosynthetic[37,149] and exocytic VEGFR2[35] turnover is linked to receptor function and VEGF-regulated cellular responses. Less is clear about the nature of the proteases involved in VEGFR processing; however, the 26S proteasome is implicated in VEGFR2 proteolysis and functional outcomes.[144,150] Interestingly, newly synthesized VEGFR3 undergoes cleavage within the sixth Ig-like domain into soluble and membrane-bound polypeptides but the nature of this ER-resident protease is unknown. A stable VEGFR3 disulfide-linked heterodimer is assembled, glycosylated and transported to the cell surface via the constitutive secretory pathway.

It is likely that both N- and O-glycosylation of either asparagine (N) or serine/threonine (S/T) occur in both VEGFRs and neuropilins (Table 1). VEGFR2 contains at least 20 kDa equivalent of N-linked glycosylation,[20] but the extent of O-glycosylation is unknown. The majority of potential N-glycosylation sites on VEGFR2 appear to be modified.[29] The extent and nature of N-glycosylation could regulate VEGFR2 interaction with VEGF ligands, and thus functional outcomes.[151] Both NRP1 and NRP2 are also N-glycosylated.[152] Interestingly, the nature of NRP1 glycosylation can affect the binding to VEGF ligand and regulate VEGFR signal transduction.[153]

A number of other PTMs on VEGF receptors have also been documented using global proteomic analyses (Table 1) but the effects of such PTMs is not known.[126] Acetylation of VEGFR2 is of functional importance. Previous work has demonstrated the VEGF-dependent acetylation occurs but peaks after tyrosine phosphorylation.[40] Interestingly, this same study showed that prior acetylation of VEGFR2 markedly increases the extent of VEGFR2 tyrosine phosphorylation and moreover this phosphorylation prevents VEGFR2 desensitization, thus prolonging downstream signaling.

6. Implications of VEGFR endocytosis in vascular physiology

Angiogenesis requires integration of signaling, trafficking and feedback mechanisms involving VEGF-VEGFR-NRP pathways. The sprouting of new blood vessels involves at least three endothelial cell populations: a leader "tip" cell which initiates sprouting and guidance, followed by stalk (proliferating) and phalanx (non-proliferating) cells. Perturbation of plasma membrane VEGFR2 endocytosis alters angiogenesis outcomes.[115] Furthermore, genetic knockout of the endocytic adaptor, epsin, affects VEGFR3 trafficking and lymphangiogenesis.[100] Other factors which modulate VEGFR function and angiogenesis includes ephrin-B2, Dab2 (Disabled homolog 2), PAR-3 and atypical PKC.[115,154,155] Such factors may have significant effects on VEGFR endocytosis and endosomal trafficking, and may thus fine tune VEGFR signaling in the context of angiogenesis. Solid tumors exhibit poorly organized and irregular vascular networks which are "leaky" and also display enhanced VEGF-VEGFR signaling properties.[155,156] It is unclear as to the extent to which endocytosis is perturbed or modulated in pathophysiological states and how that impacts on VEGF-regulated signal transduction.

6.1 Cell motility and migration

A key feature of metastatic tumors is increased cell motility and migration. Such processes require increased actin dynamics, focal adhesions and plasma membrane protrusions e.g. lamellopodia. VEGFs can stimulate increased endothelial cell migration involving activated VEGFRs and focal adhesion complexes.[157,158] The VEGFR2-pY951 epitope acts as binding sites for the adaptor protein TSAd involved in cell migration and tumor metastasis.[32,158] Both vascularization and tumor growth are reduced in *Tsad*-deficient mice.[158] FAK (focal adhesion kinase) and paxillin are recruited to focal adhesions and activate downstream signaling pathways which upregulates MMP gene expression, cell invasion and metastasis.[159,160]

Other VEGFRs are also implicated in cell migration supporting physiological and pathological angiogenesis. VEGFR1 is strongly associated with cell migration: hematopoietic progenitor cells migrate toward a VEGF gradient in a VEGFR1-dependent manner.[160] VEGFR1 neutralizing antibodies caused reduced migration of hematopoietic progenitor cells and leukocytes.[161] VEGFR1-dependent cell migration is also exhibited in tumors.[162] VEGFR1-dependent signal transduction activates the PI3K-AKT pathway leading to receptor for activated C kinase 1 (RACK1) scaffold protein activation and increased migration of macrophages and endothelial cells.[163] Calcium-regulated signaling could link both PLCγ1 and PI3K-AKT pathways to regulate macrophage migration.[164,165] Interestingly, activated VEGFR3 homodimers or VEGFR2-VEGFR3 heterodimers are linked to different signaling outcomes and lymphatic endothelial cell responses.[101]

6.2 Cell proliferation and survival

Cell proliferation is regulated by factors as cyclins and cyclin dependent kinases (CDKs). Sequestration of VEGFR2 into endothelial caveolae greatly diminishes MAPK signaling with reduced cell proliferation and endothelial tube formation.[166] VEGF-A isoforms can differentially program

VEGFR2-regulated signaling and gene expression via canonical MAPK vs. the stress-activated p38 MAPK pathways.[91] Both endothelial and lymphoendothelial cell proliferation is clearly dependent on the canonical MAPK pathway. In contrast, VEGF-A-regulated Akt activity is associated with cellular homeostasis and increased survival.

7. Conclusions

The emerging view is that integration of PTMs and trafficking on VEGF membrane receptors dictates signaling outputs in physiological processes such as vasculogenesis, angiogenesis and lymphangiogenesis. Basal (non-activated) and VEGF receptor-ligand interactions at the plasma membrane linked to PTMs and endocytosis has major outcomes on vascular and lymphatic function. However, it is beginning to emerge that PTMs occurring within the secretory pathway also influence VEGFR trafficking and plasma membrane bioavailability. This is further complicated by differences in VEGF receptor expression and function in different cell types including the endothelium, epithelium, immune and neural cell types. Pathological function of the VEGF-VEGFR axis impacts on tumor angiogenesis and age-related macular degeneration. The challenge is to better understand how the integration of VEGFR trafficking and turnover impacts on signaling in physiological and pathological processes. Such knowledge may influence a new generation of therapeutics to treat pro- and anti-angiogenic disease states.

Acknowledgments

We are grateful to members of the Endothelial Cell Biology Unit for their help and advice. This work was supported by a Leeds International Doctoral Scholarship (QS), a King Saud University PhD Scholarship (AA) and British Heart Foundation project grant PG/16/86/32474 (WRC, MAH, SP).

References

1. Adair TH, Montani JP. *Angiogenesis*. San Rafael, CA: Morgan and Claypool; 2010.
2. Klagsbrun M, D'Amore PA. Regulators of angiogenesis. *Annu Rev Physiol*. 1991;53:217–239.
3. Hanahan D, Weinberg RA. Hallmarks of cancer: the next generation. *Cell*. 2011;144:646–674.
4. Miller JW, Le Couter J, Strauss EC, Ferrara N. Vascular endothelial growth factor a in intraocular vascular disease. *Ophthalmology*. 2013;120:106–114.
5. Azizi G, Boghozian R, Mirshafiey A. The potential role of angiogenic factors in rheumatoid arthritis. *Int J Rheum Dis*. 2014;17:369–383.

6. Peach CJ, Mignone VW, Arruda MA, et al. Molecular pharmacology of VEGF-A isoforms: binding and signalling at VEGFR2. *Int J Mol Sci*. 2018;19.
7. Smith GA, Fearnley GW, Tomlinson DC, Harrison MA, Ponnambalam S. The cellular response to vascular endothelial growth factors requires co-ordinated signal transduction, trafficking and proteolysis. *Biosci Rep*. 2015;35.
8. Farhang Ghahremani M, Goossens S, Nittner D, et al. p53 promotes VEGF expression and angiogenesis in the absence of an intact p21-Rb pathway. *Cell Death Differ*. 2013;20:888–897.
9. Forsythe JA, Jiang BH, Iyer NV, et al. Activation of vascular endothelial growth factor gene transcription by hypoxia-inducible factor 1. *Mol Cell Biol*. 1996;16:4604–4613.
10. Warren RS, Yuan H, Matli MR, Ferrara N, Donner DB. Induction of vascular endothelial growth factor by insulin-like growth factor 1 in colorectal carcinoma. *J Biol Chem*. 1996;271:29483–29488.
11. Pertovaara L, Kaipainen A, Mustonen T, et al. Vascular endothelial growth factor is induced in response to transforming growth factor-beta in fibroblastic and epithelial cells. *J Biol Chem*. 1994;269:6271–6274.
12. Frank S, Hubner G, Breier G, Longaker MT, Greenhalgh DG, Werner S. Regulation of vascular endothelial growth factor expression in cultured keratinocytes. Implications for normal and impaired wound healing. *J Biol Chem*. 1995;270:12607–12613.
13. Soh EY, Sobhi SA, Wong MG, et al. Thyroid-stimulating hormone promotes the secretion of vascular endothelial growth factor in thyroid cancer cell lines. *Surgery*. 1996;120:944–947.
14. Shifren JL, Mesiano S, Taylor RN, Ferrara N, Jaffe RB. Corticotropin regulates vascular endothelial growth factor expression in human fetal adrenal cortical cells. *J Clin Endocrinol Metab*. 1998;83:1342–1347.
15. Ferrara N, Chen H, Davis-Smyth T, et al. Vascular endothelial growth factor is essential for corpus luteum angiogenesis. *Nat Med*. 1998;4:336–340.
16. Mueller MD, Vigne JL, Minchenko A, Lebovic DI, Leitman DC, Taylor RN. Regulation of vascular endothelial growth factor (VEGF) gene transcription by estrogen receptors alpha and beta. *Proc Natl Acad Sci USA*. 2000;97:10972–10977.
17. Mueller MD, Vigne JL, Pritts EA, Chao V, Dreher E, Taylor RN. Progestins activate vascular endothelial growth factor gene transcription in endometrial adenocarcinoma cells. *Fertil Steril*. 2003;79:386–392.
18. Senger DR, Galli SJ, Dvorak AM, Perruzzi CA, Harvey VS, Dvorak HF. Tumor cells secrete a vascular permeability factor that promotes accumulation of ascites fluid. *Science*. 1983;219:983–985.
19. Fearnley GW, Smith GA, Abdul-Zani I, et al. VEGF-A isoforms program differential VEGFR2 signal transduction, trafficking and proteolysis. *Biol Open*. 2016;5:571–583.
20. Vaisman N, Gospodarowicz D, Neufeld G. Characterization of the receptors for vascular endothelial growth factor. *J Biol Chem*. 1990;265:19461–19466.
21. Carmeliet P, Ferreira V, Breier G, et al. Abnormal blood vessel development and lethality in embryos lacking a single VEGF allele. *Nature*. 1996;380:435–439.
22. Ferrara N, Carver-Moore K, Chen H, et al. Heterozygous embryonic lethality induced by targeted inactivation of the VEGF gene. *Nature*. 1996;380:439–442.
23. Bellomo D, Headrick JP, Silins GU, et al. Mice lacking the vascular endothelial growth factor-B gene (Vegfb) have smaller hearts, dysfunctional coronary vasculature, and impaired recovery from cardiac ischemia. *Circ Res*. 2000;86:E29–E35.
24. Bry M, Kivela R, Holopainen T, et al. Vascular endothelial growth factor-B acts as a coronary growth factor in transgenic rats without inducing angiogenesis, vascular leak, or inflammation. *Circulation*. 2010;122:1725–1733.
25. Alitalo K, Carmeliet P. Molecular mechanisms of lymphangiogenesis in health and disease. *Cancer Cell*. 2002;1:219–227.

26. Davis-Smyth T, Chen H, Park J, Presta LG, Ferrara N. The second immunoglobulin-like domain of the VEGF tyrosine kinase receptor Flt-1 determines ligand binding and may initiate a signal transduction cascade. *EMBO J.* 1996;15:4919–4927.
27. Wu FT, Stefanini MO, Mac Gabhann F, Popel AS. A compartment model of VEGF distribution in humans in the presence of soluble VEGF receptor-1 acting as a ligand trap. *PLoS One.* 2009;4, e5108.
28. Kovacs K, Decatur C, Toro M, et al. 2-Deoxy-glucose downregulates endothelial AKT and ERK via interference with N-linked glycosylation, induction of endoplasmic reticulum stress, and GSK3beta activation. *Mol Cancer Ther.* 2016;15:264–275.
29. Chandler KB, Leon DR, Meyer RD, Rahimi N, Costello CE. Site-specific N-glycosylation of endothelial cell receptor tyrosine kinase VEGFR-2. *J Proteome Res.* 2017;16:677–688.
30. Vogel C, Bauer A, Wiesnet M, et al. Flt-1, but not Flk-1 mediates hyperpermeability through activation of the PI3-K/Akt pathway. *J Cell Physiol.* 2007;212:236–243.
31. Shaik F, Cuthbert GA, Homer-Vanniasinkam S, Muench SP, Ponnambalam S, Harrison MA. Structural basis for vascular endothelial growth factor receptor activation and implications for disease therapy. *Biomolecules.* 2020;10.
32. Zeng H, Sanyal S, Mukhopadhyay D. Tyrosine residues 951 and 1059 of vascular endothelial growth factor receptor-2 (KDR) are essential for vascular permeability factor/vascular endothelial growth factor-induced endothelium migration and proliferation, respectively. *J Biol Chem.* 2001;276:32714–32719.
33. Zachary I, Gliki G. Signaling transduction mechanisms mediating biological actions of the vascular endothelial growth factor family. *Cardiovasc Res.* 2001;49:568–581.
34. Roskoski Jr R. VEGF receptor protein-tyrosine kinases: structure and regulation. *Biochem Biophys Res Commun.* 2008;375:287–291.
35. Smith GA, Fearnley GW, Abdul-Zani I, et al. Ubiquitination of basal VEGFR2 regulates signal transduction and endothelial function. *Biol Open.* 2017;6:1404–1415.
36. Boucher JM, Clark RP, Chong DC, Citrin KM, Wylie LA, Bautch VL. Dynamic alterations in decoy VEGF receptor-1 stability regulate angiogenesis. *Nat Commun.* 2017;8:15699.
37. Zhou HJ, Xu Z, Wang Z, et al. SUMOylation of VEGFR2 regulates its intracellular trafficking and pathological angiogenesis. *Nat Commun.* 2018;9:3303.
38. Pan MR, Chang TM, Chang HC, Su JL, Wang HW, Hung WC. Sumoylation of Prox1 controls its ability to induce VEGFR3 expression and lymphatic phenotypes in endothelial cells. *J Cell Sci.* 2009;122:3358–3364.
39. Kim Y, Nam HJ, Lee J, et al. Methylation-dependent regulation of HIF-1alpha stability restricts retinal and tumour angiogenesis. *Nat Commun.* 2016;7:10347.
40. Zecchin A, Pattarini L, Gutierrez MI, et al. Reversible acetylation regulates vascular endothelial growth factor receptor-2 activity. *J Mol Cell Biol.* 2014;6:116–127.
41. Fong GH. Regulation of angiogenesis by oxygen sensing mechanisms. *J Mol Med (Berl).* 2009;87:549–560.
42. Gille H, Kowalski J, Yu L, et al. A repressor sequence in the juxtamembrane domain of Flt-1 (VEGFR-1) constitutively inhibits vascular endothelial growth factor-dependent phosphatidylinositol 3'-kinase activation and endothelial cell migration. *EMBO J.* 2000;19:4064–4073.
43. Meyer RD, Mohammadi M, Rahimi N. A single amino acid substitution in the activation loop defines the decoy characteristic of VEGFR-1/FLT-1. *J Biol Chem.* 2006;281:867–875.
44. Seetharam L, Gotoh N, Maru Y, Neufeld G, Yamaguchi S, Shibuya M. A unique signal transduction from FLT tyrosine kinase, a receptor for vascular endothelial growth factor VEGF. *Oncogene.* 1995;10:135–147.

45. Sawano A, Takahashi T, Yamaguchi S, Shibuya M. The phosphorylated 1169-tyrosine containing region of flt-1 kinase (VEGFR-1) is a major binding site for PLCgamma. *Biochem Biophys Res Commun*. 1997;238:487–491.
46. Bowler E, Oltean S. Alternative splicing in angiogenesis. *Int J Mol Sci*. 2019;20.
47. Ulyatt C, Walker J, Ponnambalam S. Hypoxia differentially regulates VEGFR1 and VEGFR2 levels and alters intracellular signaling and cell migration in endothelial cells. *Biochem Biophys Res Commun*. 2011;404:774–779.
48. Fong GH, Rossant J, Gertsenstein M, Breitman ML. Role of the Flt-1 receptor tyrosine kinase in regulating the assembly of vascular endothelium. *Nature*. 1995;376:66–70.
49. Kendall RL, Wang G, DiSalvo J, Thomas KA. Specificity of vascular endothelial cell growth factor receptor ligand binding domains. *Biochem Biophys Res Commun*. 1994;201:326–330.
50. Tripathi R, Rath G, Jain A, Salhan S. Soluble and membranous vascular endothelial growth factor receptor-1 in pregnancies complicated by pre-eclampsia. *Ann Anat*. 2008;190:477–489.
51. Takahashi T, Shibuya M. The 230 kDa mature form of KDR/Flk-1 (VEGF receptor-2) activates the PLC-gamma pathway and partially induces mitotic signals in NIH3T3 fibroblasts. *Oncogene*. 1997;14:2079–2089.
52. Katoh O, Tauchi H, Kawaishi K, Kimura A, Satow Y. Expression of the vascular endothelial growth factor (VEGF) receptor gene, KDR, in hematopoietic cells and inhibitory effect of VEGF on apoptotic cell death caused by ionizing radiation. *Cancer Res*. 1995;55:5687–5692.
53. Domingues I, Rino J, Demmers JA, de Lanerolle P, Santos SC. VEGFR2 translocates to the nucleus to regulate its own transcription. *PLoS One*. 2011;6, e25668.
54. Albuquerque RJ, Hayashi T, Cho WG, et al. Alternatively spliced vascular endothelial growth factor receptor-2 is an essential endogenous inhibitor of lymphatic vessel growth. *Nat Med*. 2009;15:1023–1030.
55. Shalaby F, Rossant J, Yamaguchi TP, et al. Failure of blood-island formation and vasculogenesis in Flk-1-deficient mice. *Nature*. 1995;376:62–66.
56. Hyde CA, Giese A, Stuttfeld E, et al. Targeting extracellular domains D4 and D7 of vascular endothelial growth factor receptor 2 reveals allosteric receptor regulatory sites. *Mol Cell Biol*. 2012;32:3802–3813.
57. Markovic-Mueller S, Stuttfeld E, Asthana M, et al. Structure of the full-length VEGFR-1 extracellular domain in complex with VEGF-A. *Structure*. 2017;25:341–352.
58. Plate KH, Breier G, Weich HA, Mennel HD, Risau W. Vascular endothelial growth factor and glioma angiogenesis: coordinate induction of VEGF receptors, distribution of VEGF protein and possible in vivo regulatory mechanisms. *Int J Cancer*. 1994;59:520–529.
59. Hartsough EJ, Meyer RD, Chitalia V, et al. Lysine methylation promotes VEGFR-2 activation and angiogenesis. *Sci Signal*. 2013;6:ra104.
60. Hughes DC. Alternative splicing of the human VEGFGR-3/FLT4 gene as a consequence of an integrated human endogenous retrovirus. *J Mol Evol*. 2001;53:77–79.
61. Chaitanya GV, Franks SE, Cromer W, et al. Differential cytokine responses in human and mouse lymphatic endothelial cells to cytokines in vitro. *Lymphat Res Biol*. 2010;8:155–164.
62. Breslin JW, Yuan SY, Wu MH. VEGF-C alters barrier function of cultured lymphatic endothelial cells through a VEGFR-3-dependent mechanism. *Lymphat Res Biol*. 2007;5:105–113.
63. Schwager S, Detmar M. Inflammation and lymphatic function. *Front Immunol*. 2019;10:308.
64. Koch S, Claesson-Welsh L. Signal transduction by vascular endothelial growth factor receptors. *Cold Spring Harb Perspect Med*. 2012;2, a006502.

65. Dumont DJ, Jussila L, Taipale J, et al. Cardiovascular failure in mouse embryos deficient in VEGF receptor-3. *Science*. 1998;282:946–949.
66. Kaipainen A, Korhonen J, Mustonen T, et al. Expression of the fms-like tyrosine kinase 4 gene becomes restricted to lymphatic endothelium during development. *Proc Natl Acad Sci USA*. 1995;92:3566–3570.
67. Wang J, Taylor A, Showeil R, et al. Expression profiling and significance of VEGF-A, VEGFR2, VEGFR3 and related proteins in endometrial carcinoma. *Cytokine*. 2014;68:94–100.
68. Staton CA, Kumar I, Reed MW, Brown NJ. Neuropilins in physiological and pathological angiogenesis. *J Pathol*. 2007;212:237–248.
69. Nakamura F, Goshima Y. Structural and functional relation of neuropilins. *Adv Exp Med Biol*. 2002;515:55–69.
70. Herzog Y, Kalcheim C, Kahane N, Reshef R, Neufeld G. Differential expression of neuropilin-1 and neuropilin-2 in arteries and veins. *Mech Dev*. 2001;109:115–119.
71. Yuan L, Moyon D, Pardanaud L, et al. Abnormal lymphatic vessel development in neuropilin 2 mutant mice. *Development*. 2002;129:4797–4806.
72. Chen H, Chedotal A, He Z, Goodman CS, Tessier-Lavigne M. Neuropilin-2, a novel member of the neuropilin family, is a high affinity receptor for the semaphorins Sema E and Sema IV but not Sema III. *Neuron*. 1997;19:547–559.
73. Giger RJ, Urquhart ER, Gillespie SK, Levengood DV, Ginty DD, Kolodkin AL. Neuropilin-2 is a receptor for semaphorin IV: insight into the structural basis of receptor function and specificity. *Neuron*. 1998;21:1079–1092.
74. Wang L, Mukhopadhyay D, Xu X. C terminus of RGS-GAIP-interacting protein conveys neuropilin-1-mediated signaling during angiogenesis. *FASEB J*. 2006;20: 1513–1515.
75. Soker S, Takashima S, Miao HQ, Neufeld G, Klagsbrun M. Neuropilin-1 is expressed by endothelial and tumor cells as an isoform-specific receptor for vascular endothelial growth factor. *Cell*. 1998;92:735–745.
76. Keyt BA, Nguyen HV, Berleau LT, et al. Identification of vascular endothelial growth factor determinants for binding KDR and FLT-1 receptors. Generation of receptor-selective VEGF variants by site-directed mutagenesis. *J Biol Chem*. 1996;271:5638–5646.
77. Caunt M, Mak J, Liang WC, et al. Blocking neuropilin-2 function inhibits tumor cell metastasis. *Cancer Cell*. 2008;13:331–342.
78. Said AM, Parker MW, Vander Kooi CW. Design, synthesis, and evaluation of a novel benzamidine-based inhibitor of VEGF-C binding to Neuropilin-2. *Bioorg Chem*. 2020;100, 103856.
79. Basagiannis D, Zografou S, Murphy C, et al. VEGF induces signalling and angiogenesis by directing VEGFR2 internalisation through macropinocytosis. *J Cell Sci*. 2016; 129:4091–4104.
80. Jopling HM, Howell GJ, Gamper N, Ponnambalam S. The VEGFR2 receptor tyrosine kinase undergoes constitutive endosome-to-plasma membrane recycling. *Biochem Biophys Res Commun*. 2011;410:170–176.
81. Zhang X, Simons M. Receptor tyrosine kinases endocytosis in endothelium: biology and signaling. *Arterioscler Thromb Vasc Biol*. 2014;34:1831–1837.
82. Mittar S, Ulyatt C, Howell GJ, et al. VEGFR1 receptor tyrosine kinase localization to the Golgi apparatus is calcium-dependent. *Exp Cell Res*. 2009;315:877–889.
83. McMahon HT, Boucrot E. Molecular mechanism and physiological functions of clathrin-mediated endocytosis. *Nat Rev Mol Cell Biol*. 2011;12:517–533.
84. Grazia Lampugnani M, Zanetti A, Corada M, et al. Contact inhibition of VEGF-induced proliferation requires vascular endothelial cadherin, beta-catenin, and the phosphatase DEP-1/CD148. *J Cell Biol*. 2003;161:793–804.

85. Roskoski Jr R. Vascular endothelial growth factor (VEGF) signaling in tumor progression. *Crit Rev Oncol Hematol.* 2007;62:179–213.
86. Takahashi T, Yamaguchi S, Chida K, Shibuya M. A single autophosphorylation site on KDR/Flk-1 is essential for VEGF-A-dependent activation of PLC-gamma and DNA synthesis in vascular endothelial cells. *EMBO J.* 2001;20:2768–2778.
87. Shibuya M. Role of VEGF-flt receptor system in normal and tumor angiogenesis. *Adv Cancer Res.* 1995;67:281–316.
88. Simons M. An inside view: VEGF receptor trafficking and signaling. *Physiology (Bethesda).* 2012;27:213–222.
89. Eichmann A, Simons M. VEGF signaling inside vascular endothelial cells and beyond. *Curr Opin Cell Biol.* 2012;24:188–193.
90. Labrecque L, Royal I, Surprenant DS, Patterson C, Gingras D, Beliveau R. Regulation of vascular endothelial growth factor receptor-2 activity by caveolin-1 and plasma membrane cholesterol. *Mol Biol Cell.* 2003;14:334–347.
91. Fearnley GW, Odell AF, Latham AM, et al. VEGF-A isoforms differentially regulate ATF-2-dependent VCAM-1 gene expression and endothelial-leukocyte interactions. *Mol Biol Cell.* 2014;25:2509–2521.
92. Basagiannis D, Christoforidis S. Constitutive endocytosis of VEGFR2 protects the receptor against shedding. *J Biol Chem.* 2016;291:16892–16903.
93. Bauer PM, Yu J, Chen Y, et al. Endothelial-specific expression of caveolin-1 impairs microvascular permeability and angiogenesis. *Proc Natl Acad Sci USA.* 2005; 102:204–209.
94. Pitulescu ME, Adams RH. Regulation of signaling interactions and receptor endocytosis in growing blood vessels. *Cell Adh Migr.* 2014;8:366–377.
95. Raikwar NS, Shibuya M, Thomas CP. VEGF-A selectively inhibits FLT1 ectodomain shedding independent of receptor activation and receptor endocytosis. *Am J Physiol Cell Physiol.* 2018;315:C214–C224.
96. Autiero M, Waltenberger J, Communi D, et al. Role of PlGF in the intra- and intermolecular cross talk between the VEGF receptors Flt1 and Flk1. *Nat Med.* 2003;9:936–943.
97. Neagoe PE, Lemieux C, Sirois MG. Vascular endothelial growth factor (VEGF)-A165-induced prostacyclin synthesis requires the activation of VEGF receptor-1 and -2 heterodimer. *J Biol Chem.* 2005;280:9904–9912.
98. Cudmore MJ, Hewett PW, Ahmad S, et al. The role of heterodimerization between VEGFR-1 and VEGFR-2 in the regulation of endothelial cell homeostasis. *Nat Commun.* 2012;3:972.
99. Wang Y, Nakayama M, Pitulescu ME, et al. Ephrin-B2 controls VEGF-induced angiogenesis and lymphangiogenesis. *Nature.* 2010;465:483–486.
100. Liu X, Pasula S, Song H, et al. Temporal and spatial regulation of epsin abundance and VEGFR3 signaling are required for lymphatic valve formation and function. *Sci Signal.* 2014;7:ra97.
101. Deng Y, Zhang X, Simons M. Molecular controls of lymphatic VEGFR3 signaling. *Arterioscler Thromb Vasc Biol.* 2015;35:421–429.
102. Salikhova A, Wang L, Lanahan AA, et al. Vascular endothelial growth factor and semaphorin induce neuropilin-1 endocytosis via separate pathways. *Circ Res.* 2008;103:e71–e79.
103. Lanahan A, Zhang X, Fantin A, et al. The neuropilin 1 cytoplasmic domain is required for VEGF-A-dependent arteriogenesis. *Dev Cell.* 2013;25:156–168.
104. Rizzolio S, Rabinowicz N, Rainero E, et al. Neuropilin-1-dependent regulation of EGF-receptor signaling. *Cancer Res.* 2012;72:5801–5811.
105. Dutta S, Roy S, Polavaram NS, et al. Neuropilin-2 regulates endosome maturation and EGFR trafficking to support cancer cell pathobiology. *Cancer Res.* 2016;76:418–428.

106. Cantuti-Castelvetri L, Ojha R, Pedro LD, et al. Neuropilin-1 facilitates SARS-CoV-2 cell entry and infectivity. *Science*. 2020;370:856–860.
107. Daly JL, Simonetti B, Klein K, et al. Neuropilin-1 is a host factor for SARS-CoV-2 infection. *Science*. 2020;370:861–865.
108. Wang HB, Zhang H, Zhang JP, et al. Neuropilin 1 is an entry factor that promotes EBV infection of nasopharyngeal epithelial cells. *Nat Commun*. 2015;6:6240.
109. Mousavi SA, Skjeldal F, Fonhus MS, et al. Receptor-mediated endocytosis of VEGF-A in rat liver sinusoidal endothelial cells. *Biomed Res Int*. 2019;2019:5496197.
110. Fearnley GW, Bruns AF, Wheatcroft SB, Ponnambalam S. VEGF-A isoform-specific regulation of calcium ion flux, transcriptional activation and endothelial cell migration. *Biol Open*. 2015;4:731–742.
111. Ikeda S, Ushio-Fukai M, Zuo L, et al. Novel role of ARF6 in vascular endothelial growth factor-induced signaling and angiogenesis. *Circ Res*. 2005;96:467–475.
112. Wu T, Zhang B, Ye F, Xiao Z. A potential role for caveolin-1 in VEGF-induced fibronectin upregulation in mesangial cells: involvement of VEGFR2 and Src. *Am J Physiol Renal Physiol*. 2013;304:F820–F830.
113. Sonveaux P, Martinive P, DeWever J, et al. Caveolin-1 expression is critical for vascular endothelial growth factor-induced ischemic hindlimb collateralization and nitric oxide-mediated angiogenesis. *Circ Res*. 2004;95:154–161.
114. Galvagni F, Anselmi F, Salameh A, Orlandini M, Rocchigiani M, Oliviero S. Vascular endothelial growth factor receptor-3 activity is modulated by its association with caveolin-1 on endothelial membrane. *Biochemistry*. 2007;46:3998–4005.
115. Nakayama M, Nakayama A, van Lessen M, et al. Spatial regulation of VEGF receptor endocytosis in angiogenesis. *Nat Cell Biol*. 2013;15:249–260.
116. Lanahan AA, Hermans K, Claes F, et al. VEGF receptor 2 endocytic trafficking regulates arterial morphogenesis. *Dev Cell*. 2010;18:713–724.
117. Ballmer-Hofer K, Andersson AE, Ratcliffe LE, Berger P. Neuropilin-1 promotes VEGFR-2 trafficking through Rab11 vesicles thereby specifying signal output. *Blood*. 2011;118:816–826.
118. Manickam V, Tiwari A, Jung JJ, et al. Regulation of vascular endothelial growth factor receptor 2 trafficking and angiogenesis by Golgi localized t-SNARE syntaxin 6. *Blood*. 2011;117:1425–1435.
119. Takahashi T, Ueno H, Shibuya M. VEGF activates protein kinase C-dependent, but Ras-independent Raf-MEK-MAP kinase pathway for DNA synthesis in primary endothelial cells. *Oncogene*. 1999;18:2221–2230.
120. Takahashi T, Yamaguchi S, Chida K, Shibuya M. A single autophosphorylation site on KDR/Flk-1 is essential for VEGF-A-dependent activation of PLC-γ and DNA synthesis in vascular endothelial cells. *EMBO J*. 2001;20:2768–2778.
121. Koch S, Tugues S, Li X, Gualandi L, Claesson-Welsh L. Signal transduction by vascular endothelial growth factor receptors. *Biochem J*. 2011;437:169–183.
122. Lamalice L, Houle F, Jourdan G, Huot J. Phosphorylation of tyrosine 1214 on VEGFR2 is required for VEGF-induced activation of Cdc42 upstream of SAPK2/p38. *Oncogene*. 2004;23:434–445.
123. Blanes MG, Oubaha M, Rautureau Y, Gratton JP. Phosphorylation of tyrosine 801 of vascular endothelial growth factor receptor-2 is necessary for Akt-dependent endothelial nitric-oxide synthase activation and nitric oxide release from endothelial cells. *J Biol Chem*. 2007;282:10660–10669.
124. Meyer RD, Sacks DB, Rahimi N. IQGAP1-dependent signaling pathway regulates endothelial cell proliferation and angiogenesis. *PLoS One*. 2008;3, e3848.
125. Meyer RD, Srinivasan S, Singh AJ, Mahoney JE, Gharahassanlou KR, Rahimi N. PEST motif serine and tyrosine phosphorylation controls vascular endothelial growth factor receptor 2 stability and downregulation. *Mol Cell Biol*. 2011;31:2010–2025.

126. Hornbeck PV, Kornhauser JM, Tkachev S, et al. PhosphoSitePlus: a comprehensive resource for investigating the structure and function of experimentally determined post-translational modifications in man and mouse. *Nucleic Acids Res.* 2012;40: D261–D270.
127. Singh AJ, Meyer RD, Band H, Rahimi N. The carboxyl terminus of VEGFR-2 is required for PKC-mediated down-regulation. *Mol Biol Cell.* 2005;16:2106–2118.
128. Igarashi K, Isohara T, Kato T, Shigeta K, Yamano T, Uno I. Tyrosine 1213 of Flt-1 is a major binding site of Nck and SHP-2. *Biochem Biophys Res Commun.* 1998;246:95–99.
129. Igarashi K, Shigeta K, Isohara T, Yamano T, Uno I. Sck interacts with KDR and Flt-1 via its SH2 domain. *Biochem Biophys Res Commun.* 1998;251:77–82.
130. Ito N, Wernstedt C, Engström U, Claesson-Welsh L. Identification of vascular endothelial growth factor receptor-1 tyrosine phosphorylation sites and binding of SH2 domain-containing molecules. *J Biol Chem.* 1998;273:23410–23418.
131. Yao J, Wu X, Zhuang G, et al. Expression of a functional VEGFR-1 in tumor cells is a major determinant of anti-PlGF antibodies efficacy. *Proc Natl Acad Sci USA.* 2011;108:11590–11595.
132. Ceci C, Atzori MG, Lacal PM, Graziani G. Role of VEGFs/VEGFR-1 signaling and its inhibition in modulating tumor invasion: experimental evidence in different metastatic cancer models. *Int J Mol Sci.* 2020;21:1388.
133. Xu WW, Li B, Lam AK, et al. Targeting VEGFR1- and VEGFR2-expressing non-tumor cells is essential for esophageal cancer therapy. *Oncotarget.* 2014;6.
134. Coso S, Zeng Y, Opeskin K, Williams ED. Vascular endothelial growth factor receptor-3 directly interacts with phosphatidylinositol 3-kinase to regulate lymphangiogenesis. *PLoS One.* 2012;7:e39558.
135. Fournier E, Dubreuil P, Birnbaum D, Borg JP. Mutation at tyrosine residue 1337 abrogates ligand-dependent transforming capacity of the FLT4 receptor. *Oncogene.* 1995;11:921–931.
136. Dixelius J, Mäkinen T, Wirzenius M, et al. Ligand-induced vascular endothelial growth factor receptor-3 (VEGFR-3) heterodimerization with VEGFR-2 in primary lymphatic endothelial cells regulates tyrosine phosphorylation sites. *J Biol Chem.* 2003;278:40973–40979.
137. Salameh A, Galvagni F, Bardelli M, Bussolino F, Oliviero S. Direct recruitment of CRK and GRB2 to VEGFR-3 induces proliferation, migration, and survival of endothelial cells through the activation of ERK, AKT, and JNK pathways. *Blood.* 2005;106:3423–3431.
138. Duval M, Bédard-Goulet S, Delisle C, Gratton JP. Vascular endothelial growth factor-dependent down-regulation of Flk-1/KDR involves Cbl-mediated ubiquitination. Consequences on nitric oxide production from endothelial cells. *J Biol Chem.* 2003;278:20091–20097.
139. Ewan LC, Jopling HM, Jia H, et al. Intrinsic tyrosine kinase activity is required for vascular endothelial growth factor receptor 2 ubiquitination, sorting and degradation in endothelial cells. *Traffic.* 2006;7:1270–1282.
140. Kobayashi S, Sawano A, Nojima Y, Shibuya M, Maru Y. The c-Cbl/CD2AP complex regulates VEGF-induced endocytosis and degradation of Flt-1 (VEGFR-1). *FASEB J.* 2004;18:929–931.
141. Bruns AF, Yuldasheva N, Latham AM, et al. A heat-shock protein axis regulates VEGFR2 proteolysis, blood vessel development and repair. *PLoS One.* 2012;7, e48539.
142. Lampugnani MG, Orsenigo F, Gagliani MC, Tacchetti C, Dejana E. Vascular endothelial cadherin controls VEGFR-2 internalization and signaling from intracellular compartments. *J Cell Biol.* 2006;174:593–604.
143. Gampel A, Moss L, Jones MC, Brunton V, Norman JC, Mellor H. VEGF regulates the mobilization of VEGFR2/KDR from an intracellular endothelial storage compartment. *Blood.* 2006;108:2624–2631.

144. Bruns AF, Herbert SP, Odell AF, et al. Ligand-stimulated VEGFR2 signaling is regulated by co-ordinated trafficking and proteolysis. *Traffic*. 2010;11:161–174.
145. Shaik S, Nucera C, Inuzuka H, et al. SCFβ-TRCP suppresses angiogenesis and thyroid cancer cell migration by promoting ubiquitination and destruction of VEGF receptor 2. *J Exp Med*. 2012;209:1289–1307.
146. Singh AJ, Meyer RD, Navruzbekov G, et al. A critical role for the E3-ligase activity of c-Cbl in VEGFR-2-mediated PLCgamma1 activation and angiogenesis. *Proc Natl Acad Sci USA*. 2007;104:5413–5418.
147. Murdaca J, Treins C, Monthouel-Kartmann MN, et al. Grb10 prevents Nedd4-mediated vascular endothelial growth factor receptor-2 degradation. *J Biol Chem*. 2004;279:26754–26761.
148. Sakaue T, Sakakibara I, Uesugi T, et al. The CUL3-SPOP-DAXX axis is a novel regulator of VEGFR2 expression in vascular endothelial cells. *Sci Rep*. 2017;7:42845.
149. Maghsoudlou A, Meyer RD, Rezazadeh K, et al. RNF121 inhibits angiogenic growth factor signaling by restricting cell surface expression of VEGFR-2. *Traffic*. 2016; 17:289–300.
150. Meissner M, Reichenbach G, Stein M, Hrgovic I, Kaufmann R, Gille J. Down-regulation of vascular endothelial growth factor receptor 2 is a major molecular determinant of proteasome inhibitor-mediated antiangiogenic action in endothelial cells. *Cancer Res*. 2009;69:1976–1984.
151. Chandler KB, Leon DR, Kuang J, Meyer RD, Rahimi N, Costello CE. N-Glycosylation regulates ligand-dependent activation and signaling of vascular endothelial growth factor receptor 2 (VEGFR2). *J Biol Chem*. 2019;294:13117–13130.
152. Fukahi K, Fukasawa M, Neufeld G, Itakura J, Korc M. Aberrant expression of neuropilin-1 and -2 in human pancreatic cancer cells. *Clin Cancer Res*. 2004;10: 581–590.
153. Shintani Y, Takashima S, Asano Y, et al. Glycosaminoglycan modification of neuropilin-1 modulates VEGFR2 signaling. *EMBO J*. 2006;25:3045–3055.
154. Suzuki A, Ohno S. The PAR-aPKC system: lessons in polarity. *J Cell Sci*. 2006;119:979–987.
155. Goldstein B, Macara IG. The PAR proteins: fundamental players in animal cell polarization. *Dev Cell*. 2007;13:609–622.
156. Carmeliet P, De Smet F, Loges S, Mazzone M. Branching morphogenesis and antiangiogenesis candidates: tip cells lead the way. *Nat Rev Clin Oncol*. 2009;6:315–326.
157. Avraham HK, Lee TH, Koh Y, et al. Vascular endothelial growth factor regulates focal adhesion assembly in human brain microvascular endothelial cells through activation of the focal adhesion kinase and related adhesion focal tyrosine kinase. *J Biol Chem*. 2003;278:36661–36668.
158. Matsumoto T, Bohman S, Dixelius J, et al. VEGF receptor-2 Y951 signaling and a role for the adapter molecule TSAd in tumor angiogenesis. *EMBO J*. 2005;24:2342–2353.
159. Mitra SK, Hanson DA, Schlaepfer DD. Focal adhesion kinase: in command and control of cell motility. *Nat Rev Mol Cell Biol*. 2005;6:56–68.
160. Poola I, DeWitty RL, Marshalleck JJ, Bhatnagar R, Abraham J, Leffall LD. Identification of MMP-1 as a putative breast cancer predictive marker by global gene expression analysis. *Nat Med*. 2005;11:481–483.
161. Luttun A, Tjwa M, Moons L, et al. Revascularization of ischemic tissues by PlGF treatment, and inhibition of tumor angiogenesis, arthritis and atherosclerosis by anti-Flt1. *Nat Med*. 2002;8:831–840.
162. Kaplan RN, Riba RD, Zacharoulis S, et al. VEGFR1-positive haematopoietic bone marrow progenitors initiate the pre-metastatic niche. *Nature*. 2005;438:820–827.

163. Wang F, Yamauchi M, Muramatsu M, Osawa T, Tsuchida R, Shibuya M. RACK1 regulates VEGF/Flt1-mediated cell migration via activation of a PI3K/Akt pathway. *J Biol Chem.* 2011;286:9097–9106.
164. Tauzin S, Chaigne-Delalande B, Selva E, et al. The naturally processed CD95L elicits a c-yes/calcium/PI3K-driven cell migration pathway. *PLoS Biol.* 2011;9, e1001090.
165. Weddell JC, Chen S, Imoukhuede PI. VEGFR1 promotes cell migration and proliferation through PLCgamma and PI3K pathways. *NPJ Syst Biol Appl.* 2018;4:1.
166. Liao WX, Feng L, Zhang H, Zheng J, Moore TR, Chen DB. Compartmentalizing VEGF-induced ERK2/1 signaling in placental artery endothelial cell caveolae: a paradoxical role of caveolin-1 in placental angiogenesis in vitro. *Mol Endocrinol.* 2009;23:1428–1444.

CHAPTER SEVEN

Endocytosis and signaling of angiotensin II type 1 receptor

Indra Mani*
Department of Microbiology, Gargi College, University of Delhi, New Delhi, India
*Corresponding author: e-mail addresses: indramanibhu@gmail.com; indra.mani@gargi.du.ac.in

Contents

Abstract

A vasoactive octapeptide angiotensin II (Ang II) hormone is the key regulator of the renin-angiotensin system (RAS). It binds with the two different plasma membrane receptors like angiotensin II type 1 (AT1) and type 2 (AT2) and consequence various biological responses occur. Further, AT1 has two subtypes such as AT1A and AT1B. These angiotensin receptors are classified to be G protein-coupled receptors (GPCRs). The main constituent of RAS is the AT1 receptor (AT1R), and its activation, signal transduction, and regulation have been extensively studied. After Ang II stimulation, the ligand-receptor complexes internalized and trafficked through the early endosome, recycling endosome, and some receptors skipped the recycling endosome and trafficked to the lysosome for metabolic degradation. Moreover, some short sequence motifs located in the carboxyl-terminus (CT) of the receptor play a vital role in the internalization, phosphorylation, subcellular trafficking, signaling, and desensitization. Furthermore, in endocytosis, the various proteins interact with the CT region of the receptor. This chapter highlights the basic mechanism of AT1 receptor internalization, trafficking and signaling in both physiological and pathophysiological conditions.

Progress in Molecular Biology and Translational Science, Volume 194
ISSN 1877-1173
https://doi.org/10.1016/bs.pmbts.2022.08.005

Abbreviations

Ang II	angiotensin II
AT1R	angiotensin II type 1 receptor
AT2R	angiotensin II type 2 receptor
CRISPR-CAS	clustered regularly interspaced short palindromic repeats (CRISPR) and CRISPR-associated (Cas) proteins
CT	carboxyl-terminus
eGFP	enhanced green fluorescent protein
GPCRs	G protein-coupled receptors
Gq	G protein coupling
HEK-293 cells	Human embryonic kidney 293 cells
PKC	protein kinase C
PLC	phospholipase C
RAS	renin-angiotensin system
RASMC	rat aortic vascular smooth muscle cells

1. Introduction

A vasoactive linear peptide angiotensin II (Ang II) hormone is composed of eight amino acids. It binds with the two different plasma membrane receptors like angiotensin II type 1 (AT1) and type 2 (AT2) and consequence various biological responses occur. Further, AT1 receptor (AT1R) has two subtypes such as AT1A and AT1B. AT1A and AT1B receptors are present in rodents, while only AT1 has been recognized in higher mammals.[1,2] In the rat, genes of AT1A, AT1B and AT2 are localized on the different chromosomes such as 17, 2 and X.[3–5] Human AT1 localized on chromosome 3.[6] Rat AT1A and AT1B receptors contain 359 amino acids,[7,8] while 359 amino acids are also present in human AT1 receptors.[9] After Ang II binding with the receptor, phosphorylation (a serine/threonine-rich domain) occurs in the carboxyl-terminus (CT)[10] and increases its association with β-arrestins.[11] However, the CT of the AT1 receptor lacks a kinase domain. For the phosphorylation of AT1, various protein kinases are involved such as the protein kinase C (PKC),[12] G protein-coupled receptor kinase type 2 (GRK2),[12] GRK4,[13] GRK5,[14] and phosphoinositide 3-kinase (PI3K).[15]

There are several heteromeric G proteins such as Gq/11, Ga12, G13, and Gi that interact with the AT1 receptors. These G protein isoforms interact with different signaling cascades, such as Gq activation, consequences in the stimulation of phospholipase C (PLC), while GαI leads to cGMP production.[16] Several studies have shown that these receptors are involved in numerous pathways of angiotensin signaling, which regulate protein tyrosine phosphorylation.[17] Irregularities in the signaling of AT1 causes various disease such as aging-related diseases, cardiovascular diseases, and albuminuria.[18,19] Further, the Ang II receptor and its subtypes help in regulating vascular functions.[20] AT1 is present in cardiomyocytes and fibroblasts in the heart but it is mainly expressed in the smooth muscle cells of the vasculature.[21]

In vivo study of AT1A knockout (AT1AKO) mice has demonstrated that the intracellular accumulation of Ang II in adrenal glands and the kidney is mediated through the AT1A receptor. Different drugs such as captopril and losartan were used to treat AT1A wild-type (AT1AWT) and AT1AKO mice for 2 weeks. After that, [^{125}I] Val^5-ANG II was infused for 1 h. Quantitative *in vivo* autoradiography data has shown that AT1A receptors mediated 80% of intracellular ANG II accumulation in the adrenal glands and kidney, while 20% of the response was mediated by AT1B receptor and others.[22] Outcome advocates that the AT1A plays an important role in these organs. Moreover, apart from agonist Ang II hormone, antagonist [Sar1, Ile8] Ang II peptide also activates AT1A receptor endocytosis *in vitro* (cell culture) while losartan (nonpeptide AT1 receptor blocker) does not internalize. An *in vivo* study in which AT1A null mice or AT1A X AT1B dual null mice and AT2-deleted were used and analysis has demonstrated that AT2 works in the direction of vasorelaxation as opposed to vasoconstriction by AT1.[23]

Ligand-stimulated AT1 receptor internalization plays a vital function in regulating cardiovascular homeostasis, whereas diminished receptor internalization may cause cardiovascular diseases. Furthermore, various hypertensive disorders such as essential hypertension, preeclampsia and malignant hypertension are associated with irregular activation of AT1R.[24] This chapter briefly discusses the basic mechanism of endocytosis and signaling of the AT1 receptor under the following headings.

2. Ligand-stimulated internalization of AT1 receptor

Seven transmembrane structure features of G protein-coupled receptors (GPCRs) are present in both AT1 and AT2 receptors. Generally, after the internalization of GPCRs, it is trafficked to the early endosome, lysosome, or recycling endosome.[25] Fig. 1 shows the major steps of the internalization, signaling and subcellular trafficking of the GPCRs.[26] After ligand binding, only AT1 receptor internalization occurs, which is a typical characteristic of many plasma membrane receptors. Ang II-AT1 receptor endocytosis progresses through clathrin-coated pits.[27] Nevertheless, the

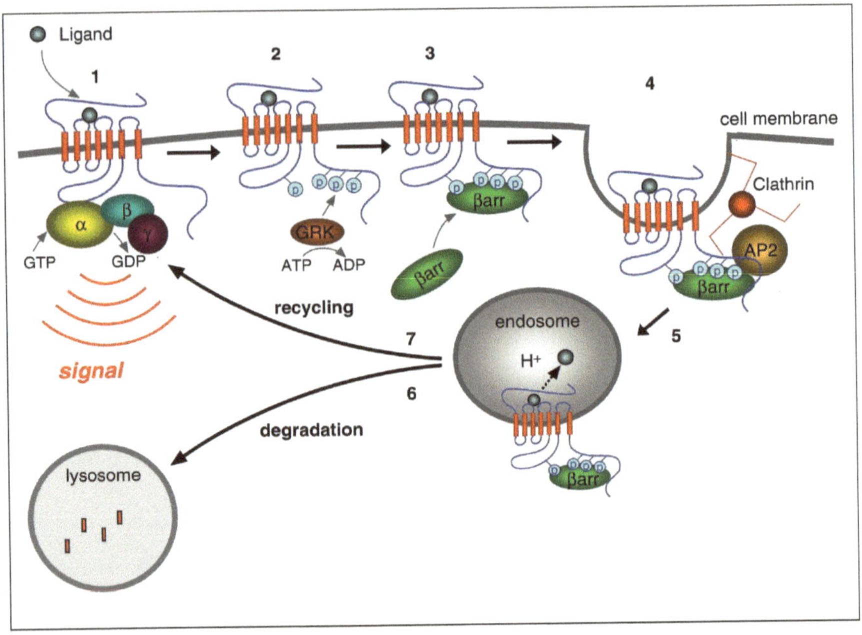

Fig. 1 Graphic depiction of internalization, subcellular trafficking, and signaling of GPCRs. All these events are shown in seven different steps: (1) After ligand binding with GPCRs, it initiates signaling by increasing guanine exchange activity, (2) several receptors are phosphorylated, (3) Arrestins bind with the phosphorylated receptor, (4) Many arrestins isoforms also bind components of endocytic lattices (AP2 alpha subunit, clathrin heavy chain, and PIP2), (5) Clathrin-dependent ligand-receptor complex is internalized, and then it trafficked to an early endosome, ligand dissociated, and receptor dephosphorylated, (6) trafficked to the lysosome for metabolic degradation, and (7) recycled back to the plasma membrane. GPCRs, G protein-coupled receptors: β-arr, β-arrestin: GRK, G protein-coupled receptor kinases: PIP2, phosphatidylinositol bisphosphate. *Figure adapted from Irannejad R, von Zastrow M. GPCR signaling along the endocytic pathway.* Curr Opin Cell Biol. *2014;27:109–116 with permission.*

exact mechanism of reduced AT1 receptor internalization is not fully understood. A general step of the endocytosis and subcellular trafficking of the receptor is shown in Fig. 2.[28] For a detailed mechanism of internalization of the AT1R process, see the excellent review by Bian et al.[28]

In 1997, Sasamura et al. showed that after Ang II binding, both AT1A and AT1B were internalized,[29] while the AT2 receptor did not undergo internalization.[30,31] Further, the study has demonstrated that the PKC inhibitor staurosporine is unable to inhibit the internalization of AT1A and AT1B receptors.[29] After ligand binding to the receptor, it activates the initiation of phospholipase C as well as Ca^{2+}/PKC signaling through

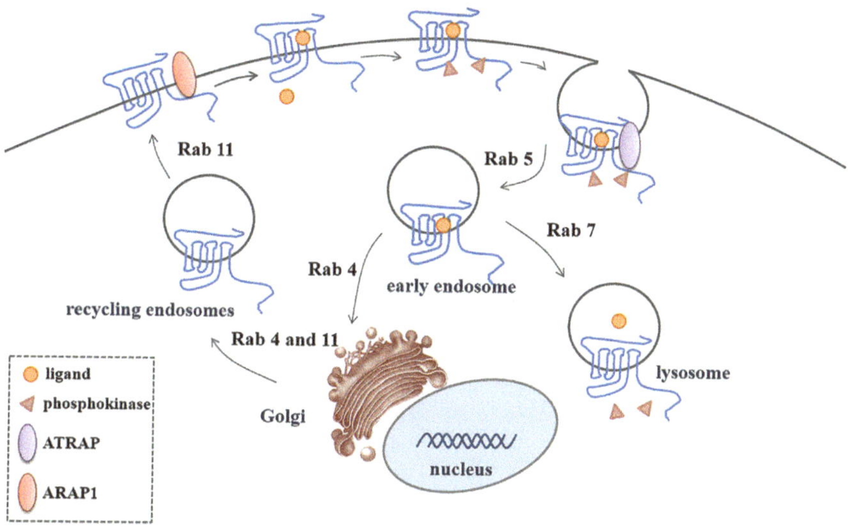

Fig. 2 Graphic illustration of the mechanism of reduced AT1 receptor internalization. Various GTP-binding proteins such as Rab5, Rab4, Rab11, and Rab7 are shown to be involved in receptor trafficking and metabolic degradation. In normal physiological conditions, after Ang II stimulation, AT1R is phosphorylated, and the ligand-receptor complex is trafficked through the early endosome, recycling endosome, and some receptors skipped the recycling endosome and trafficked to the lysosome for metabolic degradation. In AT1R internalization, AT1R-related protein (ATRAP) and AT1R-related protein 1 (ARAP1) are involved. *Figure adapted from Bian J, Zhang S, Yi M, Yue M, Liu H. The mechanisms behind decreased internalization of angiotensin II type 1 receptor.* Vascul Pharmacol. *2018;103–105:1–7 with permission.*

Gq/11 and small GTP-binding proteins help in additional signaling.[32,33] Ang II-AT1R internalization occurs as a complex with β-arrestin that roles to sterically detach the receptor from the heterotrimeric guanine-nucleotide binding protein (G protein)[34,35] and trafficked through early endosome and is not readily dephosphorylated.[36]

The co-immunoprecipitation method has been used to establish the interaction between β-arrestin1 and the AT1A receptor. It was found that interaction was agonist as well as time-dependent, and accountable upon the level of β-arrestin1 expression. However, after serial truncation of the receptor CT, the β-arrestin1 interaction was lost and diminished receptor phosphorylation. A study suggests that the CT of the receptor is vital for its internalization and interaction with β-arrestin. Moreover, receptor phosphorylation is critical for β-arrestin interaction.[11] In the CT of the AT1, T332-S338 is a key position of Ang II-induced phosphorylation.

Earlier, Kule et al. used wild-type or mutant AT1 expressing HEK-293 and CHO-K1 and demonstrated that internalization occurred in both cells through clathrin-mediated using β-arrestin. However, alanine substitution at different positions such as T332, S335, T336, and S338 prevented phosphorylation and significantly reduced AT1 internalization.[37]

A "Flag" sequence was tagged with the N-terminus of rat AT1 receptor (AT1R-NF) to stably expressed in the CHO cells. After staining with fluorescein-labeled anti-Flag monoclonal antibody, it has been analyzed through confocal microscopy. After ligand stimulation rapid internalization of AT1R-NF was observed. However, fluorescence was detected around the nucleus after 60–90 min treatment. It suggests that ligand-receptor complexes migrated to the nuclear membrane.[38] To understand the effect of Ang II and a series of substituted analogues, the enhanced green fluorescent protein (eGFP)-tagged AT1A receptor (AT1A-eGFP) fusion protein was used. Confocal microscopy is used for the visualization and quantification of chimeric receptors. A study demonstrated that nonpeptide AT1 receptor antagonist EXP3174 inhibited receptor internalization. However, after the substitution of Ang II side chains Tyr(4) and Phe(8), the internalization of the receptor was not affected but it abolished phospholipase C signaling.[39] This study suggests that specific residue of Ang II peptide involves in the stimulation of phospholipase C-mediated signaling.

To test the hypothesis of downregulation of AT1 receptor, which occurs independently of endocytosis and or G protein coupling (Gq), deleted CT of AT1A receptor has shown <5% internalized radioligand as compared to 65% for wild-type receptors. However, the ability to downregulation was not affected by truncated AT1 receptors. The study has demonstrated that endocytosis is not required for receptor degradation and get used as an alternative pathway. Further, a point mutation in the second intracellular loop was shown to impair Gq coupling. This study has also shown that after Ang II exposures, downregulation of receptors undergoes biphasic patterns such as the earliest phase (0 – 2 h) and the later phase (2–24 h).[40] Result suggests that G(q) coupling is required for the second phase of downregulation of the receptor.

3. Role of clathrin adaptor proteins in internalization of AT1R

In AT1 receptor internalization, clathrin-coated vesicles are involved in two ways; first, ligand-receptor complex internalization mediated from the cell surface and second, transports newly-synthesized receptor proteins from the *trans*-Golgi network. In both conditions, clathrin-coated vesicles contain various adaptor complexes such as HA1(AP1) and HA2 (AP2).[41,42] At physiological Ang II concentration, ligand-stimulated internalization of AT1R is β-arrestin- and dynamin-dependent internalization through clathrin-coated vesicles.[11,43,44]

Ang II stimulates cyclooxygenase2 protein expression in rat aortic vascular smooth muscle cells (RASMC). Earlier, it has been shown that it is mediated by NF-κB activation and β-arrestin-dependent receptor internalization.[45] Further, Morinelli et al. used pitstop-2 as an inhibitor to study the NF-κB activation following cyclooxygenase2 expression to be facilitated by the clathrin-dependent internalization of the activated AT1A receptor. To study the effects of pitstop-2 on Ang II binding and signaling in RASMC, various techniques have been utilized such as radioligand binding assay, immunoblotting and qRT-PCR.[46] Results have shown that the pitstop-2 significantly inhibited AT1A receptor internalization.

Moreover, various studies have utilized inhibitors and gene interventions to determine the role of clathrin in the endocytosis of the AT1 receptor.[47] For example, there is a various inhibitor which has used to inhibit clathrin-dependent endocytosis such as Chlorpromazine,[48] Phenothiazine,[49] Barbadin,[50] Eps15 mutant,[51] AP180C,[52] Dynasore,[53] Dynamin mutant, Dyn K44a,[54] Dynoles, dyngoes,[55] etc. These inhibitors have been used to inhibit the coated pit formation, invagination, and coated vesicle pinching off, consequently diminishing endocytosis, thus affecting AT1R internalization.[47,50,56] In addition, the caveolae-based endocytosis pathway involves in AT1R endocytosis.[57]

4. Subcellular trafficking of AT1 receptor

After binding Ang II to AT1, instant desensitization of the intracellular signal transduction occurred and reduced the number of cell surface receptors. Hein et al. have used immunofluorescence microscopy to demonstrate the subcellular trafficking of ligand-receptor complexes (Ang II, AT1A and AT2 receptors) in HEK-293 cells. Cells stably expressed the receptor subtypes. The study has shown that fluorescently labeled Ang II and AT1A are rapidly internalized, and trafficked into the endosome. However, after the agonist treatment, AT2 receptors were not internalized and localized in the plasma membrane. After the removal of Ang II, the AT1A receptor was recycled back on the plasma membrane, and Ang II was trafficked to the lysosome for metabolic degradation.[30] A radioligand binding assay-based study has demonstrated that after [^{3}H]Ang II binding with AT1 receptors in rat myometrial cells, ligand-receptor complex internalized and receptor recycled on the cell surface. They have treated the cells with chloroquine and monensin to understand the endosomal receptor trafficking.[58] In another study, eGFP-tagged with the CT of the receptor was used to visualize the AT1 internalization, subcellular trafficking, and interaction with the G protein. The chimeric construct of eGFP-AT1 transfected in HEK-293 cells. Confocal microscopy and immunoblotting methods were used to the study endocytosis of eGFP-AT1. The study has shown that peptide agonists induce the eGFP-AT1 internalization but not the nonpeptide ligands and recruit the endogenous Galphaq/11 protein to the membrane.[59]

Similarly, another study has used the GFP- or hemagglutinin epitope-tagged AT1 to demonstrate the internalization and trafficking of the receptor. After Ang II stimulation, ligand-receptor trafficked through the early endosome and recycling endosome. After treatment with wortmannin (WT) or LY294002 to inhibit the phosphatidylinositol (PI) 3-kinase, which caused the creation of bulky endosomal vesicles of varied Rab composition and included the Ang II-AT1 complex in their limiting membranes and minute associated vesicular structures.[60] The study indicates that after ligand binding, receptors are processed through multivesicular bodies and some receptors to the cell surface by a rapid PI 3-kinase-dependent recycling route and by a slower pathway, which was less sensitive to PI 3-kinase inhibitors.

Further, the study has shown that upon Ang II stimulation, all AT1R variants at the plasma membrane are made complexes with β-arrestin2 and internalize into endosomes. But mutant receptors have shown diverse conformations in β-arrestin2 and differentially affected mitogen-activated protein kinase (MAPK) and signaling endosomal trafficking.[61] Ferrão et al. have demonstrated the nonconventional subcellular trafficking of Ang II-AT1R/AT2R complex to the endoplasmic reticulum (ER).[62] Finding suggests that it may activate intracellular Ca^{2+} responses in ER.

5. Role of a short amino acid motif in the internalization and trafficking of AT1R

Site-directed mutagenesis (SDM) studies have shown that two regions of the cytoplasmic tail of the AT1 receptor are required for internalization, although it is an independent of G protein stimulation.[63] For the G protein coupling, the CT of the third intracellular loop of the AT1 is important.

Various studies have shown the roles of short sequence motifs in internalization, trafficking, and signaling. In addition, different proteins interact with the CT of the AT1 receptor, and these studies help to understand their role in the receptor internalization and signaling of GPCRs.[31] After Ang II binds with the AT1 receptor, rapid internalization and down-regulation occur. Serial deletions have been made in the CT of the AT1A receptor to examine the role of structural determinants in these events. Results have shown that 22 amino acids of the CT were not involved in its internalization. Remarkably, after the removal of one more amino acid residue (Leu 337), it has shown a significant reduction in receptor internalization. In addition, a 95% reduction was observed in AT1A internalization after the removal of Ser335 and Thr336.[41] The study suggested that the presence of a serine/threonine-rich region and Leu337 in the CT of the AT1 receptor is crucial for ligand-receptor complex internalization.

A deletion mutation study has shown that Ang II-induced AT1A internalization significantly reduced after 45 amino acids removal from the CT. To delineate the role of specific amino acids in this process, Thomas et al. have stably and transiently transfected wild-type and CT mutated in CHO cells. After Ang II stimulation, wild-type receptors were rapidly internalized. However, using AT1A mutated, truncated, and deleted at the CT receptors, two different regions (one membrane- distal to Lys333, within the terminal 26 amino acids and another proximal site between residues 315–329) were identified which play important role in receptor internalization. A different point mutation has been made at a different position. Further, L316F and Y319A mutants have shown a significant reduction (2.5-fold) in internalization compared to wild-type.[64] Another study has shown that deletion mutation of cytoplasmic AT1R affected receptor internalization. Becker et al. have shown that peptide sequences within the CT of the AT1 receptor play a significant role in the internalization and endosomal trafficking.[65]

A study has demonstrated that the mutant AT1 receptor undergoes normal internalization but could not activate any G-protein. In the highly conserved DRY motif of the AT1 receptor, substitution mutations of Asp 125 with an ala and Arg126 with leu have been made that were unable to stimulate G-proteins.[66] Esseltine et al. have shown that different Rab proteins such as Rab4, Rab7, and Rab11 bind to the CT of the AT1 receptor. However, Rab11 binds with receptors more effectively than Rab4 and Rab5. Whereas Rab5 binds more effectively as compared to Rab4. Further, alanine substitution analysis has demonstrated that proline 354 and cysteine 355 involve Rab protein binding, but G protein coupling was unaffected. Moreover, study has shown that increased Rab4-activity may play an important role in receptor dephosphorylation and desensitization.[67]

The NPXnY motif plays a crucial role in the internalization process of the various receptor, such as LDL receptors,[68] β2-adrenergic receptors,[69] integrin receptors,[70] and GPCRs. To know the function of a particular amino acid residue, Laporte et al. mutated Tyr302 with alanine or phenylalanine in the NPLFY site of the human AT1 receptor. To understand its role in internalization and pharmacological properties, wild-type and mutated AT1 receptors were transfected in COS-7 cells and expressed about the same amount. ^{125}I-Ang II-induced both mutant and wild-type receptors, but the rate of internalization of the mutant receptor was significantly reduced compared to the wild-type receptor. In addition, transmembrane signaling was efficiently reduced in mutant expressing cells as compared to wild-type.[71] The outcome suggests that the Tyr302 residue in the NPXnY motif of human AT1 is not pivotal for ligand binding or internalization, nevertheless, it plays a significant role in signaling.

A mutational study has shown reduced Ang II-stimulated AT1 phosphorylation after alanine and or asparagine substitution mutation of Asp236-Asp237 in the receptor. However, it was not significantly reduced inositol phosphate production, MAPK, and AT1 receptor internalization and desensitization. Moreover, dominant-negative G protein-coupled receptor kinase 2 K220M overexpression reduced Ang II-stimulated receptor phosphorylation by about 40%, however, it did not further decrease the impaired phosphorylation of diacidic mutants receptors. In addition, 30% phosphorylation was reduced after inhibiting PKC by bisindolylmaleimide in both wild-type and mutant receptors.[72]

Another mutational study has demonstrated that the NPLFY sequence motif located in AT1 significantly impaired the interaction of G proteins with receptors after the alanine substitution of Asn298. In addition, the alanine substitution of Pro299 or Tyr302 reduced the inositol phosphate response. However, the F301 mutation was not much affected. Results indicated that the NPLFY motif does not play an important role in Ang II-induced AT1 receptor internalization. Although, Phe301 and Asn298 are essential for agonist binding and important for stimulation of receptor and signal transduction.[73] In cultured rat proximal tubule cells, it has been shown that ligand was coupled to phospholipase C (PLC) through the AT1 receptor. It has also shown that the cytoskeleton-dependent internalization is required for apical Ang II AT1-mediated PLC stimulation.[74] A mutational study has shown that a polyaromatic caveolin-binding-like motif (phiXphiXXXXphiXXphi where phi is an aromatic residue) in the cytoplasmic tail of the AT1R plays an important role in receptor trafficking and signaling.[75] In addition, an aromatic amino acid residue (GDAY and FQQI) is also critical for receptor internalization, trafficking and signaling in natriuretic peptide receptor-A (NPRA).[76–78]

6. Signaling of AT1 receptor

AT1 activates epidermal growth factors (EGFRs) and tyrosine kinases and causes vascular growth, while AT2 leads to apoptosis in extreme situations by activating dephosphorylating enzymes.[23] AT1A and AT1B are joined to phospholipase C by Gq/G11 proteins and a calcium channel and negatively joined to adenyl cyclase. Irregularities in the signaling of AT1 cause various diseases such as aging-related diseases, cardiovascular diseases, and albuminuria.[18,19] A schematic representation of abnormal signaling in pathophysiological conditions is shown in Fig. 3.[79]

A mutational study was performed to understand the role of individual amino acid residue in ligand-induced stimulation of G(q/11)-dependent phosphoinositide hydrolysis. No significant effect was observed on signal transduction after alanine substitutions in positions 232–240 of the third loop. But deletion mutation at Ile (238) efficiently reduced Ang II-stimulated inositol phosphate response.[80] Another mutational study has demonstrated a mildly reduced Ang II-induced inositol phosphate signaling. Reduced signaling was observed after mutation in a conserved domain of rat AT1A receptor. The researcher has substituted alanine at positions 125 (Asp) and 126 (Arg). However, concurrent substitution at 125 (Asp) and 126 (Arg) has shown a significant decrease in both receptor internalization and inositol phosphate signaling.[81]

To understand the differential behavior of both the receptors AT1 and AT2, Hunyady et al. created a chimeric angiotensin receptor by exchanging the two different intracellular loops (IL2 and IL3) and/or the cytoplasmic tail between these receptors. Minor alternations were observed in the ligand binding properties after transient transfection in the COS-7 cells. Further, a study indicated that for normal receptor internalization, cytoplasmic tail was necessary, however, IL2 was a determinant of G protein activation. Moreover, IL3 (N-terminus) was required for the functions of both receptors. But the substitution of IL2 reduced Ang II-activated extracellular signal-regulated kinase (ERK) stimulation in AT1 receptors. Moreover, the study has demonstrated that all these substitutions from AT1 to AT2

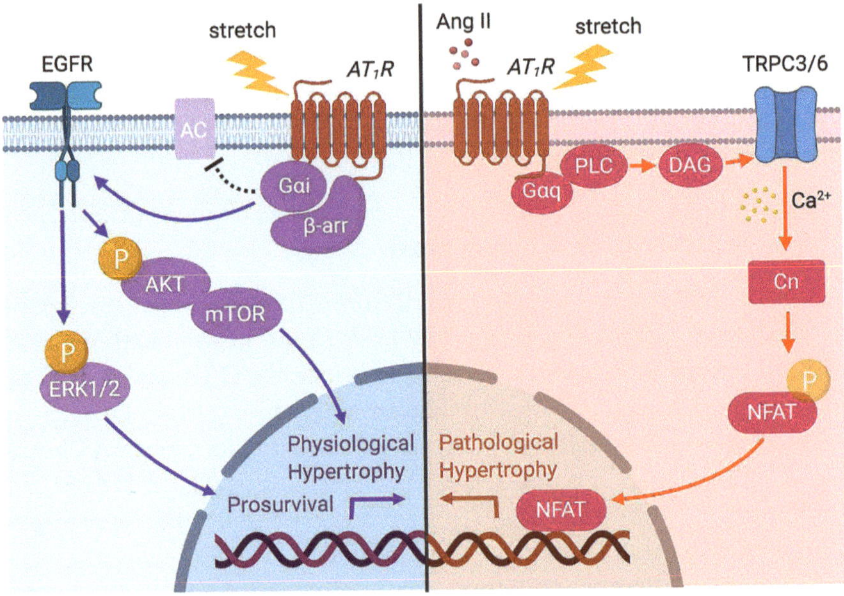

Fig. 3 Prolonged mechanical load and Ang II activate the AT1R signaling pathway. When the myocyte is subjected to load, the receptor is mainly coupled to Ga*q* protein nevertheless also coupled to Ga*i* through β-arrestin engagement. Ang II-or stretch-stimulated Ga*q* signaling activates TRPC3/6 to initiate Ca^{2+} influx, which in turn stimulates the calcineurin-NFAT pathway to trigger pathological cardiac hypertrophy (red arrows). AT1R-Ga*i* coupling stimulates Akt/ERK phosphorylation and EGFR transactivation that supports physiological hypertrophy and prosurvival signaling (blue arrows). PLC, phospholipase C; DAG, diacylglycerol; β-arr, β-arrestin; TRPC3/6, transient receptor potential canonical channel 3/6; mTOR, mammalian/mechanistic target of rapamycin; CN, calcineurin; NFAT, nuclear factor of activated T cells; ERK1/2, the extracellular signal-regulated kinase ½; EGFR, epidermal growth factor. *Figure adapted from Seo K, Parikh VN, Ashley EA. Stretch-induced biased signaling in angiotensin II type 1 and apelin receptors for the mediation of cardiac contractility and hypertrophy.* Front Physiol. *2020;11:181 with permission.*

receptors did not allow the normal receptor internalization and inositol phosphate signaling.[82] However, the result suggests that due to the deficiency of receptor endocytosis and signal generation by the AT2 receptor, it might use another activation mechanism.

7. Concluding remarks and future perspectives

After ligand binding with the AT1 and AT2 receptors, the AT1 receptors are rapidly internalized and recycled back into the plasma membrane.

Some receptors skip recycling and are trafficked to the lysosome for degradation. However, AT2 receptors are not internalized. AT1 receptor internalization and trafficking are dynamic events and continue even after desensitized inositol phosphate and Ca^{2+} signaling pathways. Various proteins bind with the cytoplasmic tail of the receptors, which activates signaling. Many short sequence motifs localized in the CT of the AT1 receptor play a significant role in receptor internalization, trafficking, and signaling. CRISPR-CAS system can use to understand the endocytosis and signaling pathway of the AT1 and AT2 receptors. Further studies are required, such as live-cell imaging and *in vivo* study of receptor endocytosis and signaling, which will help to increase our knowledge about pathophysiological conditions such as in cardio-renal and other diseases.

Conflict of interest

None.

References

1. Iwai N, Inagami T. Identification of two subtypes in the rat type I angiotensin II receptor. *FEBS Lett.* 1992;298(2–3):257–260.
2. Yoshida H, Kakuchi J, Guo DF, et al. Analysis of the evolution of angiotensin II type 1 receptor gene in mammals (mouse, rat, bovine and human). *Biochem Biophys Res Commun.* 1992;186(2):1042–1049.
3. Szpirer C, Riviere M, Szpirer J, et al. Chromosomal assignment of human and rat hypertension candidate genes: type 1 angiotensin II receptor genes and SA gene. *J Hypertens.* 1999;11:919–925.
4. Lewis JL, Serikawa T, Warnock DG. Chromosomal localization of angiotensin II type 1 receptor isoforms in the rat. *Biochem Biophys Res Commun.* 1993;194:667–682.
5. Nakajima M, Mukoyama M, Pratt RE, Horiuchi M, Dzau VJ. Cloning of cDNA and analysis of the gene for mouse angiotensin II type 2 receptor. *Biochem Biophys Res Commun.* 1993;197(2):393–399.
6. Curnow KM, Pascoe L, White PC. Genetic analysis of the human type-1 angiotensin II receptor. *Mol Endocrinol.* 1992;6(7):1113–1118.
7. Murphy TJ, Alexander RW, Griendling KK, Runge MS, Bernstein KE. Isolation of a cDNA encoding the vascular type-1 angiotensin II receptor. *Nature.* 1991;351 (6323):233–236.
8. Sandberg K, Ji H, Clark AJ, Shapira H, Catt KJ. Cloning and expression of a novel angiotensin II receptor subtype. *J Biol Chem.* 1992;267(14):9455–9458.
9. Bergsma DJ, Ellis C, Kumar C, et al. Cloning and characterization of a human angiotensin II type 1 receptor. *Biochem Biophys Res Commun.* 1992;183(3):989–995.
10. Smith RD, Hunyady L, Olivares-Reyes JA, Mihalik B, Jayadev S, Catt KJ. Agonist-induced phosphorylation of the angiotensin AT1a receptor is localized to a serine/threonine-rich region of its cytoplasmic tail. *Mol Pharmacol.* 1998;54(6):935–941.
11. Qian H, Pipolo L, Thomas WG. Association of beta-Arrestin 1 with the type 1A angiotensin II receptor involves phosphorylation of the receptor carboxyl terminus and correlates with receptor internalization. *Mol Endocrinol.* 2001;15(10):1706–1719.

12. Oppermann M, Freedman NJ, Alexander RW, Lefkowitz RJ. Phosphorylation of the type 1A angiotensin II receptor by G protein-coupled receptor kinases and protein kinase C. *J Biol Chem.* 1996;271(22):13266–13272.
13. Chen K, Fu C, Chen C, et al. Role of GRK4 in the regulation of arterial AT1 receptor in hypertension. *Hypertension.* 2014;63(2):289–296.
14. Haack KK, Engler CW, Papoutsi E, Pipinos II, Patel KP, Zucker IH. Parallel changes in neuronal AT1R and GRK5 expression following exercise training in heart failure. *Hypertension.* 2012;60(2):354–361.
15. García-Caballero A, Olivares-Reyes JA, Catt KJ, García-Saínz JA. Angiotensin AT(1) receptor phosphorylation and desensitization in a hepatic cell line. Roles of protein kinase c and phosphoinositide 3-kinase. *Mol Pharmacol.* 2001;59(3):576–585.
16. Berk BC, Corson MA. Angiotensin II signal transduction in vascular smooth muscle: role of tyrosine kinases. *Circ Res.* 1997;80(5):607–616.
17. Inagami T, Guo DF, Kitami Y. Molecular biology of angiotensin II receptors: an overview. *J Hypertens Suppl.* 1994;12(10):S83–S94.
18. Kamo T, Akazawa H, Komuro I. Pleiotropic effects of angiotensin II receptor signaling in cardiovascular homeostasis and aging. *Int Heart J.* 2015;56(3):249–254.
19. Königshausen E, Zierhut UM, Ruetze M, et al. Angiotensin II increases glomerular permeability by β-arrestin mediated nephrin endocytosis. *Sci Rep.* 2016;6:39513.
20. Matsusaka T, Ichikawa I. Biological functions of angiotensin and its receptors. *Annu Rev Physiol.* 1997;59:395–412.
21. Allen AM, Zhuo J, Mendelsohn FA. Localization and function of angiotensin AT1 receptors. *Am J Hypertens.* 2000;13(1 pt 2):31S–38S.
22. Li XC, Zhuo JL. In vivo regulation of AT1a receptor-mediated intracellular uptake of [^{125}I]Val5-ANG II in the kidneys and adrenals of AT1a receptor-deficient mice. *Am J Physiol Renal Physiol.* 2008;294(2):F293–F302.
23. Inagami T. Molecular biology and signaling of angiotensin receptors: an overview. *J Am Soc Nephrol Suppl.* 1999;11:S2–S7.
24. Lei J, Li Y, Zhang S, Wu Y, Wang P, Liu H. The prognostic role of angiotensin II type 1 receptor autoantibody in non-gravid hypertension and pre-eclampsia: a meta-analysis and our studies. *Medicine (Baltimore).* 2016;95(17):e3494.
25. Goodman Jr OB, Krupnick JG, Santini F, et al. Beta-arrestin acts as a clathrin adaptor in endocytosis of the beta2-adrenergic receptor. *Nature.* 1996;383(6599):447–450.
26. Irannejad R, von Zastrow M. GPCR signaling along the endocytic pathway. *Curr Opin Cell Biol.* 2014;27:109–116.
27. Thomas WG, Thekkumkara TJ, Baker KM. Molecular mechanisms of angiotensin II (AT1A) receptor endocytosis. *Clin Exp Pharmacol Physiol Suppl.* 1996;3:S74–S80.
28. Bian J, Zhang S, Yi M, Yue M, Liu H. The mechanisms behind decreased internalization of angiotensin II type 1 receptor. *Vascul Pharmacol.* 2018;103–105:1–7.
29. Sasamura H, Hein L, Saruta T, Pratt RE. Evidence for internalization of both type 1 angiotensin receptor subtypes (AT1a, AT1b) by a protein kinase C independent mechanism. *Hypertens Res.* 1997;20(4):295–300.
30. Hein L, Meinel L, Pratt RE, Dzau VJ, Kobilka BK. Intracellular trafficking of angiotensin II and its AT1 and AT2 receptors: evidence for selective sorting of receptor and ligand. *Mol Endocrinol.* 1997;11(9):1266–1277.
31. Guo DF, Sun YL, Hamet P, Inagami T. The angiotensin II type 1 receptor and receptor-associated proteins. *Cell Res.* 2001;11(3):165–180.
32. Hunyady L, Balla T, Catt KJ. The ligand binding site of the angiotensin AT1 receptor. *Trends Pharmacol Sci.* 1996;17(4):135–140.
33. de Gasparo M, Catt KJ, Inagami T, Wright JW, Unger T. International union of pharmacology. XXIII. The angiotensin II receptors. *Pharmacol Rev.* 2000;52(3):415–472.

34. Ferguson SS, Caron MG. Green fluorescent protein-tagged beta-arrestin translocation as a measure of G protein-coupled receptor activation. *Methods Mol Biol.* 2004; 237:121–126.
35. Krupnick JG, Benovic JL. The role of receptor kinases and arrestins in G protein-coupled receptor regulation. *Annu Rev Pharmacol Toxicol.* 1998;38:289–319.
36. Anborgh PH, Seachrist JL, Dale LB, Ferguson SS. Receptor/beta-arrestin complex formation and the differential trafficking and resensitization of beta2-adrenergic and angiotensin II type 1A receptors. *Mol Endocrinol.* 2000;14(12):2040–2053.
37. Kule CE, Karoor V, Day JN, et al. Agonist-dependent internalization of the angiotensin II type one receptor (AT1): role of C-terminus phosphorylation in recruitment of beta-arrestins. *Regul Pept.* 2004;120(1–3):141–148.
38. Merjan AJ, Kanashiro CA, Krieger JE, Han SW, Paiva AC. Ligand-induced endocytosis and nuclear localization of angiotensin II receptors expressed in CHO cells. *Braz J Med Biol Res.* 2001;34(9):1175–1183.
39. Holloway AC, Qian H, Pipolo L, et al. Side-chain substitutions within angiotensin II reveal different requirements for signaling, internalization, and phosphorylation of type 1A angiotensin receptors. *Mol Pharmacol.* 2002;61(4):768–777.
40. Modrall JG, Nanamori M, Sadoshima J, Barnhart DC, Stanley JC, Neubig RR. ANG II type 1 receptor downregulation does not require receptor endocytosis or G protein coupling. *Am J Physiol Cell Physiol.* 2001;281(3):C801–C809.
41. Hunyady L, Baukal AJ, Balla T, Catt KJ. Independence of type I angiotensin II receptor endocytosis from G protein coupling and signal transduction. *J Biol Chem.* 1994;269 (40):24798–24804.
42. Sorkin A, Carpenter G. Interaction of activated EGF receptors with coated pit adaptins. *Science.* 1996;261:612–615.
43. Gáborik Z, Szaszák M, Szidonya L, et al. Beta-arrestin- and dynamin-dependent endocytosis of the AT1 angiotensin receptor. *Mol Pharmacol.* 2001;59(2):239–247.
44. Kohout TA, Lin FS, Perry SJ, Conner DA, Lefkowitz RJ. beta-Arrestin 1 and 2 differentially regulate heptahelical receptor signaling and trafficking. *Proc Natl Acad Sci U S A.* 2001;98(4):1601–1606.
45. Morinelli TA, Kendall RT, Luttrell LM, Walker LP, Ullian ME. Angiotensin II-induced cyclooxygenase 2 expression in rat aorta vascular smooth muscle cells does not require heterotrimeric G protein activation. *J Pharmacol Exp Ther.* 2009;330(1):118–124.
46. Morinelli TA, Walker LP, Velez JC, Ullian ME. Clathrin-dependent internalization of the angiotensin II AT_1A receptor links receptor internalization to COX-2 protein expression in rat aortic vascular smooth muscle cells. *Eur J Pharmacol.* 2015;748:143–148.
47. Guo S, Zhang X, Zheng M, et al. Selectivity of commonly used inhibitors of clathrin-mediated and caveolae-dependent endocytosis of G protein-coupled receptors. *Biochim Biophys Acta.* 2015;1848(10 pt A):2101–2110.
48. Wang LH, Rothberg KG, Anderson RG. Mis-assembly of clathrin lattices on endosomes reveals a regulatory switch for coated pit formation. *J Cell Biol.* 1993;123(5): 1107–1117.
49. Kuratomi Y, Akiyama S, Ono M, et al. Thioridazine enhances lysosomal accumulation of epidermal growth factor and toxicity of conjugates of epidermal growth factor with Pseudomonas exotoxin. *Exp Cell Res.* 1986;162(2):436–448.
50. Beautrait A, Paradis JS, Zimmerman B, et al. A new inhibitor of the β-arrestin/AP2 endocytic complex reveals interplay between GPCR internalization and signalling. *Nat Commun.* 2017;8:15054.
51. Benmerah A, Lamaze C, Bègue B, Schmid SL, Dautry-Varsat A, Cerf-Bensussan N. AP-2/Eps15 interaction is required for receptor-mediated endocytosis. *J Cell Biol.* 1998;140(5):1055–1062.

52. Zhao X, Greener T, Al-Hasani H, Cushman SW, Eisenberg E, Greene LE. Expression of auxilin or AP180 inhibits endocytosis by mislocalizing clathrin: evidence for formation of nascent pits containing AP1 or AP2 but not clathrin. *J Cell Sci*. 2001;114(pt 2): 353–365.
53. Macia E, Ehrlich M, Massol R, Boucrot E, Brunner C, Kirchhausen T. Dynasore, a cell-permeable inhibitor of dynamin. *Dev Cell*. 2006;10(6):839–850.
54. van der Bliek AM, Redelmeier TE, Damke H, Tisdale EJ, Meyerowitz EM, Schmid SL. Mutations in human dynamin block an intermediate stage in coated vesicle formation. *J Cell Biol*. 1993;122(3):553–563.
55. Hill TA, Gordon CP, McGeachie AB, et al. Inhibition of dynamin mediated endocytosis by the dynoles-synthesis and functional activity of a family of indoles. *J Med Chem*. 2009;52(12):3762–3773.
56. Dutta D, Donaldson JG. Search for inhibitors of endocytosis: intended specificity and unintended consequences. *Cell Logist*. 2012;2(4):203–208.
57. Czikora I, Feher A, Lucas R, Fulton DJ, Bagi Z. Caveolin-1 prevents sustained angiotensin II-induced resistance artery constriction and obesity-induced high blood pressure. *Am J Physiol Heart Circ Physiol*. 2015;308(5):H376–H385.
58. Lázari MF, Porto CS, Freymüller E, Abreu LC, Picarelli ZP. Receptor-mediated endocytosis of angiotensin II in rat myometrial cells. *Biochem Pharmacol*. 1997;54(3):399–408.
59. Miserey-Lenkei S, Lenkei Z, Parnot C, Corvol P, Clauser E. A functional enhanced green fluorescent protein (EGFP)-tagged angiotensin II at(1a) receptor recruits the endogenous Galphaq/11 protein to the membrane and induces its specific internalization independently of receptor-g protein coupling in HEK-293 cells. *Mol Endocrinol*. 2001;15 (2):294–307.
60. Hunyady L, Baukal AJ, Gaborik Z, et al. Differential PI 3-kinase dependence of early and late phases of recycling of the internalized AT1 angiotensin receptor. *J Cell Biol*. 2002;157(7):1211–1222.
61. Cao Y, Kumar S, Namkung Y, Gagnon L, Cho A, Laporte SA. Angiotensin II type 1 receptor variants alter endosomal receptor-β-arrestin complex stability and MAPK activation. *J Biol Chem*. 2020;295(38):13169–13180.
62. Ferrão FM, Cardoso LHD, Drummond HA, et al. Luminal ANG II is internalized as a complex with AT_1R/AT_2R heterodimers to target endoplasmic reticulum in LLC-PK_1 cells. *Am J Physiol Renal Physiol*. 2017;313(2):F440–F449.
63. Hunyady L. Molecular mechanisms of angiotensin II receptor internalization. *J Am Soc Nephrol Suppl*. 1999;11:S47–S56.
64. Thomas WG, Baker KM, Motel TJ, Thekkumkara TJ. Angiotensin II receptor endocytosis involves two distinct regions of the cytoplasmic tail. A role for residues on the hydrophobic face of a putative amphipathic helix. *J Biol Chem*. 1995;270 (38):22153–22159.
65. Becker BN, Cheng HF, Hammond TG, Harris RC. The type 1 angiotensin II receptor tail affects receptor targeting, internalization, and membrane fusion properties. *Mol Pharmacol*. 2004;65(2):362–369.
66. Feng YH, Ding Y, Ren S, Zhou L, Xu C, Karnik SS. Unconventional homologous internalization of the angiotensin II type-1 receptor induced by G-protein-independent signals. *Hypertension*. 2005;46(2):419–425.
67. Esseltine JL, Dale LB, Ferguson SS. Rab GTPases bind at a common site within the angiotensin II type I receptor carboxyl-terminal tail: evidence that Rab4 regulates receptor phosphorylation, desensitization, and resensitization. *Mol Pharmacol*. 2011;79 (1):175–184.
68. Bansal A, Gierasch LM. The NPXY internalization signal of the LDL receptor adopts a reverse-turn conformation. *Cell*. 1991;67(6):1195–1201.

69. Barak LS, Tiberi M, Freedman NJ, Kwatra MM, Lefkowitz RJ, Caron MG. A highly conserved tyrosine residue in G protein-coupled receptors is required for agonist-mediated beta 2-adrenergic receptor sequestration. *J Biol Chem.* 1994;269(4): 2790–2795.
70. Bogdanović O, Delfino-Machín M, Nicolás-Pérez M, et al. Numb/Numbl-Opo antagonism controls retinal epithelium morphogenesis by regulating integrin endocytosis. *Dev Cell.* 2012;23(4):782–795.
71. Laporte SA, Servant G, Richard DE, Escher E, Guillemette G, Leduc R. The tyrosine within the NPXnY motif of the human angiotensin II type 1 receptor is involved in mediating signal transduction but is not essential for internalization. *Mol Pharmacol.* 1996;49(1):89–95.
72. Olivares-Reyes JA, Smith RD, Hunyady L, Shah BH, Catt KJ. Agonist-induced signaling, desensitization, and internalization of a phosphorylation-deficient AT1A angiotensin receptor. *J Biol Chem.* 2001;276(41):37761–37768.
73. Hunyady L, Bor M, Baukal AJ, Balla T, Catt KJ. A conserved NPLFY sequence contributes to agonist binding and signal transduction but is not an internalization signal for the type 1 angiotensin II receptor. *J Biol Chem.* 1995;270(28):16602–16609.
74. Schelling JR, Hanson AS, Marzec R, Linas SL. Cytoskeleton-dependent endocytosis is required for apical type 1 angiotensin II receptor-mediated phospholipase C activation in cultured rat proximal tubule cells. *J Clin Invest.* 1992;90(6):2472–2480.
75. Leclerc PC, Auger-Messier M, Lanctot PM, Escher E, Leduc R, Guillemette G. A polyaromatic caveolin-binding-like motif in the cytoplasmic tail of the type 1 receptor for angiotensin II plays an important role in receptor trafficking and signaling. *Endocrinology.* 2002;143(12):4702–4710.
76. Pandey KN, Nguyen HT, Garg R, Khurana ML, Fink J. Internalization and trafficking of guanylyl (guanylate) cyclase/natriuretic peptide receptor A is regulated by an acidic tyrosine-based cytoplasmic motif GDAY. *Biochem J.* 2005;388(pt 1):103–113.
77. Mani I, Garg R, Pandey KN. Role of FQQI motif in the internalization, trafficking, and signaling of guanylyl-cyclase/natriuretic peptide receptor-A in cultured murine mesangial cells. *Am J Physiol Renal Physiol.* 2016;310(1):F68–F84.
78. Mani I, Pandey KN. Emerging concepts of receptor endocytosis and concurrent intracellular signaling: mechanisms of guanylyl cyclase/natriuretic peptide receptor-A activation and trafficking. *Cell Signal.* 2019;60:17–30.
79. Seo K, Parikh VN, Ashley EA. Stretch-induced biased signaling in angiotensin II type 1 and apelin receptors for the mediation of cardiac contractility and hypertrophy. *Front Physiol.* 2020;11:181.
80. Zhang M, Zhao X, Chen HC, Catt KJ, Hunyady L. Activation of the AT1 angiotensin receptor is dependent on adjacent apolar residues in the carboxyl terminus of the third cytoplasmic loop. *J Biol Chem.* 2000;275(21):15782–15788.
81. Gáborik Z, Jagadeesh G, Zhang M, Spät A, Catt KJ, Hunyady L. The role of a conserved region of the second intracellular loop in AT1 angiotensin receptor activation and signaling. *Endocrinology.* 2003;144(6):2220–2228.
82. Hunyady L, Gáborik Z, Shah BH, Jagadeesh G, Clark AJ, Catt KJ. Structural determinants of agonist-induced signaling and regulation of the angiotensin AT1 receptor. *Mol Cell Endocrinol.* 2004;217(1–2):89–100.

CHAPTER EIGHT

B cell receptor (BCR) endocytosis

Aleah D. Roberts and Justin W. Taraska*
Biochemistry and Biophysics Center, National Heart, Lung, and Blood Institute, National Institutes of Health, Bethesda, MD, United States
*Corresponding author: e-mail address: justin.taraska@nih.gov

Contents

Abstract

The B cell receptor (BCR) interacts with foreign antigens to mediate B cell activation and secretion of antibodies. B cell activation begins with initiation of signaling pathways, such as NFAT, NF-κB, and MAPK, and endocytosis of the BCR-antigen complex. Many studies have investigated the signaling pathways associated with BCR activation, and this work has led to significant advances in drug therapies to treat cancer and autoimmune diseases that are linked to aberrant BCR signaling. Less is known, however, about the mechanisms of BCR endocytosis and the role endocytosis plays in B cell pathogenesis. This chapter will review key characteristics of the BCR, including a review of signaling pathways, and endocytic mechanisms associated with the activated BCR. We will also review recent studies investigating the role of BCR endocytosis disease pathogenesis.

Progress in Molecular Biology and Translational Science, Volume 194
ISSN 1877-1173
https://doi.org/10.1016/bs.pmbts.2022.08.003

1. Introduction to the B cell receptor

B cells are a part of the adaptive immune response that generates antigen specific antibodies. Antibodies (also known as immunoglobulins) are proteins that can bind pathogens directly to prevent infection, or bind infected cells to facilitate pathogen clearance by phagocytosis. The B cell receptor (BCR) covers the surface of mature B cells, and initiates cellular responses that lead to B cell activation and antibody production. Each B cell expresses a unique B cell receptor with specificity for a unique antigen. The BCR can bind its cognate antigen encountered directly or presented on the surface of an antigen presenting cell (APC). In either case, antigen engagement leads to BCR crosslinking and triggers activation of signaling pathways and receptor-mediated endocytosis of the BCR-antigen complex. The antigen is then degraded and inserted into major histocompatibility complex class II (MHCII) proteins that are trafficked to the plasma membrane. At the membrane, MHCII molecules present antigen to CD4+ helper T cells, which stimulate the B cell to proliferate and ultimately differentiate into an antibody producing plasma cell. In some cases, B cells may also initiate T-mediated immune responses that do not involve antibodies.

2. B cell receptor structure

Structurally, the BCR is composed of a membrane anchored immunoglobulin (mIg) that is paired with two coreceptors: CD79a and CD79b (also called Igα and Igβ; Fig. 1). CD79a and CD79b form a covalently linked heterodimer that interacts with the mIg through non-covalent interactions (charged residues) on the extracellular domains of both proteins.[1] The mIg contains the receptor's antigen interaction domain, while CD79a and CD79b mediate transduction of downstream signaling and endocytosis of the receptor-antigen complex. The extracellular portion of the mIg consists of the same basic structure as secreted antibodies: two heavy chain proteins and two light chain proteins (Fig. 1). The N-terminus region of the heavy and light chains together determine the antigen specificity of the BCR. This region is known as the variable region. Inside the variable region, there are three 5–7 amino acid sequence stretches that make up the "hypervariable regions."[2] These hypervariable regions are where somatic hypermutation occurs in activated B cells. Somatic hypermutation of the hypervariable regions allows the BCR to increase its affinity for a ligand, so that secreted antibodies are highly specific.

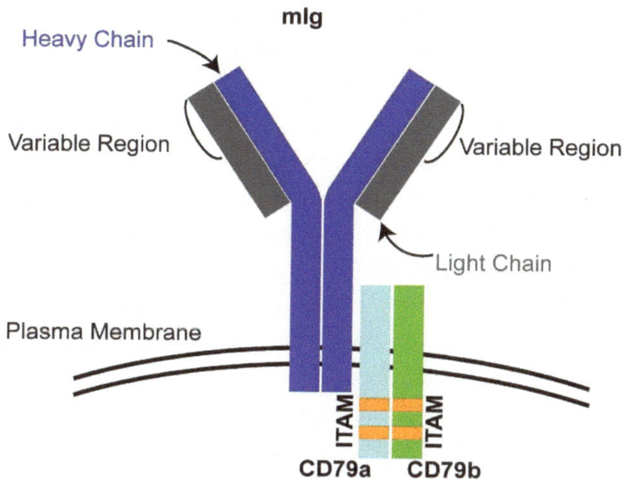

Fig. 1 Structure of the B cell receptor complex. The B cell receptor complex contains an antigen interacting domain (membrane immunoglobulin: mIg) that consists of two heavy chain and two light chain proteins. The N-terminus of the heavy and light chains contains a variable region that undergoes somatic hypermutation upon B cell activation. Central tyrosine residues in the ITAMs (immunoreceptor tyrosine-based activation motif) of CD79a/b are phosphorylated to induce receptor internalization and initiate downstream signaling pathways. The CD79a/b coreceptor is associated with all isotypes of the mIg on naïve and memory B cells.

There are multiple isotypes of the mIg on B cells. The extracellular domain of mature naïve B cell receptors may be mIgM or mIgD, while B cells that have been previously activated (memory B cells) express an irreversibly class switched mIgG. mIg isotypes vary in the number of covalent bonds between their heavy and light chain subunits.[2] While mIgM and mIgG have some differences in the structure of their extracellular domains, they function similarly in their ability to bind antigen and trigger BCR clustering.[3] The cytoplasmic tails of naïve (mIgM) and memory (mIgG) B cell receptors however, are quite different in length and function. The cytoplasmic domain of the naïve B cell receptor is only 3 amino acids long, while the cytoplasmic domain of mIgG is 28 amino acids. The longer tail sequence of mIgG is required for the increased proliferative capacity of memory B cells upon antigen stimulation.[4,5] The cytoplasmic regions of CD79a and CD79b also play an important role as phosphorylation substrates for kinases that transduce downstream signaling events and mediate endocytosis of the activated B cell Receptor.

3. B cell receptor signaling and regulation

The intracellular domains of CD79a and CD79b contain ITAM (immunoreceptor tyrosine-based activation motif) consensus sequences.

BCR aggregation triggers phosphorylation of specific tyrosines in the ITAM motifs of CD79a and CD79b by Src-family Protein Tyrosine Kinases (PTKs), including Fyn, Blk, Lyn, and Lck.[6,7] Association of Src-family PTKs with the BCR is further enhanced by lipid raft microdomains on the plasma membrane. These lipid raft domains contain increased concentrations of Src-family PTKs, which amplify BCR signaling.[8] The tyrosine kinase Syk is recruited to phosphorylated ITAM motifs through its phosphotyrosine binding domain (SH2 domain) and auto phosphorylates. Syk then acts as a docking site for other SH2 domain containing proteins, including the B cell linker protein (BLNK) and the Tec-family kinase Btk. BLNK is an adaptor protein that nucleates transcriptional activation of multiple signaling pathways that are involved in B cell activation and proliferation.[6,7,9] In brief, BLNK recruits Phospholipase C-γ2 (PLC-γ2), which is activated by Btk to generate inositol 1,4,5-trisphosphate (IP_3). IP_3 binds the endoplasmic reticulum, which triggers calcium release, and culminates in activation of the NFAT (nuclear factor of activated T cells) pathway. PLC-γ2 also contributes to activation of the NF-κB pathway by stimulating PKC (protein kinase C) activity. Finally, Diacylglycerol (DAG) produced by PLC-γ2, recruits a guanine nucleotide exchange factor protein that activates the small GTPase Ras, and leads to activation of the MAPK family of pathways.[10] A simplified overview of these signaling events is shown in Fig. 2.

The regulation of BCR signaling events is coordinated at multiple levels by accessory protein complexes. The three most well characterized accessory complexes are CD19, CD22, and FcγRIIB.[6,7] The CD19 co-receptor complex includes complement receptor 2 (CR2) and CD81, and is a positive regulator of BCR activation. Antigens that have bound complement bind CR2 in addition to the BCR, which augments BCR clustering in the early phase of B cell activation. CD81 then is a chaperone to facilitate proper segregation of the CD19 complex with the BCR.[11] CD22 and FcγRIIB are both negative regulators of BCR activation. Both receptors contain immunoreceptor tyrosine-based inhibitory motifs (ITIMs) that are phosphorylated by Lyn, a Src-family PTK.

BCR clustering and signaling is also regulated by the cortical actin cytoskeleton. Inhibition of actin remodeling using the f-actin stabilizing drug jasplakinolide for example blocked BCR aggregation and tyrosine phosphorylation, indicating that actin remodeling is critical for BCR activation.[12,13] In resting B cells, cortical actin is linked to the plasma membrane through ERM (ezrin-radixin-moesin) proteins and is thought to create a barrier that maintains the BCR in compartmentalized plasma membrane

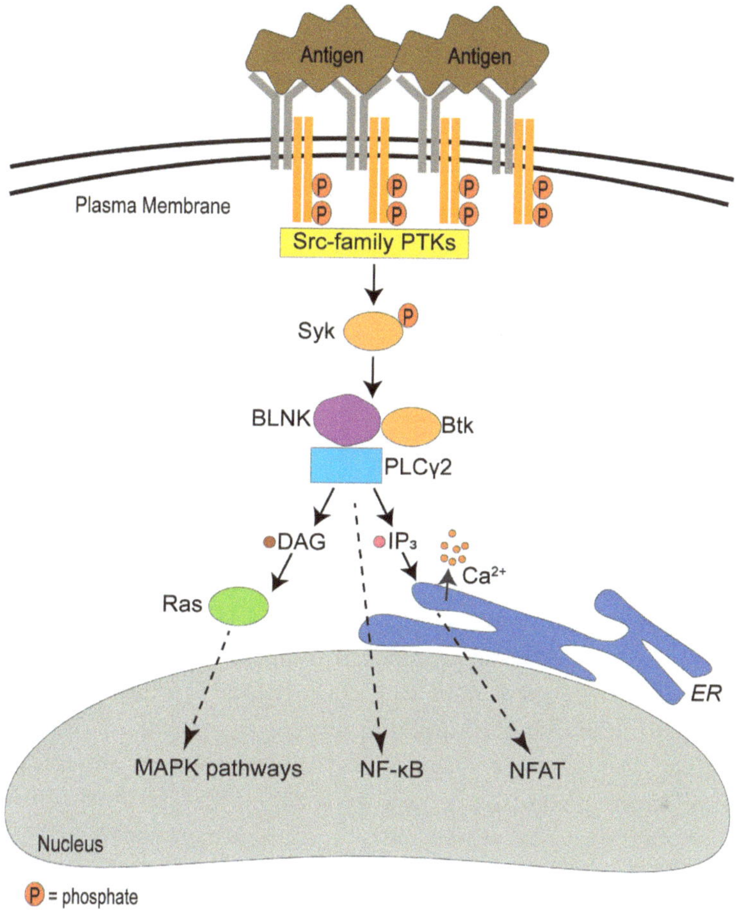

Fig. 2 Overview of B cell receptor signaling. Upon antigen stimulation the BCR clusters at the plasma membrane and triggers recruitment of Src-family protein tyrosine kinases (PTKs). Src-family PTKs phosphorylate the ITAM motifs on the BCR and recruit Syk (a tyrosine protein kinase), BLNK (B cell linker protein), and Btk (Burton's tyrosine kinase). BLNK is an adaptor protein that nucleates activation of multiple downstream signaling pathways, including MAPK (mitogen-activated protein kinase), NF-κB (nuclear Factor κ B), and NFAT (nuclear factor of activated T cells). PLCγ2—phospholipase C gamma two; DAG—diacylglycerol; IP_3—1,4,5-triphosphate.

regions.[14,15] Antigen engagement with the BCR triggers a transient depolymerization of cortical actin, which allows for increased lateral mobility of the BCR and the formation of receptor clusters. Actin then polymerizes to reform in a polarized manner around the outer edge of BCR clusters where it plays a role in facilitating endocytosis.[16,17]

Finally, antigen valency also impacts the functional response of the B cell receptor to stimulation. Specifically, monovalent antigens or Fab fragments that bind only one BCR, can stimulate calcium signaling in a Src-family PTK dependent manner.[18] BCR stimulation with monovalent antigens, however, results in reduced endocytosis, less calcium mobilization, and less efficient antigen presentation, when compared to oligomeric multivalent antigens.[19]

4. Mechanism of B cell receptor endocytosis

One of the earliest steps in the production of an effective antibody response to infection is antigen engagement with the BCR and endocytosis of the receptor-antigen complex. Upon endocytosis, the antigen moves from the early endosome to MHCII loading compartments.[16,20–23] MHCII loading compartments contain newly synthesized MHCII, and lysosomal enzymes that degrade the antigen to generate epitopes.[22] Antigen epitopes are then loaded onto MHCII by Human Leukocyte Antigen-DM.[24] Epitope loaded MHCII molecules are finally trafficked to the cell surface to be presented to CD4+ helper T cells. BCR signaling and endocytic trafficking of antigens appear to be linked. Studies have shown that activation of receptor signaling induces reorganization of antigen processing compartments that contain MHCII and fusion with vesicles containing endocytosed receptor antigen complexes.[19,25,26] B cell receptor signaling also induces polarization of the microtubule-organizing center (MTOC), which is critical for antigen processing and formation of the immune synapse between B and T cells.[21,27]

Two recent studies highlight the importance of endocytic proteins in generating specific immune responses by supporting B cell proliferation and antigen presentation functions. The first is a study of the endocytic protein endophilin A2.[28] Genetic knockdown of endophilin A2 lead to reduced antigen uptake in B cells and decreased antigen specific immune responses in mice, which was caused by a loss in the B cell's ability to proliferate in response to stimulation.[28] Another study investigated the role of dynamin in BCR-antigen endocytosis and found that expression of a dominant negative dynamin (K44A mutant) protein significantly reduced the B cell's ability to internalize the BCR-antigen complex and subsequently present antigens to T cells.[29] An overview of receptor mediated endocytic mechanisms that will be discussed in the following sections is presented in Fig. 3.

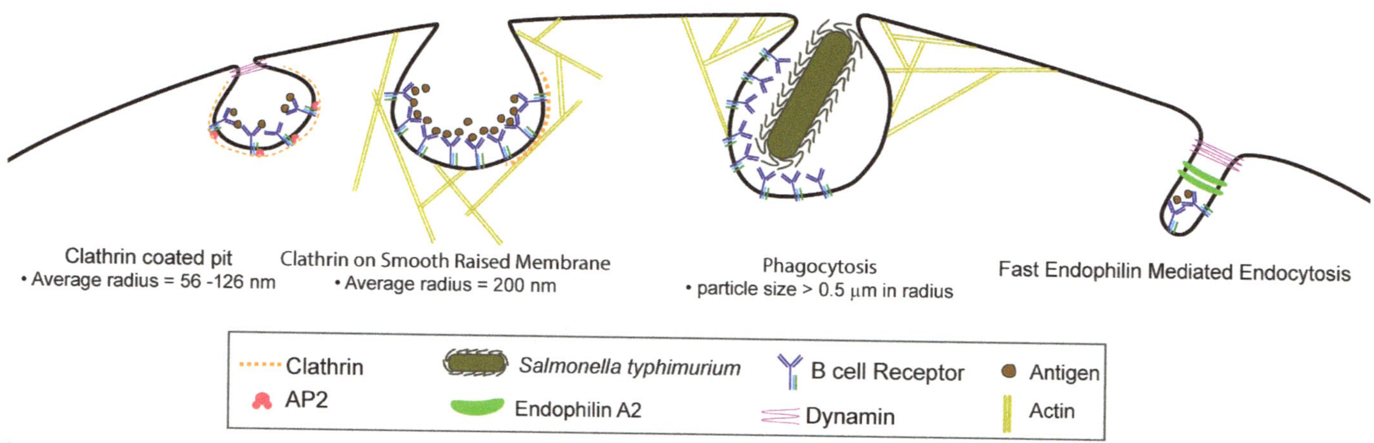

Fig. 3 Overview of B cell receptor mediated endocytosis mechanisms. The BCR may use clathrin mediated or clathrin independent endocytosis mechanisms to internalize the BCR-antigen complex. The method of endocytosis may depend on the concentration or size of the antigen. Many antigens are endocytosed by clathrin, which can typically accommodate receptor clusters between 56 and 126 nm in radius. High concentrations of antigen generate larger BCR clusters that are endocytosed by clathrin on smooth raised membrane structures, which can accommodate clusters of about 200 nm in radius. Finally, antigens or whole bacteria that are larger than 0.5 μm are internalized through phagocytosis. And fast endophilin mediated endocytosis also plays a role in BCR internalization the from the plasma membrane upon activation.

5. Clathrin mediated endocytosis

Early electron microscopy studies showed that the BCR clusters when exposed to multivalent crosslinking IgM and is then endocytosed into clathrin coated pits.[30,31] Subsequent studies have confirmed these results with fluorescence microscopy.[17,32,33] Clathrin is a structural coat protein composed of heavy and light chain triskelia. Clathrin coated vesicles nucleate at specific sites on the membrane where transmembrane cargo receptors bind adaptors including the adaptor protein complex 2 (AP2). AP2 links transmembrane receptors to clathrin and other endocytic accessory proteins. Clathrin triskelia are recruited from the cytosol to the site of concentrated adaptor proteins to generate curvature of the plasma membrane and form a cargo-loaded vesicle. Vesicle scission from the membrane occurs when the GTPase dynamin is recruited to the thin neck of the vesicle to catalyze scission. After scission, the vesicle uncoats. This allows the vesicle to fuse with endosomes.[34] Stoddart et al. showed that conditional knock down of clathrin heavy chain in B cells lead to a 70% reduction in BCR internalization after stimulation with a crosslinking antibody.[17] Early studies of BCR-antigen internalization implicated the cytoplasmic domains of both CD79a and CD79b in facilitating proper internalization of the BCR through clathrin-mediated endocytosis.[35–38]

A series of experiments in separate reports investigated the role of CD79a and CD79b in BCR internalization using truncation mutants or amino acid substitutions at central tyrosine residues that are required for phosphorylation of the ITAMs. Gazumyan et al. mutated the two central tyrosine residues in the ITAM motifs of both CD79a (CD79a$_{FF}$) and CD79b (CD79b$_{AA}$).[36] Both mutants showed significantly decreased rates of constitutive BCR internalization, with the CD79b$_{AA}$ mutant exhibiting a slightly more severe phenotype than CD79a$_{FF}$. The mutants were also stimulated with a crosslinking antibody to determine the effect of the mutated tyrosine residues on internalization. CD79b$_{AA}$ expressing cells had significantly reduced BCR internalization, while the CD79a$_{FF}$ mutant had only a modest effect on internalization after stimulation. Another study investigated the effects of truncating the cytoplasmic domain of CD79a (CD79$_{-/b}$) or CD79b (CD79$_{a/-}$) on internalization of the BCR after stimulation. In this study, the CD79a truncation mutant had a more severe effect on the cell's response to BCR crosslinking than the CD79b truncation. These studies along with others suggest that CD79a and CD79b are both involved in

BCR internalization, using different amino acid motifs to facilitate endocytosis.[35,38] Supporting this hypothesis, a study by Busman-Sahay et al. showed that the clathrin adaptor protein AP2 binds CD79b via a membrane proximal AP2 binding motif in CD79b that requires CD79a to activate AP2 binding.[39]

Although clathrin is clearly required for a component of BCR endocytosis, there is evidence that alternative mechanisms of clathrin independent endocytosis are also involved in driving BCR internalization after stimulation. First, as previously discussed, a conditional knockdown of the clathrin heavy chain protein in B cells only lead to a 70% reduction in endocytosis.[17] This suggest that the remaining fraction of endocytosing receptors must use an alternate mechanism. Additionally, a study of BCR cluster radius by Lee et al. found that BCR cluster radius ranges from about 60 nm in resting B cells to over 1 μm in stimulated cells.[40] The average radius of a clathrin coated pit is roughly between 56 and 126 nm, based on platinum replica electron microscopy studies of clathrin structures in multiple cell types.[41] The large BCR cluster radius after stimulation also suggests that an alternative mechanism of endocytosis may be required to accommodate large BCR-antigen clusters. Finally, a study of how B cells take up antigen from the APC surface found that B cells form large membrane invaginations that require clathrin to pinch off smaller antigen clusters for processing.[42] Large membrane invaginations at the APC-B cell interface are likely generated using clathrin independent mechanisms. The next section will discuss some of the clathrin independent endocytosis pathways that are active in B cells.

6. Clathrin independent endocytosis (CIE)

6.1 Phagocytosis

Phagocytosis is a process that allows cells to endocytose large particles (greater than 0.5 μm). It is classically thought to be mediated by three groups of receptors: pattern recognition receptors, opsonic receptors, or apoptotic corpse receptors.[43] Actin remodeling by Rho family GTPases is important for phagocytosis in all three groups of receptors.[43,44] Macrophages and dendritic cells are typically thought of as the primary cells responsible for phagocytosis of bacteria and antigen presentation. However, a study recently found that B cells in teleost fish (rainbow trout, specifically) are able to phagocytose bacteria.[45] This study was followed by several papers showing that human B cells are also phagocytically active,[45–49] including B cell lines

(Ramos, Raji) and primary human B cells.[48,49] Additionally, multiple B cell subtypes (B1 and follicular B cells) in mice have phagocytic activity.[46,47]

A study of BCR mediated phagocytosis in Ramos cells, which do not express the opsonic receptors (FcγRs), found that beads coated with monoclonal antibodies against the BCR were internalized within 10–20 min, while beads coated in an irrelevant antibody were not internalized.[48] This study further showed that a small proportion of primary human B cells were able to phagocytose *Salmonella typhimurium* through direct recognition of *Salmonella* antigens.[48] These data suggest that phagocytosis in B cells is mediated by the BCR and is supported by studies of B1 cells in mice.[46] B1 cells are a part of the innate immune system and contribute to early antibody responses to T-cell-independent antigens. Gao et al. found that B1 cells phagocytose *Staphylococcus aureus*, and that phagocytic activity is enhanced by antigen specific B cell receptors.[46] These studies suggest that the classical models of phagocytosis might need to be revisited in B cells.

The details of the molecular mechanisms driving BCR mediated phagocytosis in B cells are not well characterized. Several studies have shown that actin destabilizing drugs such as cytochalasin can inhibit phagocytosis in B cells. Actin is likely an important cytoskeletal driver of phagocytosis in B cells as it is in antigen presenting cells as well.[45,47–49] Additionally, a study of follicular B cells showed that Rho G is required for BCR mediated phagocytosis.[47] Rho G is a Rho family GTPase that is involved in phagocytosis via opsonic and apoptotic corpse receptors. B cells from *Rhog*$^{-/-}$ mice had fourfold less phagocytic activity when compared cells from wild type mice.[47] Future studies may determine the extent of the role B cell phagocytosis plays in natural immune responses to infection or vaccination.

6.2 Other mechanisms of CIE

Fast endophilin mediated endocytosis (FEME) is a receptor triggered mechanism of endocytosis that is mediated by endophilin proteins. Five types of endophilin proteins (A1, A2, A3, B1, and B2) in humans generate curvature by binding the plasma membrane through their BAR domains.[50] Enrichment of endophilins A1, A2, or A3 at the plasma membrane prior to receptor stimulation can mediate FEME of specific receptors.[51,52] Malinova et al. showed that knock down of endophilin A2 in Ramos cells and in primary mouse B cells lead to a reduction in antigen uptake by the BCR.[28] Furthermore, endophilin A2 colocalized with antigen and not with clathrin after receptor stimulation. Finally, genetic knock down of clathrin adaptor proteins did not affect endophilin A2 colocalization with the antigen. This study suggests that endophilin A2 mediates FEME of the BCR after antigen stimulation in a clathrin independent manner.

A study of the physical plasma membrane structures associated with crosslinked BCR clusters, identified unique Smooth Raised Membrane (SRM) structures that are significantly colocalized with large BCR clusters and may represent an additional endocytic pathway for the BCR.[32] SRM structures consistently have small clathrin lattices associated with them, and may facilitate endocytosis through a hybrid pathway that requires both clathrin and SRMs. Receptor clusters colocalized with SRMs are unlikely to be internalized by clathrin alone, however, due to their large sizes. The BCR clusters analyzed in this study were much larger than clusters typically internalized by clathrin coated pits, with an average cluster radius of about 200 nm.[32] Additionally, actin is required for efficient internalization of BCR clusters through these large membrane invaginations and forms basket like assemblies around the invaginated sites. This study suggests an alternate mechanism of BCR internalization through smooth raised membrane structures that require both actin and clathrin. Further work is required to identify additional proteins involved in this hybrid pathway of BCR internalization and to determine the role of hybrid endocytosis in BCR-antigen processing.

B cells may also internalize antigen through fluid-phase pinocytosis, similar to antigen presenting cells, although this mechanism is less efficient than receptor mediated endocytosis.[53]

6.3 Lipid rafts

The BCR has been proposed to either aggregate into nanoscale lipid raft domains or stabilize raft domains upon clustering in the plasma membrane.[54] In either case, nanoscale domains that are enriched in cholesterol and sphingolipids, and are highly resistant to detergent induced solubilization, are likely formed.[55] As previously discussed, enrichment of Src-family kinases in lipid raft domains helps to amplify BCR signaling.[8] The role of lipid rafts in internalization, however, is more complex. Early studies implicated clathrin in facilitating BCR endocytosis from lipid rafts. Experiments using B cells stimulated with F(ab′)2 crosslinking antibody fragment found that receptor stimulation led to increased phosphorylation of clathrin heavy chain (CHC) proteins localized in lipid rafts. Additionally, CHC phosphorylation was inhibited by the Src-family PTK inhibitor PP1.[33]

These data suggested that Src-family kinases, which are enriched at lipid rafts, may coordinate both signaling and endocytosis of the BCR by phosphorylating CD79a/b and clathrin proteins.[33] A second study, however, used a polyvalent antigen model to stimulate the BCR and concluded that antigen stimulation did not induce BCR uptake from lipid rafts. This study observed no effect of Methyl-β Cyclodextrin (MBCD) treatment (a cholesterol depleting drug) on the rate or extent of BCR-antigen uptake.[56] Finally, a third report directly compared the effect of crosslinking antibody or polyvalent antigen stimulation on BCR internalization.[57] BCRs stimulated with a crosslinking antibody were more strongly associated with lipid raft microdomains than antigen stimulated BCRs, and exhibited a significant decrease in endocytosis after stimulation in the presence of MBCD.[57] Taken together, these results provide a complex picture of the relationship between lipid raft microdomains and BCR internalization. Lipid rafts or microdomains may play a role in internalization of some BCR stimuli through clathrin phosphorylation at the microdomain, but not all stimuli appear to be internalized through this pathway.

7. B cell receptor signaling and endocytosis in leukemia and lymphoma

Aberrant B cell receptor signaling is a primary driver of disease pathogenesis in Chronic Lymphocytic Leukemia (CLL) and Diffuse Large B Cell Lymphoma (DLBCL). Recent advances in drug therapies that target kinases in the BCR signaling pathway have been effective at improving clinical outcomes for some patients.[58–61] However, not all subtypes of CLL or DLBCL are equally responsive to BCR pathway inhibitors.[60–62] Additionally, aggressive cancers can acquire resistance to BCR pathway inhibitors.[63] Recent studies have found that both BCR signaling and endocytosis play a role in the pathogenesis of certain subtypes of CLL and DLBCL.

8. Chronic lymphocytic leukemia

There are two major subtypes of CLL: Mutated CLL (M-CLL) and Unmutated CLL (U-CLL). M-CLL is derived from B cells that have undergone somatic hypermutation (induced by an infection) of their Immunoglobulin Heavy Chain Variable (IGHV) gene region. The BCR of malignant B cells from M-CLL patients are anergic (cannot transduce downstream signaling in response to ligation) and this subtype is often associated with better clinical outcomes than U-CLL. In U-CLL, malignant B cells have not undergone somatic hypermutation of their IGHV gene region and express BCRs that are responsive to ligation and able to activate downstream signaling pathways. U-CLL patients typically have poor clinical outcomes, but are more responsive to BCR signaling pathway inhibitor treatment than M-CLL patients.[64,65]

A recent study has shed light on the relationship between BCR signaling and endocytosis in cells from patients with both subtypes of CLL and healthy individuals.[62] This study found that B cells from M-CLL patients had significantly increased mIgM endocytic activity, compared to healthy controls, despite the anergy of the BCR and lower overall mIgM expression levels. In healthy B cells, the mIgM and CD79a/CD79b coreceptor complex are endocytosed together upon ligation of the mIg subunit. B cells from CLL patients, however, appear to endocytose mIgM independently of CD79b.[62] This study showed that in M-CLL patients, malignant B cells are able to decouple BCR signaling and endocytosis. This insight into the molecular pathogenesis of CLL is significant and could facilitate the development of new therapies targeting the aberrant endocytosis of mIgM.

9. Diffuse large B cell lymphoma

DLBCL is divided into two gene expression subtypes: activated B-cell-like (ABC DLBCL) and germinal center B-cell-like (GC DLBCL). DLBCL subtypes are distinguished by gene profiling analysis.[66] While GC DLBCL malignant cells appear to have a similar gene expression profile to other germinal center B cells, ABC DLBCL malignant cells have an expression profile that more closely resembles an activated B cell.[67] Studies have also shown that constitutive NF-κB signaling is necessary for survival of malignant cells in ABC DLBCL.[68] Tumors from ABC DLBCL patients frequently have mutations in the MYD88 oncogene, which drives continuous NF-κB activation, and also mutations in the BCR coreceptor CD79b.[66,69] As discussed in the previous section, CD79b plays an important role in BCR endocytosis. A study of mutations commonly found in the CD79b gene of ABC DLBCL patients showed that mutations in the first tyrosine residue of the ITAM motif of CD79b, lead to increased expression of the BCR.[70] In this study, CD79b mutants were also not readily phosphorylated by the Src-family kinase, Lyn, which is required for BCR internalization.[71] These studies suggest that CD79b mutations found in ABC DLBCL tumors contribute to pathogenesis by blocking endocytosis and increasing surface expression of the BCR, which could allow the receptor to maintain high levels of signaling. Several other mutations in CD79a and CD79b were presented in this study, and given that both coreceptors appear to be important for endocytosis, there may be other pathological mechanisms that contribute to altered clathrin mediated or independent endocytosis of the BCR.

10. Conclusions

B cell receptor interactions with foreign antigens are central to initiating the adaptive immune response and generating antibodies to combat infections. The B cell receptor complex mediates both activation of downstream signaling and endocytosis of the receptor-antigen complex. Signaling events associated with the BCR have been extensively studied and have led to important advances in leukemia and lymphoma treatments targeting the Src-family PTKs.[6,60] The mechanisms of BCR endocytosis, however, are not as well defined.

Clathrin mediated endocytosis plays an important role in both constitutive BCR endocytosis and endocytosis in response to receptor stimulation.[17,30,31,33] Clathrin mediated endocytosis is facilitated by the cytoplasmic domain of CD79a and CD79b which bind the clathrin adaptor protein AP2.[39] Additionally, multiple clathrin independent mechanisms appear to be active in B cells and play a role in internalization of stimulated BCR clusters. Here, along with clathrin mediated endocytosis, we highlighted a series of studies of BCR internalization though phagocytosis, FEME, a novel hybrid endocytic pathway (SRM structures), and clathrin-independent lipid raft associated mechanisms.[28,32,46–49] Future studies may investigate whether the CD79a/CD79b complex also mediates clathrin independent endocytosis and determine the role of clathrin independent endocytic pathways in B cell activation and antibody production. Improving our understanding of all the mechanisms of BCR endocytosis may uncover novel interventions for the development of new lymphoma and leukemia therapies to combat drug resistant cancers that do not respond to the current set of inhibitors. B cells also play a significant role in the pathology of inflammatory autoimmune diseases such as rheumatoid arthritis and type I diabetes. B cells are particularly efficient at internalizing and processing low concentrations of antigen for presentation to auto-reactive T cells, which exacerbates pathology in inflammatory autoimmune diseases.[72] Improving our understanding of how B cells internalize and process small amounts of antigen could lead to more targeted therapies to eliminate pathogenic B cells in inflammatory autoimmune diseases.

Acknowledgments

J. W. T. is supported by the Intramural Research Program of the National Heart, Lung, and Blood Institute, National Institutes of Health, Bethesda, MD.

References

1. Radaev S, Zou Z, Tolar P, et al. Structural and functional studies of Igαβ and its assembly with the B cell antigen receptor. *Structure.* 2010;18(8):934–943. https://doi.org/10.1016/j.str.2010.04.019.
2. Mak TW, Saunders ME, Jett BD, eds. The B cell receptor: Proteins and genes. In: *Primer to the Immune Response.* 2nd ed. Boston: Academic Cell; 2014:85–110. [Chapter 4].
3. Liu W, Meckel T, Tolar P, Won Sohn H, Pierce SK. Intrinsic properties of immunoglobulin IgG1 isotype-switched B cell receptors promote microclustering and the initiation of signaling. *Immunity.* 2010;32(6):778–789. https://doi.org/10.1016/j.immuni.2010.06.006.
4. Kaisho T, Schwenk F, Rajewsky K. The roles of γ1 heavy chain membrane expression and cytoplasmic tail in IgG1 responses. *Science.* 1997;276(5311):412–415. https://doi.org/10.1126/science.276.5311.412.
5. Martin SW, Goodnow CC. Burst-enhancing role of the IgG membrane tail as a molecular determinant of memory. *Nat Immunol.* 2002;3(2):182–188. https://doi.org/10.1038/ni752.
6. Dal Porto JM, Gauld SB, Merrell KT, Mills D, Pugh-Bernard AE, Cambier J. B cell antigen receptor signaling 101. *Mol Immunol.* 2004;41(6):599–613. https://doi.org/10.1016/j.molimm.2004.04.008.
7. Tanaka S, Baba Y. B Cell receptor signaling. In: Wang J-Y, ed. *B Cells in Immunity and Tolerance.* Singapore: Springer Singapore; 2020:23–36.
8. Pierce SK. Lipid rafts and B-cell activation. *Nat Rev Immunol.* 2002;2(2):96–105. https://doi.org/10.1038/nri726.
9. Fu C, Turck CW, Kurosaki T, Chan AC. BLNK: a central linker protein in B cell activation. *Immunity.* 1998;9(1):93–103. https://doi.org/10.1016/S1074-7613(00)80591-9.
10. Khiem D, Cyster Jason G, Schwarz John J, Black BL. A p38 MAPK-MEF2C pathway regulates B-cell proliferation. *Proc Natl Acad Sci U S A.* 2008;105(44):17067–17072. https://doi.org/10.1073/pnas.0804868105.
11. Cherukuri A, Shoham T, Sohn HW, et al. The Tetraspanin CD81 is necessary for partitioning of coligated CD19/CD21-B cell antigen receptor complexes into signaling-active lipid rafts. *J Immunol.* 2004;172(1):370. https://doi.org/10.4049/jimmunol.172.1.370.
12. Liu C, Miller H, Orlowski G, Hang H, Upadhyaya A, Song W. Actin reorganization is required for the formation of polarized B cell receptor signalosomes in response to both soluble and membrane-associated antigens. *J Immunol.* 2012;188(7):3237. https://doi.org/10.4049/jimmunol.1103065.
13. Tolar P. Cytoskeletal control of B cell responses to antigens. *Nat Rev Immunol.* 2017;17(10):621–634. https://doi.org/10.1038/nri.2017.67.

14. Song W, Liu C, Upadhyaya A. The pivotal position of the actin cytoskeleton in the initiation and regulation of B cell receptor activation. *Biochim Biophys Acta Biomembr.* 2014;1838(2):569–578. https://doi.org/10.1016/j.bbamem.2013.07.016.
15. Treanor B, Depoil D, Bruckbauer A, Batista FD. Dynamic cortical actin remodeling by ERM proteins controls BCR microcluster organization and integrity. *J Exp Med.* 2011;208(5):1055–1068. https://doi.org/10.1084/jem.20101125.
16. Brown BK, Song W. The actin cytoskeleton is required for the trafficking of the B cell antigen receptor to the late endosomes. *Traffic.* 2001;2(6):414–427. https://doi.org/10.1034/j.1600-0854.2001.002006414.x.
17. Stoddart A, Jackson AP, Brodsky FM. Plasticity of B cell receptor internalization upon conditional depletion of clathrin. *Mol Biol Cell.* 2005;16(5):2339–2348. https://doi.org/10.1091/mbc.e05-01-0025.
18. Volkmann C, Brings N, Becker M, Hobeika E, Yang J, Reth M. Molecular requirements of the B-cell antigen receptor for sensing monovalent antigens. *EMBO J.* 2016;35(21):2371–2381. https://doi.org/10.15252/embj.201694177.
19. Kim Y-M, Pan JY-J, Korbel GA, Peperzak V, Boes M, Ploegh HL. Monovalent ligation of the B cell receptor induces receptor activation but fails to promote antigen presentation. *Proc Natl Acad Sci U S A.* 2006;103(9):3327–3332. https://doi.org/10.1073/pnas.0511315103.
20. Kleijmeer MJ, Morkowski S, Griffith JM, Rudensky AY, Geuze HJ. Major histocompatibility complex class II compartments in human and mouse B lymphoblasts represent conventional endocytic compartments. *J Cell Biol.* 1997;139(3):639–649. https://doi.org/10.1083/jcb.139.3.639.
21. Yuseff M-I, Pierobon P, Reversat A, Lennon-Duménil A-M. How B cells capture, process and present antigens: a crucial role for cell polarity. *Nat Rev Immunol.* 2013;13(7):475–486. https://doi.org/10.1038/nri3469.
22. Adler LN, Jiang W, Bhamidipati K, et al. The other function: class II-restricted antigen presentation by B cells. *Front Immunol.* 2017;8:1–14. https://doi.org/10.3389/fimmu.2017.00319.
23. Avalos A, Ploegh H. Early BCR events and antigen capture, processing, and loading on MHC class II on B cells. *Front Immunol.* 2014;5:1–5. https://doi.org/10.3389/fimmu.2014.00092.
24. Sherman MA, Weber DA, Jensen PE. DM enhances peptide binding to class II MHC by release of invariant chain-derived peptide. *Immunity.* 1995;3(2):197–205. https://doi.org/10.1016/1074-7613(95)90089-6.
25. Siemasko K, Eisfelder BJ, Williamson E, Kabak S, Clark MR. Cutting edge: signals from the B lymphocyte antigen receptor regulate MHC class II containing late endosomes. *J Immunol.* 1998;160(11):5203–5208.
26. Lankar D, Hln V-S, Briken V, Yokozeki T, Ga R, Bonnerot C. Dynamics of major histocompatibility complex class II compartments during B cell receptor–mediated cell activation. *J Exp Med.* 2002;195(4):461–472. https://doi.org/10.1084/jem.20011543.
27. Wang JC, Lee JY-J, Christian S, et al. The Rap1–cofilin-1 pathway coordinates actin reorganization and MTOC polarization at the B cell immune synapse. *J Cell Sci.* 2017;130(6):1094–1109. https://doi.org/10.1242/jcs.191858.
28. Malinova D, Wasim L, Newman R, Martínez-Riaño A, Engels N, Tolar P. Endophilin A2 regulates B-cell endocytosis and is required for germinal center and humoral responses. *EMBO Rep.* 2021;22(9):e51328. https://doi.org/10.15252/embr.202051328.
29. Malhotra S, Kovats S, Zhang W, Coggeshall KM. B cell antigen receptor endocytosis and antigen presentation to T cells require Vav and dynamin. *J Biol Chem.* 2009;284(36):24088–24097. https://doi.org/10.1074/jbc.M109.014209.

30. Guagliardi LE, Koppelman B, Blum JS, Marks MS, Cresswell P, Brodsky FM. Co-localization of molecules involved in antigen processing and presentation in an early endocytic compartment. *Nature*. 1990;343(6254):133–139. https://doi.org/10.1038/343133a0.
31. Salisbury JL, Condeelis JS, Satir P. Role of coated vesicles, microfilaments, and calmodulin in receptor-mediated endocytosis by cultured B lymphoblastoid cells. *J Cell Biol*. 1980;87(1):132–141. https://doi.org/10.1083/jcb.87.1.132.
32. Roberts AD, Davenport TM, Dickey AM, Ahn R, Sochacki KA, Taraska JW. Structurally distinct endocytic pathways for B cell receptors in B lymphocytes. *Mol Biol Cell*. 2020;31(25):2826–2840. https://doi.org/10.1091/mbc.E20-08-0532.
33. Stoddart A, Dykstra ML, Brown BK, Song W, Pierce SK, Brodsky FM. Lipid rafts unite signaling cascades with clathrin to regulate BCR internalization. *Immunity*. 2002;17(4):451–462. https://doi.org/10.1016/S1074-7613(02)00416-8.
34. McMahon HT, Boucrot E. Molecular mechanism and physiological functions of clathrin-mediated endocytosis. *Nat Rev Mol Cell Biol*. 2011;12(8):517–533. https://doi.org/10.1038/nrm3151.
35. Bonnerot C, Lankar D, Hanau D, et al. Role of B cell receptor Igα and Igβ subunits in MHC class II-restricted antigen presentation. *Immunity*. 1995;3(3):335–347. https://doi.org/10.1016/1074-7613(95)90118-3.
36. Gazumyan A, Reichlin A, Nussenzweig MC. Igβ tyrosine residues contribute to the control of B cell receptor signaling by regulating receptor internalization. *J Exp Med*. 2006;203(7):1785–1794. https://doi.org/10.1084/jem.20060221.
37. Jang C, Machtaler S, Matsuuchi L. The role of Ig-α/β in B cell antigen receptor internalization. *Immunol Lett*. 2010;134(1):75–82. https://doi.org/10.1016/j.imlet.2010.09.001.
38. Siemasko K, Skaggs BJ, Kabak S, et al. Receptor-facilitated antigen presentation requires the recruitment of B cell linker protein to Igα. *J Immunol*. 2002;168(5):2127. https://doi.org/10.4049/jimmunol.168.5.2127.
39. Busman-Sahay K, Drake L, Sitaram A, Marks M, Drake JR. Cis and trans regulatory mechanisms control AP2-mediated B cell receptor endocytosis via select tyrosine-based motifs. *PLOS ONE*. 2013;8(1):e54938. https://doi.org/10.1371/journal.pone.0054938.
40. Lee J, Sengupta P, Brzostowski J, Lippincott-Schwartz J, Pierce SK. The nanoscale spatial organization of B-cell receptors on immunoglobulin M– and G–expressing human B-cells. *Mol Biol Cell*. 2016;28(4):511–523. https://doi.org/10.1091/mbc.e16-06-0452.
41. Dambournet D, Sochacki KA, Cheng AT, et al. Genome-edited human stem cells expressing fluorescently labeled endocytic markers allow quantitative analysis of clathrin-mediated endocytosis during differentiation. *J Cell Biol*. 2018;217(9):3301–3311. https://doi.org/10.1083/jcb.201710084.
42. Natkanski E, Lee W-Y, Mistry B, Casal A, Molloy JE, Tolar P. B cells use mechanical energy to discriminate antigen affinities. *Science*. 2013;340(6140):1587–1590. https://doi.org/10.1126/science.1237572.
43. Flannagan RS, Jaumouillé V, Grinstein S. The cell biology of phagocytosis. *Annu Rev Pathol Mech Dis*. 2012;7(1):61–98. https://doi.org/10.1146/annurev-pathol-011811-132445.
44. Chimini G, Chavrier P. Function of Rho family proteins in actin dynamics during phagocytosis and engulfment. *Nat Cell Biol*. 2000;2(10):E191–E196. https://doi.org/10.1038/35036454.
45. Li J, Barreda DR, Zhang Y-A, et al. B lymphocytes from early vertebrates have potent phagocytic and microbicidal abilities. *Nat Immunol*. 2006;7(10):1116–1124. https://doi.org/10.1038/ni1389.

46. Gao J, Ma X, Gu W, et al. Novel functions of murine B1 cells: active phagocytic and microbicidal abilities. *Eur J Immunol.* 2012;42(4):982–992. https://doi.org/10.1002/eji.201141519.
47. Martínez-Riaño A, Bovolenta ER, Mendoza P, et al. Antigen phagocytosis by B cells is required for a potent humoral response. *EMBO Rep.* 2018;19(9):e46016. https://doi.org/10.15252/embr.201846016.
48. Souwer Y, Griekspoor A, Jorritsma T, et al. B cell receptor-mediated internalization of Salmonella: a novel pathway for autonomous B cell activation and antibody production. *J Immunol.* 2009;182(12):7473. https://doi.org/10.4049/jimmunol.0802831.
49. Zhu Q, Zhang M, Shi M, et al. Human B cells have an active phagocytic capability and undergo immune activation upon phagocytosis of mycobacterium tuberculosis. *Immunobiology.* 2016;221(4):558–567. https://doi.org/10.1016/j.imbio.2015.12.003.
50. Casamento A, Boucrot E. Molecular mechanism of fast Endophilin-mediated endocytosis. *Biochem J.* 2020;477(12):2327–2345. https://doi.org/10.1042/BCJ20190342.
51. Boucrot E, Ferreira APA, Almeida-Souza L, et al. Endophilin marks and controls a clathrin-independent endocytic pathway. *Nature.* 2015;517(7535):460–465. https://doi.org/10.1038/nature14067.
52. Chan Wah Hak L, Khan S, Di Meglio I, et al. FBP17 and CIP4 recruit SHIP2 and lamellipodin to prime the plasma membrane for fast endophilin-mediated endocytosis. *Nat Cell Biol.* 2018;20(9):1023–1031. https://doi.org/10.1038/s41556-018-0146-8.
53. Batista FD, Neuberger MS. Affinity dependence of the B cell response to antigen: a threshold, a ceiling, and the importance of off-rate. *Immunity.* 1998;8(6):751–759. https://doi.org/10.1016/S1074-7613(00)80580-4.
54. Stone MB, Shelby SA, Núñez MF, Wisser K, Veatch SL. Protein sorting by lipid phase-like domains supports emergent signaling function in B lymphocyte plasma membranes. *eLife.* 2017;6:e19891. https://doi.org/10.7554/eLife.19891.
55. Cheng PC, Dykstra ML, Mitchell RN, Pierce SK. A role for lipid rafts in B cell antigen receptor signaling and antigen targeting. *J Exp Med.* 1999;190(11):1549–1560. https://doi.org/10.1084/jem.190.11.1549.
56. Putnam MA, Moquin AE, Merrihew M, et al. Lipid raft-independent B cell receptor-mediated antigen internalization and intracellular trafficking. *J Immunol.* 2003;170(2):905. https://doi.org/10.4049/jimmunol.170.2.905.
57. Caballero A, Katkere B, Wen X-Y, Drake L, Nashar TO, Drake JR. Functional and structural requirements for the internalization of distinct BCR-ligand complexes. *Eur J Immunol.* 2006;36(12):3131–3145. https://doi.org/10.1002/eji.200636447.
58. Byrd JC, Furman RR, Coutre SE, et al. Targeting BTK with ibrutinib in relapsed chronic lymphocytic leukemia. *N Engl J Med.* 2013;369(1):32–42. https://doi.org/10.1056/NEJMoa1215637.
59. Byrd JC, Harrington B, O'Brien S, et al. Acalabrutinib (ACP-196) in relapsed chronic lymphocytic leukemia. *N Engl J Med.* 2015;374(4):323–332. https://doi.org/10.1056/NEJMoa1509981.
60. Young RM, Shaffer AL, Phelan JD, Staudt LM. B-cell receptor signaling in diffuse large B-cell lymphoma. *Semin Hematol.* 2015;52(2):77–85. https://doi.org/10.1053/j.seminhematol.2015.01.008.
61. Wright GW, Huang DW, Phelan JD, et al. A probabilistic classification tool for genetic subtypes of diffuse large B cell lymphoma with therapeutic implications. *Cancer Cell.* 2020;37(4):551–68.e14. https://doi.org/10.1016/j.ccell.2020.03.015.
62. Eve MC, Andrea P, Silvia M, et al. In vitro and in vivo evidence for uncoupling of B-cell receptor internalization and signaling in chronic lymphocytic leukemia. *Haematologica.* 2018;103(3):497–505. https://doi.org/10.3324/haematol.2017.176164.

63. George B, Mullick Chowdhury S, Hart A, et al. Ibrutinib resistance mechanisms and treatment strategies for B-cell lymphomas. *Cancer.* 2020;12(5):1–31. https://doi.org/10.3390/cancers12051328.
64. Stevenson FK, Krysov S, Davies AJ, Steele AJ, Packham G. B-cell receptor signaling in chronic lymphocytic leukemia. *Blood.* 2011;118(16):4313–4320. https://doi.org/10.1182/blood-2011-06-338855.
65. Stevenson FK, Forconi F, Kipps TJ. Exploring the pathways to chronic lymphocytic leukemia. *Blood.* 2021;138(10):827–835. https://doi.org/10.1182/blood.2020010029.
66. Young RM, Phelan JD, Wilson WH, Staudt LM. Pathogenic B-cell receptor signaling in lymphoid malignancies: new insights to improve treatment. *Immunol Rev.* 2019;291(1):190–213. https://doi.org/10.1111/imr.12792.
67. Alizadeh AA, Eisen MB, Davis RE, et al. Distinct types of diffuse large B-cell lymphoma identified by gene expression profiling. *Nature.* 2000;403(6769):503–511. https://doi.org/10.1038/35000501.
68. Davis RE, Brown KD, Siebenlist U, Staudt LM. Constitutive nuclear factor κB activity is required for survival of activated B cell–like diffuse large B cell lymphoma cells. *J Exp Med.* 2001;194(12):1861–1874. https://doi.org/10.1084/jem.194.12.1861.
69. Ruben ALG, Anne MRS, Marie José K, Steven TP, Joost SPV. MYD88 in the driver's seat of B-cell lymphomagenesis: from molecular mechanisms to clinical implications. *Haematologica.* 2019;104(12):2337–2348. https://doi.org/10.3324/haematol.2019.227272.
70. Davis RE, Ngo VN, Lenz G, et al. Chronic active B-cell-receptor signalling in diffuse large B-cell lymphoma. *Nature.* 2010;463(7277):88–92. https://doi.org/10.1038/nature08638.
71. Niiro H, Allam A, Stoddart A, Brodsky FM, Marshall AJ, Clark EA. The B lymphocyte adaptor molecule of 32 Kilodaltons (Bam32) regulates B cell antigen receptor internalization. *J Immunol.* 2004;173(9):5601. https://doi.org/10.4049/jimmunol.173.9.5601.
72. Hampe CS. B cells in autoimmune diseases. *Scientifica.* 2012;2012:215308. https://doi.org/10.6064/2012/215308.

CHAPTER NINE

Multifunctional role of the ubiquitin proteasome pathway in phagocytosis

Angela Lalnunthangi[a], Gyaltsen Dakpa[b], and Swati Tiwari[a,*]
[a]School of Biotechnology, Jawaharlal Nehru University, New Delhi, India
[b]National Chung Hsing University, Taichung City, Taiwan
*Corresponding author: e-mail address: swati_tiwari@mail.jnu.ac.in

Contents

Abstract

Phagocytosis is a specialized form of endocytosis where large cells and particles (>0.5 μm) are engulfed by the phagocytic cells, and ultimately digested in the phagolysosomes. This process not only eliminates unwanted particles and pathogens from the extracellular sources, but also eliminates apoptotic cells within the body, and is critical for maintenance of tissue homeostasis. It is believed that both endocytosis and phagocytosis share common pathways after particle internalization, but specialized features and differences between these two routes of internalization are also likely.

Progress in Molecular Biology and Translational Science, Volume 194
ISSN 1877-1173
https://doi.org/10.1016/bs.pmbts.2022.06.023

The recruitment and removal of each protein/particle during the maturation of endocytic/phagocytic vesicles has to be tightly regulated to ensure their timely action. Ubiquitin proteasome pathway (UPP), degrades unwanted proteins by post-translational modification of proteins with chains of conserved protein Ubiquitin (Ub), with subsequent recognition of Ub chains by the 26S proteasomes and substrate degradation by this protease. This pathway utilizes different Ub linkages to modify proteins to regulate protein–protein interaction, localization, and activity. Due to its vast number of targets, it is involved in many cellular pathways, including phagocytosis. This chapters describes the basic steps and signaling in phagocytosis and different roles that UPP plays at multiple steps in regulating phagocytosis directly, or through its interaction with other phagosomal proteins. How aberrations in UPP function affect phagocytosis and their association with human diseases, and how pathogens exploit this pathway for their own benefit is also discussed.

Abbreviations

E1	Ub activating enzyme
E2	Ub conjugating enzyme
E3	Ub ligase
EE	early endosome
EEC	epididymal epithelial cells
ER	endoplasmic reticulum
ERAD	ER associated degradation
ERGIC	ER-to-Golgi intermediate compartment
ESCRT	endosomal sorting complex required for transport
FcRs	Fc receptors
GHR	growth hormone receptor
GRCs	glutaraldehyde-fixed red blood cells
ILVs	intraluminal vesicles
ITAMs	tyrosine-based activation motifs
LPS	lipopolysaccharides
MCS	membrane contact sites
MHC	major histocompatibility complex
MVBs	multivesicular bodies
PI3K	phosphatidylinositol-3-kinase
PIP_3	phosphatidylinositol 3,4,5-triphosphate
SNARE	soluble N-ethylmaleimide-sensitive factor attachment protein receptor
TAMs	tissue associated macrophages
Ub	ubiquitin
Ubl	Ub-like
UPP	ubiquitin proteasome pathway

1. Introduction

1.1 A brief history of discovery of phagocytosis

The discovery of phagocytosis is one of the milestones that finally led to our understanding of cellular immunity. Its discovery is credited to Elie

Metchnikoff (1882) who was studying the digestive process of sponges. He observed cells moving freely within transparent starfish larvae and was the first to connect it to white blood cells, thus bridging an evolutionary gap from sponges to metazoans. He proposed that these wandering cells which eat food in the body of the larvae of a starfish must eat up microbes too which was an outlandish idea at that time. To prove it, he conducted experiments in which he injected thorns and yeast spores into the body of starfish larvae and water flea *Daphnia*, respectively. He could observe the free cells inside them moving toward, surrounding and ultimately eating the thorns and yeast spores suggesting that his hypothesis was right.[1] He, along with Paul Ehrlich, received the Nobel Prize in 1908 for the discovery of cellular and humoral immunity. Carl Fredrich Claus, a zoologist, helped him in naming these wandering cells in 1906 and called them "Phagocytes"—derived from Greek word where "*phago*" means to eat and "*cytes*" means cell. Presently, phagocytosis is defined as a triggered cellular process of recognition and uptake of micron-sized particles including pathogens, larger than those internalized by endocytic pathways like clathrin-mediated endocytosis[2] into a membrane vesicle named phagosome.[3]

In this review, we describe the basic events of phagocytosis, major signaling pathways, phagosomal maturation, and current understanding of the involvement of UPP proteins at multiple steps during phagocytosis.

1.2 Phagocytosis: Multifarious pathways

Phagocytosis is an evolutionarily conserved process occurring through the remodeling of actin cytoskeleton and plasma membrane by the action of a plethora of signaling molecules.[4] It has been studied extensively ever since its discovery over a century ago. The phagosome maturation models have been mostly derived from the endosomal maturation, a much simpler and studied process.

Phagocytosis is commonly known for its role in immune defense for the removal of unwanted material from the body, and it is also used for obtaining nutrients in multicellular organisms in a process called cell cannibalism. This is not surprising given the fact that phagocytosis was first discovered in relation to its aboriginal function in feeding.[5] This non-professional phagocytosis was earlier observed in tumor cells and later also reported in normal cells, like epithelial and mesenchymal cells feeding upon neighboring live or dead cells.[5,6] Under conditions of low pH, hypoxia and low nutritional state in cancer, malignant cells cannibalize other malignant cells with metabolic alterations. This malignant cell may in turn be engulfed by another malignant cell at times, a process referred to as complex cannibalism.[7] The internalized cells undergo cell death in majority of cases, but can also divide inside the cannibalistic cell and may even escape from the outer cell.[8] Non-professional phagocytosis of necrotic neighboring cells by epithelial and mesenchymal cells has been shown to increase as cell cycle progresses, being most prominent in the G2 phase.[8] This may be explained in terms of increased plasticity of cytoskeleton during late G2 and early onset of mitosis, and the need for more essential nutrients for cell growth, similar to the need of cancer cells which is believed to lead to cell cannibalism. In cancer cells, cell cannibalism has also been thought to help in tumor immune escape by cannibalizing the lymphocyte that are supposed to kill the metastatic tumor cells.[9] Metastatic cells cannibalize sibling tumor cells, and also other cells in the tumor microenvironment without differentiating between target particles of different origins in their capacity to phagocytose them with the sole purpose to feed, survive and progress under unfavorable micro-environment.[9] It differs from professional phagocytosis in that it is non-selective unlike professional phagocytosis, and is caveolin-I dependent.[5] Additionally, it involves eating live cells and the target cell is supposed to be swallowed without much pseudopod protrusion, and target cell death is not necessary and the engulfed cell may not be killed even after entering the cytoplasm of the host cell.[5]

Macroautophagy or self-cannibalism is another highly conserved process that merges with the lysosomes. It is critical for cellular maintenance where organelles and macromolecules are degraded and breakdown products are recycled. Autophagy degrades many proteins and organelles in bulk, either constitutively at a low level, or upon induction under stress conditions. It is characterized by the recruitment of cargoes to a double membrane structure called autophagosome with the help of ubiquitin-like (Ubl) modifier Autophagy-related protein-8 (Atg8), Gamma-aminobutyric acid receptor-associated protein (GABARAB and LC3 in mammals). Atg8 conjugates with a lipid phosphatidylethanolamine (PE) and helps in expansion of the phagophore (precursor of autophagosome) after its induction and nucleation with Unc-51-like kinase family (ULK1/2) complex and a phosphatidylinositol-3-kinase (PI3K), vacuolar protein sorting-associated protein34 (Vps34), to form the autophagosome around the target. Vps34 generates phosphatidylinositol 3-phosphate (PI3P) which can be bound by several proteins including Atg18 and Atg21, known to function in macroautophagy in yeast although their precise functions in mammals are still unknown.[10] Another Ubl Atg12 also participates in elongation of autophagosome. Once the autophagosome is formed around the cargo, direct autophagosome-lysosome fusion may ensue.[11] In other cases, the autophagosome may first fuse with the endocytic vesicles viz., late endosome, to form amphisome, which may further fuse with lysosome as in phagocytosis. The contents are finally degraded and exported back to cytoplasm for reuse by the cell. Autophagy can further be regulated by ubiquitin where the cargo may be ubiquitinated and targeted to autophagosomal membrane through autophagic receptors.

2. The phagocytes

Phagocytosis is essential for host defense against pathogenic microbes, tissue remodeling, antigen cross presentation and immunological response against the pathogens. It is performed by cells called phagocytes that can be divided into two types: professional phagocytes and non-professional phagocytes, based on their functions and interactions with cells, ligands and particulates.[12] In metazoans, phagocytosis is functions in the destruction and removal of pathogens, and for removing damaged or senescent cells, and particulate matter, especially by specialized phagocytic cells like monocytes, macrophages and dendritic cells which are referred to as professional phagocytes. These cells are derived from bone marrow precursors. Epithelial cells, fibroblasts and other cells are considered non-professional phagocytes as they only have limited range of target particles since they do not have receptors that are specific for phagocytosis, and participate mostly in eliminating apoptotic cells.[12,13] This distinction was based on the observation that addition of antiserum enhanced the uptake of glutaraldehyde-fixed red blood cells (GRCs) by macrophages, while it inhibited phagocytosis by fibroblast. The antibody-coated GRCs were engulfed more efficiently compared to untreated cells. Such studies provided evidence that phagocytosis is a triggered process and likely involves some receptors, and the type of receptor present depends on the type of cell, which ultimately governs whether a phagocyte is professional or non-professional. Unlike non-professional phagocytes, the professional phagocytes release cytokines and radicals of Oxygen (O_2) and Nitrogen (N_2). While monocytes remain in blood circulation for 1 day, macrophages remain active for weeks to months and generate inflammatory responses.

3. Distinct steps of phagocytosis

Broadly, phagocytosis involves the target particle recognition and binding, downstream signaling to remodel the cytoskeletal elements, generation of phagosome and its maturation, and fusion with lysosomes to generate phagolysosome.[14] When particles, dead cells and pathogens are encountered by phagocytic cells, different signaling cascades are mediated depending on the type of receptors engaged in recognition. These events lead to the extension of phagocytic cup that is further excised from the cell membrane to form a phagosome. In lower organisms, as in amoeba, phagosomes are called food vacuoles which contain food particles. Maturation of phagosomes involves multiple steps that finally generate a destructive environment to eliminate the phagocytic prey. Accordingly, many pathogens have evolved multiple strategies to halt phagosome maturation, underscoring the importance of phagocytosis for effective immunity against pathogens.[15,16]

3.1 Target particle recognition

Phagocytosis is a receptor mediated event. Therefore, not all contacts of pathogens, microbes and particles with phagocytic cells evoke phagocytosis. It requires the active recognition of ligands on the target particle with certain receptors on the phagocytes to transduce downstream signal cascades. This process is complex as professional phagocytes can express multiple receptor types in the same cell and several receptors may simultaneously bind to the target particle and cooperate with each other to initiate internalization of the target. Phagocytic receptors are classified into two types: opsonic receptors and non-opsonic receptors. Non-opsonic receptors are phagocytic pattern recognition receptors that can directly recognize molecular patterns on the target particles. These include mannose receptor,[17] dectin-1[18,19] and scavenger receptor-A.[20] These receptors bind to mannan polysaccharides on yeast and lipopolysaccharide on gram-negative bacteria, respectively.[21] Opsonic receptors, on the other hand, do not interact directly with target pathogens and includes Fc receptors (FcRs),[22] complement receptors (CRs)[23,24] and fibronectin receptors.[15,25] They bind to host-derived molecules known as opsonins which attach to the surface of target particles.[15] FcRs bind to the constant region of immunoglobulin molecules with different affinities depending on receptor subtype involved; FcγR (FcγRIa, FcγRIIa, FcγRIIc and FcγRIIIa) interacts with IgG, FcαR with IgA, and

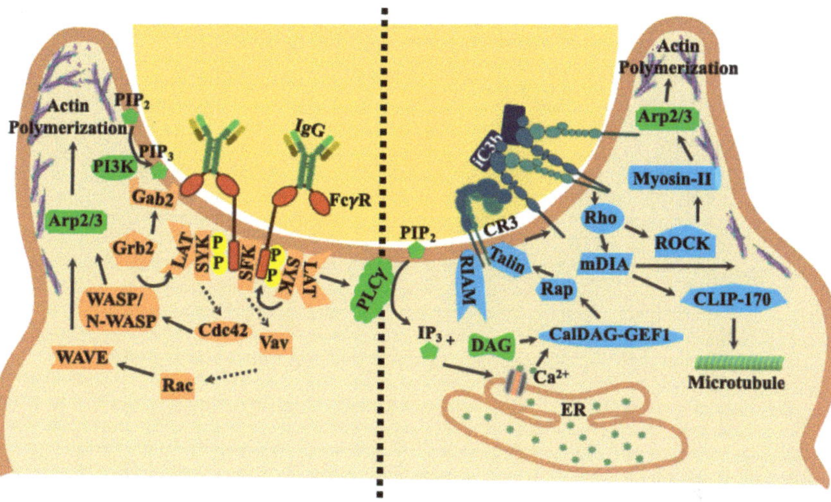

Fig. 1 Target recognition and signaling in phagocytosis. Main signaling pathways for FcγR (left) and CR3 (right)-mediated phagocytosis are shown separated by a dashed line in the center for easy understanding. Dotted arrows: unknown intermediaries/no clear proof, Peach: signaling in FcγR mediated phagocytosis, Blue: signaling in CR3 mediated phagocytosis, Green: signaling in general. Actin polymers are shown in purple.

FcεR with IgE.[26] CRs like CR3 and CR4 bind to iC3b while CR1 binds to mannan-binding lectin, C3b, C4b and C1q.[27] Different receptors generate different signaling cascades for further progression of phagocytosis (Fig. 1). Of the many receptors for phagocytosis, FcγR and Dectin-1 are considered bona fide phagocytic receptors as their ectopic expression renders phagocytic function to non-phagocytic cells.[28–30] Readers are referred to a review by Uribe-Querol and Rosales (2020) for a comprehensive summary of all known receptors involved in phagocytosis and their ligands.[31] A summary of proteins involved in signaling for pseudopod extension during FcR, CR and Dectin-1 mediated phagocytosis is provided in Table 1. Dectin-1 pathway has a cytoplasmic tail motif similar to immunoreceptor tyrosine-based activation motif (ITAM) and thus shares many signaling molecules utilized by FcR as shown in Table 1. Both FcR and CR signaling pathways have been more extensively studied compared to Dectin-1 pathway. Hence, these two pathways are described in detail in the chapter.

3.1.1 Fc receptor (FcR) signaling

FcRs are constitutively active for phagocytosis.[2] They have cytoplasmic ITAMs that lie either in the receptor tail (FcγRIIa and FcγRIIc) or in

Table 1 Different molecule requirements for signaling through Fc Receptor (FcR), Complement receptor (CR) and Dectin-1 receptor after target binding for pseudopod extension.

Signaling molecule	FcR	CR	Dectin-1
SFK	+[32]	−[32]	+[33]
SYK	+[34]	−[34]	±[19]
Gab2	+[35]	−[35]	?
PI3K	+[36]	+[36]	+[37]
Rac	+[38]	−[39]	+[19]
Cdc42	+[38]	−[39,40]	+[19]
Rho	−[39]	+[39]	+[33]
ROCK	−[41]	+[41]	+[33]
Myosin-II	−[41]	+[41]	+[33]
Vav	±[42]	+[42]	+[37]
CLIP-170	−[43]	+[43]	?
mDIA1	−[43,44]	+[43,44]	?
Arp2/3 complex	±[45,46]	+[45,46]	?
WASP/N-WASP	+[47]	±[48]	?
Actin	+[49]	+[49]	+[19]
Microtubule	+[50]	+[50]	?

+, involved; −, not involved; +/−, dispensable; ?, unknown. Numbers in superscript are the corresponding references.

the accessory FcR γ-chain (FcγRI and FcγRIIIa)[51] which become phosphorylated through receptor clustering once they bind to ligands.[26,52] Fig. 1 shows the downstream signaling pathway from the FcRs. Src family kinases (SFKs) like Hck, Fgr and Lyn phosphorylate the ITAMs upon target binding, though these kinases are dispensable.[32] Spleen tyrosine kinase (Syk) interacts with the phosphorylated receptors[53] and gets activated, further phosphorylating more ITAMs, thereby amplifying the signal. Additionally, it leads to recruitment and phosphorylation of adaptors like linker of activated T-cells (LAT).[54]

LAT allows docking of Growth factor receptor-bound protein-2 (Grb2), which in turn recruits Grb2-associated-binding protein 2 (Gab2), leading to interaction with phosphatidylinositol 3,4,5-triphosphate (PIP_3) generated by PI3K. PIP_3 is crucial for downstream signaling in FcR mediated phagocytosis.[35] Syk also activates Cell division control protein 42 homolog (Cdc42) (through unknown intermediaries), and RAC-alpha serine/threonine-protein kinase (Rac), probably through Vav,[26,55,56] which then binds to and activates Actin-related protein 2/3 complex (Arp2/3) nucleator complex to trigger actin polymerization through Wiskott-Aldrich syndrome protein/Neural Wiskott-Aldrich syndrome protein (WASP/N-WASP) and WASP family protein member (WAVE), respectively.[57,58] All these events lead to the appearance of membrane ruffling and protrusions for particle engulfment.

3.1.2 Complement receptor (CR) signaling

CRs generally require additional stimuli (e.g., Tumor necrosis factor-α (TNF-α), granulocyte-macrophage colony-stimulating factor (GM-CSF), or attachment to fibronectin-coated substrata) to engulf particles; though a case of CR-mediated phagocytosis by mature macrophages in the absence of stimuli has been reported.[59,60] Activation of CRs like CR3, an integrin family member, requires an inside-out signaling that is provided by FcR (Fig. 1). Signaling from FcR activates phospholipase-Cγ (PLCγ) through LAT that produces calcium (Ca^{2+}) and diacylglycerol (DAG), that are required to activate CalDAG-GEF1, a guanine nucleotide exchange factor.[13,61] CalDAG-GEF1 activates Rap GTPase allowing recruitment of Rap1-interacting adapter molecule (RIAM) and talin to the intracellular tail of CR3 (Fig. 1), transforming it to its active conformation.[62] The requirement of Rho family GTPases Cdc42, Rac and Rho are different for FcR- and CR3-mediated phagocytosis in that the former requires Cdc42 and Rac, while the latter requires Rho.[39] Rho is activated once CR3 binds iC3b on the target particle. Rho kinase (ROCK) is activated upon clustering of CR3, leading to phosphorylation of myosin-II, activating it to associate with Arp2/3 (Fig. 1). Separately, Rho also recruits and activates mammalian diaphanous-related formin-1 (mDIA), thus promoting actin polymerization and accumulation of microtubule associated cytoplasmic linker protein 170 (CLIP-170) at the phagocytic cup[43] (Fig. 1). SYK, whose participation in CR-mediated phagocytosis was ruled out before,[34,63] is thought to be activated indirectly by ITAM-bearing FcRγ chain and/or DNAX-activation protein 12 (DAP12)[13,64] and activates Rho probably through Vav.[42] Previously, the formation of membrane protrusion was ruled out for CR-mediated phagocytosis and it was thought that the target particle simply sinks into the cell.[2] But this concept has been disputed as subsequent research showed the formation of membrane protrusions in CR-mediated phagocytosis.[42,65]

3.2 Phagosome generation

Signaling through different types of receptors culminates in the generation of phagosomes. Receptor engagement followed by receptor clustering to induce signal transduction pathways regulates the actin cytoskeleton, so that membrane is extended to enclose target particles in a vesicle. Pinching of this vesicle from the plasma membrane forms a phagosome. The formation of phagosome involves three distinct steps: (i) binding of the target with the receptors, (ii) actin polymerization and depolymerization for membrane protrusion and (iii) closure of phagocytic cup, and finally the detachment of the phagosome from the plasma membrane (Fig. 2).[13]

3.2.1 Binding of ligand or particles with receptors

With most of the phagocytic receptors being short (around 5 nm), not easily accessible, and ligands not covering the target particle uniformly,

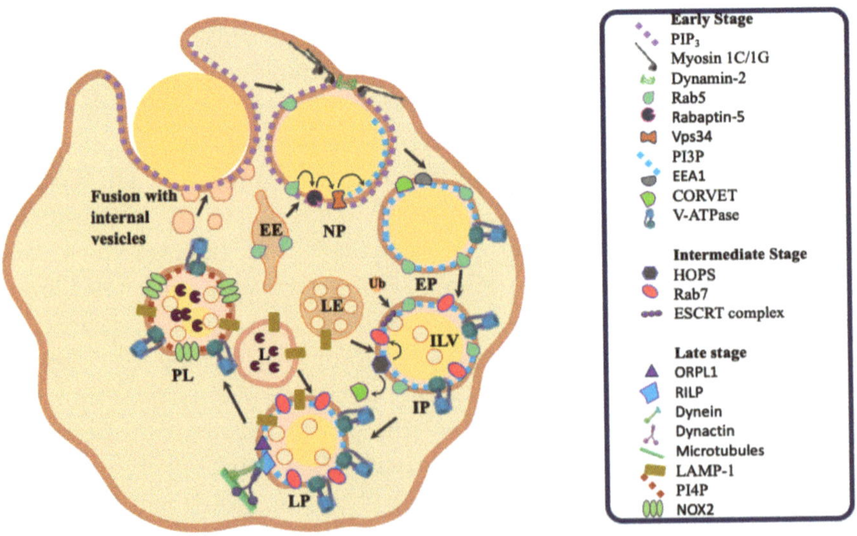

Fig. 2 Phagosome formation and maturation. Major proteins associated with scission of phagosome, and its maturation and fusion with lysosomes are shown. The key to each protein/lipid is shown on the right. Early Endosome (EE), Late Endosome (LE), Lysosome (L), Nascent Phagosomes (NP), Early Phagosome (EP), Intermediate Phagosome (IP), Phagolysosome (PL).

receptor-ligand binding becomes a challenge.[13] In a resting phagocyte, the receptors are masked by glycoproteins in a thick layer called glycocalyx.[66] In addition to reducing lateral diffusion by the glycoproteins tethered to the cytoskeleton, transmembrane phosphatases like CD45 and CD148 can dephosphorylate the activating tyrosine residues of SFKs required for phosphorylation of FcR.[28] Therefore, for efficient receptor-ligand binding to take place, accumulation of activating kinases through lateral diffusion, along with a close cell-cell contact between the phagocyte and target particle must occur to physically exclude inhibitory phosphatases.[14,67,68] The restricted diffusion of immunoreceptors can be loosened by Toll-like receptor (TLR) ligands like lipopolysaccharides (LPS) by altering their attachment to the cytoskeleton.[69] Integrins are also activated by FcR to cement a diffusion barrier to exclude CD45 through their association with actin-binding protein vinculin and talin. This requires actin polymerization.[70]

3.2.2 Phagocytic cup formation

During phagocytosis, the source of membrane for the extension of pseudopod to surround a target particle can be either from the adjacent plasma membrane, or from intracellular membrane source through membrane trafficking events.[71–74] PI3K, an enzyme that catalyzes the phosphorylation of three species of phosphatidylinositol lipid substrates viz. phosphatidylinositol-3-phosphate (PI3P) (by all PI3Ks), phosphatidylinositol-4-phosphate (PI4P) (by class-I and class-II PI3Ks), and phosphatidylinositol-4,5-bisphosphate ($PI(4,5)P_2$) (by class-I PI3K) at the 3′-hydroxyl of their D-myo-inositol head groups, is involved in membrane trafficking.[75–77] Its activity is required for optimal pseudopod extension at the initial stage of phagocytosis through exocytic membrane insertion.[36,78] PI3K exerts its effect through its downstream effector, Myosin-X, which is recruited to the phagocytic cup.[79]

Cup formation, closure and ultimately efficient phagosome formation requires the tight regulation in time and space between signaling, vesicle exocytosis and actin rearrangement. Vps-34, a class-III PI3K produces phophatidylinsitol-3-phosphate by coordinating with PIKI, a class-II PI3K to promote cup closure and phagosome maturation.[80] PI3P is transiently present on the nascent phagosomes[81] and is required for the recruitment of effector proteins including Early endosome antigen 1 (EEA1), Rabenosyn-5 and NADPH oxidase which have PI3P binding domains.[82]

For the release of membrane tension during intake of large particles, focal exocytosis of intracellular components is required in addition to rapid actin assembly-disassembly. Tight apposition between the phagocyte and target particle is established by protrusions like filopodia which provide tensile force to pull the target toward to phagocyte, and is necessary for efficient phagocytosis. After the receptors bind to target particle, the downstream signaling cascades lead to actin polymerization as well as depolymerization to form pseudopodia and phagocytic cup. This actin remodeling process takes place through different steps.[13] First, disruption of the cortical cytoskeleton which is associated with the membrane of the resting phagocyte is required. F-actin debranching protein, coronin 1, produces linear fibers which are then severed by actin severing proteins cofilin and gelsolin. PI(4,5)P_2 can sequester both cofilin and gelsolin thus preventing their association with the filament.

Breakdown of actin filaments associated with the phagocytic cup yields G-actin monomers that are then polymerized to give new filaments.[83] Arp2/3 complex initiates the nucleation of actin filaments for actin polymerization to form F-actin. This enables pseudopodia to extend and surround the target particle. In order to allow the membrane to curve around the target article and to leave space for the fusion of internal vesicles with the nascent phagosome, actin depolymerization starts from the base of the phagocytic cup.[28] ADP ribosylation factor 6 (Arf6) is responsible for the delivery of membrane to the nascent phagosome from inner pools.[84] A study showed that Arf6 might be activated by Arf6 GTPase-activating protein 2 (ACAP) recruited by Rab35, which is also suggested to co-ordinately regulate actin cytoskeleton.[85] PI(3,4,5) P_3 produced by PI3K during phagocytosis of large particles is thought to control the removal of actin by discontinuance of actin polymerization, detachment and depolymerization of the newly formed F-actin.[36,86]

3.2.3 Closure of phagocytic cup

To envelope a target particle, membrane tension has to be overcome with forces based on actin polymerization. During pseudopod extension, actin-mediated protrusions are guided and locked in place by cell-target interaction forces. As the growth of the cup continues into the cytoplasm, the closure of cup may require a mechanism to close the gap between the two ends of the cup. This contraction is likely to be mediated by myosin motor proteins which are possibly triggered by increase in membrane tension.[87] In particular myosin IC[88] and IG,[89] recruited by PI(3,4,5)

P_3 can contract the gap between the two ends to promote closure of the cup (Fig. 2). The extending phagocytic cup that surrounds the target particle then forms a phagosome at the plasma membrane that has to sink into the cytoplasm in order to proceed to the next stages for completion of phagocytosis. For this, scission of the nascent phagosome has to take place. Dynamin-2 plays an important role through its cross-talk with actin. It is known to function in separation of endocytic vesicles from the plasma membrane. It also accumulates at the site of phagosome closure (Fig. 2) and is probably involved in the closure and scission of phagosome as well.[90]

Cytokinesis, at the late stages also shares similar features with the early stages of phagocytosis. This includes the requirement of the actin cytoskeleton remodeling to form actin ring-like structure for growth of cleavage furrow (actinomycin contractile ring) during cytokinesis as well as the phagocytic cups for the internalization of particles during phagocytosis. The topologies are different in the two cases. During cytokinesis actinomyosin ring pulls the plasma membrane inwards to a pinched region while the plasma membrane is pushed out by the actin network during phagocytosis toward the completion of CR3-mediated phagocytosis. Rho family GTPases control F-actin polymerization in both cytokinesis and phagocytosis. RhoA is involved in cytokinesis and CR3-mediated phagocytosis, whose downstream effectors are formin mDia2 and mDia1, respectively, while Cdc42 and Rac1 are involved in FcR-medicated phagocytosis. As much as actin polymerization is needed, its turnover is also required for membrane remodeling during both cytokinesis and phagocytosis, where actin is severed with cofilin under the regulation of PI, particularly $PI(4,5)P_2$ hydrolysis. The delivery of membranes through vesicles from various origins is required to release the tension before abscission and phagosome closure. Both processes share few similar signaling molecules including Rab11, Rab35, and ARF6 for membrane delivery.[91]

3.3 Phagosome maturation

The operation of phagocytosis is not complete without maturation of phagosomes. During maturation phagosomes fuse with a series of endocytic organelles[92] either completely, or through transient and reversible interactions based on the "kiss and run" model,[93,94] or perhaps a combination of both.[15] Irrespective of the type of fusion, transformation of both the phagosomal membrane and its lumen occurs to create a hostile environment for target particle destruction. The specific events which mediate phagosome-endosome fusion have characteristics of endosome-endosome fusion.[95] Phagosome maturation involves several stages viz., early phagosome, intermediate phagosome, late phagosome and phagolysosome as shown in Fig. 2.

3.3.1 *Early phagosomes*

Soon after internalization, nascent phagosome starts fusing preferentially with early/sorting endosomes and starts acquiring their properties (Fig. 2).[95] Presence of Ras-related in brain 5 (Rab5) in the phagosome membrane prior to fusion with early endosomes is a pre-requisite for fusion to take place.[96] It is assumed that some Rab5 molecules that are already present on the plasma membrane are carried over to the nascent phagosome upon scission.[26,97] Fusion with early endosomes (EE) leads to additional recruitment of Rab5 to the early phagosome (Fig. 2). Rab5 recruits effectors like Rabaptin-5 which in turn recruits Vps34, which produces PI3P required for phagosome maturation.[81] PI3P accumulates on the inner leaflet of the membrane of early phagosome and engages EEA1 and Class C core vacuole/endosome tether (CORVET) (Fig. 2).[94] The pH of early phagosome is 5.4 due to the translocation of H^+ into the lumen by active V-ATPase.[98] The V-ATPase is believed to be acquired by the phagosome membrane from the plasma membrane.[99,100] This results in mild acidification of early phagosomes but they cannot digest the internalized material.

3.3.2 *Intermediate phagosomes*

As maturation of phagosome progresses, CORVET-Vps C/Homotypic protein sorting (HOPS) switch allowing Rab5 to Rab7 transition (Fig. 2).[101] HOPS functions as a GEF for Rab7 and associates with Rab5 to recruit Rab7. Rab5 is gradually lost possibly through the negative feedback from Rab7 via a Rab5-GAP. The Rab7 helps in the fusion of the early phagosome with late endosome (Fig. 2). The intermediate phagosome no longer resembles the early phagosome but has not yet been provided with the properties of late phagosome. At the same time, intra-luminal vesicles are formed through a process discussed in later sections for the degradation of proteins, notably receptors.

3.3.3 *Late phagosomes and phagolysosomes*

Late phagosomes are marked by the presence of Rab7 with complete loss of Rab5. Rab7 recruits new proteins to maturing phagosomes. Only few

Rab7 effectors have been characterized which include Oxysterol-binding protein related protein 1 (ORPL1) and Rab7-interacting lysosomal protein (RILP) that interact with dynein-dynactin complex[102] (Fig. 2). Together, these bring about the movement of the late phagosomes towards the lysosome through their interaction with the microtubules.[103] Lysosomal-associated membrane protein 1 (LAMP-1) is also recruited upon fusion with late endosomes.[104] The late phagosome become progressively acidic (~pH 5) as a consequence of additional V-ATPase in the membrane due to fusion with late endosome. Late phagosomes then fuse with lysosomes through Soluble N-ethylmaleimide-Sensitive Factor Attachment Protein Receptor (SNARE) such as vesicle-associated membrane protein 7 (VAMP7), VAMP8, syntaxin-7 and syntaxin-8.[105,106] The formed phagolysosome is highly acidic in nature with pH <4.5 due to large number of V-ATPase molecules in its membrane[15] (Fig. 2). This is required to activate enzymes such as cathepsins for the degradation of phagocytosed particles.[100] Phagolysosomes are characterized by an absence of mannose-6-phosphate receptors,[107] and have a PI4P-enriched internal membrane (Fig. 2) which is required for phagosome resolution to recover the membrane resources used in the process.[108] They acquire a number of hydrolytic enzymes. At this stage, membrane bound NOX2 NADPH oxidase forms superoxide (O_2^-). Superoxide is dismutated to hydrogen peroxide (H_2O_2) which further reacts with superoxide radicals to form reactive oxygen species (ROS) that are critical antimicrobial agents of macrophages.[109]

4. Phagocytosis: Connection with endoplasmic reticulum (ER)

The interaction of phagosome with endoplasmic reticulum (ER) has been a subject of debate. It was suggested earlier that ER, a cellular source of abundant endogenous membrane, is involved in the extension of pseudopod with the involvement of PI3K.[110] Additionally, reports of presentation of exogenous antigens by Major Histocompatibility Complex Class I (MHCI), known as cross-presentation, have invoked models suggesting involvement of the ER in the maturation of phagosomes. It is speculated that only when larger particles are to be ingested, i.e., when the need for membrane for pseudopod extension exceeds the capacity of other endosomal sources, is ER mobilized as a source of membrane.[111] The involvement of ER in phagosome formation has been debated as subsequent experiments after its first report failed to show major ER connection with phagosome in macrophage and dendritic cells.[112,113] However, the recruitment of a subset of ER resident proteins like transporter associated with antigen presentation (TAP), calnexin and calreticulin to phagocytic cups; and STIM1,[114] and especially those proteins involved in ER-associated degradation (ERAD) like AAA-type ATPase p97 and Sec61 to phagosomes as well as endosomes,[110,115–118] suggests that part of ER fuses with phagosomes. It may be through direct fusion or by membrane contact sites (MCS),[119] or through ER or ER-to-Golgi intermediate compartment (ERGIC) derived vesicles,[120] at least in the context of antigen cross-presentation and intracellular pathogen invasion.[119,121] In dendritic cells, knockdown of SNARE Sec22b, which is present both in the ER and ERGIC compartments, led to a decrease in the engagement of ER proteins with phagosomes. Sec22b knockdown also affected antigen cross-presentation without affecting phagocytosis and endocytosis. It colocalized with ERGIC marker ERGIC53, which was also found to be recruited to the phagosomes, suggesting that Sec22b is needed for the ER to contribute proteins to phagosome probably via ERGIC, and not directly form ER. Sec22b is also speculated to pair with syntaxin 4, which is present on the plasma membrane and phagosomes of dendritic cells, as both the proteins could be co-precipitated although direct evidence for the function of syntaxin 4 in trafficking from ERGIC to phagosome is still lacking.[120,122] This recruitment of ER proteins to phagosomes is thought to result in delayed lysosomal fusion thereby reducing proteolytic activity for better antigen cross presentation.[120]

In the context of *Legionella pneumophilla* infection where the organism hijacks the host's phagocytic machinery to create its intracellular niche, Sec22b has been shown to modulate the fusion of ER derived vesicles to *Legionella* containing vacuole (LCV). Sec22b was found to interact with plasma membrane-localized SNAREs including Stx2, 3, 4 and 5. *Legionella* directs this pairing through its Dot/Icm secretion system to facilitate the fusion of ER-derived vesicles to LCV.[123] The finding that ERGIC is recruited to phagosome without microbial signal in dendritic cells suggests that *Legionella* may be appropriating a preexisting mechanism to slow down or prevent lysosomal fusion for its own benefit.[120]

While the direct fusion of ER with phagosome seems to be questionable, in a study to check the role of ERAD retrotranslocation machinery in antigen cross-presentation, the detection of N-linked glycosylated peptide in phagosomes suggests the direct contact of ER with the phagosome.[124] In this study, microspheres coupled with glycosylation acceptor peptide were used as substrates for phagocytosis by dendritic cells. Isolated phagosomes could be stained with fluorescently labeled concanavalin A and detected by flow cytometry. Although these could come from ingestion of glycosylated particles, they could also be produced by the ER machinery since N-glycosylation activity is an exclusive ER property.[120,124]

The presence of syntaxin-18, a SNARE protein in the ER membrane, in J774 phagosomes also suggests the possible involvement of this protein in fusion of ER with phagosomes or plasma membrane.[125] Moreover, with a subsequent study showing that the ER contributes to 20% of the early phagosome proteome,[126] it is likely that the ER is involved during phagosome formation though its definite function remains to be determined. The fusion and connectivity of ER with phagosome could, however, be incompatible with the ionic, oxidative and proteolytic condition of the phagosome.[127] With the acute awareness of pervasive existence of MCS, which are areas of specialized functional apposition of two organelles, one of which is mostly the ER,[128] it might be possible to obviate the need for direct fusion. The remaining association of some MCS with purified phagosomes upon disruption of cells might also account for observed presence of ER proteins in the phagosome proteome and also the electron microscopy observations of ER-phagosome fusion.[110] Other than generating Ca^{2+} transients during phagosome formation and early maturation, which is likely a trigger for fusion of endosome with the nascent phagosome, the role of the MCS between ER and phagosome is largely unknown.[114,127]

5. Role of ubiquitin proteasome pathway (UPP) in phagocytosis

Phagocytosis is a significantly more complex process compared to endocytosis and utilizes diverse signaling pathways which depend on the nature of phagocyte, the target particle, and the receptors involved in the recognition of the particle. For instance, Arp2/3 complex is critical for actin rearrangement in complement mediated phagocytosis and other integrin-dependent processes, while its function is dispensable for FcR-mediated phagocytosis.[129,130] Size of the particle can also direct the signaling pathway that is activated in phagocytosis. For example, PI3K is required for phagocytosis of large particles but not for small particles.[131] For all these reasons, identification of all bona fide phagocytic proteins remains a challenge. During endocytosis, several receptors are recruited to the endosomal vesicles after being modified with ubiquitin and later the formation of multi-vesicular bodies (MVBs) also requires ubiquitin modification of the proteins of the endosomal sorting complex required for transport (ESCRT) pathway.

Though much of the molecular mechanisms involved in controlling the process of phagocytosis have been uncovered, much remains to be elucidated. The involvement of the ubiquitin-proteasome pathway (UPP) in phagocytosis is one such aspect of phagocytosis, whose role is only recently being appreciated, and needs more extensive analysis and research. While UPP is involved at multiple steps in phagocytosis, the underlying molecular mechanisms are still mostly obscure.

Ubiquitin (Ub) modification of cellular proteins plays a major role in a number of cellular processes. Post-translational modification of substrate proteins with Ub is a highly selective process requiring a cascade of three classes of enzymes. The first step is the activation of Ub by Ub activating enzyme (E1) in an ATP-dependent reaction, then transfer of Ub to Ub conjugating enzymes (E2s) which constitute a family with dozens of members, and finally its conjugation to the substrate by a large number of Ub ligases (E3s) of which there are several classes. Ub chains can be linked via various internal lysine residues that serve different purposes. While a major function of ubiquitination is to degrade old and misfolded proteins by the 26S proteasomes, a protein's localization, interactions and activity are also regulated by diverse ubiquitin mediated signals. Therefore, the UPP is critical for maintaining homeostasis in the cell. In addition to protein quality control, it plays a crucial role in DNA repair, transcription, signal transduction, cell cycle progression (see the following excellent reviews for details of this pathway[132–134]). Hence, defects in this system is correlated with various human diseases including malignancies,[135] neurodegenerative diseases,[136] immune and inflammatory response.[137] It is a recognized therapeutic target for cancer, neurodegenerative, autoimmune, inflammatory and parasitic diseases.[138,139]

Involvement of UPP in receptor-mediated endocytosis and in the formation of multivesicular bodies (MVBs) during endosomal maturation is well established. During endocytosis, ubiquitination is required to internalize and downregulate cargo receptors like epidermal growth factor receptor (EGFR), growth hormone receptor (GHR), epithelial sodium channel (ENaC), colony-stimulating factor 1 (CSF1) receptors, as well as FcgRIIA receptor,[140–142] thereby playing diverse roles in regulating the activity of plasma membrane proteins.[143] The first direct evidence showing that ubiquitination is required for endosomal sorting of proteins to luminal vesicles of MVBs came from the study of *Saccharomyces cerevisiae* surface protein, Ste2p, which is a G-protein coupled receptor, whose mutation at its single lysine residue abrogated its mono-ubiquitination at the cell surface and also its internalization.[143–145]

Ubiquitination is required for intracellular sorting of cargos like EGFR, and major histocompatibility complex (MHC)-II into multivesicular bodies (MVB) during endosomal maturation in endocytosis.[146,147] These MVBs have intralumenal vesicles (ILVs) formed by the inward budding of vesicles that is promoted by endosomal sorting complexes required for transport (ESCRT) complexes.[141,148,149] In general, MVB formation requires the ubiquitination of cargo protein for the recognition by ESCRT-0, which recruits ESCRT-I that interacts weakly with the membrane. ESCRT-0 then hands over the ubiquitinated cargo to ESCRT-I, which then activates ESCRT-II, recruiting it to the membrane, handing over the cargo to ESCRT-II and producing membrane invaginations. Additionally, both ESCRT-0 and ESCRT-II bind to PI3P which provides endosomal and membrane recruitment specificity. The next ESCRT complex, ESCRT-III, is then activated after it is bound by ESCRT-II but, unlike the other ESCRT complexes, it does not bind to ubiquitinated cargo. It recruits deubiquitination machinery for removing ubiquitin from the cargo, packing the cargo into ILVs, drives vesicle budding and regulates scission and release of ILVs. The vesicles then bud off inward from the membrane and ESCRT-III complex is then finally disassembled by an AAA-ATPase, Vps4, so that it can be reused again.[150–152]

Endocytosis and phagocytosis have fundamental differences with distinct requirements, e.g., the dependence of the former on clathrin, while the latter being dependent on localized F-actin assembly at sites of particle uptake. Nonetheless, they make use of several common protein components during cargo internalization as well as during maturation processes, including dynamin II and amphiphysin.[153] Moreover, endocytic vesicles and phagosomes go through similar maturation steps after internalization, acquiring different markers as they mature.[140] A number of studies provide the evidence that UPP plays an important role at multiple levels in phagocytosis as described below.

5.1 UPP in phagosome formation and maturation

Though it was previously believed that ubiquitination is not essential for phagocytosis and phagosome maturation,[140] subsequent studies have shown that both ubiquitination and proteasome function is required for efficient phagocytosis. Phagosome associated proteins have been reported to be ubiquitinated, and sorting of the cargo into MVBs is inhibited if mono- and polyubiquitination is impaired. Additionally, sorting of the cargo to MVBs is also inhibited in the absence of proteasome activity.[149] Proteasomes are recruited to the phagosomes during antigen-cross presentation,[154] and may process phagosomal membrane proteins, thus playing additional roles in phagosomal functions. A direct molecular association between the alpha-7 subunit of proteasomes and Rab7, which governs early to late endosomal transport, has been reported.[155] Proteasome inhibition also arrested sorting of FcγRIIA from phagosome to MVB.[149] Proteasomes thus probably function by degrading a negative regulator of MVB formation or may trim polyubiquitinated receptor so that they may be recognized by the endocytic pathway machinery.[149,156] The precise mechanisms by which ubiquitination and proteasomes regulate phagocytosis remains obscure as neither the specific ubiquitin conjugating enzymes, the ubiquitin ligases, nor the substrates acted upon by them, are known.

As mentioned earlier, UPP is involved in a number of cellular pathways due to its principal role in protein modification. Given the fact that phagosome formation and maturation involve the integrated action of many signaling molecules, it is expected that UPP also has additional roles in phagocytosis other than in MVB formation, and may even have a role at the very early step of binding to the target. Upon binding to the target, the FcγRs cluster together, leading to cytoskeletal remodeling due to activation of tyrosine kinases to initiate downstream signaling. One such signaling protein was found to be c-Cbl, which is an E3.

Compared to wild-type control, an unexpected increase in binding with IgG-opsonized targets by $Cbl^{-/-}$ macrophages was observed, in addition to enhanced tyrosine phosphorylation and overall phagocytosis, suggesting that target binding is not a passive process. Both c-Cbl and its isoform, Cbl-b, were phosphorylated by tyrosine kinases upon FcγR stimulation and were found to be recruited to the phagocytic cups in bone marrow-derived macrophage (BMDM). Both $Cbl\text{-}b^{-/-}$ and $c\text{-}Cbl^{-/-}$ murine BMDMs showed enhanced phagocytosis when challenged with opsonized target. The increase in binding capacity could not be explained in terms of surface FcγR expression. Rather, it was attributed to the regulation of the ratio between inhibitory and activating FcγRs by Cbl proteins wherein the former is favored to inhibit maximal binding (Fig. 3A). It was

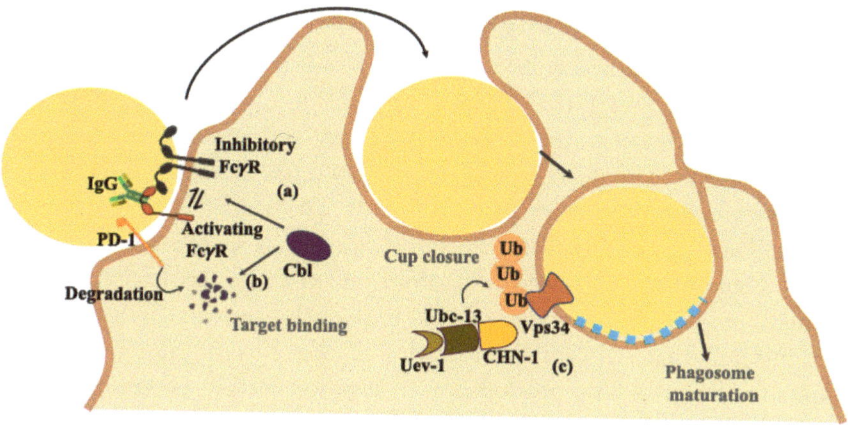

Fig. 3 UPS in phagosome formation and maturation. (A) Negative regulation of phagocytosis by Cbl E3. (B) Enhancement of tumor phagocytosis by c-Cbl. (C) UPP in phagosomal maturation.

speculated that FcγRI may have a lower activation threshold in macrophages that do not have Cbl proteins as a result of its pre-association with plasma membrane microdomains containing Src family proteins. Cbl may also function as an adapter to negatively regulate phagocytosis.[157] The opposite effect of Cbl has been found during tumor phagocytosis (Fig. 3B), where Cbl was shown to enhance phagocytosis through binding and ubiquitination of Programmed cell death-1 (PD-1), targeting it for proteasomal degradation. This leads to down regulation of PD-1 surface expression, thus initiating tumor cell phagocytosis and ultimately resulting in tumor growth suppression.[157,158] Whether the ubiquitin ligase function of Cbl is required for these effects on target binding is not known.

Further, the phagosome maturation process is also likely to be regulated by participation of the UPP as reported in *Caenorhabditis elegans*. In this organism, an E2 UBC-13, in conjunction with a non-catalytic E2 variant UEV-1, and a Ub ligase CHN-1, were shown to ubiquitinate and stabilize Vps-34 through K-63 linked polyubiquitination (Fig. 3C). As discussed before, Vps-34 is a PI3K whose activity is required for the maturation process. The deletion and expression of loss-of-function mutant forms of UBC-13, UEV-1 and CHN-1 led to decreased association of Rab5 and PI3P in early stage phagosomes, as well as Rab7 and Lysosomal Lysine/Arginine Transporter (LAAT)-1 in late stage phagosomes, thus deregulating phagosome maturation.[80]

5.2 UPP in innate immunity and antigen cross presentation

UPP also plays important roles in regulating innate immunity. Monoclonal nonspecific suppressor β (MNSFβ), a Ubl, was found to regulate Dectin-1 mediated phagocytosis by conjugating covalently with endophilin II (Fig. 4A). Double knockdown of both MNSFβ and endophilin II using siRNA resulted in increased phagocytosis of zymosan by RAW 264.7 cells through Dectin-1 receptor to the same amount as endophilin II siRNA alone. This shows that MNSFβ alone is not the primary player in Dectin-1 mediated phagocytosis, as previously shown.[159] It was also demonstrated that MNSFβ/Endophylin II complex negatively regulates zymosan-induced TNFα production. In addition to this, IκBα degradation was enhanced with knockdown of endophilin II and also MNSFβ, indicating that MNSFβ/Endophylin II complex inhibits signaling pathway upstream of IKK activation. It was proposed that MNSFβ does not directly regulate degradation of IκBα. Rather, it changes the interactions of the modified protein with other cellular proteins.[160]

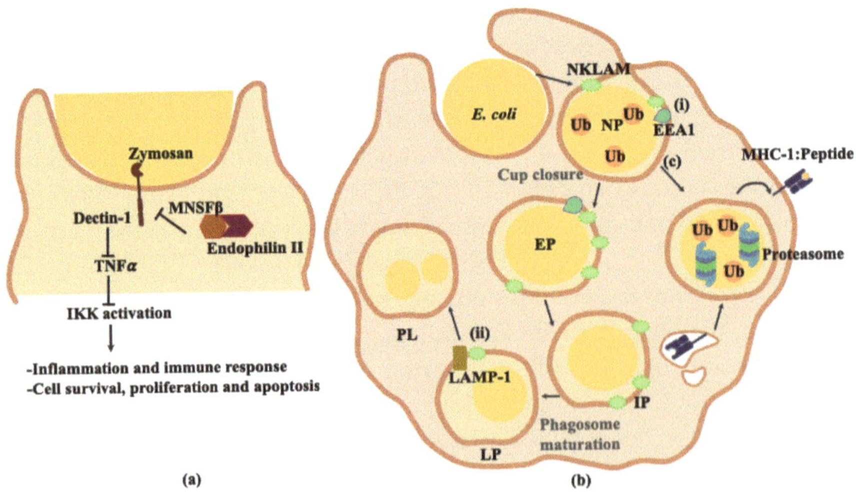

Fig. 4 UPS in innate immunity and antigen cross presentation. (A) Regulation of innate immunity by the post-translational modification of Endophilin II with the ubiquitin-like protein Monoclonal nonspecific suppressor β (MNSFβ). (B) Time dependent association of the ubiquitin ligase, NKLAM with maturing phagosome. (i) Natural killer lytic-associated molecule (NKLAM) colocalization with EEA1 and (ii) LAMP-1. (C) Antigen cross-presentation from the phagosomes. NP, Nascent Phagosome; EP, Early Phagosome; IP, Intermediate Phagosome; LP, Late Phagosome; PL, Phagolysosome.

Additionally, natural killer lytic-associated molecule (NKLAM), a RING-between-RING (RBR) E3 enzyme, has also been observed to translocate to phagosomes in bone marrow-derived macrophages early during maturation, where its translocation increases 35 min post-ingestion and then decreases with maturation (Fig. 4B). It was found to colocalize with phagosome proteins EEA1 and LAMP-1.[161] It is involved in *E. coli* killing inside phagosomes and its overexpression increased the killing in J774 cells as compared to control-transfected cells. However, it is not involved in bacterial uptake and phagosomal maturation process.[162] The identity of phagosome targets of NKLAM is unknown and exact mechanism of how NKLAM functions in the bactericidal activity is still unclear. Its localization with the phagosome coincides with an increase in phagosome associated ubiquitinated proteins. Whether its catalytic RING domain has access to phagosomal cargo or cytosolic targets, is not known at present.[161] NKLAM has also been found to interact with three members of PI3K complex according to BioGRID database.[161,163]

Phagocytosis also links innate immunity to adaptive immunity through antigen presentation. UPP has been implicated in antigen cross presentation. For instance, active proteasomes that are present inside the phagosomes of BMDCs act to generate peptides which stabilize MHCI on the cell surface for antigen cross-presentation (Fig. 4C). Labeled immunoproteasome subunit LMP2, and three different antibodies specific for S2 subunit located in the 19S cap, α5 and β5 components of outer and inner rings respectively of the 20S core particle of the proteasome were observed in LAMP-1 labeled membrane bound compartments of endolysosome. Ubiquitination of phagosomal cargo was also observed with GFP-tagged ubiquitin.[164] This suggests that an active UPP is present in phagosomes and involved in antigen cross presentation. It is unknown, however, whether this association is required only in the context of antigen cross-presentation, or is also required in the basic mechanism of phagocytosis. The involvement and function of ubiquitination in phagocytosis may be restricted by the type of phagocyte involved and also the type of target particle ingested.

Therefore, it is not known if the same effect would be observed if the size or nature of the target particle is changed.

5.3 UPP in removal of apoptotic body and maintenance of tissue homeostasis

With its vast target range, UPP may also be involved in other non-immunological functions of phagocytosis. A forward genetics study to

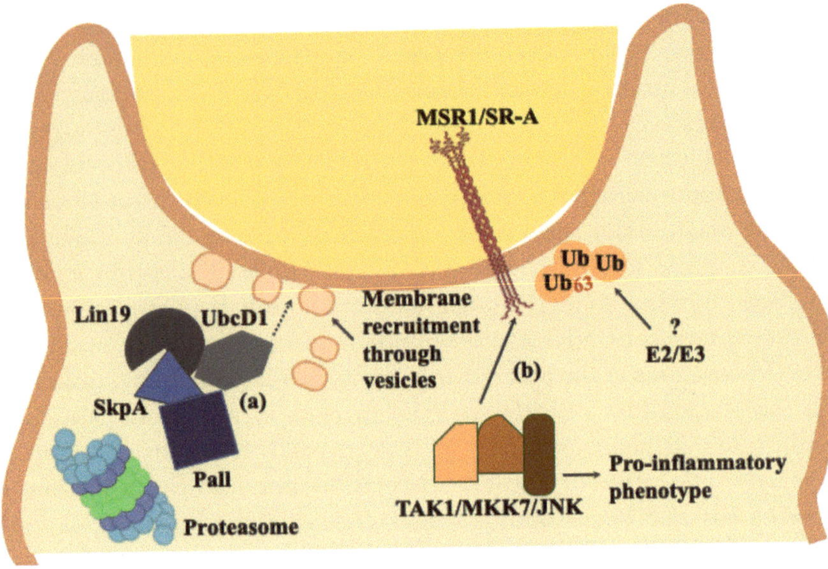

Fig. 5 UPS in removal of apoptotic body and maintenance of tissue homeostasis. (i) Hypothetical representation of UbcD1 and SCF complex consisting of SkpA, Lin19 and Pall proteins in membrane recruitment and subsequent membrane enlargement in removal of apoptotic corpse in *Drosophila melanogaster.* (ii) Phenotypic change of anti-inflammatory M2 macrophage into pro-inflammatory M1 macrophage by association of TAK1/MKK7/JNK signaling complex to Macrophage scavenger receptor-1/Scavenger receptor A (MSR1/SR-A) via K63-linked polyubiquitination on MSR1/SR-A. The E2/E3 involved in this process is unknown. Dotted arrow: unknown process.

identify genes involved in the removal of apoptotic corpse by phagocytosis in *Drosophila* embryonic macrophages showed the involvement of an E2, Effete (or UbcD1), and an E3 SCF complex, and proteasomes in the process (Fig. 5A). The identified SCF complex consisted of Skp1 homolog SkpA, Cullin-1 homolog Lin19, and an F-box protein Pall. Loss-of-function mutant of *pall*, null mutants of *skpA* and *lin-19* and loss-of-function mutant of *effete* genes led to similar defects in phagocytosis.[165,166] Pall physically interacts with SkpA and acts as an engulfment specific subunit of the SCF complex.[166] Also, the expression of temperature sensitive dominant-negative subunits of 26S proteasome led to strong phagocytic defects under the restrictive temperature in the drosophila study. However, Pall-SCF complex was found to be neither essential for maturation and degradation, nor for phagosome formation. It was hypothesized to be involved in bringing about membrane recruitment and subsequently macrophage enlargement since phagocytosis was not fully eliminated, but instead, engulfment was markedly reduced. Since *pall* mutant macrophages did not show defects in endocytosis of scavenger receptor, Pall-SCF appears to be specific for phagocytosis. This activity is thought to be essential only for clearance or phagocytosis of apoptotic cells or large particles, and appears to be unnecessary for phagocytosis of extracellular particles or organisms, as was shown for *Escherichia coli* and *Staphylococcus aureus*.[167] But, further experimental evidence is required to draw a final conclusion.[165]

Tissue homeostasis is another role played by phagocytosis. Macrophages can polarize into many different phenotypes in response to environmental cues due to their high plasticity. Interleukin 4 (IL-4) can induce a shift from an anti-inflammatory to a pro-inflammatory phenotype. Lysine-63-linked polyubiquitin chains function as a scaffold for assembly of signaling complexes in this process. The Macrophage scavenger receptor-1/ Scavenger receptor A (MSR1/SR-A) enables the phenotypic shift of macrophages via TAK1/MKK7/JNK signaling complex in alternatively activated M2 macrophages. MSR1 is a marker for tissue activated macrophages (TAMs) that are associated with tumor progression. Polyubiquitination of MSR1 by K63-linked ubiquitin chains by an unknown E2/E3 ligase is necessary to recruit the TAK1/MKK7/JNK signaling complex to phagosomes on activation with IL-4 (Fig. 5B). Increased expression of K63-polyubiquitinated MSR1, as well as increased activation of JNK pro-inflammatory signaling, is also observed in TAMs. This indicates that TAMs of human cancers have JNK signaling downstream of K-63 linked polyubiquitinated MSR1 and could potentially be involved in tumor promotion.[168]

5.4 UPP and marking of the phagocytic target

In addition to its functions inside the phagocyte, ubiquitination may also affect the phagocytosis of target particles directly. For instance, ubiquitination was found to play a role in marking defective sperm for phagocytosis by the epididymal epithelial cells (EEC).[169] The ubiquitin-dependent sperm phagocytosis was reconstituted in vitro. Sperm killed by enzymatic digestion was also present in the isolated cell suspensions. A quantitative study was done and demonstrated that compared to controls, ubiquitin-coated beads had higher affinity for cultured EEC. Possibly both normal and defective sperm intrinsically may have ubiquitinated substrates. However, during epididymal passage, only defective sperms become surface ubiquitinated.[169] It was hypothesized that epididymal ubiquitination at the intersection, or at the common end-point of many apoptotic mechanisms may recognize damaged sperm accessory structures or structural damage of sperm DNA and eliminated by phagocytosis. The phagocytosed sperm is likely to be digested by a process similar to that for endocytosed receptors by utilizing the lysosomal machinery.[142,169]

5.5 UPP proteins with unknown roles in phagocytosis

Many proteins of the UPP have also been detected in the proteome of phagosome, some of which have also been found to regulate phagocytosis.[170,171] A mass spectrometry approach to compare phagosome proteomes between RAW 264.7 and bone marrow derived macrophage (BMDM) identified 3535 phagosome-associated proteins. Quantitation showed 2510 proteins that were found in two of the three replicates. This included a substantial amount of UPP related proteins.[170] Among them are ubiquitin activating enzyme (E1) like UBA1, E2s like Ube2g2 (fragment) and Ube2j1 and E3s like CBL, ITCH, MGRN1, RNF13, RNF 25, RNF128, RNF130, RNF149, RNF213, RNF220, Synoviolin, TRIPC, WWP1, WWP2, ZNRF1 and ZNRF2. Other than CBL, MGRN1 and ITCH which function in endocytic trafficking,[172–174] the functions of other proteins in phagocytic process remain unknown. Nevertheless, association of a large number of proteins related to UPP pathway indicates a close connection of this pathway with phagocytosis. In addition to this, another study using magnetic bead and genome-wide CRISPR analysis to screen genes regulating phagocytosis identified *UBE2L3, UBE2D3, STUB1, USP22, KCTD5* and *SUPT20H* which code for proteins of the ubiquitin machinery along with other genes involved in various other cellular pathways. The knockout of the aforementioned genes could negatively or positively regulate phagocytosis depending on the substrate.[171] The observations made in some of these organisms have not been tested in mammalian systems. It would be worth determining if the E2–E3 complexes associated with phagocytosis in other organisms play the same roles in mammals to fill the gaps in understanding the underlying role of UPP in phagocytosis.

6. Aberrations of UPP in phagocytosis and human diseases

β-Amyloid (Aβ), forms aggregates to produce amyloid plaques in the brain in Alzheimer's disease (AD). It is phagocytosed by macrophages in

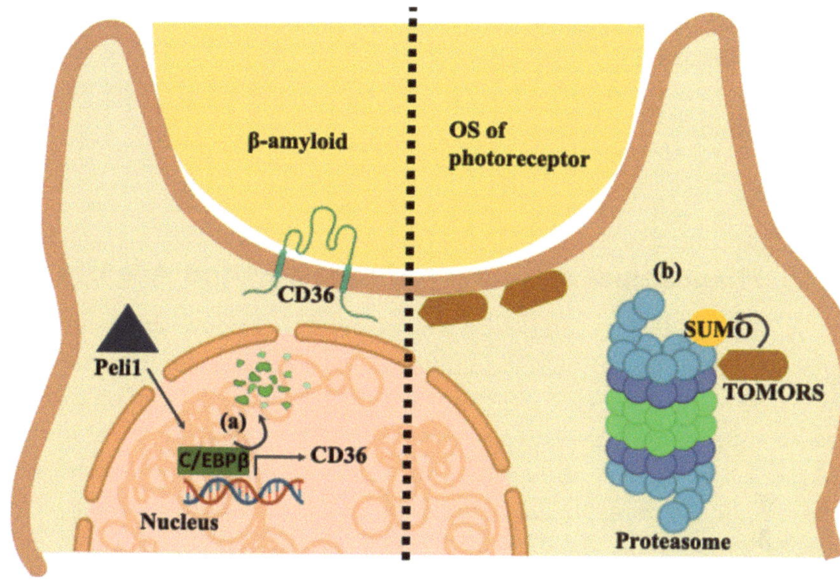

Fig. 6 Aberrations of UPS in phagocytosis and human diseases. (i) Pellino1 (Peli1) E3 ligase in regulation of β-amyloid (Aβ) phagocytosis by scavenger receptor CD36. (ii) Clearance of outer segment (OS) of photoreceptor cells in retinal pigment epithelium (RPE) via dual E3 ubiquitin and SUMO1 ligase, Topoisomerase-1 binding arginine/serine-rich protein (TOMORS).

normal subjects while its phagocytosis is deficient in most AD patients.[175] An E3 ligase Pellino1 (Peli1), negatively regulates phagocytosis of Aβ by microglia. Peli1 was shown to inhibit the expression of scavenger receptor CD36 by directly degrading C/EBPβ, a major transcription factor for CD36 expression (Fig. 6A). In AD brain, Peli1 has been found to be induced, further impairing phagocytosis of Aβ by microglia during disease pathogenesis.[176]

Retinal pigment epithelium (RPE) is one of the most active phagocytic cell groups in the body and functions to engulf outer segments (OS) of photoreceptor cells. The failure of this process can lead to retinopathies.[177] Mutations of Topoisomerase-1 binding arginine/serine-rich protein (TOMORS), a dual E3 ubiquitin and SUMO1 ligase, leads to autosomal dominant retinal pigmentosa and also autosomal dominant pericentral retinal dystrophy.[178,179] TOMORS was shown to physically interact with P26s4, the 26S proteasome regulatory subunit 4 (Fig. 6B). The interaction mostly involved the domain of TOMORS which contains the SUMO1 ligase activity suggesting that P26s4 may be a substrate for SUMOylation for TUMORS. Interestingly, immunostaining of adult mouse retina shows P26s4 to be localized between photoreceptor OS and RPE. This is suggestive of P26s4 involvement in RPE phagocytosis of photoreceptor and RPE homeostasis.[180]

7. Phagosomal machinery of parasites and role of UPP

A number of organisms have evolved strategies to make the destructive phagosome into a conducive environment for survival and growth. Though the exact mechanism and pathways are yet unknown, ubiquitination definitely plays an important role in phagosome functions and modulating its environment.[172] This importance can be demonstrated by the specific targeting of the ubiquitination machinery during infection by *Legionella pneumophila*. After phagocytosis, *L. pneumophila* replicates in a *Legionella* containing vesicle (LCV). The efficient replication of the bacteria requires the UPP of the host. The LCV is covered with ubiquitinated proteins which is achieved by effector proteins of the bacteria that either mimic ubiquitin ligases or by other novel mechanisms. SidC, is one such novel effectors that has a novel fold with an E3 ligase activity. SidC helps in polyubiquitin conjugate formation around the bacterial phagosome along with the recruitment of ER components.[181] Thus, the pathogen manipulates the host secretory pathway by utilizing ubiquitin ligase activity of its effector proteins.[181]

Initial steps of phagocytosis in *Entamoeba* have been elucidated in some detail and utilize a set of unique proteins. Expression of enzymatically inactive form of EhUbc7, an orthologue of human Ube2g2 and yeast Ubc7, has been shown to result in reduced number of phagocytic cups and phagosomes without complete abrogation of phagocytosis, and without any effect on the locomotion of the parasite.[182] This suggests a ubiquitination dependent process, separate from the actin requiring process involved in the extension of pseudopod around the target particle for the formation of phagocytic cups. In addition to this, unlike its orthologues, EhUbc7 was found to colocalize in the plasma membrane and in phagocytic cups with calreticulin, an *E. histolytica* receptor for C1q that facilitates phagocytosis.[183] Whether this effect of Ubc7 on phagocytosis is specific for *E. histolytica* or conserved in other organisms is unknown.

8. Future perspectives

Phagocytosis is a complex process, not only in terms of the shapes and sizes of the targets that are engulfed, but also in terms of the types of cells that can function as phagocytes, and the various receptors and their associated downstream signaling processes that are involved. While we now understand many different aspects of this complex process, a lot still needs to explored and understood. With a large number of ubiquitinated phagosomal proteins and UPP proteins found to be present in the phagosomes, there is a need to identify the ubiquitin conjugating enzymes and ubiquitin ligases, and other Ubls that function in the pathway. There is also a need to understand the direct vs indirect effects of inhibition or mutation of the involved UPP proteins on phagocytosis. With the increasing advancement in both proteomic and microscopic techniques, more discoveries on the specific phagocytic functions of the UPP proteins will hopefully soon ensue. This can further pave the way for a clearer understanding of the process of phagocytosis and thus its manipulation and identification of novel drug targets.

Acknowledgments

The authors acknowledge the support of Indian Council of Medical research (ICMR) grant 5/8-1(56)/2013-14-ECD-II to S.T., Department of Biotechnology (DBT), Government of India for Master's in Biotechnology fellowship support (to A.L. and G.D.), and DBT Research fellowship (DBT/JRF/BET-18/1/2018/AL/175) to A.L.

References

1. Tan SY, Dee MK. Elie Metchnikoff (1845-1916): Discoverer of phagocytosis. *Singapore Med J*. 2009;50(5):456–457.
2. Aderem A, Underhill DM. Mechanisms of phagocytosis in macrophages. *Annu Rev Immunol*. 1999;17(1):593–623. https://doi.org/10.1146/annurev.immunol.17.1.593.
3. Hanawa-Suetsugu K, Itoh Y, Ab Fatah M, et al. Phagocytosis is mediated by two-dimensional assemblies of the F-BAR protein GAS7. *Nat Commun*. 2019;10 (1):1–13. https://doi.org/10.1038/s41467-019-12738-w.
4. Boulais J, Trost M, Landry CR, et al. Molecular characterization of the evolution of phagosomes. *Mol Syst Biol*. 2010;6(1):1–14. https://doi.org/10.1038/msb.2010.80.
5. Fais S, Overholtzer M. Cell-in-cell phenomena in cancer. *Nat Rev Cancer*. 2018;18 (12):758–766. https://doi.org/10.1038/s41568-018-0073-9.
6. Seeberg JC, Loibl M, Moser F, et al. Non-professional phagocytosis: A general feature of normal tissue cells. *Sci Rep*. 2019;9(1):1–8. https://doi.org/10.1038/s41598-019-48370-3.
7. Gupta N, Jadhav K, Shah V. Emperipolesis, entosis and cell cannibalism: Demystifying the cloud. *J Oral Maxillofac Pathol*. 2017;21(1):92–98. https://doi.org/10.4103/0973-029X.203763.

8. Hofmann A, Putz F, Büttner-Herold M, Hecht M, Fietkau R, Distel LV. Increase in non-professional phagocytosis during the progression of cell cycle. *PLoS One*. 2021;16 (2):1–16. https://doi.org/10.1371/journal.pone.0246402.
9. Fais S. Cannibalism: A way to feed on metastatic tumors. *Cancer Lett*. 2007;258 (2):155–164. https://doi.org/10.1016/j.canlet.2007.09.014.
10. Parzych KR, Klionsky DJ. An overview of autophagy: Morphology, mechanism, and regulation. *Antioxid Redox Signal*. 2014;20(3):460–473. https://doi.org/10.1089/ars.2013.5371.
11. Gómez-Díaz C, Ikeda F. Roles of ubiquitin in autophagy and cell death. *Semin Cell Dev Biol*. 2019;93:125–135. https://doi.org/10.1016/j.semcdb.2018.09.004.
12. Rabinovitch M. Professional and non-professional phagocytes: An introduction. *Trends Cell Biol*. 1995;5(3):85–87. https://doi.org/10.1016/S0962-8924(00)88955-2.
13. Rosales C, Uribe-Querol E. Phagocytosis: A fundamental process in immunity. *Biomed Res Int*. 2017;2017:1–18. https://doi.org/10.1155/2017/9042851.
14. Levin R, Grinstein S, Canton J. The life cycle of phagosomes: Formation, maturation, and resolution. *Immunol Rev*. 2016;273(1):156–179. https://doi.org/10.1111/imr.12439.
15. Flannagan RS, Cosío G, Grinstein S. Antimicrobial mechanisms of phagocytes and bacterial evasion strategies. *Nat Rev Microbiol*. 2009;7(5):355–366. https://doi.org/10.1038/nrmicro2128.
16. Uribe-Querol E, Rosales C. Control of phagocytosis by microbial pathogens. *Front Immunol*. 2017;8:1–23. https://doi.org/10.3389/fimmu.2017.01368.
17. Ezekowitz RA, Sastry K, Bailly P, Warner A. Molecular characterization of the human macrophage mannose receptor: Demonstration of multiple carbohydrate recognition-like domains and phagocytosis of yeasts in Cos-1 cells. *J Exp Med*. 1990;172 (6):1785–1794. https://doi.org/10.1084/jem.172.6.1785.
18. Brown GD, Gordon S. A new receptor for β-glucans. *Nature*. 2001;413:6851. https://doi.org/10.1038/35092620.
19. Herre J. Dectin-1 uses novel mechanisms for yeast phagocytosis in macrophages. *Blood*. 2004;104(13):4038–4045. https://doi.org/10.1182/blood-2004-03-1140.
20. Peiser L, Gough PJ, Kodama T, Gordon S. Macrophage class A scavenger receptor-mediated phagocytosis of Escherichia coli: Role of cell heterogeneity, microbial strain, and culture conditions in vitro. *Infect Immun*. 2000;68(4): 1953–1963. https://doi.org/10.1128/IAI.68.4.1953-1963.2000.
21. Rosales C. Control of phagocytosis by microbial pathogens. *Front Immunol*. 2017;8 (October):1–23. https://doi.org/10.3389/fimmu.2017.01368.
22. Anderson CL, Shen L, Eicher DM, Wewers MD, Gill JK. Phagocytosis mediated by three distinct fc gamma receptor classes on human leukocytes. *J Exp Med*. 1990;171 (4):1333–1345. https://doi.org/10.1084/jem.171.4.1333.
23. van Lookeren CM, Wiesmann C, Brown EJ. Macrophage complement receptors and pathogen clearance. *Microbiology*. 2007;9(9):2095–2102. https://doi.org/10.1111/j.1462-5822.2007.00981.x.
24. Ross GD, Reed W, Dalzell JG, Becker SE, Hogg N. Macrophage cytoskeleton association with CR3 and CR4 regulates receptor mobility and phagocytosis of iC3b-opsonized erythrocytes. *J Leukoc Biol*. 1992;51(2):109–117. https://doi.org/10.1002/jlb.51.2.109.
25. Blystone SD, Graham IL, Lindberg FP, Brown EJ. Integrin alpha v beta 3 differentially regulates adhesive and phagocytic functions of the fibronectin receptor alpha 5 beta 1. *J Cell Biol*. 1994;127(4):1129–1137. https://doi.org/10.1083/jcb.127.4.1129.
26. Flannagan RS, Jaumouillé V, Grinstein S. The cell biology of phagocytosis. *Annu Rev Pathol Mech Dis*. 2012;7(1):61–98. https://doi.org/10.1146/annurev-pathol-011811-132445.

27. Brown EJ. Complement receptors and phagocytosis. *Curr Opin Immunol.* 1991;3 (1):76–82. https://doi.org/10.1016/0952-7915(91)90081-B.
28. Freeman SA, Grinstein S. Phagocytosis: Receptors, signal integration, and the cytoskeleton. *Immunol Rev.* 2014;262(1):193–215. https://doi.org/10.1111/imr.12212.
29. Indik Z, Kelly C, Chien P, Levinson AI, Schreiber AD. Human fc gamma RII, in the absence of other fc gamma receptors, mediates a phagocytic signal. *J Clin Invest.* 1991;88 (5):1766–1771. https://doi.org/10.1172/JCI115496.
30. Strijbis K, Tafesse FG, Fairn GD, et al. Bruton's tyrosine kinase (BTK) and Vav1 contribute to Dectin1-dependent phagocytosis of Candida albicans in macrophages. *PLoS Pathog.* 2013;9(6), e1003446. https://doi.org/10.1371/journal.ppat.1003446.
31. Uribe-Querol E, Rosales C. Phagocytosis: our current understanding of a universal biological process. *Front Immunol.* 2020. https://doi.org/10.3389/fimmu.2020.01066.
32. Fitzer-Attas CJ, Lowry M, Crowley MT, et al. Fcγ receptor–mediated phagocytosis in macrophages lacking the Src family tyrosine kinases Hck, Fgr, and Lyn. *J Exp Med.* 2000;191(4):669–682. https://doi.org/10.1084/jem.191.4.669.
33. Choraghe RP, Kołodziej T, Buser A, Rajfur Z, Neumann AK. RHOA mediated mechanical force generation through Dectin-1. *J Cell Sci.* 2020;133(5):1–11. https://doi.org/10.1242/jcs.236166.
34. Kiefer F, Brumell J, Al-Alawi N, et al. The Syk protein tyrosine kinase is essential for Fcγ receptor signaling in macrophages and neutrophils. *Mol Cell Biol.* 1998;18 (7):4209–4220. https://doi.org/10.1128/MCB.18.7.4209.
35. Gu H, Botelho RJ, Yu M, Grinstein S, Neel BG. Critical role for scaffolding adapter Gab2 in FcγR-mediated phagocytosis. *J Cell Biol.* 2003;161(6):1151–1161. https://doi.org/10.1083/jcb.200212158.
36. Cox D, Tseng C-C, Bjekic G, Greenberg S. A requirement for phosphatidylinositol 3-kinase in pseudopod extension. *J Biol Chem.* 1999;274(3):1240–1247. https://doi.org/10.1074/jbc.274.3.1240.
37. Shah VB, Ozment-Skelton TR, Williams DL, Keshvara L. Vav1 and PI3K are required for phagocytosis of β-glucan and subsequent superoxide generation by microglia. *Mol Immunol.* 2009;46(8–9):1845–1853. https://doi.org/10.1016/j.molimm.2009.01.014.
38. Massol P, Montcourrier P, Guillemot J-C, Chavrier P. Fc receptor-mediated phagocytosis requires CDC42 and Rac1. *EMBO J.* 1998;17(21):6219–6229. https://doi.org/10.1093/emboj/17.21.6219.
39. Caron E, Hall A. Identification of two distinct mechanisms of phagocytosis controlled by different rho GTPases. *Science.* 1998;282(5394):1717–1721. https://doi.org/10.1126/science.282.5394.1717.
40. Park H, Cox D. Cdc42 regulates fc γ receptor-mediated phagocytosis through the activation and phosphorylation of Wiskott-Aldrich syndrome protein (WASP) and neural-WASP. Parent C, ed. *Mol Biol Cell.* 2009;20(21):4500–4508. https://doi.org/10.1091/mbc.e09-03-0230.
41. Olazabal IM, Caron E, May RC, Schilling K, Knecht DA, Machesky LM. Rho-kinase and myosin-II control phagocytic cup formation during CR, but not FcγR, phagocytosis. *Curr Biol.* 2002;12(16):1413–1418. https://doi.org/10.1016/S0960-9822(02)01069-2.
42. Hall AB, Gakidis MAM, Glogauer M, et al. Requirements for vav guanine nucleotide exchange factors and rho GTPases in FcγR- and complement-mediated phagocytosis. *Immunity.* 2006;24(3):305–316. https://doi.org/10.1016/j.immuni.2006.02.005.
43. Lewkowicz E, Herit F, Le CC, Bourdoncle P, Perez F, Niedergang F. The microtubule-binding protein CLIP-170 coordinates mDia1 and actin reorganization during CR3-mediated phagocytosis. *J Cell Biol.* 2008;183(7):1287–1298. https://doi.org/10.1083/jcb.200807023.

44. Colucci-Guyon E, Niedergang F, Wallar BJ, Peng J, Alberts AS, Chavrier P. A role for mammalian diaphanous-related Formins in complement receptor (CR3)-mediated phagocytosis in macrophages. *Curr Biol.* 2005;15(22):2007–2012. https://doi.org/10.1016/j.cub.2005.09.051.
45. May RC, Caron E, Hall A, Machesky LM. Involvement of the Arp2/3 complex in phagocytosis mediated by FcγR or CR3. *Nat Cell Biol.* 2000;2(4):246–248. https://doi.org/10.1038/35008673.
46. Rotty JD, Brighton HE, Craig SL, et al. Arp2/3 complex is required for macrophage integrin functions but is dispensable for FcR phagocytosis and in vivo Motility. *Dev Cell.* 2017;42(5):498–513.e6. https://doi.org/10.1016/j.devcel.2017.08.003.
47. Lorenzi R, Brickell PM, Katz DR, Kinnon C, Thrasher AJ. Wiskott-Aldrich syndrome protein is necessary for efficient IgG-mediated phagocytosis. *Blood.* 2000;95(9):2943–2946. https://doi.org/10.1182/blood.V95.9.2943.009k17_2943_2946.
48. Leverrier Y, Lorenzi R, Blundell MP, et al. Cutting edge: The Wiskott-Aldrich syndrome protein is required for efficient phagocytosis of apoptotic cells. *J Immunol.* 2001;166(8):4831–4834. https://doi.org/10.4049/jimmunol.166.8.4831.
49. Rougerie P, Miskolci V, Cox D. Generation of membrane structures during phagocytosis and chemotaxis of macrophages: Role and regulation of the actin cytoskeleton. *Immunol Rev.* 2013;256(1):222–239. https://doi.org/10.1111/imr.12118.
50. Khandani A, Eng E, Jongstra-Bilen J, et al. Microtubules regulate PI-3K activity and recruitment to the phagocytic cup during Fcγ receptor-mediated phagocytosis in nonelicited macrophages. *J Leukoc Biol.* 2007;82(2):417–428. https://doi.org/10.1189/jlb.0706469.
51. Goodridge HS, Underhill DM, Touret N. Mechanisms of fc receptor and Dectin-1 activation for phagocytosis. *Traffic.* 2012;13(8):1062–1071. https://doi.org/10.1111/j.1600-0854.2012.01382.x.
52. Daëron M. Fc receptor biology. *Annu Rev Immunol.* 1997;15(1):203–234. https://doi.org/10.1146/annurev.immunol.15.1.203.
53. Crowley MT, Costello PS, Fitzer-Attas CJ, et al. A critical role for Syk in signal transduction and phagocytosis mediated by Fcγ receptors on macrophages. *J Exp Med.* 1997;186(7):1027–1039. https://doi.org/10.1084/jem.186.7.1027.
54. Tridandapani S, Lyden TW, Smith JL, Carter JE, Coggeshall KM, Anderson CL. The adapter protein LAT enhances Fcγ receptor-mediated signal transduction in myeloid cells. *J Biol Chem.* 2000;275(27):20480–20487. https://doi.org/10.1074/jbc.M909462199.
55. Deckert M, Tartare-Deckert S, Couture C, Mustelin T, Altman A. Functional and physical interactions of Syk family kinases with the vav proto-oncogene product. *Immunity.* 1996;5(6):591–604. https://doi.org/10.1016/S1074-7613(00)80273-3.
56. Patel JC, Hall A, Caron E. Vav regulates activation of Rac but not Cdc42 during FcγR-mediated phagocytosis. *Mol Biol Cell.* 2002;13(4):1215–1226. https://doi.org/10.1091/mbc.02-01-0002.
57. Cox D, Greenberg S. Phagocytic signaling strategies: Fcγreceptor-mediated phagocytosis as a model system. *Semin Immunol.* 2001;13(6). https://doi.org/10.1006/smim.2001.0330.
58. Cox D, Chang P, Zhang Q, Reddy PG, Bokoch GM, Greenberg S. Requirements for both Rac1 and Cdc42 in membrane ruffling and phagocytosis in leukocytes. *J Exp Med.* 1997;186(9):1487–1494. https://doi.org/10.1084/jem.186.9.1487.
59. García-García E, Rosales C. Adding complexity to phagocytic signaling: Phagocytosis-associated cell responses and phagocytic efficiency. In: *Molecular Mechanisms of Phagocytosis.* Springer US; 2006:58–71. https://doi.org/10.1007/978-0-387-28669-3_5.
60. Newman SL, Becker S, Halme J. Phagocytosis by receptors for C3b (CR 1), iC3b (CR 3), and IgG (fc) on human peritoneal macrophages. *J Leukoc Biol.* 1985;38(2):267–278. https://doi.org/10.1002/jlb.38.2.267.

61. Stolla M, Stefanini L, André P, et al. CalDAG-GEFI deficiency protects mice in a novel model of FcγRIIA-mediated thrombosis and thrombocytopenia. *Blood.* 2011;118(4):1113–1120. https://doi.org/10.1182/blood-2011-03-342352.
62. Lagarrigue F, Kim C, Ginsberg MH. The Rap1-RIAM-Talin axis of integrin activation and blood cell function. *Blood.* 2016;128(4):479–487. https://doi.org/10.1182/blood-2015-12-638700.
63. Allen L-AH, Aderem A. Mechanisms of phagocytosis. *Curr Opin Immunol.* 1996;8 (1):36–40. https://doi.org/10.1016/S0952-7915(96)80102-6.
64. Mócsai A, Ruland J, Tybulewicz VLJ. The SYK tyrosine kinase: A crucial player in diverse biological functions. *Nat Rev Immunol.* 2010;10(6):387–402. https://doi.org/10.1038/nri2765.
65. Patel PC, Harrison RE. Membrane ruffles capture C3bi-opsonized particles in activated macrophages. *Mol Biol Cell.* 2008;19(11):4628–4639. https://doi.org/10.1091/mbc.e08-02-0223.
66. Springer TA. Adhesion receptors of the immune system. *Nature.* 1990;346 (6283):425–434. https://doi.org/10.1038/346425a0.
67. Bakalar MH, Joffe AM, Schmid EM, Son S, Podolski M, Fletcher DA. Size-dependent segregation controls macrophage phagocytosis of antibody-opsonized targets. *Cell.* 2018;174(1):131–142.e13. https://doi.org/10.1016/j.cell.2018.05.059.
68. Goodridge HS, Reyes CN, Becker CA, et al. Activation of the innate immune receptor Dectin-1 upon formation of a 'phagocytic synapse. *Nature.* 2011;472(7344):471–475. https://doi.org/10.1038/nature10071.
69. Freeman SA, Jaumouillé V, Choi K, et al. Toll-like receptor ligands sensitize B-cell receptor signalling by reducing actin-dependent spatial confinement of the receptor. *Nat Commun.* 2015;6(1):1–16. https://doi.org/10.1038/ncomms7168.
70. Freeman SA, Goyette J, Furuya W, et al. Integrins form an expanding diffusional barrier that coordinates phagocytosis. *Cell.* 2016;164(1–2):128–140. https://doi.org/10.1016/j.cell.2015.11.048.
71. Huynh KK, Kay JG, Stow JL, Grinstein S. Fusion, fission, and secretion during phagocytosis. *Phys Ther.* 2007;22(6):366–372. https://doi.org/10.1152/physiol.00028.2007.
72. Lennartz MR, Yuen AF, Masi SM, Russell DG, Buttle KF, Smith JJ. Phospholipase A2 inhibition results in sequestration of plasma membrane into electronlucent vesicles during IgG-mediated phagocytosis. *J Cell Sci.* 1997;110(17):2041–2052. https://doi.org/10.1242/jcs.110.17.2041.
73. Swanson JA. Shaping cups into phagosomes and macropinosomes. *Nat Rev Mol Cell Biol.* 2008;9(8):639–649. https://doi.org/10.1038/nrm2447.
74. Bajno L, Peng X-R, Schreiber AD, Moore H-P, Trimble WS, Grinstein S. Focal exocytosis of Vamp3-containing vesicles at sites of phagosome formation. *J Cell Biol.* 2000;149(3):697–706. https://doi.org/10.1083/jcb.149.3.697.
75. Jean S, Kiger AA. Classes of phosphoinositide 3-kinases at a glance. *J Cell Sci.* 2014;127 (5):923–928. https://doi.org/10.1242/jcs.093773.
76. Shepherd PR, Reaves BJ, Davidson HW. Phosphoinositide 3-kinases and membrane traffic. *Trends Cell Biol.* 1996;6(3):92–97. https://doi.org/10.1016/0962-8924(96)80998-6.
77. Vanhaesebroeck B, Guillermet-Guibert J, Graupera M, Bilanges B. The emerging mechanisms of isoform-specific PI3K signalling. *Nat Rev Mol Cell Biol.* 2010;11 (5):329–341. https://doi.org/10.1038/nrm2882.
78. Bhattacharya S, McElhanon KE, Gushchina LV, Weisleder N. Role of phosphatidylinositol-4,5-bisphosphate 3-kinase signaling in vesicular trafficking. *Life Sci.* 2016;167:1–17. https://doi.org/10.1016/j.lfs.2016.10.018.
79. Cox D, Berg JS, Cammer M, et al. Myosin X is a downstream effector of PI(3)K during phagocytosis. *Nat Cell Biol.* 2002;4(7):469–477. https://doi.org/10.1038/ncb805.

80. Liu J, Li M, Li L, Chen S, Wang X. Ubiquitination of the PI3-kinase VPS-34 promotes VPS-34 stability and phagosome maturation. *J Cell Biol*. 2018;217(1):347–360. https://doi.org/10.1083/jcb.201705116.
81. Vieira OV, Botelho RJ, Rameh L, et al. Distinct roles of class I and class III phosphatidylinositol 3-kinases in phagosome formation and maturation. *J Cell Biol*. 2001;155(1):19–26. https://doi.org/10.1083/jcb.200107069.
82. Vieira OV, Botelho RJ, Grinstein S. Phagosome maturation: Aging gracefully. *Biochem J*. 2002;366(3):689–704. https://doi.org/10.1042/bj20020691.
83. Bernstein BW, Bamburg JR. ADF/cofilin: A functional node in cell biology. *Trends Cell Biol*. 2010;20(4):187–195. https://doi.org/10.1016/j.tcb.2010.01.001.
84. Niedergang F, Colucci-Guyon E, Dubois T, Raposo G, Chavrier P. ADP ribosylation factor 6 is activated and controls membrane delivery during phagocytosis in macrophages. *J Cell Biol*. 2003;161(6):1143–1150. https://doi.org/10.1083/jcb.200210069.
85. Egami Y, Fukuda M, Araki N. Rab35 regulates phagosome formation through recruitment of ACAP2 in macrophages during FcγR-mediated phagocytosis. *J Cell Sci*. 2011;124(21):3557–3567. https://doi.org/10.1242/jcs.083881.
86. Schlam D, Bagshaw RD, Freeman SA, et al. Phosphoinositide 3-kinase enables phagocytosis of large particles by terminating actin assembly through Rac/Cdc42 GTPase-activating proteins. *Nat Commun*. 2015;6(1):1–12. https://doi.org/10.1038/ncomms9623.
87. Vorselen D, Labitigan RLD, Theriot JA. A mechanical perspective on phagocytic cup formation. *Curr Opin Cell Biol*. 2020;66:112–122. https://doi.org/10.1016/j.ceb.2020.05.011.
88. Swanson JA, Johnson MT, Beningo K, Post P, Mooseker M, Araki N. A contractile activity that closes phagosomes in macrophages. *J Cell Sci*. 1999;112(3):307–316.
89. Dart AE, Tollis S, Bright MD, Frankel G, Endres RG. The motor protein myosin 1G functions in FcγR-mediated phagocytosis. *J Cell Sci*. 2012;125(24):6020–6029. https://doi.org/10.1242/jcs.109561.
90. Marie-Anaïs F, Mazzolini J, Herit F, Niedergang F. Dynamin-actin cross talk contributes to phagosome formation and closure. *Traffic*. 2016;17(5):487–499. https://doi.org/10.1111/tra.12386.
91. Deschamps C, Echard A, Niedergang F. Phagocytosis and cytokinesis: Do cells use common tools to cut and to eat? Highlights on common themes and differences. *Traffic*. 2013;14(4):355–364. https://doi.org/10.1111/tra.12045.
92. Desjardins M, Huber LA, Parton RG, Griffiths G. Biogenesis of phagolysosomes proceeds through a sequential series of interactions with the endocytic apparatus. *J Cell Biol*. 1994;124(5):677–688. https://doi.org/10.1083/jcb.124.5.677.
93. Desjardins M. Biogenesis of phagolysosomes: The "kiss and run" hypothesis. *Trends Cell Biol*. 1995;5(5):183–186. https://doi.org/10.1016/S0962-8924(00)88989-8.
94. Pauwels A, Trost M, Beyaert R, Hoffmann E. Patterns, receptors, and signals: Regulation of phagosome maturation. *Trends Immunol*. 2017;38(6):407–422. https://doi.org/10.1016/j.it.2017.03.006.
95. Mayorga LS, Bertini F, Stahl PD. Fusion of newly formed phagosomes with endosomes in intact cells and in a cell-free system. *J Biol Chem*. 1991;266(10):6511–6517. https://doi.org/10.1016/S0021-9258(18)38147-X.
96. Duclos S, Diez R, Garin J, et al. Rab5 regulates the kiss and run fusion between phagosomes and endosomes and the acquisition of phagosome leishmanicidal properties in RAW 264.7 macrophages. *J Cell Sci*. 2000;113(19):3531–3541.
97. Chavrier P, Parton RG, Hauri HP, Simons K, Zerial M. Localization of low molecular weight GTP binding proteins to exocytic and endocytic compartments. *Cell*. 1990;62(2):317–329. https://doi.org/10.1016/0092-8674(90)90369-P.

98. Hackam DJ, Rotstein OD, Zhang W-J, et al. Regulation of Phagosomal acidification. *J Biol Chem*. 1997;272(47):29810–29820. https://doi.org/10.1074/jbc.272.47.29810.
99. Brisseau GF, Grinstein S, Hackam DJ, et al. Interleukin-1 increases vacuolar-type H +-ATPase activity in murine peritoneal macrophages. *J Biol Chem*. 1996;271(4):2005–2011. https://doi.org/10.1074/jbc.271.4.2005.
100. Kinchen JM, Doukoumetzidis K, Almendinger J, et al. A pathway for phagosome maturation during engulfment of apoptotic cells. *Nat Cell Biol*. 2008;10(5):556–566. https://doi.org/10.1038/ncb1718.
101. Rink J, Ghigo E, Kalaidzidis Y, Zerial M. Rab conversion as a mechanism of progression from early to late endosomes. *Cell*. 2005;122(5):735–749. https://doi.org/10.1016/j.cell.2005.06.043.
102. Jordens I, Fernandez-Borja M, Marsman M, et al. The Rab7 effector protein RILP controls lysosomal transport by inducing the recruitment of dynein-dynactin motors. *Curr Biol*. 2001;11(21):1680–1685. https://doi.org/10.1016/S0960-9822(01)00531-0.
103. Harrison RE, Bucci C, Vieira OV, Schroer TA, Grinstein S. Phagosomes fuse with late endosomes and/or lysosomes by extension of membrane protrusions along microtubules: Role of Rab7 and RILP. *Mol Cell Biol*. 2003;23(18):6494–6506. https://doi.org/10.1128/MCB.23.18.6494-6506.2003.
104. Pitt A, Mayorga LS, Stahl PD, Schwartz AL. Alterations in the protein composition of maturing phagosomes. *J Clin Invest*. 1992;90(5):1978–1983. https://doi.org/10.1172/JCI116077.
105. Antonin W. A SNARE complex mediating fusion of late endosomes defines conserved properties of SNARE structure and function. *EMBO J*. 2000;19(23):6453–6464. https://doi.org/10.1093/emboj/19.23.6453.
106. Luzio JP, Pryor PR, Bright NA. Lysosomes: Fusion and function. *Nat Rev Mol Cell Biol*. 2007;8(8):622–632. https://doi.org/10.1038/nrm2217.
107. Griffiths G, Hoflack B, Simons K, Mellman I, Kornfeld S. The mannose 6-phosphate receptor and the biogenesis of lysosomes. *Cell*. 1988;52(3):329–341. https://doi.org/10.1016/S0092-8674(88)80026-6.
108. Levin-Konigsberg R, Grinstein S. Phagosome-endoplasmic reticulum contacts: Kissing and not running. *Traffic*. 2020;21(1):172–180. https://doi.org/10.1111/tra.12708.
109. Babior BM. NADPH oxidase. *Curr Opin Immunol*. 2004;16(1):42–47. https://doi.org/10.1016/j.coi.2003.12.001.
110. Gagnon E, Duclos S, Rondeau C, et al. Endoplasmic reticulum-mediated phagocytosis is a mechanism of entry into macrophages. *Cell*. 2002;110(1):119–131. https://doi.org/10.1016/S0092-8674(02)00797-3.
111. Becker T, Volchuk A, Rothman JE. Differential use of endoplasmic reticulum membrane for phagocytosis in J774 macrophages. *Proc Natl Acad Sci*. 2005;102(11):4022–4026. https://doi.org/10.1073/pnas.0409219102.
112. Touret N, Paroutis P, Grinstein S. The nature of the phagosomal membrane: Endoplasmic reticulum versus plasmalemma. *J Leukoc Biol*. 2005;77(6):878–885. https://doi.org/10.1189/jlb.1104630.
113. Touret N, Paroutis P, Terebiznik M, et al. Quantitative and dynamic assessment of the contribution of the ER to phagosome formation. *Cell*. 2005;123(1):157–170. https://doi.org/10.1016/j.cell.2005.08.018.
114. Nunes P, Cornut D, Bochet V, et al. STIM1 juxtaposes ER to phagosomes, generating Ca2+ hotspots that boost phagocytosis. *Curr Biol*. 2012;22(21):1990–1997. https://doi.org/10.1016/j.cub.2012.08.049.

115. Guermonprez P, Saveanu L, Kleijmeer M, Davoust J, van Endert P, Amigorena S. ER–phagosome fusion defines an MHC class I cross-presentation compartment in dendritic cells. *Nature*. 2003;425(6956):397–402. https://doi.org/10.1038/nature01911.
116. Garin J, Diez R, Kieffer S, et al. The phagosome proteome. *J Cell Biol*. 2001;152 (1):165–180. https://doi.org/10.1083/jcb.152.1.165.
117. Zehner M, Marschall AL, Bos E, et al. The Translocon protein Sec61 mediates antigen transport from endosomes in the cytosol for cross-presentation to CD8+ T cells. *Immunity*. 2015;42(5):850–863. https://doi.org/10.1016/j.immuni.2015.04.008.
118. Muller-Taubenberger A. Calreticulin and calnexin in the endoplasmic reticulum are important for phagocytosis. *EMBO J*. 2001;20(23):6772–6782. https://doi.org/10.1093/emboj/20.23.6772.
119. Nunes-Hasler P, Demaurex N. The ER phagosome connection in the era of membrane contact sites. *Biochim Biophys Acta - Mol Cell Res*. 2017;1864(9):1513–1524. https://doi.org/10.1016/j.bbamcr.2017.04.007.
120. Cebrian I, Visentin G, Blanchard N, et al. Sec22b regulates Phagosomal maturation and antigen Crosspresentation by dendritic cells. *Cell*. 2011;147(6):1355–1368. https://doi.org/10.1016/j.cell.2011.11.021.
121. Desjardins M. ER-mediated phagocytosis: A new membrane for new functions. *Nat Rev Immunol*. 2003;3(4):280–291. https://doi.org/10.1038/nri1053.
122. Dingjan I, Linders PTA, Verboogen DRJ, Revelo NH, ter Beest M, van den Bogaart G. Endosomal and Phagosomal SNAREs. *Physiol Rev*. 2018;98(3):1465–1492. https://doi.org/10.1152/physrev.00037.2017.
123. Arasaki K, Roy CR. Legionella pneumophila promotes functional interactions between plasma membrane Syntaxins and Sec22b. *Traffic*. 2010;11(5):587–600. https://doi.org/10.1111/j.1600-0854.2010.01050.x.
124. Ackerman AL, Giodini A, Cresswell P. A role for the endoplasmic reticulum protein Retrotranslocation machinery during Crosspresentation by dendritic cells. *Immunity*. 2006;25(4):607–617. https://doi.org/10.1016/j.immuni.2006.08.017.
125. Hatsuzawa K, Tamura T, Hashimoto H, et al. Involvement of Syntaxin 18, an endoplasmic reticulum (ER)-localized SNARE protein, in ER-mediated phagocytosis. *Mol Biol Cell*. 2006;17(9):3964–3977. https://doi.org/10.1091/mbc.e05-12-1174.
126. Campbell-Valois FX, Trost M, Chemali M, et al. Quantitative proteomics reveals that only a subset of the endoplasmic reticulum contributes to the phagosome. *Mol Cell Proteomics*. 2012;11(7):1–13. https://doi.org/10.1074/mcp.M111.016378.
127. Levin-Konigsberg R, Grinstein S. Phagosome-endoplasmic reticulum contacts: Kissing and not running. *Traffic*. 2020;21(1):172–180. https://doi.org/10.1111/tra.12708.
128. Scorrano L, De Matteis MA, Emr S, et al. Coming together to define membrane contact sites. *Nat Commun*. 2019;10(1):1–11. https://doi.org/10.1038/s41467-019-09253-3.
129. Linder S. Arp2/3: Not absolutely required after all. *Dev Cell*. 2017;42(5):436–438. https://doi.org/10.1016/j.devcel.2017.08.018.
130. Motility IV, Rotty JD, Brighton HE, et al. Arp2 / 3 complex is required for macrophage integrin functions but is dispensable for FcR phagocytosis article Arp2 / 3 complex is required for macrophage integrin functions but is dispensable for FcR phagocytosis and in vivo Motility. *Dev Cell*. 2017;42(5):498–513. https://doi.org/10.1016/j.devcel.2017.08.003.
131. Schlam D, Bagshaw RD, Freeman SA, et al. Phosphoinositide 3-kinase enables phagocytosis of large particles by terminating actin assembly through Rac/Cdc42 GTPase-activating proteins. *Nat Commun*. 2015;6(1):8623. https://doi.org/10.1038/ncomms9623.

132. Brinkmann K, Schell M, Hoppe T, Kashkar H. Regulation of the DNA damage response by ubiquitin conjugation. *Front Genet.* 2015;6:1–15. https://doi.org/10.3389/fgene.2015.00098.
133. Dang F, Nie L, Wei W. Ubiquitin signaling in cell cycle control and tumorigenesis. *Cell Death Differ.* 2021;28(2):427–438. https://doi.org/10.1038/s41418-020-00648-0.
134. Geng F, Wenzel S, Tansey WP. Ubiquitin and proteasomes in transcription. *Annu Rev Biochem.* 2012;81(1):177–201. https://doi.org/10.1146/annurev-biochem-052110-120012.
135. Tu Y, Chen C, Pan J, Xu J, Zhou ZG, Wang CY. The ubiquitin proteasome pathway (UPP) in the regulation of cell cycle control and DNA damage repair and its implication in tumorigenesis. *Int J Clin Exp Pathol.* 2012;5(8):726–738.
136. Zheng Q, Huang T, Zhang L, et al. Dysregulation of ubiquitin-proteasome system in neurodegenerative diseases. *Front Aging Neurosci.* 2016;8:1–10. https://doi.org/10.3389/fnagi.2016.00303.
137. Lecker SH, Goldberg AL, Mitch WE. Protein degradation by the ubiquitin–proteasome pathway in Normal and disease states. *J Am Soc Nephrol.* 2006;17(7):1807–1819. https://doi.org/10.1681/ASN.2006010083.
138. Nakamura N. Ubiquitin System. *Int J Mol Sci.* 2018;19(4):1–5. https://doi.org/10.3390/ijms19041080.
139. Wang J, Maldonado MA. The ubiquitin-proteasome system and its role in inflammatory and autoimmune diseases. *Cell Mol Immunol.* 2006;3(4):255–261.
140. Booth JW, Kim M-K, Jankowski A, Schreiber AD, Grinstein S. Contrasting requirements for ubiquitylation during fc receptor-mediated endocytosis and phagocytosis. *EMBO J.* 2002;21(3):251–258. https://doi.org/10.1093/emboj/21.3.251.
141. Katzmann DJ, Odorizzi G, Emr SD. Receptor downregulation and multivesicular-body sorting. *Nat Rev Mol Cell Biol.* 2002;3(12):893–905. https://doi.org/10.1038/nrm973.
142. Strous GJ, van Kerkhof P, Govers R, Ciechanover A, Schwartz AL. The ubiquitin conjugation system is required for ligand-induced endocytosis and degradation of the growth hormone receptor. *EMBO J.* 1996;15(15):3806–3812. https://doi.org/10.1002/j.1460-2075.1996.tb00754.x.
143. Hicke L, Riezman H. Ubiquitination of a yeast plasma membrane receptor signals its ligand-stimulated endocytosis. *Cell.* 1996;84(2):277–287. https://doi.org/10.1016/S0092-8674(00)80982-4.
144. Hochstrasser M. Ubiquitin-dependent protein degradation. *Annu Rev Genet.* 1996;30(1):405–439. https://doi.org/10.1146/annurev.genet.30.1.405.
145. Terrell J, Shih S, Dunn R, Hicke L. A function for Monoubiquitination in the internalization of a G protein–coupled receptor. *Mol Cell.* 1998;1(2):193–202. https://doi.org/10.1016/S1097-2765(00)80020-9.
146. Felder S, Miller K, Moehren G, Ullrich A, Schlessinger J, Hopkins CR. Kinase activity controls the sorting of the epidermal growth factor receptor within the multivesicular body. *Cell.* 1990;61(4):623–634. https://doi.org/10.1016/0092-8674(90)90474-S.
147. Kleijmeer M, Ramm G, Schuurhuis D, et al. Reorganization of multivesicular bodies regulates MHC class II 11 antigen presentation by dendritic cells. *J Cell Biol.* 2001;155(1):53–63. https://doi.org/10.1083/jcb.200103071.
148. Katzmann DJ, Babst M, Emr SD. Ubiquitin-dependent sorting into the multivesicular body pathway requires the function of a conserved endosomal protein sorting complex, ESCRT-I. *Cell.* 2001;106(2):145–155. https://doi.org/10.1016/S0092-8674(01)00434-2.
149. Lee WL, Kim M-K, Schreiber AD, Grinstein S. Role of ubiquitin and proteasomes in phagosome maturation. *Mol Biol Cell.* 2005;16(4):2077–2090. https://doi.org/10.1091/mbc.e04-06-0464.

150. Henne WM, Buchkovich NJ, Emr SD. Review the ESCRT pathway. *Dev Cell.* 2011;21(1):77–91. https://doi.org/10.1016/j.devcel.2011.05.015.
151. Piper RC, Dikic I, Lukacs GL. Ubiquitin-dependent sorting in endocytosis. *Cold Spring Harb Perspect Biol.* 2014;6(1):1–22. https://doi.org/10.1101/cshperspect.a016808.
152. Piper RC, Katzmann DJ. Biogenesis and function of multivesicular bodies. *Annu Rev Cell Dev Biol.* 2007;23(1):519–547. https://doi.org/10.1146/annurev.cellbio.23.090506.123319.
153. Gold ES, Morrissette NS, Underhill DM, Guo J, Bassetti M, Aderem A. Amphiphysin IIm, a novel Amphiphysin II isoform, is required for macrophage phagocytosis. *Immunity.* 2000;12(3):285–292. https://doi.org/10.1016/S1074-7613(00)80181-8.
154. Houde M, Bertholet S, Gagnon E, et al. Phagosomes are competent organelles for antigen cross-presentation. *Nature.* 2003;425(6956):402–406. https://doi.org/10.1038/nature01912.
155. Dong J, Chen W, Welford A, Wandinger-Ness A. The proteasome α-subunit XAPC7 interacts specifically with Rab7 and late endosomes. *J Biol Chem.* 2004;279(20):21334–21342. https://doi.org/10.1074/jbc.M401022200.
156. van Kerkhof P, dos Santos CMA, Sachse M, Klumperman J, Bu G, Strous GJ. Proteasome inhibitors block a late step in lysosomal transport of selected membrane but not soluble proteins. *Mol Biol Cell.* 2001;12(8):2556–2566. https://doi.org/10.1091/mbc.12.8.2556.
157. Dale BM, Traum D, Erdjument-Bromage H, Tempst P, Greenberg S. Phagocytosis in macrophages lacking Cbl reveals an unsuspected role for Fcγ receptor signaling and actin assembly in target binding. *J Immunol.* 2009;182(9):5654–5661. https://doi.org/10.4049/jimmunol.0803942.
158. Lyle C, Richards S, Yasuda K, et al. C-Cbl targets PD-1 in immune cells for proteasomal degradation and modulates colorectal tumor growth. *Sci Rep.* 2019;9(1):1–14. https://doi.org/10.1038/s41598-019-56208-1.
159. Nakamura M, Shimosaki S. The ubiquitin-like protein monoclonal nonspecific suppressor factor β conjugates to endophilin II and regulates phagocytosis. *FEBS J.* 2009;276(21):6355–6363. https://doi.org/10.1111/j.1742-4658.2009.07348.x.
160. Nakamura M, Watanabe N. Ubiquitin-like protein MNSFβ/endophilin II complex regulates Dectin-1-mediated phagocytosis and inflammatory responses in macrophages. *Biochem Biophys Res Commun.* 2010;401(2):257–261. https://doi.org/10.1016/j.bbrc.2010.09.045.
161. Lawrence DW, Willard PA, Cochran AM, Matchett EC, Kornbluth J. Natural killer lytic-associated molecule (NKLAM): An E3 ubiquitin ligase with an integral role in innate immunity. *Front Physiol.* 2020;11:1–13. https://doi.org/10.3389/fphys.2020.573372.
162. Lawrence DW, Kornbluth J. E3 ubiquitin ligase NKLAM is a macrophage phagosome protein and plays a role in bacterial killing. *Cell Immunol.* 2012;279(1):46–52. https://doi.org/10.1016/j.cellimm.2012.09.004.
163. Huttlin EL, Bruckner RJ, Paulo JA, et al. Architecture of the human interactome defines protein communities and disease networks. *Nature.* 2017;545(7655):505–509. https://doi.org/10.1038/nature22366.
164. Sengupta D, Graham M, Liu X, Cresswell P. Proteasomal degradation within endocytic organelles mediates antigen cross-presentation. *EMBO J.* 2019;38(16):1–15. https://doi.org/10.15252/embj.201899266.
165. Ferrandon D. Ubiquitin-proteasome: Pallbearer carries the deceased to the grave. *Immunity.* 2007;27(4):541–544. https://doi.org/10.1016/j.immuni.2007.10.003.
166. Silva E, Au-Yeung HW, Van Goethem E, Burden J, Franc NC. Requirement for a drosophila E3-ubiquitin ligase in phagocytosis of apoptotic cells. *Immunity.* 2007;27(4):585–596. https://doi.org/10.1016/j.immuni.2007.08.016.

167. Stuart LM, Boulais J, Charriere GM, et al. A systems biology analysis of the drosophila phagosome. *Nature*. 2007;445(7123):95–101. https://doi.org/10.1038/nature05380.
168. Guo M, Härtlova A, Gierliński M, et al. Triggering MSR1 promotes JNK-mediated inflammation in IL-4-activated macrophages. *EMBO J*. 2019;38(11):1–15. https://doi.org/10.15252/embj.2018100299.
169. Sutovsky P, Moreno R, Ramalho-Santos J, Dominko T, Thompson WE, Schatten G. A putative, ubiquitin-dependent mechanism for the recognition and elimination of defective spermatozoa in the mammalian epididymis. *J Cell Sci*. 2001;114(9):1665–1675. https://doi.org/10.1242/jcs.114.9.1665.
170. Guo M, Härtlova A, Dill BD, Prescott AR, Gierliński M, Trost M. High-resolution quantitative proteome analysis reveals substantial differences between phagosomes of RAW 264.7 and bone marrow derived macrophages. *Proteomics*. 2015;15(18):3169–3174. https://doi.org/10.1002/pmic.201400431.
171. Haney MS, Bohlen CJ, Morgens DW, et al. Identification of phagocytosis regulators using magnetic genome-wide CRISPR screens. *Nat Genet*. 2018;50(12):1716–1727. https://doi.org/10.1038/s41588-018-0254-1.
172. Dean P, Heunis T, Härtlova A, Trost M. Regulation of phagosome functions by post-translational modifications: A new paradigm. *Curr Opin Chem Biol*. 2019;48:73–80. https://doi.org/10.1016/j.cbpa.2018.11.001.
173. Kim BY, Olzmann JA, Barsh GS, Chin L-S, Li L. Spongiform neurodegeneration-associated E3 ligase Mahogunin ubiquitylates TSG101 and regulates endosomal trafficking. Lippincott-Schwartz J, ed. *Mol Biol Cell*. 2007;18(4):1129–1142. https://doi.org/10.1091/mbc.e06-09-0787.
174. Piper RC, Dikic I, Lukacs GL. Ubiquitin-dependent sorting in endocytosis. *Cold Spring Harb Perspect Biol*. 2014;6(1). https://doi.org/10.1101/cshperspect.a016808.
175. Fiala M, Lin J, Ringman J, et al. Ineffective phagocytosis of amyloid-β by macrophages of Alzheimer's disease patients. *J Alzheimers Dis*. 2005;7(3):221–232. https://doi.org/10.3233/JAD-2005-7304.
176. Xu J, Yu T, Pietronigro EC, et al. Peli1 impairs microglial Aβ phagocytosis through promoting C/EBPβ degradation. *PLoS Biol*. 2020;18(10), e3000837. https://doi.org/10.1371/journal.pbio.3000837.
177. Kevany BM, Palczewski K. Phagocytosis of retinal rod and cone photoreceptors. *Phys Ther*. 2010;25(1):8–15. https://doi.org/10.1152/physiol.00038.2009.
178. Chakarova CF, Papaioannou MG, Khanna H, et al. Mutations in TOPORS cause autosomal dominant retinitis pigmentosa with perivascular retinal pigment epithelium atrophy. *Am J Hum Genet*. 2007;81(5):1098–1103. https://doi.org/10.1086/521953.
179. Selmer KK, Grøndahl J, Riise R, et al. Autosomal dominant pericentral retinal dystrophy caused by a novel missense mutation in the TOPORS gene. *Acta Ophthalmol*. 2009;88(3):323–328. https://doi.org/10.1111/j.1755-3768.2008.01465.x.
180. Czub B, Shah AZ, Alfano G, Kruczek PM, Chakarova CF, Bhattacharya SS. TOPORS, a dual E3 ubiquitin and Sumo1 ligase, interacts with 26 S protease regulatory subunit 4, encoded by the PSMC1 gene. *PLoS One*. 2016;11(2), e0148678. https://doi.org/10.1371/journal.pone.0148678.
181. Qiu J, Luo Z-Q. Hijacking of the host ubiquitin network by legionella pneumophila. *Front Cell Infect Microbiol*. 2017;7(DEC):1–12. https://doi.org/10.3389/fcimb.2017.00487.
182. Kumari R, Gupta P, Tiwari S. Ubc7/Ube2g2 ortholog in Entamoeba histolytica: Connection with the plasma membrane and phagocytosis. *Parasitol Res*. 2018;117(5):1599–1611. https://doi.org/10.1007/s00436-018-5842-6.
183. Vaithilingam A, Teixeira JE, Miller PJ, Heron BT, Huston CD. Entamoeba histolytica cell surface calreticulin binds human C1q and functions in amebic phagocytosis of host cells. *Infect Immun*. 2012;80(6):2008–2018. https://doi.org/10.1128/IAI.06287-11.

CHAPTER TEN

Receptors of immune cells mediates recognition for tumors

Vandana Anang[a], Aayushi Singh[b], Sarat Kumar Kottarath[c,*], and Chaitenya Verma[d,*]
[a]International Center for Genetic Engineering and Biotechnology (ICGEB), New Delhi, India
[b]Invictus Oncology Pvt. Ltd, New Delhi, India
[c]Department of Experimental Therapeutics, MD Anderson Cancer Center, Huston, TX, United States
[d]Department of Pathology, Wexner Medical Center, Ohio State University, Columbus, OH, United States
*Corresponding authors: e-mail addresses: skottarath@mdanderson.org; chaitenya999@gmail.com; chaitenya.verma@osumc.edu

Contents

Progress in Molecular Biology and Translational Science, Volume 194
ISSN 1877-1173
https://doi.org/10.1016/bs.pmbts.2022.09.009

Abstract

Over the last few decades, the immune system has been steered toward eradication of cancer cells with the help of cancer immunotherapy. T cells, B cells, monocytes/macrophages, dendritic cells, T-reg cells, and natural killer (NK) cells are some of the numerous immune cell types that play a significant part in cancer cell detection and reduction of inflammation, and the antitumor response. Briefly stated, chimeric antigen receptors, adoptive transfer and immune checkpoint modulators are currently the subjects of research focus for successful immunotherapy-based treatments for a variety of cancers. This chapter discusses ongoing investigations on the mechanisms and recent developments by which receptors of immune cells especially that of lymphocytes and monocytes/macrophages regulate the detection of immune system leading to malignancies. We will also be looking into the treatment strategies based on these mechanisms.

Abbreviations

ADCC antibody dependent cell cytotoxicity
ADCP antibody dependent cell phagocytosis
AICL activation-induced C type lectin
AIDS acquired immunodeficiency syndrome
ALL lymphoblastic leukemia
AOM azoxymethane
AP-1 activator protein-1
ATP adenosine triphosphate
ATR ataxia telangiectasia and Rad3-related protein
BAG-6 Bcl-2 associated athanogene-6
BCR B cell receptor
BTK Bruton tyrosine kinase
CAC colitis-associated cancer
CAR chimeric antigen receptors
CCA cholangiocarcinoma
CCL C—C chemokine ligand
CCL2 C—C motif chemokine ligand 2
CCL20 C—C motif chemokine ligand 20
CCR C—C chemokine receptors
CDK4 cyclin-dependent kinase 4
CLP common lymphoid progenitor
CMV cytomegalovirus
CSF-1 colony stimulating factor-1
CSTB cystatin B
CTL cytotoxic T-lymphocytes

DAMP	danger-associated molecular patterns
DCs	dendritic cells
DEC1	chondrocyte expressed gene 1
DNA	deoxyribonucleic acid
DSS	dextran sodium sulfate
EGF	epidermal growth factor
GM-CSF	granulocyte macrophage-colony stimulating factor
H_2O_2	hydrogen peroxide
HCC	hepatocellular carcinoma
HGF	hepatocytes growth factor
HOCL	hypochlorous acid
HSCs	hematopoietic stem cells
ICAM	intracellular adhesion molecule
IEC	intestinal epithelial cells
IL-1β	interleukin-1 beta
IL-6	interleukin-6
iNOS	inducible nitric oxide synthase
ITAM	immunoreceptor tyrosine based activation motif
KLF-4	Kruppel-like factor-4
Klrk1	killer cell lectin-like receptor subfamily member 1
LAT	linker for the activation of T cells
MCSF-1	macrophage colony stimulating factor-1
MHC	major histocompatibility complex
MMP	matrix metalloproteinases
MoAb	monoclonal antibody
NCR	natural cytotoxicity receptors
NF-κB	nuclear factor kappa B
NK	natural killer
NKG2D	natural killer group 2D
NLRP3	NOD, LRR and pyrin domain containing protein 3
NLRs	NOD-like receptors
NO	nitric oxide
NTAL	non-T cell activation linker
PAMPS	pathogen-associated molecular patterns
PDL-1	programmed death ligand-1
PI3K	phosphoinositide 3-kinase phosphoinositide 3-kinase
PMS2	post meiotic segregation increased
PRR	pattern recognition receptors
SAP	SLAM-associated proteins
SDF-1	stromal cell derived factor-1
SLAM	signaling lymphocyte activating molecules
STAT3	signal transducer and activator of transcription 3
SYK	spleen tyrosine kinase
TAA	tumor associated antigen
TAMs	tumor-associated macrophages
TCR	T cell receptor
TGF-β	tumor growth factor-beta

TIL	tumor-infiltrating lymphocytes
TLRs	toll-like receptors
TME	tumor microenvironment
TNF-α	tumor necrosis factor alpha
TSA	tumor-specific antigen
VCAM	vascular cell adhesion protein
VEGF	vascular endothelial growth factor

1. Introduction: Inflammation in tumor microenvironment and cancer

Rudolf Virchow was the first to discover an association between cellular inflammation and the development of cancer in 1863.[1] After a century of gradual development in studies inspired by Virchow, epidemiological data and numerous mechanistic studies have successfully correlated and reinforced the relationship between inflammation and carcinogenesis.[2] Given that the majority of cancers occur in older population and that cell senescence is associated with the characteristics of both inflammation and cancer conditions, it is likely that the accumulation of inflammatory mediators during the chronic stage is one of the mediators that causes tumor growth. Industrial irritants, silica, and cigarette smoke are all capable of initiating tumor formation as well as eliciting an acute inflammatory response.[3,4] Long-term exposure to these irritants has the potential to cause chronic inflammation as well as tumor formation.[5]

Inflammation caused by viral and bacterial infections altering the function of the innate immune system via the NOD and TLR receptors, which promotes the growth of various cancers.[6] *Helicobacter pylori* infection, for example, continues to be one of the top causes of persistent gastrointestinal cancer[7] while Schistosomiasis is associated to the development of bladder and colon cancer and hepatitis C infection is connected with an increased risk of liver cancer.[8,9] A trematode parasite called *Opisthorchis viverrini*, which causes a food-borne illness known as liver fluke also leads to the development of cholangiocarcinoma.[10] *Schistosoma haematobium* is another example of parasitic infection leading to carcinoma of squamous cells.[11] Some carcinogens produced by infections have been reported to accelerate the tumor formation and inflammation associated with stomach cancer. Similarly, obesity has been linked to an alarmingly high cancer prevalence and has been shown to promote the development of cancer in the liver, pancreas, and colon.[12,13]

The infiltration of leukocytes into the growing tumor microenvironment, as well as the mitogenic properties of certain irritants, formed the foundation of Virchow's theory that inflammation is linked to cancer.[14] We now know that increased cell proliferation alone is not sufficient to trigger neoplastic transformation.[15] The presence of DNA damaging factors, the inflammatory microenvironment (inflammatory cells, cytokines), and growth factors is a third important factor in cancer development and tumor growth stimulation.[16] The functional link between inflammation and cancer is mediated by a dynamic between pro-oncogenic signaling processes and paracrine communication between tumor and inflammatory cells. A decrease in inflammation during cancer progression can thus be explained by focusing on both the initiation and progression of the carcinogenic process.

2. Tumor initiation

Reoccurrence of tissue damage, regeneration of peroxonitrite (also called peroxynitrite) and reactive oxygen/nitrogen species released from a pro-inflammatory cell that leads to point mutations, deletion, or rearrangement. Chronic inflammatory bowel disease and viral hepatitis can both have elevated levels of inducible nitric oxide synthase (iNOS) and free radical nitric oxide production (NO).[17] Further, these NO intermediates have potential nitrosating products that produce mutagenic and carcinogenic nitrosamines.[18,19] Aside from reactive nitrogen species, chronic inflammation and activated myeloid cells contain a variety of reactive oxygen species, including superoxide, hypochlorous acid, and hydrogen peroxide.[20,21] Secondary, metal-catalyzed degradation of hydrogen peroxide produced hydroxyl radical (OH') are well-established mutation causing species capable of oxidizing DNA.[22] P53 mutations have been found in chronic inflammatory diseases such as inflammatory bowel disease and rheumatoid arthritis, and the frequency of mutations in cancer is comparable. These findings emphasize the mutagenic nature of chronic inflammation.[23] Despite being a low-risk carcinogen, dextran sodium sulphate (DSS) may cause DNA damage that promotes the growth of colonic adenomas.[24] A pro-inflammatory environment might thus promote AID expression, increasing the chance of mutations and other genetic instability in critical cancer-related genes such as Bcl-6, Tp53, and c-Myc in tumor cells. Furthermore, these genes contribute to the development of liver cancer and gastric lymphomas.[25,26] Inflammatory mediators alter the DNA repair systems, and the cell cycle checkpoints.[27] Epigenetic alteration by the JmjC domain major protein, encoded by the NF-κB gene target, has lately been linked to inflammation.[26]

Few enterocytes lose their ability to perform two MLH1 and PMS2 mismatch repairs in Gia2 knockout mice, which is impacted by the inhibition of the DEC-1-mediated MLH1 promoter gene and histone deacetylase (HDAC) gene. This results in colon inflammation and subsequently cancer.[28] Another epigenetic mechanism which highlights tumor initiation includes the DNA methylation and microRNA-based silencing of tumor suppressors like INK4 and APC.[29] However, the importance of these inflammation-inducing epigenetic mechanisms in tumor growth development has yet to be shown. Diethyl-nitrosamine (DEN) is one of the most common causes of hepatocellular cancer. DNA damage, as is well known, accelerates necrotic cell death, which leads to tumor formation and a variety of inflammatory responses.[30,31] Furthermore, a number of oncoproteins, such as Myc, RET, and Ras, activate signaling pathways that can lead to the release of pro-inflammatory cytokines/chemokines such as IL-6, IL-18, CCL2 and CCL20.[32] Additionally, stress induced NKG2D ligand binds to NK and γδ-T cells, eradicating stressed cells or local inflammatory responses.[33]

Likewise, CD11b+ Gr1+ enriched myeloid cells in the skin results from the mosaic structural deletion of DNA-repair genes such as Tp53 and ATR showing as a part of an archetypal immune response.[34] Fen1 exonuclease deficiency causes a defect in the DNA repair mechanism, resulting in a large amount of tumor-promoting inflammatory reaction; pattern recognition receptors may play an important role in initiating this inflammatory reaction.[35]

3. Tumor promotion

Inflammation promotes cancer directly by influencing the survival and anti-apoptotic pathways, reconstruction of the micro-environment, angiogenesis, and metastasis.[36] As shown in Fig. 1, this involves various cellular and soluble mediators of inflammation.

A classical succession of inflammatory events in a malfunctioned tissue is the infiltration by peripheral mononuclear cells (PMCs) like neutrophils and macrophages, followed by the release of various soluble mediators like chemokines prostaglandins, cytokines, proteases, angiogenesis promoters and growth factors.[34] Differential inflammatory immune cells activate

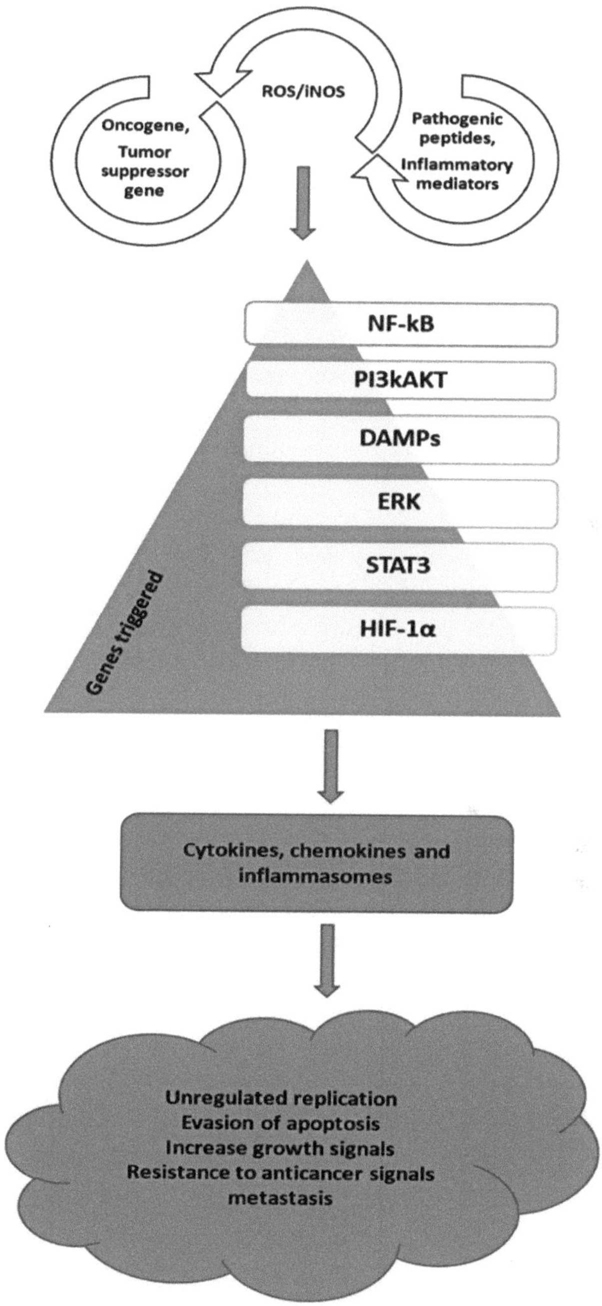

Fig. 1 Inflammation has a vital role in cancer from initiation, proliferation and, eventually, metastatic illness.

transcription factors such as AP-1, STAT-3, and NF-κB in early malignant states and begin producing too many promoting cytokines such as tumor necrosis factor-α (TNF-α) to induce genes that can alter multiple cellular processes such as proliferation, survival, cellular angiogenesis, invasiveness, motility, chemokines, and cytokine production.[37,38] Pathogenic components such as pathogen associated molecular patterns (PAMPs) and damage associated molecular patterns can also activate oncogenic transcription factors via pattern recognition receptors (DAMPs).[39] TNF and TNF-R1 signaling, in particular, are required for AP-1 activation in cutaneous carcinoma,[40] whereas the activation of signal transducer and activator of transcription 3 (STAT3) in malignant cells is mostly dependent on cellular production of IL-6, IL-11, IL-22, epidermal growth factor (EGF), hepatocyte growth factor (HGF), and an oncogenic tyrosine kinase such as Src and c-Met.[34,37,41,42]

Cytokines such as IL-1, TNF-α, and IL-6, which regulate the function of immune cells such as macrophages and T cells, have been linked to colitis-associated cancer (CAC),[43] other cytokines such as IL-6, IL-23 and TNF-α contributes into obesity associated tumor promotion in hepatocellular carcinoma (HCC).[44] IL-6 is known to play a key role in the promotion of HCC-related tumors. Mice lacking IL-6 had a reduced HCC response to pro-carcinogenic DEN.[40] STAT3 induces a high level of circulatory IL-6 as a tumor-promoting effect in CAC and HCC, while specific cell type deactivation among enterocytes and hepatocytes suppresses the formation of CAC and HCC tumors, even in mice treated with DSS, azoxymethane (AOM), and DEN.[34,42,45] High level of circulatory IL-6 is coupled with HCC risk factors. In addition, the development of CAC in mice is entirely reliant on IKKβ-mediated NF-κB activation in enterocytes. This model's principal goal is to enhance the endurance of early cancerous cells.[45]

4. Role of pattern recognition receptors (PRRs) in tumor progression

In several immunopathological states, prolonged activation of innate immune PRRs is associated with tissue damage caused by bacterial and viral infection and metabolic abnormalities that contribute to chronic inflammation.[46] This intrinsic immune function and its inflammatory role in cancer initiation and development are well established.[47] Innate immunity serves as the primary line of defense against internal tissue injury or infection at all times. It accomplishes this by using PRRs to discriminate between well-conserved sets of molecular structures known such as PAMPs and DAMPs.[48] Moreover, PRRs employ Toll-like receptors (TLR) and NOD-like receptors (NLR) to promote downstream communication and elicit a variety of pro-inflammatory responses, resulting in more effective antigen-specific immunity.[49] Additionally, TLR signals promote dendritic cell (DC) maturation, efficient antigen presentation, and CD8+ cytotoxic effector activity, all of which are required for successful tumor immunity.[50] TLRs serve an important role in tissue homeostasis by inducing a pathogenic immune response and modulating the immune response against helpful bacteria. Interfering with TLR function also impairs protective gut walls and triggers a range of proliferative signals. Changes in homeostasis caused by abnormal TLR expression or chronic activation might amplify the negative regulation of antitumor immunity. A biased immunosuppressive activity has been reported to promote the proliferation of regulatory T cells.[51] Similarly, NLR activation resulted in the release of mature versions of IL-1 and IL-18, which trigger multiple inflammatory pathways, resulting in an inflamed tumor microenvironment.[52] DAMPs such as HMGB1, ATP, and microbial/bacterial macromolecules such as dsDNA and dsRNA function as intercellular and extracellular ligands for these innate PPRs.[53] During metabolic disruption dur to tissue injury, obesity, and infection; TLRs and NLRs act as essential sensors to promote restoration of cellular homeostasis.[54] Thus, NLRs and TLRs monitor a coordinated signaling system that detects pathogenic infiltration or tissue disturbance using multiple molecular markers such as PAMPs/DAMPs. This accurately signals the expression of IL-6, TNF-α, and IL-1 genes.[55] PPRs may not be confined to immune cells; clinical investigations indicate that they are also elevated in many human epithelial malignancies. Chronic TLR activation induces pro-inflammatory responses that promote carcinogenesis, as well as increases proliferative, anti-apoptotic, pro-fibrogenic signals in the tumor microenvironment and tumor cells.[47] Thus, PRR homeostasis and atypical expression provide an immune enriching function that boosts anticancer immunity, while atypical and dysregulated expression generates a tumor-promoting effect.

5. Immune evasion and angiogenesis

Tumor progression is always accompanied by immune evasion as a result of the angiogenic process. TLRs along with NF-κB intercellular signaling, regulate integrin B1 expression and immersion in the tumor cells.[56]

LPS induced TLR4 activation has been shown to stimulate epithelial-mesenchymal transition (EMT) and invasive phenotypes in human breast cancer cell line like MDA-MB-231 and MCF-7.[57] Metastasis incursion requires the chemokine-mediated dynamic transport of metastatic cells through the sub-capsular sinus epithelium.[58] TLR activation induces CCR7 and CCR8 expression; these key chemokines are implicated in tumor mediated lymphatic remodeling of peri-tumoral lymph vessels and darning lymph-nodes, thereby boosting metastasis. Several metastatic tumors seem to use the expression of these chemokines to colonize a new tissue or organ. In contrast, substandard expression of homing receptors like CXCR3, CXCR4, CXCR7, and CCR6 has been observed in differential organ metastasis of colon cancer.[59] Additionally, metastatic CRC patients observed with higher expression of CXCL19, CCL20 and SDF-1 receptors ligands in liver and lung region.[60] Integrin signaling is an additional metastatic process modulated by the TLR activation. Melanoma and other tumor cells play a particular role in accelerating coordinated immune cell trafficking.[61,62] Over-expression of TLR, PD-L1, CCL4, TGF-β, and CCL5 has enabled the invasion and proliferation of such leukocytic tumor cells within the lymph-nodes. Lymphocytic tumor appears with increased E/P-selectins, L-selectins ligand, α4B-1, ICAM/VACM-1 required for trafficking and homing to distal tissue.[62] Moreover, IL-1β and IL-18 are inflammatory mediators and effector pro-inflammatory cytokines that upregulated adhesion molecules expression and played an important role in hepatic metastasis.[63]

5.1 Immune evasion

TLRs also partially aid tumor cells in invading immune systems, increasing tumor aggressiveness and TLR expression in numerous malignancies, which appears to hold advantages for both immune evasion and immunological suppression.[64] Key immunosuppressive cytokines like IL-10, TGF-β are produced in higher concentration while TLR signaling gets altered.[65] Moreover, activation of TLR along with increased expression of HLA-G and PD-L1 an suppressive co-stimulatory molecule, promotes development of immune tolerance.[66] This suppressive immune mechanism is quite beneficial during inflammatory autoimmune disorders. TLR-4 signaling lead to the enhanced secretion of suppressive co-stimulatory markers B7-H1/H2 and inducible iNOS by providing tolerance against cytotoxic T cells (CTL) damage.[67] Supernatants from TLR stimulated tumor cells has been observed to suppress both NK cell and T cell proliferation. Whereas, blocking TLR pathways cause delays in tumor growth and inverts tumor induced inhibition of NK cell activity and T cell proliferation.[68]

5.2 Angiogenesis

IL-1β and IL-18 plays a decisive role in tumor angiogenesis, they get matured in the presence of NLRP3 inflammatory mediators and facilitates metastasis process and cancer progression. Upregulated IL-1β in melanoma cells promoting angiogenesis in both means autocrine and paracrine and thus promoting to invasion and metastasis.[63] In addition, IL-1β induces the expression of angiogenic factors such as IL-18 and vascular endothelial growth factor (VEGF), which arouse tumor growth.[69] Similarly, Lewis lung carcinoma (LLC) cells, which also overexpress IL-1β, show VEGF associated with angiogenesis, enhanced macrophage inflammatory protein-2 (MIP-2/CXCL2) secretion, and faster tumor growth. At the same time, IL-1β knockdown mice showed decline in angiogenesis and growth in tumor size.[70] Further, DAMPs activate NLRP3 in the tumor environment is also observed to constantly produce activated IL-1β, thereby contributing to neovascularization.[71] IL-1β signals stimulate various angiogenic factors like IL-8 and VEGF from tumor and stromal cells, thus further promoting tumor growth.[72]

6. Macrophages function in tumor recognition

Macrophages are found in all vertebrate tissue and have dual roles in both host defense and tissue damage. Macrophages are mononuclear cells that play a substantial role in immunity and can phagocytose microbes. They consist of two populations: first; tissue-resident macrophages and second; postnatal monocytes-derived macrophages; independent of origin, they are pliable. According to the local environmental stimuli, they are being able to change their phenotype. Macrophages are generally polarized into M1 or M2, TAMs, and ATMs like phenotype. Macrophages ingest and degrade tumor cells, debris, dead cells, and other foreign materials. In addition, they regulate the activation and proliferation of lymphocytes. Macrophages are the effector cellular component of the innate immune system that release both pro/anti-inflammatory mediators.[73] Moreover, they play an essential role in eliminating damaged cells through programmed cell death.[72]

6.1 Macrophages in tumor-tissue resident macrophages

Macrophages present in tumors are heterogeneous in population and have altered developmental origins. Initially, it was considered that these macrophages are responsible for antitumor response, but later experimental and clinical data indicated that macrophages participate in tumor initiation, development, and metastasis.[74] Together with other immune cells, macrophages produce inflammatory cytokines including IFN-γ, IL-6, and TNF-α in response to ongoing infection, which sustains the chronic inflammation that promotes the growth of tumors. Furthermore, after establishing tumors, they instruct the tumor-associated macrophages (TAMs) left from an immunologically active state to accept the immunosuppressive phenotype responsible for tumor progression to malignancy.[75]

Many types of cancer cells abundantly release extracellular mediators such as chemokines, cytokines, and several other growth factors, such as IL-6, TGF-β, TNF-α, chemokine (C-C motif) ligand-2 (CCL-2), matrix metalloproteinases (MMPs), and granulocyte macrophage-colony stimulating factor (GM-CSF).[76] These immunogenic proteins play significant role in changing the phenotypes of macrophages. Additionally, the number and quantity of cytokines produced in the tumor microenvironment by cancer cells determine how macrophages react, which may promote or repress immune responses and, as a result, slow or prolong cancer progression, respectively. Based on the origin, macrophages are divided into two distinct linages, which are, monocyte derived macrophages and tissue resident macrophages.

6.2 Monocytes derived macrophages

Monocytes derived macrophages comes from adult hematopoietic stem cells in the bone marrow. These macrophages are circulated to tissues as monocytes, which are able to differentiate into macrophages depending differential signals from different organs and other pathogenic conditions. Monocytes-derived macrophages play important role in tumor antigen presentation, although it does not directly affect tumor progression. They accumulate in high proportion during lung injury.[77] The recruitment and differentiation of CCR2+ inflammatory monocytes are facilitated by CCL2, which is also known to be increased in the microenvironment of cancer. Additionally, they transform into F4/80+/CD11b+ macrophage and promote breast cancer spread through VEGF-dependent mechanisms.[78] CCL2 incubated with human CD11b+ monocytes, triggers CD14/CD206/CD11b population. These macrophages produce pro-tumor cytokines and accompany a tumor promoting cellular phenotypes.[79] Tissue-resident macrophages are independent of adult hematopoiesis process. They initiate from progenitor cells during embryonic development[80] and have self-renewal ability with distinct features. The names of tissue-resident macrophages vary on the basis of the organ where they reside. Tissue resident macrophages are exceptionally diverse and display varying levels of surface markers and transcription factors compared to monocyte-derived macrophages.[81] Reports shows that tissue resident macrophages are engaged in tissue remodeling instead of chronic inflammation.[82]

Further, some studies suggested that tissue resident macrophages can self-renew because of tumor derived CSF-1 (Colony Stimulating Factor-1) promote tumor progression. Moreover, CSF1/2 regulates cardiac tissue-resident macrophages' angiogenic and proliferation capacity by controlling KFL4 (Kruppel-like factor-4) levels. However, macrophage features in each organ are diverse and vary in distinct tumor microenvironments.[83]

6.3 Macrophages recruitment to the tumor site

In the tumor microenvironment, the immune cell displays a critical role. A detailed understanding of the immune cells will therefore help the cancer metastasis study. Among all immune cells, tumor associated macrophages (TAMs) are one of the most abundant infiltrates in solid tumor, which perform a substantial role in tumor invasion and development, leading to tumor cells metastasis. TAMs were derived from circulating monocytes; they differentiate in two distinct M1 and M2 phenotypes. Naturally, activated M1 like TAMs suppress cancer progression, while the M2 type promotes it.[84] However, the specific phenotypes of TAMs depend on the later stage of tumor progression. In early stages of tumor progression, TAMs adopt M-1 like phenotypes to inhibit angiogenesis in combination with stimulation of tumor immunity.

TAMs, on the other hand, enter the M2 state and release a variety of chemicals that stimulate tumor growth. The most comprehensive explanation for how M2 TAMs promotes cancer metastasis is that they emit features that favor metastasis and the proliferation of pre-malignant cells. The components include epidermal growth factor (EGF), macrophage colony stimulating factor-1 (MCSF-1), matrix metallopeptidase 2 (MMP-2), MMP-7, MMP-9, WNT5a, and VEGF.[85] Additionally, Cystatin B (CSTB) and WNT5A stimulate cancer cell proliferation and dissemination through the SMAD and c-Jun pathways.[86] According to these findings, TAMs are becoming a more prominent therapeutic target for limiting cancer spread and tumor development.

6.4 Polarization of macrophages

Macrophage polarization is essential for tissue healing and maintenance of homeostasis. The M1/M2 macrophages paradigm plays a significant role in tumor advancement. Traditionally, M1 macrophages are thought to be antitumor, but M2-polarized macrophages, also known as tumor, associated macrophages (TAMs), play a significant role in several pro-tumorigenic outcomes in cancer through immune suppression, hypoxia induction, tumor cell proliferation, and metastasis. These pro-tumorigenic effects are made possible by the tumor microenvironment, which can influence macrophage polarization and recruitment. Therefore, investigating TME-stimulated macrophage polarization is essential for improving knowledge of the pro-tumor function of TAMs and perhaps creating new therapeutic approaches.

M1 macrophages are proinflammatory and have a higher antigen presentation magnitude, generating IL-13 and an aggressive type-1 T cell response.[87,88] This serves to limit cell growth by releasing pro-inflammatory cytokines and nitric oxide, which cause tissue damage. Furthermore, Type 1 helper T cells release cytokines such IL-1 and IFN-γ[89–92] while M2 macrophages induced by high IL-4, IL-10 and IL-13, decreased IL-12 leads to an immune suppressive environment.[88] These cells stimulate tissue repair, cell proliferation, angiogenesis, and phagocytosis in order to block Th1 inhibitors and reduce inflammation.[93] However, because macrophage polarization is frequently characterized at specific stages of development, it is important to note that these phenotypic markers are repeatedly present on many subtypes of macrophages in variable levels of expression, for example, M2 macrophage can express M1 phenotypes to some extent but at a lower level than M1 macrophage and vice versa.[94]

6.5 Macrophages based immunotherapies

Depletion of tumor-associated macrophages is a separate therapeutic category that has been developed to battle tumor growth on several fronts. CSF-1 or its receptor expression in TME is linked to cancer prognosis. Because CSF-1 is important for monocyte/macrophage proliferation and survival, its interaction with the CSF-1 receptor might be an appealing target for lowering the quantity of TAMs in TME. Similarly, in a pre-clinical mouse research, the monoclonal antibody emactuzumab significantly reduces the amount of TAM while increasing the cytotoxic T cell ratio in TME by targeting CSF-1R.[95] Also, the small drug molecule PLX3397, which is intended to disrupt the CSF-1-CSF-1R link, showed increased cytotoxic T cell infiltration in the TME in a number of tumor mice models.[96] PI3K pathways may be activated momentarily by temporarily blocking CSF-1R, which is a striking target of CSF-1/CSF-1R. Thus, PI3K inhibition paired with CSF-1R inhibition has shown good results in pre-clinical investigations[97]; however, this treatment has become less effective with time. The tertiary approach in TAM cancer treatment, on the other hand, is the production of M1-like macrophages by reprogramming of M2 TAMs. TLRs play an important role in M1 programming because ligand interaction to these receptors activates macrophages and causes them to express M1.[98] Several studies have been conducted on TLRs especially TLR-3, TLR-7, TLR-8, and TLR-9 in recent years. Imiquimod is the only FDA-approved medicine that promotes TLR-7 expression and is used topically to treat squamous and basal cell carcinoma.[99] Poly I:C stimulation also increases TLR-3 expression and efficiently reprograms TAMs from M2 to M1. In an in-vitro investigation, nanoparticles loaded with poly I:C were utilized to target M2 tumor macrophages, and enhanced production of TNF-α, NO, and iNOS was seen, albeit treatment has not yet been established.[100]

Macrophages function differentially according to their different origin, changing metabolic environment, and pathological conditions.[101] On the basis of macrophages function and various immune settings they were classified in two major compartments, first, those which have the capability to kill through iNOS mechanism by responding to TLR-4, INF-α, and IFN-γ activation. IL-4 and IL-13 machinery response in macrophages as a secondary defense against parasitic infection and asthma.[102] Later, Mills et.al. categorized macrophages in-vitro and characterized M1 "activated" and M2 "alternative."[103] These findings imply that TAMs may serve as tumor killers through M1 macrophages or tumor promoters via M2 macrophages.[104] However, such kind of extreme definition of macrophage polarization plausibly cannot exist, particularly in complicated tumor microenvironment since macrophages were well described in multiple phenotypes in term of TAM activity especially in different pathological settings with their differential biological functions. Alternatively, studies on different tumor types demonstrated various other macrophages accompanying with varied phenotypes advocates for abundance of macrophage population. Additionally, most of the large-scale transcriptome analysis observed vast variety of mixed macrophages expressing both M1 and M2 phenotypes.[105] Lack of conclusive experiments to ablation of macrophage in M1 or M2 population, thus their definite role in tumor microenvironment is remain inconclusive.

7. Lymphocytes driven immune surveillance

Lymphocytes are white blood cells that mostly consist of T cells, B cells, and NK cells. They have a consistent appearance yet perform a variety of biological tasks. They control immune response by producing antibodies against harmful peptides and directly destroying virus-infected or tumor cells. T and B cells are capable of developing antigen-specific immunological memory through V(D)J somatic gene rearrangement utilizing a specifically expressed recombination-activating gene (RAG) protein. NK cells, on the other hand, employ RAG-independent hapten-based contact sensitizers to elicit antigen-specific responses. Cancer growth and progression are linked to leucocyte infiltration in developing tumors.[106] Furthermore, it was considered that initial immune response against neoplasm mimics response to tissue injury, later sequential infiltration myeloid cells and then lymphocytes confer tumor development.[107] However, since the dynamics of neoplastic cells and tumor formation modify the local immune milieu, distinguishing between an immune response to infection and tumor development is difficult.

7.1 T cell receptor (TCR) in tumor

T cells have the capacity to detect changes in peptides, even single amino acids, through their TCR receptors. Thus, activated T cells may be better able to recognize cancer cells through peptide linked to MHC-I and MHC-II molecules, ultimately leading to cell death. Studies in mouse models have revealed rejection of mutagen-induced tumors, lending credence to this assertion. T cell-dependent immunity turns against antigens after being exposed to a solid mutagen. T cell infiltration is a clear and effective anti-cancer immune response. TCRs are composed of two separate protein chains, namely α and β, which are produced on T cells as a result of productive gene rearrangement. A small fraction of matured T cells also express γ and δ chain, however the percentage is small.[108]

αβTCR mediated antigen recognition is foremost function of adaptive immune system. αβTCR binds antigen-presenting cells with peptide loaded in MHC-I or MHC-II (pMHC). The interaction between pMHC molecules and αβTCR executed due to T cells ability to distinguish between rare foreign particle and abundant self MHC molecules.[109] In cancer patients, essential adaptive immune response governs through crucial role of CD8+ T cells, and they activated by TCR recognition of specific peptide epitope. Primarily, MHC-1 present on surface of tumor cells and endogenously generates peptide epitopes for antigenic presentation.[110] These MHC-I proteins are class of membrane protein that expressed on all nucleated cells and encoded with several HLA-A/B/C human leukocytes antigen gene.[111] Although HLA gene expression was significantly increased by interferon signaling, however, in case of tumor their presentation is remarkably decreased. The loss of HLA gene expression associated with immune invasion and disease progression among cancer patients.[112,113] T cell activation requires antigenic peptide processed through MHC complex, processed antigen presented by MHC molecules to the TCR and then to intercellular signaling pathways.[114–116] Studies conducted in-vivo and in-vitro suggests cytokine expression from T cells determined against the dose of antigen presented by the MHC complex.[117,118]

7.2 Structure of T cell receptor

T cell receptor expressed on T cells and express somatically arranged natural occurring antigen receptors on their surface. TCRs has heterodimer consisting with 95% α and β chains receptor and remain 5% is γ and δ chain, TCR receptor's hold abilities to bind variable region of antigen binding; neither chain has intrinsic capacity to signal. For T cell activation, receptors depend on accessory signaling molecules that CD3 binds non-covalently to. Further, two α and β TCR chains utilizing conserved cysteine residue form constant region of both chain and turn covalent to link through a single disulfide bond.[119]

Comparable to the variable region of MoAb, the variable domain of each TLR chain is made up of the following regions; a germline encoded sequences that is recombined from highly polymorphic family of variable (V) and the diversity (D) alleles and a series of the non-consecutive hypervariable loops that are referred to as complementary determining regions (CDRs), in addition to the V/D alleles, the β chain has an additional allele known as the junctional (J) allele, which allows for an even broader range of possible outcomes. The CDR loops extend outward from the TLR chain and make direct physical contact with parts of the MHC molecule, either by themselves or in combination with the peptide. Because they are the most hypervariable upon somatic rearrangement, the centrally positioned CDR3 loops dominates the interaction with the peptide and, as results, often contribute to the fine specificity of TCR for a particular peptide. The outward facing and germline encoded CDR1 and CD2 loops, on the other hand, provide a baseline level of TLR affinity for generic MHC molecules via highly conserved interactions. CDRq, on the other hand, has the potential to engage peptides and contribute to their specificity. The TLR creates a non-covalent oligomeric complex with three dimeric CD3 signaling molecules to trigger signaling farther down the line. These are CD3 homodimers, as well as CD3 and CD3 heterodimers. In contrast to the overwhelming majority of second-generation CAR signaling domains, the peptide/MHC -TLR complex has ten ITAMs.[120]

7.3 TCR-based immunotherapies/CAR therapies

7.3.1 *Chimeric antigen receptors (CARs) T cell therapy*

Strategies for cancer therapy are continually changing to address new issues in medical research. Adoptive cell therapy can be used as a last resort for patients who have failed to stop certain types of cancer with conventional treatments. In adoptive cell therapy, human T cells are combined with technological advances to cure illnesses. An artificial T cell receptor may be attached to these cells as one method of modification. TCRs bind to antigens on lymphocytes to recognize them. T cell therapies that alter the receptors to treat cancer include CAR T cell therapy and engineered TCR therapy.

Chimeric antigen receptors (CARs) are a surprisingly adaptable and fully synthetic single chain receptor that consists of various modules such as an antigen recognition domain, hinge region, transmembrane domain, and T cell signaling domain. CAR recognition domains are created from MoAb variable regions and then connected together as a single chain variable fragment (scFv).[121] Currently, existing CAR T cell treatments are tailored to specific individuals. CAR T cells are made mostly from T cells obtained from patients' blood and are genetically modified to create proteins termed chimeric antigens on their surface. Furthermore, CARs may detect and attach to particular antigens/proteins on the surface of cancer cells.[122] These modified T cells were grown in vitro before being reinfused into the same patient. These CAR T cells will multiply in TME indefinitely if T cell engineering is effective. In order to deliver successful treatment, these CAR T cells have a target antigen on their surface that allows them to detect and destroy cancer cells (Fig. 2).

CAR T cell Therapy was first developed to treat children with acute lymphoblastic leukemia (ALL). B cell related ALL-cancer alone accounts for 80% of all malignancies diagnosed in children. Despite the fact that it may be treated with intense chemotherapy or stem cell transplantation, the proportion of non-responders and recurrence cases remains a substantial obstacle for a successful treatment approach. Tisagenlecleucel, the first

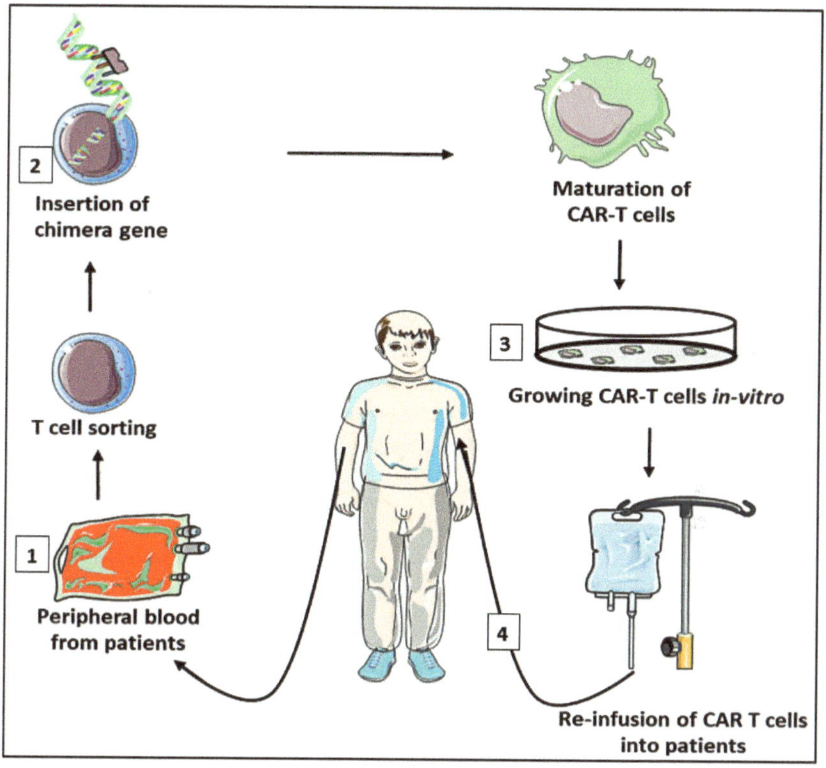

Fig. 2 A type of therapy called CAR-T cell therapy involves genetically modifying a patient's T cells in a lab so they will bind to and kill particular proteins (antigens) on cancer cells. (1) Blood is drawn from the patient and T cells were collected. (2) Next, in the lab, T cells are given the genes for a unique receptor known as the chimeric antigen receptor (CAR). The patient's T cell expresses the designed CAR protein from the gene, resulting in the development of a CAR T cell. (3) Countless CAR T cells were raised in the lab. (4) They are then administered as an intravenous infusion to the patient, CAR T cells attach to the cancer cells' antigens and destroy them.

CAR T cell treatment, was presented in 2017 following extensive clinical research and FDA clearance (Kymriah^{TM}). The findings are promising since it has been found to remove cancer in children with relapsed ALL cases. CAR T cell treatment advancements in children with ALL are encouraging.

7.3.2 Adoptive cell transfer (ACT)

Adoptive cell therapy was first published in 1988 and quickly became the most effective therapeutic technique in individuals with metastatic melanoma. Patients with metastatic melanoma who underwent ACT treatment saw an unbiased response, with some achieving a full response. ACT—immunotherapy is divided into four categories: tumor infiltrating lymphocytes (TIL) treatment, endogenous T cell therapy, chimeric antigen receptors T cell therapy, and TCR transduced T cell therapy. All of these immunotherapies include the collection of immune cells, which are then multiplied to generate a high quantity before being returned to the same patient. T cells were produced from the tumor itself for TIL treatment, while tumor specific T cells were created from the patient's blood for endogenous T cell therapy. CAR T, on the other hand, is generated by chimeric antibody/T cell receptor gene infused in peripheral T cells and in TCR transduced T cell therapy, designed T cell receptor gene from a third party with tumor recognition capacity infused in patient's peripheral T cells.[123]

For the treatment of recurrent acute ALL and diffuse large B cell lymphoma, an FDA-approved ACT that identifies CD19 (CART-19) is present in certain B cell neoplasia and even on normal B cells. Furthermore, CAR-T cell technology is being evaluated against a broad spectrum of hematological and solid cancers, not only B cell maturation antigen (BCMA).[117] CAR-T 20 is being evaluated in pre-clinical trials for cancers such as non-Hodgkin lymphoma, mesothelioma, non-small cell lung cancer, pancreatic adenocarcinoma, HER-2 positive breast cancer, and gliomas.[124,125] Potent T-cell immunity after ACT therapy is dependent on a number of processes, the most important of which are the permanence of transfected cells and the persistence of anti-tumor activity. The capacity of transplanted cells to differentiate into effector memory (TEM), central memory (TCM), stem cell memory (TSEM), and terminal effector memory (TTE) has been associated to increased ACT treatment endurance. Since several decades, investigations have indicated that certain TCR-Vα/Vβ T-cell clones are present in the peripheral blood and tumor-infiltrating lymphocytes (TIL) of melanoma patients.[126,127] Particular single chain TR VP T cells grown in vitro have specialized killing effects, and these clonal T-cells exhibit specific anti-melanoma cytotoxicity. In addition, Fujio et al. have shown that the retrovirus-generated TCRαβ gene transduction system may restore the antigen-specific immunity of peripheral T cells in mice.[127]

7.3.3 Immune checkpoint inhibitors

An important immune system regulator is the immunological checkpoint. Their primary function is to stop the immune reaction from being so potent that it completely eradicates immune cells from the body. Immune checkpoints are activated when immune cells bind to partner proteins on the surface of other cells, such as tumor cells. These proteins are known as immune checkpoint proteins. When the checkpoint and partner proteins bind together, they send a "off signal" to the cells. As a consequence, the immune system is unable to destroy the cancerous cells. Immune checkpoint inhibitors are medications or small compounds used in immunotherapy to block checkpoint proteins from interacting with their partner proteins. As a consequence, since the "off signal" is not given, immune cells are able to kill cancer cells.

Cancer cells may evade being detected and eliminated by host lymphocytes by changing their surface antigens in the tumor microenvironment. An important aspect in tumor-induced immune suppression is increased synthesis of ligands that bind inhibitory T cell receptors.[128–130] These ligands, also known as immunological checkpoints, work physiologically to prevent the development of autoimmunity at different phases of the immune response. The key processes in T cell regulation are the suppression of potential autoreactive naive T cells (marked by TCR oriented towards self-antigen) in lymph nodes during the early stages or the deactivation of T cells in peripheral tissues later in the process. CTLA-4 and PD-1 immunological checkpoints play important roles in this process, which is known as peripheral tolerance. Tumor cells have evolved techniques to take advantage of peripheral tolerance and avoid immune detection by generating a defective immunological check point expression by T cells[131–133] (Fig. 3).

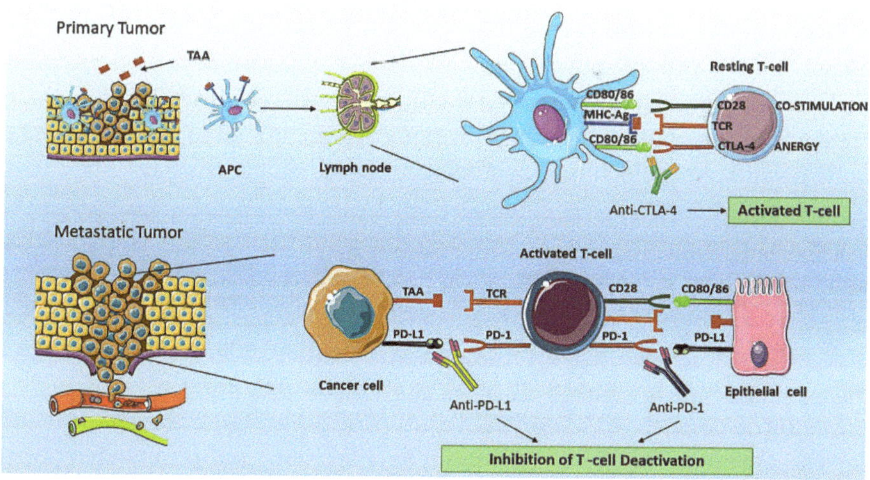

Fig. 3 Antigen presenting cells (APCs) process tumor associated antigen (TAA) and attach them to MHC molecules in the tumor microenvironment. APC moved to T cell-dependent regions of tumor-draining lymph nodes and presented TAA to naive or quiescent T cells to activate them. *Adopted from Franzin R, Netti GS, Spadaccino F, Porta C, Gesualdo L, Stallone G, Castellano G, Ranieri E. The use of immune checkpoint inhibitors in oncology and the occurrence of AKI: where do we stand?.* Front Immunol. *2020;11:574271. https://doi.org/10.3389/fimmu.2020.574271. eCollection 2020. Review. PubMed PMID: 33162990.*

8. Role of B cell receptor in tumor

B lymphocytes are the key cellular constituents of the humoral compartment of adaptive immunity. After birth and numerous rounds of development, hematopoietic stem cells (HSCs) reach the common lymphoid progenitor (CLP) stage and generate T, B, or NK cells. Throughout fetal life, they develop mostly in the liver. The B cell linages include B lymphocytes of fetal liver heritage.

They are a kind of B lymphocyte that lives in body cavities and may self-renew and produce restricted-specificity, secret IgM antibodies (natural antibodies). B cell growth and adaptive immunological responses are dependent on B cell development and adaptive immune responses are strongly reliant on the transmembrane receptor produced by it, known as the B cell receptor (BCR). Pre-BCR maturation is required for B cells to mature into functional B cells capable of recognizing a wide range of antigens.

A wide range of foreign antigens are detected by mature B cells through their individual BCR, which is responsible for antibody responses specific to these antigens and promotes the transition to plasma cells and memory B cells. Activation of B cells upon stimulation is achieved by signal transduction via the activation of membrane-proximal kinases such as PI3K, notably PI3K8, a p110 isoform widely expressed in lymphocytes, Bruton tyrosine kinase (BTK), and spleen tyrosine kinase (SYK). Kinase inhibitors are presently being developed to target these kinases. B-cell receptors (BCR), which are made up of auxiliary signaling proteins that are linked to transmembrane isoforms of IgM molecules, begin to accumulate on growing B cells. These are composed of two heavy chains formed by recombination between the Ig heavy chain locus's (VH, DH, and JH). Each of them, when joined with a light chain formed by recombination between variable regions (VL and JL) of IgK or λ light (L) chain loci, results in a light chain.[134] About five million of B cell clones, each of which expressed a distinct BCR that recognized a different epitope, are produced as a result of this procedure.[135]

Antigen contact with the BCR and subsequent signal transduction are the earliest phases of B cell activation, proliferation, and ultimately plasma cell differentiation In their study of the function of BCR signaling in antitumor immunity, Bing Yang et al. focused on the importance of a human immunoglobulin G1 (IgG1) variant specific to Asia with a GLy 396 to Arg

396 substitution (hIgG1-G396R) in colorectal cancer. An epidemiological investigation discovered a link between higG1-396R and progression-free survival in colorectal cancer. Plasma cells, rather than B lymphocytes, were seen infiltrating the tumor microenvironment in human samples and colorectal cancer models in animals. Individuals with the hIgG1- G396R variant had a larger number of CD8+ T cells, dendritic cells, and tertiary lymphoid structures. These variations may decrease carcinogenesis by increasing B cell responses, and it has been proposed that changing BCR signaling might improve the efficacy of immunotherapy in cancer. Effective cancer immunotherapies may result from the development of monotherapies or combinations of therapies that modulate B cells and BCR signaling; mice with B cell deficiency or immunized with tumor antigen lacking the BCR-specific epitope failed to produce data on the importance of B cells in antitumor immunity.

8.1 BCR-based immunotherapies

An antigen-specific BCR which is membrane-anchored is expressed by both naive and memory B cells. After the BCR and B cells have activated and differentiated into plasma cells and immunoglobulins in response to the antigen recognition. These antibodies, which are specific to tumor antigens, can recognize cancer cells through their variable regions and mediate tumor killing by complement, CD8+ T cells, or NK cells through their constant regions recognized by Fc receptors on immune cells, or by phagocytosis (antibody-dependent cell phagocytosis; ADCP; antibody-dependent cell cytotoxicity; ADCC).

Over the past 10 years, inhibitors of SYK tyrosine kinase, Bruton tyrosine kinase (BTK4), and PI3K6[136,137] have been created. Ibrutinib, a BTK inhibitor, is now authorized for the treatment of people with CLL, MCL, WM, and MZL. Idelalisib, a PI3k inhibitor, is authorized for use in patients with FL and relapsed CLL (when combined with rituximab, a monoclonal antibody that targets the B lymphocyte antigen CD20). Due to their superior pre-clinical and clinical activity characterization, two first-in-class agents have been chosen. Additionally, both medications have exceptional therapeutic success in individuals suffering from CLL4,[136] MCLS,[138] FL,[137] MZL,[139] and WM.[140] Responses, however, are rare or limited to certain patient populations with other B cell cancers.[6]

Additionally, there are differences in the overall response rates (ORRs) and longevity of responses to BTK and PI3K inhibitors across various B cell malignancies. The majority of individuals with WM[140] and CLL[141–143] react effectively to both kinds of inhibitors. The median progression-free survival (PFS) for patients treated with ibrutinib and idelalisib is 13.9 months and 3.7 months, respectively, which is significantly less than that for patients with CLI treated with these drugs. Both BTK and PI3K6 inhibitors achieve sizeable initial response rates in relapsed and refractory MCL. Responses are only seen in the ABC subtype of DLBCL patients, which, unlike GCB-DLBCL, requires on BCR signaling and NF-k pathway activation for survival. However, even in this subtype, response rates are low and response times seldom last longer than a few months.[144]

The activation of pathways that can overcome BTK or PI3K6 inhibition is believed to depend on both the genetic background and the degree of genetic instability. One way to look at the large variations in response rates and response times among the different B cell malignancies is to consider how they differ in their reliance on BC signaling and their interactions with the milieu. An illustration of this sort of escape mechanism is the proliferation of ibrutinib-resistant MCL cells that is mediated by cyclin-dependent kinase 4 (CDK4), which may be inhibited in pre-clinical mice models by using PD0332991 (Palbociclib).[145] It should come as no surprise that the therapeutic function of BCR signaling pathway inhibitors varies depending on the kind of B cell malignancy, in part due to the biochemical variations outlined above and the various clinical needs. Chemotherapy can cure the indolent lymphomas, which include CLL, FL, MZL, and the majority of MCL patients. BCR signaling inhibitors have emerged as a promising treatment option for patients with these forms of lymphoma who have relapsed following first-line therapy, which is often chemotherapy with a CD209 monoclonal antibody (chemo-immunotherapy). Additionally, chemotherapy as usual now includes BCR signaling inhibitors. based regimens, or even in the creation of curative regimens, to lengthen the duration of response.

Approaches to treating patients with CLL, MCL, WM, MZL, and FL is fundamentally altered by targeted therapy using BTK and PI3K inhibitors. These results have brought attention to the function of BCR signaling in leukemia and B cell lymphoma. Indeed, the kinases BTK and PI3K5, which are the targets of these drugs, are crucial for BCR signaling. Therefore, the major targets of these drugs may be BCR-driven proliferation and survival of malignant B cells. The effects of these substances on malignant B cells and the microenvironment, however, are also becoming more and more clear. The function of additional cell surface receptors, such as those that signal chemokine receptors[146] and those that collaborate to control the tissue homing of B lymphocytes,[147–150] is similarly impacted. Second, they target BTK-containing macrophages in the microenvironment and modify their activity.[151,152] Third, they target many off-target kinases in addition to their intended kinase, such as BTK or PI3K, as shown by the off-target suppression of T or B lymphocyte kinase (BLK) with ibrutinib.[153,154] T and B cell activity is modulated by this. The effects on homing receptor function and signaling are particularly noteworthy and clinically significant because the migration of malignant B cells from the lymphoid tissue into the peripheral blood causes rapid onset redistribution lymphocytosis, frequently dramatic shrinkage of lymph nodes, and enlargement of the spleen. Thus, it is appealing to think that ibrutinib and idelalisib's alternate mode of action is to induce anoikis[155] by removing CLL cells and other malignant B cells from a supportive tissue environment that triggers a variety of pro-survival signals. Instead of oncogenic transformation, B cell malignancies were treated by targeting nonspecific signaling pathways that are triggered by interactions between malignant B cells and their tissue milieu. It is believed that the success of BTK and PI3K inhibitors will spur interest in this field of study, where investigations of the intricate interactions between malignant cells and their microenvironment are expected to quickly identify new effective treatment targets.

9. Natural killer (NK) cells and receptors

A group of innate lymphoid cells known as natural killer cells are essential to the host immune system defense against infection and tumor development. A complicated network of inhibitory and activating signals controls the intense cytotoxic action of these cells. NK cells have been shown to be crucial in regulating tumor development and providing a potent anti-metastatic impact in a variety of experimental animal cancer models. In several human solid tumors, a high density of tumor-infiltrating NK cells has been associated with a favorable prognosis. However, due to the existence of an immunosuppressive microenvironment that affects NK cell function, some cancers seem to be resistant to NK cell-mediated death. Several studies have demonstrated the significance of NK cells in the defense against various cancer types.[156] Tumor development and metastasis are accelerated when NK cells are eliminated by antibodies or have their cytotoxic abilities compromised.[157,158] Clinical research has also repeatedly demonstrated a link between a larger tumor-infiltrating NK cell count and a better prognosis in solid tumors, including lungs, breast, neck, and colorectal cancer.[159,160] The crucial importance of NK cell-mediated cytotoxicity in cancer immunosurveillance was highlighted by an 11-year follow-up research that found reduced NK cell cytotoxicity activity is linked to a greater risk of cancer occurrence.[161,162] Additionally, recent research revealed that individuals with breast and prostate cancer who had NK cells that express more activating receptors do considerably better.[163,164] Functionally, NK-derived IFN-γ also causes M1 macrophages to become polarized and activated, which, along with the activated NK cells, kills tumor cells.[165] Also, IFN-γ upregulates the production of CXC19 and CXCL10 by tumor cells, which in turn enhances the recruitment of cytotoxic CD27^{high} NK cells in a CXCR3-dependent pathway.[166] More crucially, NK cells have recently been discovered to have antigen-specific memory features like T cells and B cells, despite previously being known as short-lived cytotoxic killer cells. This enables NK-mediated immunosurveillance for up to three months.[165–167]

IFN-γ, TNF-α, GM-CSF, and chemokines are among the many cytokines that are highly active NK cells and T cells may release.[168] IFN-γ is expected to have powerful anticancer effects, including increasing MHC1 expression and making tumor cells more susceptible to CD8+ T cell eradication. In subcutaneous tumor transplantation models, γδ-T cells are recognized as a crucial source of IFN-γ, and IFN-derived from NK cells is favorably correlated with patient survival in various malignancies.[169] TNF-α directly influences cytotoxicity activity by inducing apoptosis mediated by caspase-8.[170] Senescence is also brought on by TNF-α and IFN-γ together.[171] As a result, cytokines released by innate cells may directly combat tumor growth. Through the production of cytokines or interactions between receptors and ligands, NK cells can also control the activity of other leucocytes, including T cells and dendritic cells.[172,173] However, collaboration between various leucocytes has not been thoroughly investigated by tumor immunology. According to a recent study, NK-cell-derived IFN-γ may drive tumor cells and polarize macrophages towards a tumoricidal "M1" state, which protects against carcinogen-induced sarcoma.[165]

9.1 NK cell receptor

NK cells communicate with other cells often while patrolling. The balance of signals from activating and inhibitory receptors on the NK cell's surface determines whether or not the NK cell kills these cells. Inhibitory receptors monitor NK cell death whereas activating receptors turn on NK cells when they identify chemicals expressed on the surface of cancer cells and infected cells. The majority of healthy, normal NK cells have MHC1 receptors, designating them as "self." The cognate MHC1 is recognized by inhibitory receptors on the surface of NK cells, which turns off the NK cells and stops them from killing. When the NK cell decides to kill, it releases cytotoxic granules containing perforin and granzymes, causing the target cells to lyse. Individuals differ greatly in their MHC1 genes as well as the inhibitory receptors that the NK cell uses to distinguish them. The alleles of these genes have been connected to possibility of developing certain autoimmune diseases as well as the capacity to defend against HIV infection. Chronic viral infections like cytomegalovirus have an impact on NK cell types, which also alter with aging (CMV). Cancer immunotherapies are interested in NK cells as potential targets. Some therapeutic monoclonal antibodies depend on the destruction of NK cells. Additionally, scientists are evaluating the therapeutic potential of genetically altered live NK cells and designing therapies to activate NK cells using small chemicals or cytokines.

9.2 Role of natural killer group 2D (NKG2D)

In the NK gene complex, the gene Kirk1 (Killer Cell Lectin-Like Receptor Subfamily Member 1) is responsible for encoding the NKG2D receptor, which is a lectin-like type-2 transmembrane receptor.[174] NKG2D is incapable of independently transmitting activating signals due to the narrow intracellular domain it has.[175] Instead, it is the charged residues in the transmembrane region of NKG2D that allow it to link with the signaling adaptor proteins DAP 10.[176–178] This pairing is crucial for NKG2D surface expression as well as subsequent signaling. Only a certain splice isoform of NKG2D is capable of pairing with DAP12, and this isoform is only seen in active mouse NK cells and not in resting NK cells.[176] DAP10 is responsible for recruiting the p85 subunit of phosphoinositide 3-kinase, often known as P13K, as well as growth factor receptor bound protein 2.[179,180] DAP2 contains an immunoreceptor tyrosine-based activation motif (ITAM) at the same time, which allows it to engage spleen tyrosine kinase (SYK) and zeta chain associated protein kinase 70 (ZAP70).[178] When NKG2D is engaged, the downstream activation of PI3K and GRB2 is initiated in mouse NK cells, as well as SYK and ZAP7O, takes place. Killing, generation of cytokines, activation of immune cells, and proliferation of immune cells are all stimulated by these signaling biomolecules. Nearly all NK cells and CD8+ T cells, CD4+ T cells and γδ-T cells, express the NKG2D gene. It has not been established whether or whether not an ordinary CD4+ T cells express NKG2D in humans; nonetheless, NKG2D is expressed by all $CD8^+$ T cells. NKG2D is able to identify a number of ligands that are linked to MHC-I, and it has been demonstrated to be elevated in a variety of cancer cells.[176,181–183] NKG2D ligands in humans may be broken down into three categories: MHC class-I chain-related proteins A and B, as well as UL16-binding proteins. There is evidence to suggest that NKG2D ligands are expressed in a wide variety of human malignancies, some of which include melanoma, leukemia, myeloma, prostate, glioma, and colon cancer.[184–187] There is an increasing number of data indicating that the expression of NKG2D is essential for the destruction of tumor cells in both in vitro and in vivo settings.[181,182,188] It was shown that the use of NKG2D antibody blockades led to an increase in the number of MCA-induced tumors, while there was no such increase in NKG2D knockout (KO) mice.[183,189,190]

9.3 Natural cytotoxicity receptors

In order to recognize tumor cells, certain receptors, such as NKp46 (also known as NCR1 and CD335), NKp30 (as NCR3, CD337), and NKp44 (as NCR2, CD336), are thought to be important. These receptors, referred as "Natural Cytotoxicity Receptors" (NCR) since their role in natural killing is not more significant than the role of NKG2, NKp46, NKp30, and NKp44, are all expressed on human NK cells, however NKp44 is only expressed on human NK cells that have been activated. NCRs are connected non-covalently with adaptor proteins such as DAP12 (NKp44), CD32-FceRIY (NKp30), and FceRiy-CD34 (NKp46), and they are members of the immunoglobulin (Ig) class of receptors.[191–195] NCRs also feature extracellular I domains.[196] These adaptor proteins include TAMs, which are necessary for enlisting SYK and ZAP70 and activating them. Phosphorylation of transmembrane adaptors by these kinases, such as the linker for the activation of T cells (LAT) and the Non-T-cell Activation linker (NTAL), leads to the activation of PI3K and phospholipase.[197–201] The calcium flux that arises as a consequence of these signaling events ultimately results in the remodeling of the cytoskeleton, the production of cytokines, and the cytotoxicity of the cell.[191-193,202] Studies conducted in vitro have shown that these receptors have a role in the death of particular tumor cells by the natural killer (NK) system.[192,196,203–206] In addition, the nuclear protein Bcl-2 associated athanogene 6 (BAG-6), which is also known as BAT3, was hypothesized to function as a cellular ligand for NKp30 and was connected with in vivo tumor recognition.[207,208]

9.4 NKp80 (KLRF1-killer cell lectin-like receptor subfamily F1)

NKp80 is a homo-dimeric receptor that has the properties of a C-type lectin. This receptor is present not just on human NK cells but also on $CD3^{+}CD56^{+}$ T cells.[209] The unique tyrosine-based motifs known as hem-ITAMs may be found in the cytoplasmic tail of the NKp50 protein. These motifs trigger the SYK kinase pathway and induce NKp80-mediated cytotoxicity.[210,211] Activation-Induced C-type Lectin is a kind of activating receptor that is most often expressed on myeloid cells (AIL). This receptor binds to AIL, which then causes it to become activated. The link between AICL and NKp50 is what stimulates NK cytotoxicity and plays a role in the identification of myeloid tumor cells. This connection is responsible for both effects.[209]

9.5 Signaling lymphocytic activation molecule (SLAM)-related receptors

Macrophages, mast cells, dendritic cells, and natural killer (NK) cells they all express the SLAM (Signaling Lymphocyte Activating Molecules) associated receptors on their surfaces. The 2B4 (CS244), CD54, CRACC (CD319), and SLAM (CD150) receptors are all members of this family of receptors. The majority of SLAM receptors participate in homotypic interactions with other cells that have the same proteins. The binding of SLAM receptors triggered the activation of signals via the coordinated efforts of a number of receptor-associated adaptor proteins that belonged to the family of SAP (SLAM-associated proteins).[212] In the absence of the SAP family adaptors that may suppress activation produced by other receptors like NKG2D, the SLAM family receptors transmit powerful inhibitory signals. This is the case because these adaptors are not readily accessible.[213–216] NK-mediated death of hematopoietic target cells is dependent on SLAM receptors, whereas the recognition of non-hematopoietic tumor cells does not need SLAM receptors. SLAM receptors play a crucial role in NK-mediated killing of hematopoietic target cells.[213] NK cells lose some of their capacity to destroy hematopoietic target cells in people and mice that are lacking in the SAP gene.[217–220]

9.6 Adhesion molecules and DNA-1

NK cells generate an immunological synapse in order to attach with target cells. Within this synapse, activating and inhibitory receptors temporally cluster in order to maximize the efficiency of intracellular signaling. When it comes to the formation of immunological synapse, adhesion molecules like the integrin LFA-1 are necessary.[221,222] The polarization of cytotoxic granules in the direction of the immunological synapse is caused by the interaction of LFA-1 with ligands that are present on target cells.[223–225] It is necessary for the interaction of NK cells and T-cells with a wide variety of target cells. The DNAM-1 cytoplasmic domain is subsequently phosphorylated by Fyn kinase as a result of the Ig superfamily receptor DNA-1 (CD226) with LFA-1 on the cell surface.[226,227] Numerous types of leukocytes, including monocytes, platelets, T cells, B cells, and NK cells, are responsible for the expression of DNA-1, which is common in cancer cells.[228–231] Nectin-2 and PVR are widely dispersed throughout the body's various tissues, including neurons, epithelial cells, endothelial cells, and fibroblasts, among others.[232] In some circumstances, the interaction is necessary for the proper activation of NK cell-mediated cytotoxicity and the generation of cytokines.[226,229,230,233] The destruction of neuroblastoma cells by NK cells came from a variety of patients and was shown to have a strong correlation with the expression of DNA1 ligands on the tumor cells.[234] However, activation with DNAM-1 is not sufficient to initiate the death of target cells; rather, it synergistically activates NK cells, which are also activated through the 2B4, NKp46, and perhaps additional receptors. This results in the destruction of target cells.[224] Notably, Dnam1$^{-/-}$ animals showed faster development of fibrosarcoma and papilloma in reaction to chemical carcinogens, and they had decreased rejection of transplanted tumor cells expressing DNAM-1 ligand. Both of these results were in response to chemical carcinogens.[235,236] According to these results, DNAM-1 likely plays an important part in the immune surveillance of tumors.

9.7 MHC associated NK cell inhibitory receptors

NK cells are stimulated by the MHC-I molecules that are produced by target cells and they express a variety of receptors that are able to send inhibiting signals to other cells. These receptors include those belonging to the Ly49 and KIR families, which detect MHC class-I molecules in a direct manner. Mice express Ly49 receptors but not KIRs, in contrast to humans. On the other hand, Ly49 receptors are not expressed in humans, only KIRs.

The CD94/NKG2A heterodimeric receptor's third common inhibitory receptor recognizes a peptide made from the class 1a leader sequence that is made available by a specific nonclassical MHC I molecule (HLA-E in humans and Qa-1 in mice).[235–239] The cytoplasmic tails of inhibitory receptors include immunoreceptor tyrosine-based inhibitory motifs, also known as ITIMs. These motifs are responsible for the inhibition of signaling by the receptors. ITIMs undergo tyrosine phosphorylation and recruitment of the protein tyrosine phosphatases SHP-1 and SHP-2 in response to the engagement of their respective receptors.[240–242] It is suspected that the phosphatases may impede activation by dephosphorylating one or more crucial signaling molecules, particularly VAV-1.[243–246]

Only a small percentage of natural killer (NK) cells have the KIR, Ly49, or CD94/NKG2A receptors expressed. As a consequence, the NK repertoire may recognize host cells that have lost only a single class la allele.[239,247–249] Because the selective loss of MHC-1 expression is a characteristic of many tumor cells,[250] the diverse KIR expression by NK cells makes it possible to recognize and eliminate tumor cells or other cells that have such alterations in MHC-I expression and is termed 'missing self-recognition'. This is true whether the cells in question are tumor cells or other cells.[251] Studies that showed RMA-S, a lymphoma cell line chosen for lack of MHC-I expression, was susceptible to NK cells but RMA cells, an MHC+ counterpart to RMA-S, was not led to the discovery of missing self-recognition.[252] Later research demonstrated that NK cells are capable of destroying even normal cells even if they do not have MHC-I.[253] The lack of MHC I molecules on RMA-S cells or MHC-I deficient normal cells is not a sufficient requirement for the destruction of the cells; the susceptible cells must also express ligands that activate NK cells, although to a lesser extent. This is necessary for the destruction of the cells. Recent research reveals that in order for this kind of killing to occur, particular activating receptors on NK cells, namely those belonging to the SLAM family, need to be activated.[213] This result is supported in part by the fact that, NK cells that lacked the SAP family adapters for SLAM receptors, were unable to kill MHC-deficient normal cells or RMA-S cells in either in vivo or in vitro studies. Hematopoietic cells are the only cells in the body that are able to produce SLAM proteins, which are notable because they also act as ligands for SLAM receptors. Because of this, the signals that are produced by the activation of SLAM receptors are inadequate to counteract the inhibition that is imposed by MHC-I specific receptors. This explains why normal MHC+ cells are not sensitive to being killed by NK-mediated death.

9.8 NK cell-based therapy

An up-regulation of activated NK cell ligands as well as a down-modulation of their activity has been revealed. Activating ligands have been shown to express less of themselves on the surface of cancer cells as the disease progresses, with metastatic lesions revealing lower quantities of these molecules in comparison to their original counterparts.[254,255] This reduced expression is often ascribed to the production of a soluble version of the molecule, the shedding of ligands, and release in the extracellular space. The NKG2D ligands known as MICs and ULBPs are two of the well-defined ligands that are secreted by melanoma cells. Additionally, the shedding of B7-H6, which is engaged by NKp30, has been observed. As was to be predicted, the surface ligands decrease with the course of the illness, however the shed ligands exhibit the opposite trend compared to the surface equivalent.[256–258] Although it has also been identified as a crucial method for downregulating NK cell cytotoxicity, the over-expression of activating ligands is a factor that must be considered.[259,260] Both ligand shedding and over-expression will ultimately result in the receptor being hyper-stimulated, at which point it will be internalized to prevent excessive activation. This will lead to NK cells becoming desensitized and having less cytotoxic capability.[256] It is a well-established fact that NK cells in melanoma patients usually have a decreased expression of activating receptors and a diminished capacity to react to malignant cells.[261–263] Damage sensor kinases ATR and ATM are responsible for recognizing DNA damage in the early stages of carcinogenesis. This recognition stimulates a kinase cascade that includes downstream mediators including the checkpoint kinases CHK1 and CHK2 as well as the tumor suppressor p53. This mechanism is a DNA damage response that induces both apoptosis and senescence, hence inhibiting the development of cancer cells.[264] This stress response also disrupts the capacity of tumor cells to be eliminated by NK cells by promoting the production of NKG2D ligands, which is necessary for this process.[265] When used against melanoma cells, the lethal effects of DIC and INZ, as well as those of cisplatin, are dependent on the capacity of these agents to create adducts within DNA and to trigger the stress response, therefore exerting a corresponding immunomodulatory impact.[266]

In a model of murine melanoma, it was discovered that DTIC may upregulate the expression of NKG2D ligands. In addition, for the medicine to be able to exert its anticancer activity, a functioning immune system is required.[267] Both the perforin/granzyme B-dependent killing of melanoma cells and the IFN-γ secretion by natural killer (NK) cells were enhanced by the expression of NKG2D ligands, which was increased by DTIC. The expression of MHC-I molecules and endogenous antigen presentation on melanoma cells, in turn, were upregulated by IFN-γ. This allowed for the melanoma cells to be recognized and eliminated by cytotoxic T cells, most likely through both perforin granzyme and Fas-mediated pathways. Melanoma cells were able to be recognized and eliminated by cytotoxic T cells.[268]

As a result, the targeting of melanoma cells that have been sensitized by alkylating agents by natural killer (NK) cells would be essential for inducing both innate and adaptive cytotoxic anti-cancer responses within the TME. This would ultimately result in an effective inhibition of the progression of the tumor. Melanoma patients that are treated with DTIC see an overall increase in NK cells that express NKp16.[269] In addition to this, patients who had a response to DTIC showed a greater level of NK cell activation and cytolytic activity against melanoma cells.[270] Cisplatin has a capacity comparable to that of increasing NKG2D ligand expression, which is one of the mechanisms by which the medication might boost the cytotoxicity of NK cells. Studies have also been done on the cisplatin-induced over-expression of additional activating molecules such as ICAM-1, B7-H6, and Fas.[271–276] Furthermore, cisplatin has the capacity to sensitize tumor cells to granzyme B by raising the penetrability of tumor cells and by elevating the expression of the granzyme-target caspase-3, which makes the execution phase of apoptosis easier.[277,278]

It is necessary to have a properly functioning cytoskeleton in order to facilitate the particular aggregation of signaling receptors and adhesion molecules as well as the polarization program of cytotoxic granules at the immunological synapse in order for there to be successful killing.[279] Although PTX eventually impairs microtubule dynamics, it is hypothesized that it will impede NK cell cytotoxicity, as various authors have shown.[280–282] On the other hand, there are findings that contradict these data and suggest that PTX could improve the elimination of tumor cells by NK cells by boosting NK cell cytotoxicity, promoting the expression of ICAM-1 and MIC-B, and sensitizing melanoma cells to the possibility of being killed.[283,284] Since low doses of chemotherapeutics normally stimulate immunomodulation and large dosages are immunosuppressors, these conflicting findings may be a consequence of the varying quantities of PTX that were utilized in the various experimental scenarios.[285]

The immunological characteristics that are generated by DTX mostly coincide with those that are observed for PTX. The effects of DTX on NK cell-mediated killing are diverse, with some investigators finding a reduction while others demonstrated an increased NK cell cytotoxicity.[286] On the other hand, the essential molecular pathways seemed to be shared with PTX. These mechanisms included an increase in the expression of ICAM-1 and NKG2D ligand.[287] However, up-regulation of Fas by DTX has also been reported, indicating that the two taxanes may activate different pathways.[288]

The capacity of medications used in melanoma chemotherapy to upregulate the expression of NK cell triggering ligands on melanoma cells may be caused by an immunomodulatory activity that may contribute, along with direct cytotoxicity, to the anti-tumoral effects of chemotherapeutics.

This would improve the ability of NK cells to recognize and kill melanoma cells. Furthermore, it would seem that the primary targets of this up-regulation are NKG2D ligands; nevertheless, it has been shown that certain medications may promote other engaging ligands, such as, B7-H6, Fas, and ICAM-1. These non-overlapping actions have the potential to be successfully employed to produce patient-tailored treatments in which the individual melanoma immune phenotype dictates the choice of the chemotherapeutic.

10. Concluding remarks

Recent studies have focused on immune cells including macrophages, B cells, T cells, and NK cells. Reduced angiogenesis by targeting Nrp1 and ANG 2 might potentially lead to a reduction in macrophage recruitment. Depletion of the tumor-associated macrophages by utilizing inhibitors or antibodies against them or some of their ligands could also suppress macrophage recruitment. CSF1 inhibitors may prevent the development of new macrophages because CSF1 provides a signal for macrophage differentiation. Furthermore, RP-182 activates CD206 on tumor-associated macrophages, reprogramming them and enhancing their anti-tumor activity.[289]

Cancer immunotherapy that concentrates on B cells and T cells has become a popular strategy in the fight against cancer. The ability of TCR-T- ACT to treat cancer is promising. For T cell reactivity, immunological responses, and its clinical impact on the foreign antigens, TCR tuning is essential. The effectiveness of immunotherapy is also increased by modified TCRs that target specific antigens. Additionally, crucial is the security of genetically altered T cells for

ACT. The two inhibitory checkpoints that are often seen on activated T-cells are CTLA-4 and PD-1, may prove to be the most effective targets for the therapy of malignancies.[290] A number of human malignancies may be treated by blocking these inhibitory receptors on T cells with different antibodies. Studies of the intricate connections between malignant cells and their microenvironment are projected to quickly identify new effective therapeutic targets in this field of study, and the popularity of BTK and PI3K inhibitors is predicted to enhance interest in it. NK cells in the melanoma TME may directly destroy tumor cells and encourage other immune cells to engage in anti-tumor activity. The rising field of cancer immunotherapy continues to grow for novel druggable targets.[291]

Reference

1. Balkwill F, Mantovani A. Inflammation and cancer: back to Virchow? *Lancet*. 2001;357:539–545.
2. Aggarwal BB, Vijayalekshmi RV, Sung B. Targeting inflamma-tory pathways for prevention and therapy of cancer: short-term friend, longterm foe. *Clin Cancer Res*. 2009;15:425–430.
3. Ershler WB, Keller ET. Age-associated increased interleukin-6 gene expression, late-life diseases, and frailty. *Annu Rev Med*. 2000;51:245–270.
4. Rodier F, Coppe JP, Patil CK, et al. Persistent DNA damage signalling triggers senescence-associated inflammatory cytokine secretion. *Nat Cell Biol*. 2009;11: 973–979.
5. Takahashi H, Ogata H, Nishigaki R, Broide DH, Karin M. Tobacco smoke promotes lung tumorigenesis by triggering IKKbeta and JNK1-dependent inflammation. *Cancer Cell*. 2010;17:89–97.
6. Kolb R, Liu GH, Janowski A, Sutterwala F, Zhang W. Inflammasomes in cancer: a double-edged sword. *Protein & Cell*. 2014;5(1):12–20.
7. Blaser MJ, Chyou PH, Nomura A. Age at establishment of Helicobacter pylori infection and gastric carcinoma, gastric ulcer, and duodenal ulcer risk. *Cancer Research*. 1995;55:562–565.
8. Ferreri AJ, Ernberg I, Copie-Bergman C. Infectious agents and lymphoma development: molecular and clinical aspects. *J Intern Med*. 2009;265:421–438.
9. Kuper H, Adami HO, Trichopoulos D. Infections as a major preventable cause of human cancer. *J Intern Med*. 2000;248:171–183.
10. Sripa B, Brindley PJ, Mulvenna J, et al. The tumorigenic liver fluke Opisthorchis viverrini—multiple pathways to cancer. *Trends Parasitol*. 2012;28:395–407.
11. Ernst PB, Gold BD. The disease spectrum of Helicobacter pylori: the immunopathogenesis of gastroduodenal ulcer and gastric cancer. *Annu Rev Microbiol*. 2000;54: 615–640.
12. Khasawneh J, Schulz MD, Walch A, et al. Inflammation and mitochondrial fatty acid beta-oxidation link obesity to early tumor promotion. *Proc Natl Acad Sci USA*. 2009;106:3354–3359.
13. Park EJ, Lee JH, Yu GY, et al. Dietary and genetic obesity promote liver inflammation and tumorigenesis by enhancing IL-6 and TNF expression. *Cell*. 2010;140:197–208.
14. Reuter S, Gupta SC, Chaturvedi MM, Aggarwal BB. Oxidative stress, inflammation, and cancer: how are they linked? *Free Rad Biol Med*. 2010;49:1603–1616.

15. Schetter AJ, Heegaard NH, Harris CC. Inflammation and cancer: interweaving microRNA, free radical, cytokine and p53 pathways. *Carcinogenesis*. 2010;31:37–49.
16. Xu K, Shu HK. EGFR activation results in enhanced cyclooxygenase-2 expression through p38 mitogen-activated protein kinase-dependent activation of the Sp1/Sp3 transcription factors in human gliomas. *Cancer Res*. 2007;67:6121–6129.
17. Marletta MA. Mammalian synthesis of nitrite, nitrate, and N-nitrosating agents. *Chem Res Toxicol*. 1988;1:249–257.
18. Bartsh HE, Hietanen E, Malaveille C. Carcinogenic nitrosamines: free radical aspects of their action. *Free Rad Biol Med*. 1989;7:637–644.
19. Babbs CF. Oxygen radicals in ulcerative colitis. *Free Rad Biol Med*. 1992;13:169–181.
20. Phull PS, Green CJ, Jacyna MR. A radical review of the stomach: the role of oxygen-derived free radicals and antioxidants in gastroduodenal disease. *Eur J Gastroenterol Hepatol*. 1995;7:265–274.
21. Wiseman H, Halliwell B. Damage to DNA by reactive oxygen and nitrogen species: role in infammatory disease and progression to cancer. *Biochem J*. 1996;313:17–29.
22. Sies H. Oxidative stress: oxidants and antioxidants. *Exp Physiol*. 1997;82(2):291–295.
23. Robles AI, Harris CC. Clinical outcomes and correlates of TP53 mutations and cancer. *Cold Spring Harb Perspect Biol*. 2010;2(3), a001016.
24. Meira LB, Bugni JM, Green SL, et al. DNA damage induced by chronic infammation contributes to colon carcinogenesis in mice. *J Clin Invest*. 2008;118:2516–2525.
25. Takai A, Toyoshima T, Uemura M, et al. A novel mouse model of hepatocarcinogenesis triggered by AID causing deleterious p53 mutations. *Oncogene*. 2009;28:469–478.
26. Edwards RA, Witherspoon M, Wang K, et al. Epigenetic repression of DNA mismatch repair by infammation and hypoxia in infammatory bowel disease-associated colorectal cancer. *Cancer Res*. 2009;69:6423–6429.
27. De Santa F, Totaro MG, Prosperini E, Notarbartolo S, Testa G, Natoli G. The histone H3 lysine-27 demethylase Jmjd3 links inflammation to inhibition of polycomb-mediated gene silencing. *Cell*. 2007;130:1083–1094.
28. Cooper CS, Foster CS. Concepts of epigenetics in prostate cancer development. *Br J Cancer*. 2009;100:240–245.
29. Maeda S, Kamata H, Luo JL, Leffert H, Karin M. IKKbeta couples hepatocyte death to cytokine-driven compensatory proliferation that promotes chemical hepatocarcinogenesis. *Cell*. 2005;121:977–990.
30. Sakurai T, He G, Matsuzawa A, et al. Hepatocyte necrosis induced by oxidative stress and IL-1 alpha release mediate carcinogen-induced compensatory proliferation and liver tumorigenesis. *Cancer Cell*. 2008;14:156–165.
31. Mantovani A, Allavena P, Sica A, Balkwill F. Cancer-related infammation. *Nature*. 2008;2008(454):436–444.
32. Strid J, Roberts SJ, Filler RB, et al. Acute upregulation of an NKG2D ligand promotes rapid reorganization of a local immune compartment with pleiotropic effects on carcinogenesis. *Nat Immunol*. 2008;9:146–154.
33. Ruzankina Y, Schoppy DW, Asare A, Clark CE, Vonderheide RH, Brown EJ. Tissue regenerative delays and synthetic lethality in adult mice after combined deletion of Atr and Trp53. *Nat Genet*. 2009;41:1144–1149.
34. Grivennikov SI, Greten FR, Karin M. Immunity, inflammation, and cancer. *Cell*. 2010;140(6):883–899.
35. Zheng L, Jia J, Finger LD, Guo Z, Zer C, Shen B. Functional regulation of FEN1 nuclease and its link to cancer. *Nucleic Acids Res*. 2011;39(3):781–794.
36. Lewis CE, Pollard JW. Distinct role of macrophages in different tumor microenvironments. *Cancer Res*. 2006;66:605–612.
37. Yu H, Pardoll D, Jove R. STATs in cancer inflammation and immunity: a leading role for STAT3. *Nat Rev Cancer*. 2009;9:798–809.

38. Rakoff-Nahoum S, Medzhitov R. Toll-like receptors and cancer. *Nat Rev Cancer*. 2009;9:57–63.
39. Balkwill F. Tumor necrosis factor and cancer. *Nat Rev Cancer*. 2009;9:361–371.
40. Bollrath J, Phesse TJ, von Burstin VA, et al. gp130-mediated Stat3 activation in enterocytes regulates cell survival and cell-cycle progression during colitis-associated tumorigenesis. *Cancer Cell*. 2009;15:91–102.
41. Pandey S, Singh S, Anang V, Bhatt AN, Natarajan K, Dwarakanath BS. Pattern recognition receptors in cancer progression and metastasis. *Cancer Growth Metastasis*. 2015;8:25–34. https://doi.org/10.4137/CGM.S24314.
42. Joyce JA, Pollard JW. Microenvironmental regulation of metastasis. *Nat Rev Cancer*. 2009;9:239–252.
43. Sun B, Karin M. Obesity, inflammation, and liver cancer. *J Hepatol*. 2012;56 (3):704–713.
44. Malgulwar PB, Sharma V, Tomar AS, et al. Transcriptional co-expression regulatory network analysis for Snail and Slug identifies IL1R1, an inflammatory cytokine receptor, to be preferentially expressed in ST-EPN-RELA and PF-EPN-A molecular subgroups of intracranial ependymomas. *Oncotarget*. 2018;9(84):35480–35492.
45. Kumar H, Kawai T, Akira S. Pathogen recognition by the innate immune system. *Int Rev Immunol*. 2011;30(1):16–34.
46. Takeuchi O, Akira S. Pattern recognition receptors and inflammation. *Cell*. 2010;140 (6):805–820.
47. Kawai T, Akira S. TLR signaling. *Semin Immunol*. 2007;19(1):24–32. Review.
48. Fukata M, Vamadevan AS, Abreu MT. Toll-like receptor (TLRS) and Nod-like receptors (NLRS) in inflammatory disorders. *Semin Immunol*. 2009;21(4):242–253.
49. Takeda K, Kaisho T, Akira S. Toll-like receptors. *Annu Rev Immunol*. 2003;21: 335–376. https://doi.org/10.1146/annurev.immunol.21.120601.141126. Epub 2001 Dec 19. Review.
50. Yokota S, Okabayashi T, Rehli M, Fujii N, Amano K. Helocobacter pylori lipopolysaccharides upregulate toll-like receptor 4 expression and proliferation of gastric epithelial cells via the MEK1/2-ERK1/2 mitogen-activated protein kinase pathway. *Infect Immun*. 2010;78(1):468–476.
51. Sato Y, Goto Y, Narita N, Hoon DS. Cancer cells expressing toll-like receptors and the tumor microenvironment. *Cancer Microenviron*. 2009;2(suppl 1):205–214.
52. Boone BA, Lotze MT. Targeting damage-associated molecular pattern molecules (DAMPS) and DAMP receptors in melanoma. *Methods Mol Biol*. 2014;1102: 537–552.
53. Gombault A, Baron L, Couillin I. ATP release and purinergic signaling in NLRP3 inflammasome activation. *Front Immunol*. 2013;3:414.
54. Scheede-Bergdahl C, Watt HL, Trutschnigg B, et al. Is IL-6 the best proinflammatory biomarker of clinical outcomes of cancer cachexia. *Clin Nutr*. 2011;31(1):85–88.
55. Pradere JP, Dapito DH, Schwabe RF. The Yin and Yang of toll- like receptors in cancer. *Oncogene*. 2014;33(27):3485–3495.
56. Tafani M, Russo A, Di Vito M, et al. Up-regulation of pro-inflammatory genes as adaptation to hypoxia in MCF-7 cells and in human mammary invasive carcinoma microenvironment. *Cancer Sci*. 2010;101(4):1014–1023.
57. Ben-Baruch A. Organ selectivity in metastasis: regulation by chemokines and their receptors. *Clin Exp Metastasis*. 2008;25(4):345–356.
58. Ghadjar P, Coupland SE, Na IK, et al. Chemokine receptor CCR6 expression level and liver metastases in colorectal cancer. *J Clin Oncol*. 2006;24(12):1910–1916.
59. Iwasa S, Yanagawa T, Fan J, Katoh R. Expression of CXCR4 and its ligand SDF-1 in intestinal-type gastric cancer is associated with lymph node and liver metastasis. *Anticancer Res*. 2009;29(11):4751–4758.

60. Dunn JH, Ellis LZ, Fujita M. Inflammasomes as molecular mediators of inflammation and cancer: potential role in melanoma. *Cancer Lett*. 2012;314(1):24–33.
61. Lefrancois L, Parker CM, Olson S, Muller W, Wagner N, Puddington L. The role of ß-7 integrins in CD8 T cell trafficking during an antiviral immune response. *J Exp Med*. 1999;189(10):1631–1638.
62. Maker AV, Katabi N, Qin LX, et al. Cyst fluid interleukin 1-ß levels predict the risk of carcinoma in intraductal papillary mucinous neoplasms of the pancreas. *Clin Cancer Res*. 2011;17(6):1502–1508.
63. Weng PH, Huang YL, Page JH, et al. Polymorphism of an innate immune gene, toll-like receptor 4, and aggressive prostate cancer risk: a systematic review and meta- analysis. *PLoS One*. 2014;9(10):e110569.
64. Jin Y, Wi HJ, Choi MH, Hong ST, Bae YM. Regulation of anti-inflammatory cytokines IL-10 and TGF-ß in mouse dendritic cells through treatment with Clonorchis sinensis crude antigen. *Exp Mol Med*. 2014;46:e74.
65. Schreibelt G, Tel J, Sliepen KH, et al. Toll-like receptor expression and function in human dendritic cell subsets: implications for dendritic cell-based anti-cancer immunotherapy. *Cancer Immunol Immunother*. 2010;59(10):1573–1582.
66. Qian Y, Deng J, Geng L, et al. TLR4 signaling induces B7-H1 expression through MAPK pathways in bladder cancer cells. *Cancer Invest*. 2008;26(8):816–821.
67. Trinchieri G, Sher A. Coorporation of toll-like receptor signals in innate immune defence. *Nat Rev Immunol*. 2007;7(3):179–190.
68. Carmi Y, Dotan S, Rider P, et al. The role of IL-1ß in the early tumor cell- induced angiogenic response. *J Immunol*. 2013;190(7):3500–3509.
69. Saijo Y, Tanaka M, Miki M, et al. Proinflammatory cytokine IL-1ß promotes tumor growth of Lewis lung carcinoma by induction of angiogenic factors: in vivo analysis of tumor-stromal interaction. *J Immunol*. 2002;169(1):469–475.
70. Halle A, Hornung V, Petzold GC, et al. The NALP3 inflammasome is involved in the innate immune response to amyloid ß. *Nat Immunol*. 2008;9(8):857–865.
71. Vidal-Vanaclocha F, Fantuzzi G, Mendoza L, et al. IL-18 regulates IL-1ß dependent hepatic melanoma metastasis via vascular cell adhesion molecule-1. *Proc Natl Acad Sci USA*. 2000;97(2):734–739.
72. Hirayama D, Iida T, Nakase H. The phagocytic function of macrophage-enforcing innate immunity and tissue homeostasis. *Int J Mol Sci*. 2017;29:19(1). https://doi.org/10.3390/ijms19010092. Review.
73. Kumari N, Choi SH. Tumor-associated macrophages in cancer: recent advancements in cancer nanoimmunotherapies. *J Exp Clin Cancer Res*. 2022;41(1):68. https://doi.org/10.1186/s13046-022-02272-x. Review.
74. Tanaka T, Narazaki M, Kishimoto T. IL-6 in inflammation, immunity, and disease. *Cold Spring Harb Perspect Biol*. 2014;6(10):a016295. https://doi.org/10.1101/cshperspect.a016295. Review.
75. Kano A. Tumor cell secretion of soluble factor(s) for specific immunosuppression. *Sci Rep*. 2015;9(5):8913.
76. Geissmann F, Manz MG, Jung S, Sieweke MH, Merad M, Ley K. Development of monocytes, macrophages, and dendritic cells. *Science*. 2010;327(5966):656–661.
77. Dong Q, Nelson PJ, Zhao Y. Editorial: cancer cell metabolism and immunomodulation in the context of tumor metastasis. *Front Oncol*. 2021;11:803213.
78. Kienzl P, Polacek R, Reithofer M, et al. The cytokine environment influence on human skin-derived T cells. *FASEB J*. 2019;33(5):6514–6525.
79. Casanova-Acebes M, Dalla E, Leader AM, et al. Tissue-resident macrophages provide a pro-tumorigenic niche to early NSCLC cells. *Nature*. 2021;595(7868):578–584.
80. Italiani P, Boraschi D. New insights into tissue macrophages: from their origin to the development of memory. *Immune Netw*. 2015;15(4):167–176.

81. Davies LC, Jenkins SJ, Allen JE, Taylor PR. Tissue-resident macrophages. *Nat Immunol.* 2013;14(10):986–995.
82. Corliss BA, Azimi MS, Munson JM, Peirce SM, Murfee WL. Macrophages: an InflammatoryLinkBetweenAngiogenesisandLymphangiogenesis. *Microcirculation.* 2016; 23(2):95–121.
83. Yang Q, Guo N, Zhou Y, Chen J, Wei Q, Han M. The role of tumor-associated macrophages (TAMs) in tumor progression and relevant advance in targeted therapy. *Acta Pharm Sin B.* 2020;10:2156–2170.
84. Sangro B, Sarobe P, Hervás-Stubbs S, Melero I. Advances in immunotherapy for hepatocellular carcinoma. *Nat Rev Gastroenterol Hepatol.* 2021;18(8):525–543.
85. Kasprzak A. Angiogenesis-related functions of Wnt signaling in colorectal carcinogenesis. *Cancers (Basel).* 2020;12(12).
86. Verreck FAW, de Boer T, Langenberg DML, et al. Human IL-23-producing type 1 macrophages promote but IL-10-producing type 2 macrophages subvert immunity to (Myco) bacteria. *Proc Natl Acad Sci USA.* 2004;101. 4560–456.
87. Varikuti S, Verma C, Holcomb E, et al. MicroRNA-21 deficiency promotes the early Th1 immune response and resistance toward visceral leishmaniasis. *J Immunol.* 2021;207 (5):1322–1332.
88. Bouhlel MA, Derudas B, Rigamonti E, et al. PPARγ activation primes human monocytes into alternative M2 macrophages with anti-inflammatory properties. *Cell Metabol.* 2007;6:137–143.
89. Wang N, Liang H, Zen K. Molecular mechanisms that influence the macrophage M1–M2 polarization balance. *Front Immunol.* 2014;5.
90. Verma C, Ankush KR, Anang V, et al. Calcium dynamics regulate protective responses and growth of Staphylococcus aureus in macrophages. *Biomol Concepts.* 2020;11(1): 230–239.
91. Kottarath SK, Bhat M, Verma C, et al. Folate receptor-β targeted cholesterol-chitosan nanocarrier for treatment of rheumatoid arthritis: an animal study. *J Drug Deliv Sci Technol.* 2020;60:101946. https://doi.org/10.1016/j.jddst.2020.101946.
92. Yunna C, Mengru H, Lei W, Weidong C. Macrophage M1/M2 polarization. *Eur J Pharmacol.* 2020;15(877):173090.
93. Orecchioni M, Ghosheh Y, Pramod AB, Ley K. Macrophage polarization: different gene signatures in M1 (LPS+) vs. classically and M2 (LPS–) vs. alternatively activated macrophages. *Front Immunol.* 2019;10:1084.
94. Boimel PJ, Smirnova T, Zhou ZN, et al. Contribution of CXCL12 secretion to invasion of breast cancer cells. *Breast Cancer Res.* 2012;14:R23.
95. Ries CH, Cannarile MA, Hoves S, et al. Targeting tumor-associated macrophages with anti-CSF-1R antibody reveals a strategy for cancer therapy. *Cancer Cell.* 2014;25: 846–859.
96. Xu J, Escamilla J, Mok S, et al. CSF1R signaling blockade stanches tumor-infiltrating myeloid cells and improves the efficacy of radiotherapy in prostate cancer. *Cancer Res.* 2013;73:2782–2794.
97. Quail DF, Bowman RL, Akkari L, et al. The tumor microenvironment underlies acquired resistance to CSF-1R inhibition in gliomas. *Science.* 2016;352:aad3018.
98. Papadavid E, Stratigos AJ, Falagas ME. Imiquimod: an immune response modifier in the treatment of precancerous skin lesions and skin cancer. *Exp Opin Pharmacother.* 2007;8:1743–1755.
99. Maeda A, Digifico E, Andon FT, Mantovani A, Allavena P. Poly(I:C) stimulation is superior than imiquimod to induce the antitumoral functional profile of tumor-conditioned macrophages. *Eur J Immunol.* 2019;49:801–811.
100. Wynn TA, Chawla A, Pollard JW. Macrophage biology in development, homeostasis and disease. *Nature.* 2013;496:445–455.

101. Gordon S. Alternative activation of macrophages. *Nat Rev Immunol.* 2003;3:23–35.
102. Mills CD. M1 and M2 macrophages: oracles of health and disease. *Crit Rev Immunol.* 2012;32:463–488.
103. Sica A, Larghi P, Mancino A, et al. Macrophage polarization in tumor progression. *Semin Cancer Biol.* 2008;18:349–355.
104. Qian BZ, Pollard JW. Macrophage diversity enhances tumor progression and metastasis. *Cell.* 2010;141:39–51.
105. Hanahan D, et al. *Cell.* 2011. PMID: 21376230 Review.
106. Ruffell B, DeNardo DG, Affara NI, Coussens LM. Lymphocytes in cancer development: polarization towards pro-tumor immunity. *Cytokine Growth Factor Rev.* 2010;21(1):3–10. https://doi.org/10.1016/j.cytogfr.2009.11.002. Epub 2009 Dec 11. Review.
107. Germain RN, Stefanova I. The dynamics of T cell receptor signaling: complex orchestration and the key roles of tempo and cooperation. *Annu Rev Immunol.* 1999;17: 467–522.
108. Phan GQ, Rosenberg SA. Adoptive cell transfer for patients with metastatic melanoma: the potential and promise of cancer immunotherapy. *Cancer Control.* 2013;20:289–297.
109. Brown SD, Warren RL, Gibb EA, et al. Neo-antigens predicted by tumor genome metaanalysis correlate with increased patient survival. *Genome Res.* 2014;24:743–750.
110. Agrawal S, Kishore MC. MHC class I gene expression and regulation. *J Hematother Stem Cell Res.* 2000;9:795–812.
111. Leone P, Shin EC, Perosa F, Vacca A, Dammacco F, Racanelli V. MHC class I antigen processing and presenting machinery: organization, function, and defects in tumor cells. *J Natl Cancer Inst.* 2013;105:1172–1187.
112. Zhang N, Bevan MJ. CD8(+) T cells: foot soldiers of the immune system. *Immunity.* 2011;35:161–168.
113. Obst R. The timing of T cell priming and cycling. *Front Immunol.* 2015;6:56.
114. Singh A, Anang V, Kumar Rana A, et al. Deciphering the role of calcium homeostasis in T cells functions during mycobacterial infection. *Cell Immunol.* 2020;357:104198.
115. Corse E, Gottschalk RA, Allison JP. Strength of TCR-peptide/MHC interactions and in vivo T cell responses. *J Immunol.* 2011;186:5039–5045.
116. Tkach KE, Barik D, Voisinne G, et al. T cells translate individual, quantal activation into collective, analog cytokine responses via time-integrated feedbacks. *Elife.* 2014;3:e01944.
117. Zhao L, Cao YJ. Engineered T cell therapy for cancer in the clinic. *Front Immunol.* 2019;10:2250.
118. Hwang JR, Byeon Y, Kim D, Park SG. Recent insights of T cell receptor-mediated signaling pathways for T cell activation and development. *Exp Mol Med.* 2020;52 (5):750–761.
119. Chandran SS, Klebanoff CA. T cell receptor-based cancer immunotherapy: emerging efficacy and pathways of resistance. *Immunol Rev.* 2019;290(1):127–147.
120. Alhabbab RY. Targeting cancer stem cells by genetically engineered chimeric antigen receptor T cells. *Front Genet.* 2020;11:312.
121. Garcia KC, Adams JJ, Feng D, Ely LK. The molecular basis of TCR germline bias for MHC is surprisingly simple. *Nat Immunol.* 2009;10(2):143–147.
122. Chen L, Qiao DJ, Wang JT, Tian G, Wang MJ. Cancer immunotherapy with lymphocytes genetically engineered with T cell receptors for solid cancers. *Immunol Lett.* 2019;216:51–62.
123. Zhang YK, Li YQ. T cell receptor-engineered T cells for leukemia immunotherapy. *Cancer Cell Int.* 2019;19.
124. Forsberg EMV, Lindberg MF, Jespersen H, et al. HER2 CAR-T cells eradicate uveal melanoma and T-cell therapy-resistant human melanoma in IL2 transgenic NOD/SCID IL2 receptor knockout mice. *Cancer Res.* 2019;79(5):899–904.

https://doi.org/10.1158/0008-5472.CAN-18-3158. Epub 2019 Jan 8. PubMed PMID: 30622115.
125. Ye L, Lou Y, Lu L, Fan X. Mesothelin-targeted second generation CAR-T cells inhibit growth of mesothelin-expressing tumors *in vivo*. *Exp Ther Med.* 2019;17(1):739–747. https://doi.org/10.3892/etm.2018.7015. Epub 2018 Nov 27. PubMed PMID: 30651858; PubMed Central PMCID: PMC6307389.
126. Jager E, Maeurer M, Hohn H, et al. Clonal expansion of Melan A-specific cytotoxic T lymphocytes in a melanoma patient responding to continued immunization with melanoma-associated peptides. *Int J Cancer.* 2000;86(4):538–547.
127. Fujio K, Misaki Y, Setoguchi K, et al. Functional reconstitution of class II MHC-restricted T cell immunity mediated by retroviral transfer of the alpha beta TCR complex. *J Immunol.* 2000;165(1):528–532.
128. Das S, Johnson DB. Immune-related adverse events and anti-tumor efficacy of immune checkpoint inhibitors. *J Immunother Cancer.* 2019;7(1):306. https://doi.org/10.1186/s40425-019-0805-8. Review. PubMed PMID: 31730012; PubMed Central PMCID: PMC6858629.
129. Jenkins RW, Barbie DA, Flaherty KT. Mechanisms of resistance to immune checkpoint inhibitors. *Br J Cancer.* 2018;118(1):9–16. https://doi.org/10.1038/bjc.2017.434. Epub 2018 Jan 2. Review. PubMed PMID: 29319049; PubMed Central PMCID: PMC5765236.
130. Perazella MA, Shirali AC. Immune checkpoint inhibitor nephrotoxicity: what do we know and what should we do? *Kidney Int.* 2020;97(1):62–74. https://doi.org/10.1016/j.kint.2019.07.022. Epub 2019 Aug 23. Review. PubMed PMID: 31685311.
131. Schneider H, Downey J, Smith A, et al. Reversal of the TCR stop signal by CTLA-4. *Science.* 2006;313(5795):1972–1975. https://doi.org/10.1126/science.1131078. Epub 2006 Aug 24. PubMed PMID: 16931720.
132. Perez VL, Van Parijs L, Biuckians A, Zheng XX, Strom TB, Abbas AK. Induction of peripheral T cell tolerance in vivo requires CTLA-4 engagement. *Immunity.* 1997;6(4):411–417. https://doi.org/10.1016/s1074-7613(00)80284-8. PubMed PMID: 9133420.
133. Franzin R, Netti GS, Spadaccino F, et al. The use of immune checkpoint inhibitors in oncology and the occurrence of AKI: where do we stand? *Front Immunol.* 2020;11, 574271. https://doi.org/10.3389/fimmu.2020.574271. eCollection 2020. Review. PubMed PMID: 33162990; PubMed Central PMCID: PMC7580288.
134. Jung D, Alt FW. Unraveling V(D)J recombination; insights into gene regulation. *Cell.* 2004;116:299–311.
135. Pieper K, Grimbacher B, Eibel H. B-cell biology and development. *J Allergy Clin Immunol.* 2013;131:959–971.
136. Furman RR, et al. Idelalisib and rituximab in relapsed chronic lymphocytic leukemia. *N Engl J Med.* 2014;370:997–1007.
137. Gopal AK, et al. PI3Kδ inhibition by idelalisib in patients with relapsed indolent lymphoma. *N Engl J Med.* 2014;370:1008–1018.
138. Kahl BS, et al. A phase 1 study of the PI3Kδ inhibitor idelalisib in patients with relapsed/refractory mantle cell lymphoma (MCL). *Blood.* 2014;123:3398–3405.
139. Noy A, et al. Targeting Bruton tyrosine kinase with ibrutinib in relapsed/refractory marginal zone lymphoma. *Blood.* 2017;129:2224–2232.
140. Treon SP, et al. Ibrutinib in previously treated Waldenstrom's macroglobulinemia. *N Engl J Med.* 2015;372:1430–1440.
141. Byrd JC, et al. Targeting BTK with ibrutinib in relapsed chronic lymphocytic leukemia. *N Engl J Med.* 2013;369:32–42.
142. Byrd JC, et al. Three-year follow-up of treatmentnaive and previously treated patients with CLL and SLL receiving single-agent ibrutinib. *Blood.* 2015;125:2497–2506.

143. Brown JR, et al. Idelalisib, an inhibitor of phosphatidylinositol 3-kinase p110δ, for relapsed/refractory chronic lymphocytic leukemia. *Blood.* 2014;123:3390–3397.
144. Wilson WH, et al. Targeting B cell receptor signaling with ibrutinib in diffuse large B cell lymphoma. *Nat Med.* 2015;21:922–926.
145. Chiron D, et al. Cell-cycle reprogramming for PI3K inhibition overrides a relapse-specific C481S BTK mutation revealed by longitudinal functional genomics in mantle cell lymphoma. *Cancer Discov.* 2014;4:1022–1035.
146. Ponader S, et al. The Bruton tyrosine kinase inhibitor PCI-32765 thwarts chronic lymphocytic leukemia cell survival and tissue homing in vitro and in vivo. *Blood.* 2012;119:1182–1189.
147. Spaargaren M, et al. The B cell antigen receptor controls integrin activity through Btk and PLCγ2. *J Exp Med.* 2003;198:1539–1550.
148. de Gorter DJ, et al. Bruton's tyrosine kinase and phospholipase Cγ2 mediate chemokine-controlled B cell migration and homing. *Immunity.* 2007;26:93–104.
149. de Rooij MF, et al. The clinically active BTK inhibitor PCI-32765 targets B-cell receptor- and chemokine-controlled adhesion and migration in chronic lymphocytic leukemia. *Blood.* 2012;119:2590–2594.
150. de Rooij MF, et al. Ibrutinib and idelalisib target B cell receptor- but not CXCL12/CXCR4-controlled integrin-mediated adhesion in Waldenstrom macroglobulinemia. *Haematologica.* 2016;101:e111–e115.
151. Gunderson AJ, et al. Bruton tyrosine kinase-dependent immune cell cross-talk drives pancreas cancer. *Cancer Discov.* 2016;6:270–285.
152. Ni Gabhann J, et al. Btk regulates macrophage polarization in response to lipopolysaccharide. *PLoS One.* 2014;9:e85834.
153. Dubovsky JA, et al. Ibrutinib is an irreversible molecular inhibitor of ITK driving a Th1-selective pressure in T lymphocytes. *Blood.* 2013;122:2539–2549.
154. Kim E, et al. Ibrutinib inhibits pre-BCR+ B-cell acute lymphoblastic leukemia progression by targeting BTK and BLK. *Blood.* 2017;129:1155–1165.
155. Frisch SM, Screaton RA. Anoikis mechanisms. *Curr Opin Cell Biol.* 2001;13:555–562.
156. Malmberg K-J, Carlsten M, Björklund A, Sohlberg E, Bryceson YT, Ljunggren H-G. Natural killer cell-mediated immu- nosurveillance of human cancer. *Semin Immunol.* 2017;31:20–29. https://doi.org/10.1016/j.smim.2017.08.002.
157. Gorelik E, Wiltrout RH, Okumura K, Habu S, Herberman RB. Role of NK cells in the control of metastatic spread and growth of tumor cells in mice. *Int J Cancer.* 1982;30:107–112. https://doi.org/10.1002/ijc.2910300118.
158. Smyth MJ, Thia KY, Cretney E, et al. Perforin is a major contributor to NK cell control of tumor metastasis. *J Immunol.* 1999;162:6658–6662.
159. Attri S, Sharma V, Kumar A, Verma C, Gahlawat SK. Dissecting role of founder mutation p.V727M in GNE in Indian HIBM cohort. *Open Med (Wars).* 2021;16(1):1733–1744.
160. Zhang S, Liu W, Hu B, et al. Prognostic significance of tumor-infiltrating natural killer cells in solid tumors: a systematic review and meta-analysis. *Front Immunol.* 2020;11. https://doi.org/10.3389/fimmu.2020.01242.
161. Nersesian S, Schwartz SL, Grantham SR, et al. NK cell infiltration is associated with improved overall survival in solid cancers: a systematic review and meta-analysis. *Transl Oncol.* 2021;14:100930.
162. Imai K, Matsuyama S, Miyake S, Suga K, Nakachi K. Natural cytotoxic activity of peripheral-blood lymphocytes and cancer incidence: an 11-year follow-up study of a general population. *Lancet.* 2000;356:1795–1799.
163. Ascierto ML, Idowu MO, Zhao Y, et al. Molecular signatures mostly associated with NK cells are predictive of relapse free survival in breast cancer patients. *J Transl Med.* 2013;11:145. https://doi.org/10.1186/1479-5876-11-145.

164. Pasero C, Gravis G, Granjeaud S, et al. Highly effective NK cells are associated with good prognosis in patients with metastatic prostate cancer. *Oncotarget*. 2015;6: 14360–14373. https://doi.org/10.18632/oncotarget.3965.
165. O'Sullivan T, Saddawi-Konefka R, Vermi W, et al. Cancer immunoediting by the innate immune system in the absence of adaptive immunity. *J Exp Med*. 2012;209: 1869–1882.
166. Saha PK, Sharma PK, Singh BK, Verma C. Role of CXCR3+CCR5+ Th1 cells in pulmonary tuberculosis patients: at pathogenic site. *Biomed Biotechnol Res J*. 2019; 3:80–86.
167. Paust S, Gill HS, Wang B-Z, et al. Critical role for the chemokine receptor CXCR6 in NK cell-mediated antigen-specific memory of haptens and viruses. *Nat Immunol*. 2010;11:1127–1135. https://doi.org/10.1038/ni.1953.
168. Vivier E, Raulet DH, Moretta A, et al. Innate or adaptive immunity? The example of natural killer cells. *Science*. 2011;331:44.
169. Menard C, Blay JY, Borg C, et al. Natural killer cell IFN-gamma levels predict long-term survival with imatinib mesylate therapy in gastrointestinal stromal tumor-bearing patients. *Cancer Res*. 2009;69:3563.
170. Peter ME, Krammer PH. The CD95(APO-1/Fas) DISC and beyond. *Cell Death Diff*. 2003;10:26.
171. Braumuller H, Wieder T, Brenner E, et al. T-helper-1-cell cytokines drive cancer into senescence. *Nature*. 2013;494:361–365.
172. Martin-Fontecha A, Thomsen LL, Brett S, et al. Induced recruitment of NK cells to lymph nodes provides IFN-gamma for T(H)1 priming. *Nat Immunol*. 2004; 5:1260.
173. Varikuti S, Singh B, Volpedo G, et al. Ibrutinib treatment inhibits breast cancer progression and metastasis by inducing conversion of myeloid-derived suppressor cells to dendritic cells. *Br J Cancer*. 2020;122(7):1005–1013.
174. Raulet DH. Roles of the NKG2D immunoreceptor and its ligands. *Nat Rev Immunol*. 2003;3:781.
175. Houchins JP, Yabe T, McSherry C, Bach FH. DNA sequence analysis of NKG2, a family of related cDNA clones encoding type II integral membrane proteins on human natural killer cells. *J Exp Med*. 1991;173:1017.
176. Diefenbach A, Tomasello E, Lucas M, et al. Selective associations with signaling proteins determine stimulatory versus costimulatory activity of NKG2D. *Nat Immunol*. 2002;3:1142.
177. Rosen DB, Araki M, Hamerman JA, Chen T, Yamamura T, Lanier LL. A Structural basis for the association of DAP12 with mouse, but not human, NKG2D. *J Immunol*. 2004;173:2470.
178. Zompi S, Hamerman JA, Ogasawara K, et al. NKG2D triggers cytotoxicity in mouse NK cells lacking DAP12 or Syk family kinases. *Nat Immunol*. 2003;4:565.
179. Sutherland CL, Chalupny NJ, Schooley K, VandenBos T, Kubin M, Cosman D. UL16-binding proteins, novel MHC class I-related proteins, bind to NKG2D and activate multiple signaling pathways in primary NK cells. *J Immunol*. 2002;168:671.
180. Wu J, Song Y, Bakker AB, et al. An activating immunoreceptor complex formed by NKG2D and DAP10. *Science*. 1999;285:730.
181. Bauer S, Groh V, Wu J, et al. Activation of NK cells and T cells by NKG2D, a receptor for stress-inducible MICA. *Science*. 1999;285:727.
182. Cerwenka A, Bakker AB, McClanahan T, et al. Retinoic acid early inducible genes define a ligand family for the activating NKG2D receptor in mice. *Immunity*. 2000;12:721–727.
183. Guerra N, Tan YX, Joncker NT, et al. NKG2D-deficient mice are defective in tumor surveillance in models of spontaneous malignancy. *Immunity*. 2008;28:571.

184. Friese MA, Platten M, Lutz SZ, et al. MICA/NKG2D-mediated immunogene therapy of experimental gliomas. *Cancer Res.* 2003;63:8996.
185. Salih HR, Antropius H, Gieseke F, et al. Functional expression and release of ligands for the activating immunoreceptor NKG2D in leukemia. *Blood.* 2003;102:1389.
186. Vetter CS, Groh V, thor Straten P, Spies T, Brocker EB, Becker JC. Expression of stress-induced MHC class I related chain molecules on human melanoma. *J Investig Dermatol.* 2002;118:600.
187. Watson NF, Spendlove I, Madjd Z, et al. Expression of the stress-related MHC class I chain-related protein MICA is an indicator of good prognosis in colorectal cancer patients. *Int J Cancer.* 2006;118:1445.
188. Diefenbach A, Jensen ER, Jamieson AM, Raulet DH. Rae1 and H60 ligands of the NKG2D receptor stimulate tumor immunity. *Nature.* 2001;413:165.
189. Crowe NY, Smyth MJ, Godfrey DI. A critical role for natural killer T cells in immunosurveillance of methylcholanthrene-induced sarcomas. *J Exp Med.* 2002; 196:119.
190. Smyth MJ, Crowe NY, Godfrey DI. NK cells and NKT cells collaborate in host protection from methylcholanthrene-induced fibrosarcoma. *Int Immunol.* 2001;13:459.
191. Cantoni C, Bottino C, Augugliaro R, et al. Molecular and functional characterization of IRp60, a member of the immunoglobulin superfamily that functions as an inhibitory receptor in human NK cells. *Eur J Immunol.* 1999;29:3148.
192. Pende D, Parolini S, Pessino A, et al. Identification and molecular characterization of NKp30, a novel triggering receptor involved in natural cytotoxicity mediated by human natural killer cells. *J Exp Med.* 1999;190:1505.
193. Pessino A, Sivori S, Bottino C, et al. Molecular cloning of NKp46: a novel member of the immunoglobulin superfamily involved in triggering of natural cytotoxicity. *J Exp Med.* 1998;188:953.
194. Sivori S, Vitale M, Morelli L, et al. p46, a novel natural killer cell-specific surface molecule that mediates cell activation. *J Exp Med.* 1997;186:1129.
195. Vitale M, Bottino C, Sivori S, et al. NKp44, a novel triggering surface molecule specifically expressed by activated natural killer cells, is involved in non-major histocompatibility complex-restricted tumor cell lysis. *J Exp Med.* 1998;187:2065.
196. Moretta A, Bottino C, Vitale M, et al. Activating receptors and coreceptors involved in human natural killer cell-mediated cytolysis. *Annu Rev Immunol.* 2001;19:197.
197. Billadeau DD, Brumbaugh KM, Dick CJ, Schoon RA, Bustelo XR, Leibson PJ. The Vav-Rac1 pathway in cytotoxic lymphocytes regulates the generation of cell-mediated killing. *J Exp Med.* 1998;188:549.
198. Cella M, Fujikawa K, Tassi I, et al. Differential requirements for Vav proteins in DAP10- and ITAM-mediated NK cell cytotoxicity. *J Exp Med.* 2004;200:817.
199. Spaggiari GM, Carosio R, Pende D, et al. NK cell-mediated lysis of autologous antigen-presenting cells is triggered by the engagement of the phosphatidylinositol 3-kinase upon ligation of the natural cytotoxicity receptors NKp30 and NKp46. *Eur J Immunol.* 2001;31:1656.
200. Tassi I, Presti R, Kim S, Yokoyama WM, Gilfillan S, Colonna M. Phospholipase C-gamma 2 is a critical signaling mediator for murine NK cell activating receptors. *J Immunol.* 2005;175:749.
201. Upshaw JL, Schoon RA, Dick CJ, Billadeau DD, Leibson PJ. The isoforms of phospholipase C-gamma are differentially used by distinct human NK activating receptors. *J Immunol.* 2005;175:213.
202. Singh A, Anang V, Verma C, et al. Bcl2 negatively regulates protective immune responses during mycobacterial infection. *Biomol Concepts.* 2021;12(1):94–109.
203. Arnon TI, Achdout H, Lieberman N, et al. The mechanisms controlling the recognition of tumor- and virus-infected cells by NKp46. *Blood.* 2004;103:664.

204. Costello RT, Sivori S, Marcenaro E, et al. Defective expression and function of natural killer cell-triggering receptors in patients with acute myeloid leukemia. *Blood*. 2002; 99:3661.
205. Fauriat C, Just-Landi S, Mallet F, et al. Deficient expression of NCR in NK cells from acute myeloid leukemia: evolution during leukemia treatment and impact of leukemia cells in NCRdull phenotype induction. *Blood*. 2007;109:323.
206. Brandt CS, Baratin M, Yi EC, et al. The B7 family member B7-H6 is a tumor cell ligand for the activating natural killer cell receptor NKp30 in humans. *J Exp Med*. 2009;206:1495.
207. Kaifu T, Escaliere B, Gastinel LN, Vivier E, Baratin M. B7-H6/NKp30 interaction: a mechanism of alerting NK cells against tumors. *Cell Mol Life Sci*. 2011;68:3531.
208. Pogge von Strandmann E, Simhadri VR, von Tresckow B, et al. Human leukocyte antigen-B-associated transcript 3 is released from tumor cells and engages the NKp30 receptor on natural killer cells. *Immunity*. 2007;27(965).
209. Welte S, Kuttruff S, Waldhauer I, Steinle A. Mutual activation of natural killer cells and monocytes mediated by NKp80-AICL interaction. *Nat Immunol*. 2006.
210. Dennehy KM, Klimosch SN, Steinle A. Cutting edge: NKp80 uses an atypical hemi-ITAM to trigger NK cytotoxicity. *J Immunol*. 2011;186:657.
211. Ruckrich T, Steinle A. Attenuated natural killer (NK) cell activation through C-type lectin-like receptor NKp80 is due to an anomalous hemi-immunoreceptor tyrosine-based activation motif (HemITAM) with impaired Syk kinase recruitment capacity. *J Biol Chem*. 2013;288:17725.
212. Veillette A. Immune regulation by SLAM family receptors and SAP-related adaptors. *Nat Rev Immunol*. 2006;6:56.
213. Dong Z, Cruz-Munoz ME, Zhong MC, Chen R, Latour S, Veillette A. Essential function for SAP family adaptors in the surveillance of hematopoietic cells by natural killer cells. *Nat Immunol*. 2009;10:973.
214. Lee JC, Lee KM, Kim DW, Heo DS. Elevated TGF-beta1 secretion and Down-modulation of NKG2D underlies impaired NK cytotoxicity in cancer patients. *J Immunol*. 2004;172:7335.
215. McNerney ME, Guzior D, Kumar V. 2B4 (CD244)—CD48 interactions provide a novel MHC class I-independent system for NK cell self-tolerance in mice. *Blood*. 2005.
216. Vaidya SV, Stepp SE, McNerney ME, et al. Targeted disruption of the 2B4 gene in mice reveals an in vivo role of 2B4 (CD244) in the rejection of B16 melanoma cells. *J Immunol*. 2005;174:800.
217. Benoit L, Wang X, Pabst HF, Dutz J, Tan R. Defective NK cell activation in X-linked lymphoproliferative disease. *J Immunol*. 2000;165:3549.
218. Bloch-Queyrat C, Fondaneche MC, Chen R, et al. Regulation of natural cytotoxicity by the adaptor SAP and the Src-related kinase Fyn. *J Exp Med*. 2005;202:181.
219. Nakajima H, Cella M, Bouchon A, et al. Patients with X-linked lymphoproliferative disease have a defect in 2B4 receptor-mediated NK cell cytotoxicity. *Eur J Immunol*. 2000;30:3309.
220. Tangye SG, Phillips JH, Lanier LL, Nichols KE. Functional requirement for SAP in 2B4-mediated activation of human natural killer cells as revealed by the X-linked lymphoproliferative syndrome. *J Immunol*. 2000;165:2932.
221. Long EO, Sik Kim H, Liu D, Peterson ME, Rajagopalan S. Controlling natural killer cell responses: integration of signals for activation and inhibition. *Annu Rev Immunol*. 2013;31:227.
222. Verma C, Sharma SK, Natarajan K, et al. Immunological alterations in tuberculosis associated immune reconstitution inflammatory syndrome in HIV infected patients. *Indian J Exp Biol*. 2019;57:786–795.

223. Barber DF, Faure M, Long EO. LFA-1 contributes an early signal for NK cell cytotoxicity. *J Immunol.* 2004;173:3653.
224. Bryceson YT, March ME, Barber DF, Ljunggren HG, Long EO. Cytolytic granule polarization and degranulation controlled by different receptors in resting NK cells. *J Exp Med.* 2005;202:1001.
225. Mentlik AN, Sanborn KB, Holzbaur EL, Orange JS. Rapid lytic granule convergence to the MTOC in natural killer cells is dependent on dynein but not cytolytic commitment. *Mol Biol Cell.* 2010;21:2241.
226. Shibuya A, Campbell D, Hannum C, et al. DNAM-1, a novel adhesion molecule involved in the cytolytic function of T lymphocytes. *Immunity.* 1996;4:573.
227. Shibuya K, Lanier LL, Phillips JH, et al. Physical and functional association of LFA-1 with DNAM-1 adhesion molecule. *Immunity.* 1999;11:615.
228. Shibuya K, Shirakawa J, Kameyama T, et al. CD226 (DNAM-1) is involved in lymphocyte function-associated antigen 1 costimulatory signal for naive T cell differentiation and proliferation. *J Exp Med.* 2003;198:1829.
229. Burns GF, Triglia T, Werkmeister JA, Begley CG, Boyd AW. TLiSA1, a human T lineage-specific activation antigen involved in the differentiation of cytotoxic T lymphocytes and anomalous killer cells from their precursors. *J Exp Med.* 1985;161:1063.
230. Bottino C, Castriconi R, Pende D, et al. Identification of PVR (CD155) and Nectin-2 (CD112) as cell surface ligands for the human DNAM-1 (CD226) activating molecule. *J Exp Med.* 2003;198:557.
231. Tahara-Hanaoka S, Shibuya K, Onoda Y, et al. Functional characterization of DNAM-1 (CD226) interaction with its ligands PVR (CD155) and nectin-2 (PRR-2/CD112). *Int Immunol.* 2004;16:533.
232. Fuchs A, Cella M, Giurisato E, Shaw AS, Colonna M. Cutting edge: CD96 (tactile) promotes NK celltarget cell adhesion by interacting with the poliovirus receptor (CD155). *J Immunol.* 2004;172:3994.
233. Lakshmikanth T, Burke S, Ali TH, et al. NCRs and DNAM-1 mediate NK cell recognition and lysis of human and mouse melanoma cell lines in vitro and in vivo. *J Clin Investig.* 2009;119:1251.
234. Castriconi R, Dondero A, Corrias MV, et al. Natural killer cell-mediated killing of freshly isolated neuroblastoma cells: critical role of DNAX accessory molecule-1-poliovirus receptor interaction. *Cancer Res.* 2004;64:9180.
235. Gilfillan S, Chan CJ, Cella M, et al. DNAM-1 promotes activation of cytotoxic lymphocytes by nonprofessional antigen-presenting cells and tumors. *J Exp Med.* 2008;205:2965.
236. Iguchi-Manaka A, Kai H, Yamashita Y, et al. Accelerated tumor growth in mice deficient in DNAM-1 receptor. *J Exp Med.* 2008;205:2959.
237. Braud VM, Allan DS, O'Callaghan CA, et al. HLA-E binds to natural killer cell receptors CD94/NKG2A, B and C. *Nature.* 1998;391:795.
238. Lee N, Llano M, Carretero M, et al. HLA-E is a major ligand for the natural killer inhibitory receptor CD94/NKG2A. *Proc Natl Acad Sci USA.* 1998;95:5199.
239. Vance RE, Kraft JR, Altman JD, Jensen PE, Raulet DH. Mouse CD94/NKG2A is a natural killer cell receptor for the nonclassical MHC class I molecule Qa-1b. *J Exp Med.* 1998;188:1841.
240. Lanier LL. NK cell receptors. *Annu Rev Immunol.* 1998;16:359.
241. Long E. Regulation of immune responses through inhibitory receptors. *Annu Rev Immunol.* 1999;17:875.
242. Ravetch JV, Lanier LL. Immune inhibitory receptors. *Science.* 2000;290:84.
243. Binstadt BA, Brumbaugh KM, Dick CJ, et al. Sequential involvement of Lck and SHP-1 with MHC-recognizing receptors on NK cells inhibits FcR-initiated tyrosine kinase activation. *Immunity.* 1996;5:629.

244. Burshtyn D, Scharenberg A, Wagtmann N, et al. Recruitment of tyrosine phosphatase HCP by the NK cell inhibitory receptor. *Immunity*. 1996;4:77.
245. Campbell KS, Dessing M, Lopez-Botet M, Cella M, Colonna M. Tyrosine phosphorylation of a human killer inhibitory receptor recruits protein tyrosine phosphatase 1C. *J Exp Med*. 1996;184:93.
246. Olcese L, Lang P, Vely F, et al. Human and mouse killer-cell inhibitory receptors recruit Ptp1c and Ptp1d protein tyrosine phosphatases. *J Immunol*. 1996;156:4531.
247. Moretta A, Bottino C, Vitale M, et al. Receptors for HLA class-I molecules in human natural killer cells. *Annu Rev Immunol*. 1996;14:619.
248. Raulet DH, Vance RE, McMahon CW. Regulation of the natural killer cell receptor repertoire. *Annu Rev Immunol*. 2001;19:291.
249. Shilling HG, Young N, Guethlein LA, et al. Genetic control of human NK cell repertoire. *J Immunol*. 2002;169:239.
250. Garrido F, Ruiz-Cabello F, Cabrera T, et al. Implications for immunosurveillance of altered HLA class I phenotypes in human tumors. *Immunol Today*. 1997;18:89.
251. Ljunggren HG, Karre K. In search of the 'missing self': MHC molecules and NK cell recognition. *Immunol Today*. 1990;11:237.
252. Karre K, Ljunggren HG, Piontek G, Kiessling R. Selective rejection of H-2-deficient lymphoma variants suggests alternative immune defense strategy. *Nature*. 1986; 319:675.
253. Bix M, Liao NS, Zijlstra M, Loring J, Jaenisch R, Raulet D. Rejection of class I MHC-deficient haemopoietic cells by irradiated MHC-matched mice. *Nature*. 1991; 349:329.
254. Cagnano E, Hershkovitz O, Zilka A, et al. Expression of ligands to NKp46 in benign and malignant melanocytes. *J Invest Dermatol*. 2008;128(4):972–979. https://doi.org/10.1038/sj.jid.5701111.
255. Lakshmikanth T, Burke S, Ali TH, et al. NCRs and DNAM-1 mediate NK cell recognition and lysis of human and mouse melanoma cell lines in vitro and in vivo. *J Clin Invest*. 2009;119(5):1251–1263. https://doi.org/10.1172/JCI36022.
256. Groh V, Wu J, Yee C, Spies T. Tumor-derived soluble MIC ligands impair expression of NKG2D and T-cell activation. *Nature*. 2002;419(6908):734–738. https://doi.org/10.1038/nature01112.
257. Paschen A, Sucker A, Hill B, et al. Differential clinical significance of individual NKG2D ligands in melanoma: soluble ULBP2 as an indicator of poor prognosis superior to S100B. *Clin Cancer Res*. 2009;15(16):5208–5215. https://doi.org/10.1158/1078-0432.CCR-09-0886.
258. Schlecker E, Fiegler N, Arnold A, et al. Metalloprotease-mediated tumor cell shedding of B7-H6, the ligand of the natural killer cell-activating receptor Nkp30. *Cancer Res*. 2014;74(13):3429–3440. https://doi.org/10.1158/0008-5472.CAN-13-3017.
259. Oppenheim DE, Roberts SJ, Clarke SL, et al. Sustained localized expression of ligand for the activating NKG2D receptor impairs natural cytotoxicity in vivo and reduces tumor immunosurveillance. *Nat Immunol*. 2005;6(9):928–937. https://doi.org/10.1038/ni1239.
260. Coudert JD, Scarpellino L, Gros F, Vivier E, Held W. Sustained NKG2D engagement induces cross-tolerance of multiple distinct NK cell activation pathways. *Blood*. 2008;111(7):3571–3578. https://doi.org/10.1182/blood2007-07-100057.
261. Cristiani CM, Turdo A, Ventura V, et al. Accumulation of circulating CCR7+ natural killer cells marks melanoma evolution and reveals a CCL19-dependent metastatic pathway. *Cancer Immunol Res*. 2019;7(5):841–852.
262. Fregni G, Messaoudene M, Fourmentraux-Neves E, et al. Phenotypic and functional characteristics of blood natural killer cells from melanoma patients at different clinical stages. *PLoS One*. 2013;8(10), e76928. https://doi.org/10.1371/journal.pone.0076928.

263. Mirjačić Martinović KM, Lj Babović N, Džodić RR, Jurišić VB, Tanić NT, Konjević GM. Decreased expression of NKG2D, NKp46, DNAM-1 receptors, and intracellular perforin and STAT-1 effector molecules in NK cells and their dim and bright subsets in metastatic melanoma patients. *Melanoma Res.* 2014;24(4):295–304.
264. Lord CJ, Ashworth A. The DNA damage response and cancer therapy. *Nature.* 2012;481(7381):287–294. https://doi.org/10.1038/nature10760.
265. Raulet DH, Gasser S, Gowen BG, Deng W, Jung H. Regulation of ligands for the NKG2D activating receptor. *Annu Rev Immunol.* 2013;31:413–441. https://doi.org/10.1146/annurev-immunol-032712-095951.
266. Dasari S, Tchounwou PB. Cisplatin in cancer therapy: molecular mechanisms of action. *Eur J Pharmacol.* 2014;740:364–378. https://doi.org/10.1016/j.ejphar.2014.07.025.
267. Hervieu A, Rébé C, Végran F, et al. Dacarbazine-mediated upregulation of NKG2D ligands on tumor cells activates NK and CD8 T cells and restrains melanoma growth. *J Invest Dermatol.* 2013;133(2):499–508. https://doi.org/10.1038/jid.2012.273.
268. Yang S, Haluska FG. Treatment of melanoma with 5-fluorouracil or dacarbazine in vitro sensitizes cells to antigen-specific CTL lysis through perforin/granzyme- and Fas-mediated pathways. *J Immunol.* 2004;172(7):4599–4608. https://doi.org/10.4049/jimmunol.172.7.4599.
269. Fregni G, Perier A, Pittari G, et al. Unique functional status of natural killer cells in metastatic stage IV melanoma patients and its modulation by chemotherapy. *Clin Cancer Res.* 2011;17(9):2628–2637. https://doi.org/10.1158/1078-0432.CCR-10-2084.
270. Mignot G, Hervieu A, Vabres P, et al. Prospective study of the evolution of blood lymphoid immune parameters during dacarbazine chemotherapy in metastatic and locally advanced melanoma patients. *PLoS One.* 2014;9(8):e105907.
271. Kim H, Sin JI. Electroporation driven delivery of both an IL-12 expressing plasmid and cisplatin synergizes to inhibit B16 melanoma tumor growth through an NK cell mediated tumor killing mechanism. *Hum Vaccin Immunother.* 2012;8(11):1714–1721.
272. Cao G, Wang J, Zheng X, Wei H, Tian Z, Sun R. Tumor therapeutics work as stress inducers to enhance tumor sensitivity to natural killer (NK) cell cytolysis by up-regulating NKp30 ligand B7-H6. *J Biol Chem.* 2015;290(50):29964–29973. https://doi.org/10.1074/jbc.M115.674010.
273. Shi L, Lin H, Li G, et al. Cisplatin enhances NK cells immunotherapy efficacy to suppress HCC progression via altering the androgen receptor (AR)-ULBP2 signals. *Cancer Lett.* 2016;373(1):45–56. https://doi.org/10.1016/j.canlet.2016.01.017.
274. Okita R, Yukawa T, Nojima Y, et al. MHC class I chain-related molecule A and B expression is upregulated by cisplatin and associated with good prognosis in patients with non-small cell lung cancer. *Cancer Immunol Immunother.* 2016;65(5):499–509. https://doi.org/10.1007/s00262-016-1814-9.
275. Matsuzaki I, Suzuki H, Kitamura M, Minamiya Y, Kawai H, Ogawa J. Cisplatin induces Fas expression in esophageal cancer cell lines and enhanced cytotoxicity in combination with LAK cells. *Oncology.* 2000;59(4):336–343. https://doi.org/10.1159/000012192.
276. Sundelin K, Roberg K, Grénman R, Håkansson L. Effects of cisplatin, alphainterferon, and 13-Cis retinoic acid on the expression of Fa intercellular adhesion molecule-1 (ICAM-1), and epidermal growth factor receptor (EGFR) in oral cancer cell lines. *J Oral Pathol Med.* 2007;36(3):177–183.
277. Ramakrishnan R, Assudani D, Nagaraj S, et al. Chemotherapy enhances tumor cell susceptibility to CTL-mediated killing during cancer immunotherapy in mice. *J Clin Invest.* 2010;120(4):1111–1124. https://doi.org/10.1172/JCI40269.
278. Bae SH, Park YJ, Park JB, Choi YS, Kim MS, Sin JI. Therapeutic synergy of human papillomavirus E7 subunit vaccines plus cisplatin in an animal tumor model: causal involvement of increased sensitivity of cisplatin treated tumors to CTL-mediated killing

in therapeutic synergy. *Clin Cancer Res.* 2007;13(1):341–349. https://doi.org/10.1158/1078-0432.CCR-06-1838.
279. Martín-Cófreces NB, Sánchez-Madrid F. Sailing to and docking at the immune synapse: role of tubulin dynamics and molecular motors. *Front Immunol.* 2018;9:1174. https://doi.org/10.3389/fimmu.2018.01174.
280. Puente J, Diaz M, Salas MA, et al. Studies of natural killer cell activity in a drug-free, healthy population. Response to a challenge with taxol, estramustine and lipopolysaccharide. *Int J Clin Pharmacol Ther.* 1995;33(8):457–461.
281. Loubani O, Hoskin DW. Paclitaxel inhibits natural killer cell binding to target cells by down-regulating adhesion molecule expression. *Anticancer Res.* 2005;25(2A):735–741.
282. Markasz L, Stuber G, Vanherberghen B, et al. Effect of Frequently Used Chemotherapeutic Drugs on the Cytotoxic Activity of Human Natural Killer Cells. *Mol Cancer Ther.* 2007;6(2):644–654. https://doi.org/10.1158/1535-7163.MCT-06-0358.
283. Kubo M, Morisaki T, Matsumoto K, et al. Paclitaxel probably enhances cytotoxicity of natural killer cells against breast carcinoma cells by increasing perforin production. *Cancer Immunol Immunother.* 2005;54(5):468–476. https://doi.org/10.1007/s00262-004-0617-6.
284. Law KS, Chen HC, Liao SK. Non-cytotoxic and sublethal paclitaxel treatment potentiates the sensitivity of cultured ovarian tumor SKOV-3 cells to lysis by lymphokine-activated killer cells. *Anticancer Res.* 2007;27(2):841–850.
285. Shurin MR, Naiditch H, Gutkin DW, Umansky V, Shurin GV. ChemoImmuno Modulation: immune regulation by the antineoplastic chemotherapeutic agents. *Curr Med Chem.* 2012;19(12):1792–1803.
286. Di Modica M, Sfondrini L, Regondi V, et al. Taxanes enhance trastuzumab-mediated ADCC on tumor cells through NKG2D-mediated NK cell recognition. *Oncotarget.* 2016;7(1):255–265. https://doi.org/10.18632/oncotarget.6353.
287. Grünberg E, Eckert K, Maurer HR. Docetaxel treatment of HT-29 colon carcinoma cells reinforces the adhesion and immunocytotoxicity of peripheral blood lymphocytes in vitro. *Int J Oncol.* 1998;12(4):957–963. https://doi.org/10.3892/ijo.12.4.957.
288. Symes JC, Kurin M, Fleshner NE, Medin JA. Fas-mediated killing of primary prostate cancer cells is increased by mitoxantrone and docetaxel. *Mol Cancer Ther.* 2008;7(9):3018–3028.
289. Albini A, Bruno A, Noonan DM, Mortara L. Contribution to tumor angiogenesis from innate immune cells within the tumor microenvironment: implications for immunotherapy. *Front Immunol.* 2018;9:527. https://doi.org/10.3389/fimmu.2018.00527. eCollection 2018. Review.
290. Manfredi F, Cianciotti BC, Potenza A, et al. TCR redirected T cells for cancer treatment: achievements, hurdles, and goals. *Front Immunol.* 2020;11:1689. https://doi.org/10.3389/fimmu.2020.01689. eCollection 2020. Review.
291. Hu S, Gokhale S, Jung J, et al. Multifaceted immunomodulatory effects of the BTK inhibitors ibrutinib and acalabrutinib on different immune cell subsets—beyond B lymphocytes. *Front Cell Dev Biol.* 2021;9:727531. https://doi.org/10.3389/fcell.2021.727531. eCollection 2021. Review.

CHAPTER ELEVEN

Role of lymphocytes, macrophages and immune receptors in suppression of tumor immunity

Aayushi Singh[a], Vandana Anang[b], Komal Kumari[c], Sarat Kumar Kottarath[d,*], and Chaitenya Verma[e,*]

[a]Invictus Oncology Pvt. Ltd, New Delhi, India
[b]International Center for Genetic Engineering and Biotechnology (ICGEB), New Delhi, India
[c]Department of Biotechnology, Central University of South Bihar, Bihar, India
[d]Department of Experimental Therapeutics, MD Anderson, Cancer Center, Huston, TX, United States
[e]Department of Pathology, Wexner Medical Center, Ohio State University, Columbus, OH, United States
*Corresponding authors: e-mail addresses: skottarath@mdanderson.org; chaitenya999@gmail.com; chaitenya.verma@osumc.edu

Contents

Progress in Molecular Biology and Translational Science, Volume 194
ISSN 1877-1173
https://doi.org/10.1016/bs.pmbts.2022.10.002

Abstract

Cancer is now the leading cause of mortality across the world. Inflammatory immune cells are functionally important in the genesis and progression of tumors, as demonstrated by their presence in human tumors. Numerous research has recently been conducted to determine if the innate and adaptive immune systems' cytotoxic cells can inhibit tumor growth and spread. Majority of cancers, when growing into identifiable tumors use multiple strategies to elude immune monitoring by lowering tumor immunity. Immunological suppression in the tumor microenvironment is achieved through interfering with antigen-presenting cells and effector T cells. Treatment of cancer requires managing both the tumor as well as tumor microenvironment (TME). Most patients will not be able to gain benefits from immunotherapy because of the immunosuppressive tumor microenvironment. The actions of many stromal myeloid and lymphoid cells are regulated to suppress tumor-specific T lymphocytes. These frequently exhibit inducible suppressive processes brought on by the same anti-tumor inflammatory response the immunotherapy aims to produce. Therefore, a deeper comprehensive understanding of how the immunosuppressive environment arises and endures is essential. Here in this chapter, we will talk about how immune cells, particularly macrophages and lymphocytes, and their receptors affect the ability of tumors to mount an immune response.

Abbreviations

AIDS	acquired immune deficiency syndrome
AML	acute myeloid leukemia
APCs	antigen presenting cells
ATG	autophagy
BCSCs	breast cancer stem cells
Cbl-b	Casitas-B-lineage lymphoma
CD	cluster of differentiation
CDK	cyclin-dependent kinases
CRC	colorectal cancer
CSF1	colony stimulating factor-1
CTL	cytotoxic T lymphocyte
CTLA4	cytotoxic T-lymphocyte associated antigen 4
CXCL	chemokines
DCs	dendritic cells
ECM	extracellular matrix
EGF	epidermal growth factor
EMT	epithelial-to-mesenchymal transition
ER	estrogen receptor
ERK	MEK/extracellular signal-regulated kinase
FGF	fibroblast growth factor
GM-CSF	granulocyte-monocyte colony stimulating factor
HER-2	herceptin-2
HL	Hodgkin's lymphoma
HLA	human leukocyte antigen
IDO	indoleamine-2,3-dioxygenase

IFN-γ	interferon-γ
IgSF	immunoglobulin superfamily member
ILs	interleukins
IPMN	intraductal papillary mucinous neoplasm
IRs	inhibitory receptors
ITIMs	immunoreceptor tyrosine-based inhibitory motifs
JAK	Janus kinases
LAG3	lymphocyte activation gene 3
LPS	lipopolysaccharide
MAPK	mitogen-activated protein kinase
MARCO	macrophage receptor with collagenous structure
MCS-F	macrophage colony-stimulating factor
MDSCs	myeloid-derived suppressor cells
MHC	major histocompatibility complex
MMP	matrix metalloproteinase
MR	mannose receptor
MSCs	melanoma stem cells
NF-κB	nuclear factor-κB
NKT	natural killer T cells
NSCLC	non-small cell lung cancer
PA	plasminogen activator
PD-1/PD-L1	programmed cell death protein 1/programmed death ligand 1
PDAC	pancreatic ductal adenocarcinoma
PDGF	platelet-derived growth factor
PKD-2	protein kinase D isoform 2
RR	relative risk
STA-3	signal transducer and activator of transcription 3
TAMCs	tumor-associated myeloid cells
TAMs	tumor-associated macrophages
TANs	tumor associated neutrophils
TCR	T-cell receptor
TEMs	Tie2-expressing monocytes/macrophages
TGF-β	transforming growth factor beta
Th	T helper
TLR	toll-like receptor
TME	tumor microenvironment
TNF	tumor necrosis factor
VEGF	vascular endothelial growth factor
ZAP70	zeta-associated protein of 70 kDa
γδT cells	gamma delta T cells

1. Introduction

Cancer is represented by an accumulation of cellular anomalies that lead to unregulated cell growth, invasion of healthy tissue and organs, and metastasis (the spread of cancer throughout the body). During all stages

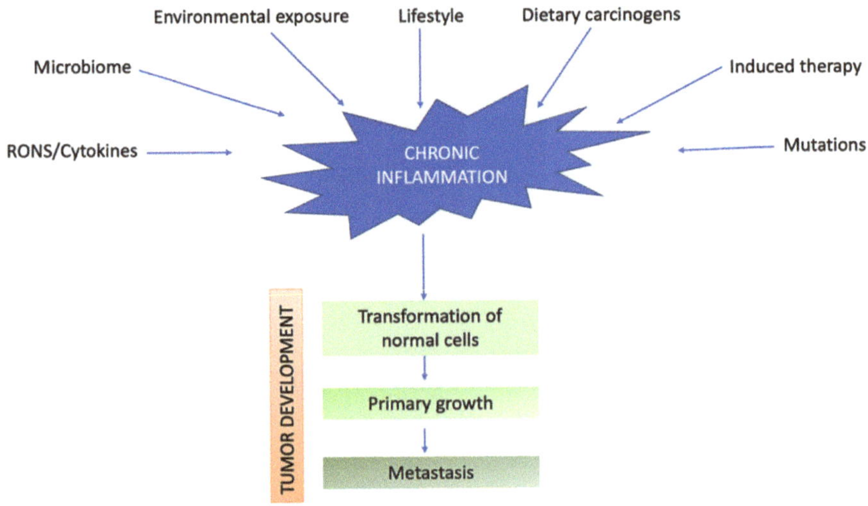

Fig. 1 Chronic inflammation is the driver for tumor development and promotion. Various inflammatory conditions (Microbiome, RONS/Cytokines, Environmental Exposure, Lifestyle, dietary carcinogens, Chemical induced therapy and mutations) can lead to transformation of otherwise healthy cells to trigger primary growth ultimately leading to metastasis.

of the disease, from tumor genesis through tumor development and metastasis (Fig. 1), the cancer cells must contend with diverse types of immune and stromal cells.[1,2] In-depth studies conducted over the course of many decades have shown the tissue microenvironment (TME) playing a vital role in influencing tumor behavior. TME's biological components are diverse and include local stromal cells, essential immune cells, and endothelial cells that have invaded the tissue.

Crosstalk between accessory cells and cancer cells influences and fuels the growth of tumors.[3] It has been shown over the past few decades that immune cells play an important role in inflammation associated with cancer. Since then, research has also centered on figuring out how immune cells affect tumors' development in various illnesses. Aberrant innate and adaptive immune responses are now thought to contribute to carcinogenesis through induction of immunocompromised environment and encouraging proliferation of cancer cells and metastasis, hence it is crucial to avoid the destruction of cancer cells. It was observed that cytotoxic CD8+ immune cells and natural killer (NK) cells can find and kill more immunogenic cancer cells at initial stages of expansion.[4] However, they let out different cells, like tumor-associated macrophages (TAM's) and myeloid derived suppressor cells (MDSC), including cytokines IL-6, TNF-α, and IL-1 (Interleukin-1),[5] as well as other soluble compounds that may be critical in speeding the growth of tumors.

It has been observed that during chronic inflammation, immune cells are thought to serve a distinct purpose, since adaptive immunological responses may cause the overactivation of the innate immune system (1, 2), tissue antibody accumulation, which draws more innate immune cells, and T-lymphocyte malfunction as opposed to activation (3). The development of both spontaneous and virally driven malignancies can be predisposed to immunodeficiency, and existing tumors frequently produce immunosuppressive microenvironments that can inhibit effective anti-tumor immunity. The interaction between a tumor and its host immune system is considered to evolve in three phases; first elimination, in which immune response eradicated newly formed tumor cells, second equilibrium, when immune response maintains tumor growth and metastasis; and third escape occurs when tumor cells have developed acquired immunity to the host's immune system, called "immune-editing" process.

One of the many escape mechanisms that have been discovered is alteration in the expression and/or functionality of human leukocyte antigens (HLAs) expressed by tumor cells. These changes are important because they may significantly affect the interaction among tumor cells and host immune cells. Understanding the pathophysiology of malignancies linked to inflammation and developing more potent cancer treatments can benefit from identifying these processes. The primary markers and receptors expressed that impact cancer immune suppression will be the topic of this chapter's discussion of tumor-associated immune cells.

1.1 Role of macrophages in tumors

Macrophages, the most typical immune cells infiltrating tumors, make for around 50% of the blood-derived cells in TME. CCL-2, CSF-1 (colony stimulating factor), and other cytokines, which operate as signaling molecules, draw circulating monocytes and other tissue-resident macrophages to tumor locations. Macrophages that participate in the early stages of cancer have anti-tumor activity, but when linked to tumors, they develop certain phenotypes in the relevant microenvironment. TME has been found to control tumor associated macrophages (TAMS), which then become pro-tumor and in addition accelerate tumor growth.[2,6] This could happen as a result of the initiation and stimulation of the activation of numerous growth factors and signaling molecules, including the immunosuppressive chemokines and interleukins (IL-6, IL-10, and CXCL-8), the macrophage colony-stimulating factor (M-CSF), Transforming growth factor (TGF), and vascular endothelial growth factor (VEGF). Most of the studies reports the tumor-promoting role of TAMs within the primary TME.[7,8]

Unexpectedly, macrophages which naturally possessed phagocytic and cytotoxic activities and were previously thought to be a sign of anticancer immunity, represents an essential part of the immune cell infiltrate in the TME.[7,9] Contrary to widespread assumption, macrophages do not act as hosts' protectors among inflammatory microenvironments. Once they are exposed to TME signals, macrophages undergo functional reprogramming and may change its phenotypes to react to either pro- or anti-inflammatory environmental stimuli.[10] Importantly, the TME, which is composed of macrophages, tumor cells, and endothelial cells, is another tumor-promoting structure for metastasis that may be seen at metastatic sites and has been shown to be predictive of likelihood metastasis in human breast cancer malignancies[11]; this result is significant for cancer cell survival because of TAMs' role in immunosuppression, invasion, metastasis, and angiogenesis. TME is inundated with cytokines and growth factors when benign invasive cancer transforms into invasive malignant cancer. TAMs can maintain the "steaming inflammation" in the TME that required for the pro-tumor activity because they have tardy and defective NF-κB activation in response to signals like LPS (Lipopolysaccharide) and TNF-α (Tumor Necrosis Factor- α).[12]

2. Differentiation and polarization of tumor-associated macrophages

According to the knowledge that is currently available, TAMs that invade existing tumors pick up phagocytic characteristics similar to those of M2 macrophages and phenotypes that promote angiogenesis, tumor development, tissue remodeling, and also inhibition of anti-tumor immunity. Similar to the Th1 and Th2 classification, macrophage can be categorized in two polarization states which are M1 (activated) and M2 (alternatively activated) functional status in the TME.[13,14] The activation of M2 macrophage can be induced by IL-4, IL-6, IL-10, IL-13, IL-33 and Notch, whereas microbial stimuli or IFN-γ alone or in combination of with TNF-α and GM-CSF cytokines triggers classically activated M1 macrophages and immune complex.[15,16] Alveolar macrophages stimulated to take on the M2 phenotype by IL-13 are further amplified by IL-33.[17] In response to signals from TMEs, TAMs may encourage tumor invasion and metastasis by creating activating factors for stromal and cancer cells TGF-β, PDGF (platelet derived growth factor), VEGF, EGF (epidermal growth factor), FGF (fibroblast growth factor), matrix metalloproteases, serine proteases, and cathepsins.[18–20] The word "activation" instead of "polarization" is preferred for the classification of the functional status of TAM in recently rising efforts to create a standard vocabulary for defining the characteristics of the macrophages under investigation.[17] TAMs are the collective name for macrophages seen in human TME that cannot be universally categorized as either M1 or M2 macrophages.[21] However, M1 and M2 macrophage are two ends of spectrum that exist in a variety of functional states (Fig. 2).[23–25] In addition, recent research using highly standardized stimulation of human macrophages shown that the existing M1 and M2 polarization paradigm may be stretched to "spectrum model" including approximately nine distinct macrophage activation programs. It is now known that dynamic changes in macrophage phenotype take place during development,

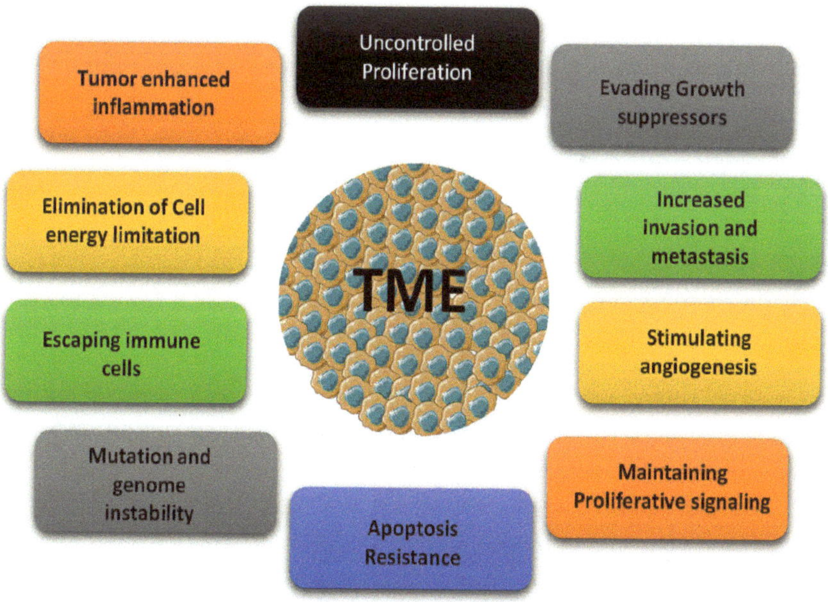

Fig. 2 The tumor microenvironment and characteristics of cancer. This figure describes the main factors of cancer that widely influence tumor microenvironments.[22] *Adapted from Wang M, et al. Role of tumor microenvironment in tumorigenesis.* J Cancer. *2017;8 (5):761–773. doi: 10.7150/jca.17648.*

genesis, and metastasis of tumors and that different TAM subpopulations are in charge of various tumor-promoting actions.[6,26,27] According to the extensive transcriptome investigation, macrophages exhibit a mixed phenotype and express both M1 and M2-like markers. Notably, TAMs distribution and function within neoplastic tissue vary considerably, and tumors have a wide range of illnesses. The activation of TAMs and prognosis of the tumor as a whole appear to be influenced by various signals from specific areas in the TME. TAMs can take many different forms when studying the microanatomical structure, including the accumulation of pro-tumor cells in the hypoxic regions and changes in inflammatory pathways and components across tumors originating in various anatomical places. TAMs are also known to have proangiogenic activity, and tumors typically have significant vascular densities when macrophages are infiltrated there. The bulk of M2-like TAMs are identified in hypoxic tumor areas, demonstrating greater proangiogenic activity *in-vivo* where they have proliferated with tumor progression. Many TAMs display a number of angiogenesis modulating factors, such as βFGF, TNF-α, IL-1β, CXCL-8, cyclooxygenase-2, plasminogen activator (PA), VEGF, PDGF-β, MMP-7, MMP-9 and IL-12.[28] As the tumor progresses, the immune microenvironment's makeup itself and TAMs' activity become highly favorable for tumor growth and thus the role of macrophages functionally gets altered from tumor initiation to progression.[29]

Concurrently, M2-like macrophage can convert to M1-like cells and reduction of immunosuppressive effect from M2 population been evident following IFN-γ treatment in TAMs providing M1 phenotypes.[30,31] These findings suggest the possibility of TAM activation being reversed and offer new treatment approaches, such as TAM re-education. Strategies for re-education may be facilitated by discovering the genetic and epigenetic factors underpinning the diversity of macrophages in tissues and their many modes of activation.[29] TAMs, which exhibit an M2-like polarization and block immunostimulatory signals while demonstrating little cytotoxic action, aid in the growth and survival of tumors (Table 1).[32]

TAMs can be considered corrupted cells that promote the formation of tumors and are resistant to traditional anticancer treatments like chemotherapy and radiotherapy and the most advanced immunotherapies like anti-PD1. These make TAMs interesting targets for cutting-edge anti-tumor therapies. Transforming immune-suppressive, pro-tumor macrophage (M-like) into immunostimulatory, anti-tumor cytotoxic effector cells (M1-like) is the aim of reprogramming TAMs. If this change is successful and long-lasting,

Table 1 Prognostic significance of TAMs and their markers in different human cancer types.[32]

Type of cancers	TAMs markers	Prognostic impact
Glioma	CD68, CD163, CD204, IBA	Bad
	CD163/CCL3 ratio	Bad (if ratio high)
	CD 68, CD163/AIF ratio	Bad (if ratio high)
Mesothelioma	CD68, CD163	Bad
Lung cancer (NSCLC)	CD68	Good in tumor islets/bad in tumor stroma
Pancreatic cancer	CD68/iNOS (for M1)	Bad (if M2 > M1)
	CD68/CD163 (For M2)	
	CD68, CD204, CD163	Bad
Breast cancer	CD68, CD163	Bad
Colorectal cancer	iNOS	Good (if together with high CD8+ cells)
	CD68	Good (at the invasive front)
	Wnt 5a, CD68	Bad
Melanoma	CD68, CD163	Bad (CD163 at the tumor stroma and CD68 at the invasive front)
Bladder cancer	CD68	Bad

it should restore the patients' reactive immune system, which is proficient of battling and removing the tumor.

Tumor vasculature is an important pathway for promoting metastasis of malignant tumors. A mechanism known as the "angiogenic switch" is activated by several other mechanisms once solid tumors grow up to a specific size in order to trigger the formation of high density vasculature for the provision of nutrients and elimination of waste products.[33] TAMs are important participants in regulating "angiogenic switch". As angiogenesis and metastasis hotspots, TAMs from clusters in the intra-tumoral zones and invasive fronts. On the contrary, absence of TAMs significantly reduces the vessel density by 40%.[34,35] Additionally, TAMs also contribute to lymphoangiogenesis which serves as the crucial pathway for the tumor cell dissemination into locally nodes and distant metastatic sites. Taken together, TAMs influence the development of new tumor vessels as well as the modification of existing vasculature into a tortuous and leaky shape to promote dissemination of tumors.[36]

3. Role of TAM receptors in suppressing tumor immunity

3.1 Mannose receptor

One of the earliest immunological adhesion molecules, mannose receptor (MR), also referred as CD206, related to C-type lectin receptor family (CTLR). It is often found on the surface of antigen-presenting cells such as dendritic cells and macrophages.[37] Whereas, utilizing endocytosis, MR perform a critical role in bridging acquired and innate immune responses. It is inappropriately expressed in pancreatic, breast, and colorectal cancer and regulates cell homeostasis. In addition, it has been noted that TAMs express MR. The mutual crosslinking of tumor-like mucoprotein-mediated MRs infiltrating TAMs suppresses tumor immunity and aids tumor progression.[38] Additionally, it has been established that MRs, which are exhibited by the endothelial cells of lymphatic vessels, contribute to the dissemination of tumors along the lymphatic system.[39] Recent studies show that MR+ liver sinusoidal vessels help colon cancer cells spread to the liver through a mechanism involving IL-1 induced upregulation.[5] Mucins found in the TME can activate MR on human TAM. This interaction strengthens their immunosuppressive character even more and can be viewed as an additional method of immune evasion by tumors.[37,38,40]

3.2 CD68 and CD163 receptors

M2 macrophages are crucial for the suppression of tumor immunity because they regulate inflammatory reactions and adaptive Th1 immunity to support angiogenesis, tissue remodeling, and repair. According to recent research, CD68 and CD163 are the most prevalent TAM markers. Kuffer cells, germinal center, splenic macrophages, and granulocyte precursors are a few examples of many tissue macrophages that recognize epitopes when tagged with the pan-macrophages marker CD68. CD163, a hemoglobin scavenger receptor, was initially identified while looking for specific markers of mononuclear phagocyte differentiation. It has since been established that CD163 is a phenotypic marker of M2 macrophages. According to Minami et al. research, malignancies such as hepatocellular, breast, bladder, and ovarian cancer have poor patient prognosis when CD68+ and CD163+ cells are present in tumors.[41] In their investigation of the relationship among CD163/68 ratio and patients prognosis in hepatocellular carcinoma, Komohara et al. discovered that allowing CD163/68 ratio was connected to poor prognosis.[42] These findings might also show how different and heterogenous TAMs are in their ability to influence the prognostic impact of macrophages.[41]

3.3 Macrophage receptor with collagenous structure

Macrophage receptor with collagenous structure (MARCO), a scavenger receptor class-A protein is expressed on the surface of naive tissue-resident macrophages.[43] In non-small cell lung cancer which is a solid tumor, the quantity of TAMs expressing MARCO is correlated with an increase in regulatory T cells, effector T cells, and a reduction of NK cells.[44] TAMs with MARCO expression are unable to activate NK cells or produce cytokines, which reduces their ability to eradicate tumors. In human breast cancer, metastatic melanoma, periampullary adenocarcinoma, and pancreatic cancer, the expression of MARCO is associated with a poor prognosis, and considerable infiltration of TAMs expressing MARCO is associated with treatment resistance.[45] Clinical studies data suggest that tumor infiltration with macrophages expressing high CD163 and MARCO indicates lower survival rates. Patients with a poor response can receive anti-MARCO medication along with anti-CTLA-4 and anti-PD1/PD-L1 to enhance ineffective checkpoint treatments in NSCLC patients.

3.4 MerTK and Axl

MerTK and Axl are receptor tyrosine kinase that are expressed in multiple cells and tissues of the body.[46] Approximately 50–70% of the patients with non-small cell lung cancer,[47] breast cancer,[48] ovarian cancer,[49] melanoma,[50] colorectal cancer[51] and leukaemia[52] are positive for the expression of MerTK and Axl receptors. Gas6 (Growth Arrest Specific Protein 6) is the ligand for MerTK, and ProS1 (Protein S) is the ligand for both the receptors.[53] Signaling is mostly related to efferocytosis, where TAM receptors produced by macrophages, via ligands ProS1/Gas6 bind PtdSer are exposed to cell debris or to cells undergoing apoptosis. The ensuing signaling instructs macrophages to adopt an M2 phenotype, ensuring the tissue healing and quiet clearance of dead cells.[54] These TAM receptors have a crucial role in controlling and regulating inflammatory immune responses.[55]

4. Role of lymphocytes promotion and suppression of tumor immunity

Leukocyte infiltration into growing tumors is now thought to be a hallmark of the genesis of cancer.[56] The immune system reacts to early cancer in the same way as the body responds to an acute tissue damage, with successive infiltration of diverse myeloid populations leading to lymphocyte infiltration.[57] Furthermore, it is difficult to make conclusions regarding the association between immune response to injury or infection and tumor formation since tumor growth kinetics and neoplastic cells themselves change the local immunological milieu. Regardless, suppose the pre-cancer cells are notely removed and initial acute inflammatory reaction does not go away. In that case, a condition of chronic inflammation will undoubtedly arise inside the local tissue. Myeloid-lineage cells, in particular TAMs, MDSCs, tumor-associated neutrophils (TANs), tumor-associated dendritic cells that can act as suppressors, Tie2-expressing monocytes, and monocytes that express the angiopoietin-2 receptor, promote the early development of cancer through a variety of mechanisms.[13,31] The immunological microenvironments of a neoplastic tissue comprise the composition of invasive leukocytes as well as the bio-effector function played by these cells inside the tissue. The expression of tissue-specific cytokines, chemokines, and other immune mediators, as well as the presence of cells inside tumors, are important determinants of whether an immune response is pro- or anti-tumor.[3,13]

Myeloid cells that have infiltrated tumors respond immediately to soluble and insoluble signals from the neoplastic environment, rather than simply to inflammatory stimuli that induce tissue damage. Responses manifest as radically altered gene expression that modifies the immune cells' bio-effector functions. These typically result in increased production of components or mediators that promote the development and survival of neoplastic cells, the induction and maintenance of angiogenic responses, the advances of tissue remodeling, and the inhibition of anti-tumor immune response.[12] An infection or autoimmune condition that causes tissue to become chronically inflamed can change the likelihood that cancer will develop by fostering the survival and subsequent proliferation of preneoplastic cells that have already begun to grow, as well as by producing compounds that damage DNA, such as reactive oxygen and nitrogen species, which raise the frequency of mutation.[58] While infiltrating immune cells can control all of these features of the growth of solid tumors, this article will concentrate on the components of carcinogenesis that are controlled by infiltrating lymphocytes.

4.1 T-lymphocytes

T cells grow in the thymus from a common lymphoid progenitor and are distinguished by the production of distinct antigen binding molecules on their surface, known as T cells receptors.[59,60] T cells can only recognize antigen when it is provided by MHC molecules, also referred to as human leukocytes antigen or HLA, as opposed to membrane-bound antibodies on B cells that can do so on their own,[61] T cells are further subdivided into two different subpopulation of T cell that recognize peptides provided by MHC-1 or MHC-II molecules, respectively, known as CD4+ T helper and CD8+ cytotoxic lymphocytes (CTL).[59] Th cells recognize and engage with an antigen-MHC-II molecular complex, activating them to become effector cells that release different growth factors known as cytokines. The released cytokines are crucial for triggering the activation of B cells and T cells, macrophage, and other immune response-producing cells. Different forms of immunological response are brought on by variations in the cytokine production pattern of activated Th cells.

Cytotoxic T cells that recognize antigen MHC-I complex molecules multiply and develop into effector CTLs under the influence of cytokines released by Th cells. The CTL typically demonstrate cell killing, cytotoxic activity, and produces fewer cytokines than the cytotoxic T cell. CTLs are crucial in keeping track of the body's cells and removing any antigen express, like virus infected cells, tumor cells, and cells from an outside of tissue graft. CTLs are directed against changed self-cells, defined as cells expressing foreign antigen complex with a class I MHC molecules.

Th cells are further subdivided into Th1 cells that express IFN-γ and TNF-α, and Th2 cells that express IL-4, IL-5, and IL-13. This compact model of the T cells compartment has been enhanced by the identification of a number of other subtypes, including follicular helper T cells (fTh), IL-17 producing Th cells (Th17), and regulatory T cells (Treg). For the CD4+ T cells compartment, type-1, type-2, and CD8+ T cells, as well as regulatory CD8+ cells, have been described. There are two other T cell subsets that may be activated by the TCR or cytokines and natural killer T lymphocytes get glycolipids through the unconventional MHC protein CD1d (NKT),[62] in contrast γδT cells are not restricted to MHC molecules and may recognize a wide variety of other molecules including soluble non protein antigen.[63] Each of these T lymphocytes subset have unique role in directing in immune response and have been studied for their significant role.[59]

4.2 Cytotoxic T lymphocytes

In addition to spontaneous tumors and lymphomas, carcinogen-induced sarcomas are more common in mice with specific immune-based genetic abnormalities.[64,65] Numerous studies have demonstrated that both CD8+ CTLs and NK cells are essential mediators of the anti-tumor response because of their capacity to generate IFN-γ and kill target cells directly.[64] γδT cells, which resemble CTLs and NK cells in terms of their properties, are also implicated in the anti-tumor response in epithelial tissue such as skin, where they can be predominant T cells population.[63] The proportional significance of CTLs, NK cells, and T cells depends greatly on the malignancy being treated.

According to epidemiological studies cancer risk varies substantially by organ and cause, as shown by studies of cancer in AIDS patients and organ transplant recipients.[65,66] Infectious or viral reactivation in the absence of T cells is related with an increase in viral-associated malignancies, such as Herpes Virus 8 -associated Kaposi's sarcoma, Epstein-Barr virus-associated Non-lymphoma, Hodgkin's and HPV -associated squamous carcinoma.[67,68] Some types of cancer are more prevalent in populations with suppressed immune systems for reasons unrelated to immune suppression, such as long-term exposure to toxins associated to tobacco use.[69,70] However, for the most common types of epithelial-based solid tumors that are not virally associated, such as breast and prostate cancers, the risk ratio (RR) is reduced in immune suppressed individuals, with some of them having RRs smaller than 1.0.[65,66]

Tumor infiltrating T cells are capable of eliciting an immune response against the tumor because they express activation signals, recognize tumor antigens (including immunogenic ones), and kill tumor cells.[63] Neo- and overexpressed c-myc, HER-2/neu, and p53 antigens are examples of tumor antigens,[69] along with host stromal cell-derived antigen specific to a particular tumor. Autoantibodies against extracellular matrix (ECM) substances such non-collagen Type I, III, and V as well as anti-fibronectin antibodies have been shown in clinical investigations to accumulate in patients with lung cancer in their nasopharynx.[70] It's crucial to remember that lymphocytes collaborate with one another in order to function. Many of their effector functions depend on their capacity to create cytokines and bind inhibitory and activating receptors to ligands released by other leukocytes, stromal cells, and even malignant cells. For instance, cytokines produced by mature DCs and activated T cells are required for increasing NK cell effector function.

Robust regulation of CD8+ T cell responses is enabled by IFN-γ, which is secreted by NK cells and functions as a maturation signal for tissue-resident DCs and a driver of CD8+ T cell effector activity.[71] A proper immune response must be initiated, directed, maintained, and ultimately shut down by these interconnected regulatory mechanisms. This mechanism also prevents an unnecessary immune response; while peripheral tolerance, which is important to prevent autoimmune disorders, is mediated by cytokines, inhibitory receptors, and immune regulatory cell types, central tolerance is achieved through the negative selection of self-reactive cells. Given the regularity with which cancer cells identify themselves as "self," it is not unexpected that they employ comparable techniques to inhibit anti-tumor immunity. As a consequence of this, T cells are absolutely necessary for controlling both pro- and anti-tumor immunity, in addition to chronic inflammation in the microenvironment of the tumor.[59]

4.3 T helper cells

The conventional development of CD4+ T cells into Th1 and Th2 cells is mediated by IL-12 and IL-4, respectively. Recent research has demonstrated that a unique T helper cell subset known as the Th17 lineage is involved in this classical differentiation process. Th17 cells are produced in response to TGF-β and IL-6; these cells subsequently exercise their effects by secreting cytokines including IL-17, IL-21, IL-22, and IL-17F. Most of these cytokines cause inflammation by making mice's autoimmunity worse, but they also seem to play a key role in fighting off some infections from the outside. This may be achieved by mediating leukocyte recruitment by inducing chemokine expression,[72,73] which would then promote leukocyte recruitment.

T cells have been seen in patients with malignancies of the ovary, prostate, and stomach. It has also been revealed that a high number of IL-17-producing cells in patients with hepatocellular carcinoma is associated with a poor prognosis. In 4T1 mammary cancer cells, inhibiting IL-17 production leads to decreased survival and the formation of tumors, while transgenic production of IL-17 by cells leads to enhanced tumor growth in mice and enhances angiogenesis in these animals. Depletion of IL-17 also has the effect of reducing the rate at which chemically induced papillomas develop. It seems that this impact is achieved by a mechanism in which IL-17 serves to stimulate the production of IL-6 by both neoplastic and stromal cells, which then activates STAT3 in turn. This could cause a feedback loop that is bad for the body's health because it controls the production of IL-17 and genes that weaken the immune system and make tumors grow.

Inflammation caused by eFdIL-17 may have both beneficial and detrimental consequences, depending on the context. The development of a tumor may be affected by a variety of variables, which depends upon distinct tumor models. For instance, MC38 colon cancer cells grow rapidly in animals lacking IL-17,[74] but tumor regression may be induced by adoptive transfer of *in-vitro* polarized Th17 cells that are specific for the B16 melanoma antigen.[75] Regression was prevented by blocking antibodies, and tumors missing IL-17 exhibited a reduced frequency of IFN-γ, suggesting that IFN-γ is necessary for tumor growth. Confusion is increased by the fact that both IL-21 and IL-22 may activate STAT3 and encourage Th17 differentiation. Clinical studies are now being done on IL-21 therapy as a way to treat cancer. This is because CD8+ T cells that have been primed with IL-21 *in-vitro* have a stronger anti-tumor response after being adopted.[76]

It should not come as a surprise that reducing the number of CD4+ T cells and eliminating the population of Th1, Th2, Th17, and Treg had different impacts in a variety of animal cancer models.[77] It is often believed that polarization of the Th1 immune system has anti-tumor effects, while polarization of the Th2 immune system encourages tumor growth.[78] Directly concentrating on Th1 production and effector function through IL-12 and IFN-γ implies a role in Th1 tumor rejection. Loss of IFN-γ also increases the risk of lung adenocarcinomas and spontaneous lymphomas. When there is a genetic abnormality in IFN-γ, IFN-γR, or when anti-IFN-γ antibody is employed in therapy, MCA-induced sarcomas are more frequent. Exogenous IL-12 therapy, like IL-12 genetic deficiency, has the opposite impact, increasing the prevalence of chemically-induced sarcomas and papillomas.[79] Anti-INF antibodies were remarkable in their ability to inhibit IL-12-dependent sarcoma cell rejection. Though IFN-γ is produced by Th1 cells, CD8+ T cells, T cells, NK cells, and NKT cells, IL-12 is still required for this process.[64] It was shown that in IFN-γ deficient mice, MCA-induced carcinogenesis was amplified by T cells' inability to generate IFN-γ.[80] As a result, Th1 cells might not be important, at least not in the skin, where there are a greater number of T cells.

The greatest evidence for a specific function for the Th1 phenotype may be seen in STAT-6 deficient mice, which exhibits enhanced Th1 polarization attributed to a block in IL-4 signaling.[64] These mice had a greater resistance to the growth of the 4T1 mammary cancer cell line and were able to reject tumors that were created through the injection of an astrocytoma cell line that developed permissively in normal mice. Also, these mice were more resistant to the growth of the 4T1 mammary cancer cell line than other mice.[81] After secondary stimulation, lymphocytes from mice with tumors caused by STAT-6 deficiency generated IFN-γ, and their spleens showed enhanced cell death. The barrier in IFN-γ signaling causes higher Th2 polarization in STAT-1 defective animals, who are also more prone to developing tumors. These findings align with the theory that Th1 polarization boosts IFN-γ production, which improves CTL responses to produce more potent anti-tumor immunity. In addition, reduction of IFN-γ sensitivity decreased the immunogenicity of tumors. Direct effects of IFN-γ have been reported on suppressing proliferation, inducing apoptosis, and decreasing angiogenesis. Interestingly research has shown that IFN-γ increases MHC1 expression. It also indicates a distinct manner by which IFN-γ generation by Th1 and other cells may enhance tumor immunity by increasing CTL destruction on MCA-induced sarcomas.[59,82]

4.4 Natural killer T cells

Natural killer (NK) cells exhibit a wide variety of phenotypic traits, depending on where in the maturation process, they are found. These fundamental components of the innate immune system have been shown to possess potent antitumor and antimetastatic capabilities. In order to escape being destroyed by NK cells, breast cancer cells downregulate their activation receptors while they are dormant. The interaction of the MHC-1 with NK cells that includes immunoreceptor tyrosine-based inhibitory motif (ITIMs) helps in the licensing of NK cells and inhibits the killing of healthy cells.[83,84] It's interesting to note that tumor cells express little or no MHC-I to avoid CD8+ T cell-mediated cytotoxicity, which reduces tumor immunity.

In order to NKT cells to maintain their viability and proliferation, a variety of cytokines and transcription factors must be generated. They provide a vital regulatory function in guiding a Th1 or Th2 polarized immune response by rapidly producing IFN-γ, TNF-α, IL-4, and IL-13 in response to stimulation. This allows them to play a role in the immune response.

This is shown in mice that are deficient in NKT cells due to the fact that infections, particularly those caused by bacteria or parasites, are often more severe in these animals.[85–88] The TCR α and β chain are expressed by thymic NKT cells, just as they are expressed by conventional CTL and T helper cells. T cells, which are NKT cells, recognize glycoprotein antigens presented by CD1d, a non-classical member of the MHC family, rather than peptides presented by MHC class I or II molecules.[62] Type-II NKT cells display a range of TCR chains, are also capable of recognizing CD1d, and are triggered into action by glycolipids that are yet poorly understood.

Peripheral blood CD4+ NKT cells in people produced both Th1 and Th2 type cytokines upon stimulation, but CD4 NKT cells only expressed IFN-γ and TNF-α.[89,90] It's also impossible to completely rule out the possibility that NKT cells have lethal effector activities. Human NKT cells have been shown by Song and colleagues to directly kill human neuroblastoma cell lysate-pulsed monocytes.[91] Using immunocompromised mice, researchers discovered that humans compromised with NKT cells exhibit decreased monocytes in tumor and inhibit xenograft growth, demonstrating NKT cells also block pro-tumor immunity by killing TAMs.[88] Therefore, tumor location, the type of tumor may have a significant impact on how NKT cells express cytokines and whether they help or hinder the anti-tumor response.

4.5 Gamma delta (γδ) T cells

γδ-T cells in the tumor microenvironment may function differentially, ranging from cells that suppress to those that promote inflammation, depending on the tumor stage of development and the peripheral regulation. While it is known that some subgroups of γδ-T cells have features that promote the growth of tumors, others have the potential to serve as therapeutic targets.

γδ-T cells are the predominant inflammatory cells that may be found in the tumor microenvironment of mice. These cells are largely responsible for the production of the pro-inflammatory cytokines IL-8, GM-CSF, and IL-2, as well as the immunosuppressive cytokine IL-17.[92] γδ-T cells may exhibit as T regulatory phenotype and can adopt a Th2 phenotype by generating more suppressive cytokines. This promotes the growth of tumors and suppresses the immune system's ability to combat tumors. The accumulation of γδ-T cells expressing IL-10 and TGF-β reduced the immune responses of both adaptive and natural killer cells against tumors in mice.[93]

In addition, research indicates that the suppressive characteristics of γδ-T cells that produce IL-10 may impede the rejection of tumors by adaptive and anti-tumor γδ-T cells that are involved in immunosurveillance.[94]

Tumor-infiltrating CD39+ T lymphocytes in humans with colorectal cancer secrete large amounts of adenosine, demonstrating potent immunosuppressive action.[95] Cytokines produced by pro-tumor γδ-T cells enhance tumor development and metastasis by inducing angiogenesis and suppressing the immune system. This action is analogous to Treg/Th2 activity and, by interfering with DC effector function, directly suppresses anti-tumor adaptive immunity.[92]

4.6 B lymphocytes

B lymphocytes develop in the bone marrow, and they have a specific antigen-binding receptor that only they have. This antibody molecule that is attached to the cell membrane acts as a B cell or antigen-binding receptor when it is exposed to an antigen. A naive B cell divides quickly, and its progeny become memory and effector B cells, which are also called plasma cells. Memory B cells live longer than naive B cells and make the same antibody bound to the membrane as their parent B cells. Both plasma cells and the B cells from which they come make membrane-bound antibodies. Plasma cells make a form of the antibody that can be secreted, with little or no antibody attached to the cell membrane. B cells make up more than 25% of the cells in most tumors. Also, in breast cancers, B cells make up about 40% of the lymphocytes that move into the tumor. Antibodies made by B cells can change how antigenic targets of tumor cells work. They can also start the complement cascade, opsonize tumor cells to stop cross-presentation of tumor antigen by other antigen-presenting cells like DCs, and help NK cells kill tumor cells by binding to antibodies.[96]

To activate Fc-receptors on resident and invading myeloid cells, particular immune complexes deposit in pre-malignant tissue in a genetic mouse model of squamous cell carcinoma.[97] Additionally, they stimulate the complement system, which in turn promotes angiogenesis and the growth and spread of tumors.[64,98] Extracellular vesicles (EVs) generated from B16-F10 melanoma tumor cells were used in recent research. These EVs include membrane and luminal proteins that may trigger the host's adaptive immune response and induce antibodies that promote carcinogenesis.[97]

5. Suppression of anti-tumor immunity via pro-tumorigenic factors and cytokine production by B cells

Angiogenesis of lymphatic vessels, which promotes tumor growth and the spread of cancerous cells throughout the body, is one characteristic of cancer.[99] This emphasizes the vital role that lymphatic veins play in the growth of malignancies. It is possible for inflammation to promote the development of new lymphatic channels, which would make it easier for immune cells to enter the lymphatic system. The interaction of stromal, tumor, and immune cells may be essential for triggering lymph angiogenesis. CXCL13, which is produced by B cells that have infiltrated the tumor, draws additional B cells to the tumor location.[4,100] After that, these tumor-infiltrating cells secrete lymphotoxin, which triggers NF-κB and STAT-3 signaling in other cancer cells to promote the growth and development of tumors.[101,102] Lymphotoxin is a survival factor that induces angiogenesis in cancers. B cells are able to inhibit effector cells such as CD8+ cytotoxic T lymphocytes and NK-T cells, which enables them to decrease the immune response against tumors. B cells also have the ability to produce a variety of cytokines that are involved in immunoregulation (Fig. 3). Regulatory B cells, also known as B-reg, are a diverse population of B cells that may also produce TGF-β and IL-35, despite the fact that they are operationally characterized as cells that release IL-10.[103]

Additionally, it appears that these B-reg cells have different tumor-evoked B-regs and B cells that block CD4+ T cell response, as well as phenotypic and functional variations. Thus inflammation is supported by B-regs that generate IL-10, which also aids in accelerating the progression of cancer. This immune-modulatory activity may reduce the host's ability to identify and eradicate mutated cancer cells, thereby allowing cancer cells to survive and proliferate in an environment that is chronically inflammatory. While B cells may be beneficial to autoimmune disease by suppressing a Th1 type response during chronic inflammation, this activity may also reduce the host's ability to identify and eradicate mutated cancer cells.

However, the issue that has to be asked is the same as before: how do B cells get involved in regulation? The neoplasm itself is responsible for the method through which B cells acquire regulatory potential. The tumor exerts long-range control to drive the recruitment of cells and mediators to support its development. B-reg can be produced by the action of

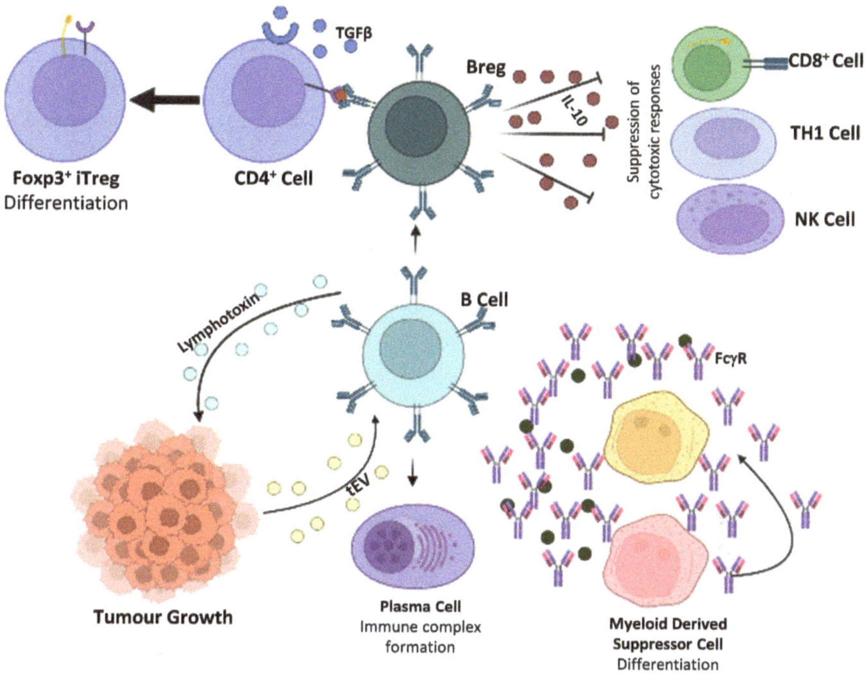

Fig. 3 Suppression of the anti-tumor responses by B cells. B cells generate lymphotoxin, which increases tumor angiogenesis. Tumor-produced extracellular vesicles (tEVs) may drive B cells to produce antibodies, which can bind to antigens to form immunological complexes (ICs). These circulating ICs may activate Fc receptors on myeloid cells, leading them to become myeloid-derived suppressor cells that limit CD4+ and CD8+ T cell responses. Regulatory B cells (B-regs), a subset of B cells, may secrete immunoregulatory cytokines such IL-10, which suppresses CD4+ Th1 cells, NK cells, and CD8+ cytotoxic T cells, and TGF, which induces CD4+ T cells to become Foxp3+ CD4+ T regulatory cells (Tregs).[96] *Adapted from Yuen GJ, Demissie E, Pillai S. B lymphocytes and cancer: a love–hate relationship.* Trends Cancer. *2016;2(12):747–757. doi: 10.1016/j.trecan.2016.10.010. Permission license 5379550489353.*

5-lipoxygenase metabolites, such as leukotriene B4, which are produced by the tumor and activate PPAR-α in B cells to start the differentiation process into B-reg. B-reg can be generated by the action of 5-lipoxygenase metabolites, such as leukotriene B4, which are produced by the tumor.[98] TGF-β that is produced by B-regs has the potential to convert naive CD4+ T cells into FoxP3+ T cells, which inhibits the capacity of NK cells and effector CD8+ T cells to prevent the spread of malignant cells.[99,104] The fact that tumor-evoked B-regs initiate a cascade of events that results in immune suppression emphasizes the need of developing a way to control and eradicate these cells during therapeutic intervention. They will most likely continue to be induced until the cancer cells exist.

Moreover, immune cells in the vicinity oflymph nodes that drain tumors may also induce a B-reg phenotype. It has been shown that B cells become more "regulatory" in character due to the synthesis of IL-21 by Tregs, since IL-21 promotes the overexpression of IL-10, IDO, and granzyme-B through B cells. Granzyme-B, a serine protease, is often a crucial effector protein produced by cytotoxic NK cells, CD8+ and CD4+ CTLs, but it also has an immunomodulatory role. Granzyme B has reportedly been transferred to T cells by B-regs. This inhibits anti-tumor T cell activity by destroying the zeta (ζ) chain of T cell receptors without causing these cells to undergo apoptosis.[100]

In one of the mouse models of B16-F10 melanoma, phenotypic analysis of B lymphocytes that are preferentially aggregated in the tumor draining lymph nodes revealed the precise subset that is driving the remodeling of the tumor-draining lymph nodes and encouraging metastasis.[101] As a result, regulatory B cells form a significant portion of the lymphocyte population engaged in inhibiting anti-tumor immune responses. By generating TGF-β and turning resting CD4+ T cells into Tregs, these B cells either directly suppress anti-tumor immunity by secreting IL-10 or indirectly weaken immunological responses to tumors. There is growing evidence that B-regs are present in human tumors; while anti-CD20 and other chemicals can be used to deplete B cells more broadly, there are presently no treatment methods available to remove B-regs exclusively. Additionally, there are conflicting data on B cell depletion's ability to stop tumor growth in mice models. In one investigation, a solid tumor model's ability to grow was inhibited by B cell elimination through anti-CD20 antibody.[102] However, Rituximab for individuals with melanoma and renal cell carcinoma has not been shown to have any further effects in other investigations. Additionally, evidence from studies on mice suggests that depleting anti-CD20 can hinder anti-tumor immunity mediated by T cells[103] and potentially facilitate cancer escape because B-reg induced by tumors frequently express low levels of CD20.[105] By blocking Bruton's tyrosine kinase, ibrutinib reduces the number of B cells and their ability to survive. Reducing breast cancer's ability to spread and develop new tumors is one of ibrutinib's main effects.[106]

The presence of inhibitory receptors on the surface of T cells helps to control the immune system and prevent the development of autoimmune disorders. Tumors, however, can use these "checkpoints" as a means of protection from the immune system. Exhausted and nonresponsive tumor-specific T cells display high levels of inhibitory receptors such as CTLA4, PD1, and LAG3. Tumor-derived regulatory T cells are immunosuppressive and express a wide variety of checkpoint receptors. Therefore, modern advances in cancer immunotherapy are focused on overcoming this inhibitory receptor-mediated immune tolerance against cancer development.

6. T-lymphocyte inhibitory receptors and their role in suppressing tumor immunity

Signals like antigen recognition by T cells must first be recognized by the TCR and co-stimulation by variety of receptors that bind with corresponding ligands on APCs to begin an adaptive immune response. Under homeostatic conditions, costimulatory receptor activity must be balanced by signaling via inhibitory receptors (IRs) to guarantee a calibrated response that, without regulation, would lead to aggravated activation and emergence of autoimmune diseases. However, when cancer progresses, tumor-specific T lymphocytes exhibit increased, persistent expression of several inhibitory receptors, including but not limited to PD1, LAG3, and TIM3, which impairs their responsiveness and exhausts their functional capacity. These worn-out CD8+ TILs (tumor-infiltrating lymphocytes) are incapable of proliferating in response to antigen and minimal essential capabilities including cytotoxicity and cytokines production. The following immunological tolerance creates many barriers to tumor elimination, including infiltration of Treg cells into the tumor, co-inhibitory signaling through IRs, and production of suppressive cytokines such IL-10, TGF-β, and IL-35.

6.1 PD-1/PDL-1 pathway in suppressing tumor immunity

The PD-1/PD-L1 relationship regulates the establishment and maintenance of immune tolerance in the tumor microenvironment. Degeneration of the anti-tumor immune response in cancer is produced by the PD-1 receptor and its ligands, PD-L1 and PD-L2. This occurs because of the proliferation, activation, and release of cytotoxic T cells. Cancerous cells proliferate because PD-L1 suppresses the immune system's defense mechanism and is highly expressed on T cells that attack malignancies. Different signals generated by cancer cells may influence the PD-1/PD-L1 pathway, which plays a pivotal role in tumor development. Therefore, it is crucial to learn about this signaling network to enhance the current knowledge of the

PD-1/PD-L1 pathway's function in tumor formation and progression. Important for dampening immune responses and encouraging self-tolerance, PD-1 regulates T cell activity, prevents regulatory T cell apoptosis, and triggers the death of antigen-specific T cells. By dampening the body's immune response to tumor cells, PD-L1 also plays an important role in a variety of cancers.

Breast, colon, and squamous cell carcinomas are one of several type of tumor where PD-L1 overexpression dampens the anticancer immune response of T cells. *In-vivo* studies proven that that transgenic expression of PD-L1 promotes tumorigenesis and tumor invasion. Increased PD-L1 expression in malignancies makes them even more resistant to *in-vitro* specific CD8+ T cells-mediated lysis. There may be a relationship between a shift in PD-1 expression on CD4+ and CD8+ T lymphocytes and T effector cell exhaustion in the tumor microenvironment. In addition, cancer cells' ability to elude detection may be facilitated by innate and adaptive immune resistance.[107] PD-L1 may become activated in cancer cells as well as many immune cells, such as activated T cells, natural killer cells, myeloid suppressor cells, dendritic cells, macrophages, and B lymphocytes, when they are present in an inflammatory tumor microenvironment.[108]

By attaching to its receptors and turning on signaling pathways for proliferation and survival, PD-L1 works as a pro-tumorigenic factor in cancer cells. This observation added more support to the hypothesis that PD-L1 will play a vital role in the progression of tumors in the forthcoming years. In addition, it has been shown that the proliferative activities of PD-non-immune-L1 may have an effect on a variety of different kinds of tumor cells. For example, in renal cancer cells, PD-L1 triggered epithelial to mesenchymal transition (EMT), which indicates that the presence of the intrinsic pathway of PD-L1 promotes the progression of kidney cancer. Stem cell-like phenotypes were also triggered by PD-L1, which shows that PD-L1 is involved in the progression of kidney cancer.[109] Given that it may be both helpful and detrimental, PD-1 performs two conflicting tasks. Regarding its advantageous effects, it is crucial for minimizing the control of immunological responses that are damaging or ineffectual as well as for preserving immune tolerance. On the other hand, PD-1 inhibits the protective immune response, which in turn encourages malignant cells to expand and become more visible.

CD8+ T cells, production of Th1 cytokines and chemical factors, interferons, and certain gene expression patterns are all associated with high PD-L1 levels.[108] Tumor cells produce PD-L1 as a "adaptive immune strategy" to evade anti-tumor response.[107] In ovarian cancer cells, it has been found that IFN-γ, upregulates PDL-1, which is important for the course of the illness. Conversely, it has been found that blocking IFN-γ, receptor 1 in mice models of AML (acute myeloid leukemia) can weaken the expression of PD-L1.[110] IFN-γ in turn induces protein kinase D2 (PKD2), which plays a crucial role in modulating PD-L1 expression. Abridged PD-L1 expression with PKD2 inhibition leads to an effective anti-tumor immune response. The IFN-γ secreted by natural killer cells and T lymphocytes seems to increase the production of PD-L1 on the surface of targeted cells, including tumor cells.[111]

Agents that block PD1/PD-L1 can improve tumor immunity by focusing on the immunological changes that a tumor's microenvironment elicits. In particular, PD-1/PD-L1 inhibitors augmented the cytolytic activity of T cells that are specific for tumor and decreased the production of the suppressive cytokine IL-10. On the other hand, this production of pro-inflammatory cytokines, promoted the presence of T-effs, and reduced the quality of Tregs as well as their ability to suppress the growth of tumors.

6.1.1 Effect of signaling pathway on PD-1/PD-L1 expression in cancer

The PD-1/PD-L1 axis, which plays a crucial role in the process of carcinogenesis, may be modulated in cancer cells by a diverse range of signals. Therefore, in order to facilitate our present level of comprehension, it is necessary to understand the complex signaling network that exists inside of the cells.

6.1.1.1 JAK/STAT pathways

Certain cytokines, interferons, growth factors, and other related chemicals employ a signaling mechanism called the JAK/STAT system, which is highly conserved throughout evolution.[112] Extracellular influences may regulate gene expression using this route. Thus, it may serve as a simple illustration of how cells interpret information from their surroundings to control their development and differentiation.[113] Toshifumi et al. found that blocking the JAK/STAT pathway reduced PD-L1 expression at the mRNA and protein levels, indicating the significance of these pathways in regulating PD-L1 expression. Interferon, produced by NK cells via the JAK and STAT pathway, increases expression of PD-L1 on the surface of tumor cells.[114] Furthermore, studies have shown that JAK/STAT-IRF1 driven IFN-γ generating T cells may regulate expression of ligand PD-L1 in melanoma cells. The activation of JAK/STAT3 signaling and upregulation of

PD-L1 expression *in-vitro* were also linked to FGFR2 signaling and over-expression of theses signaling mouse model of colorectal cancer led to elevated levels of the ligand PD-L1 and a growth in tumor size.

6.1.1.2 MAPK signaling pathway

An essential part of the signal transduction system, the mitogen-activated protein kinase (MAPK) signaling pathway is involved in the translation of extracellular signals into cellular responses. Phosphorylation activation of the MAPK pathway has a crucial role in regulating cell proliferation, differentiation, invasion, metastasis, and death.[115] Recent research has placed a greater emphasis on determining the nature of the relationship that exists between the MAPK pathway and the PD-1/PD-L1 axis.[116] In another study, researchers found that inhibiting the MAPK pathway prevented the mRNA and protein production of PD-L1 in lung cancer cells, which was activated by the growth hormones transforming growth factor beta (TGF-β) and interferon beta (IFN-β). The use of anti-PD-L1 antibody was also shown to decrease the generation of p-P38 and p-ERK in all HL lines, and this antibody was found to be linked to components involved in MAPK signaling in Hodgkin's lymphoma (HL) cells by Jalali et al. Blocking the MAPK pathway molecule MEK1/2 has a similar effect on PD-L1 expression in renal cell cancer.[117]

6.1.1.3 NF-κB signaling pathway

Recent research has shown that the TLRs or IFN-γ, driven nuclear factor NF-κB can induce PD-L1 gene expression. Together, anti-CTLA4 checkpoint suppression treatment and along with NF-κB inhibitor able to suppress the proliferation of breast cancer, colon carcinoma, and melanoma cell lines, suggesting that NF-κB inhibition may target both tumor immune checkpoints, tumor cell proliferation, and survival.[118] Studies indicate that, probably NF-κB has role in LMP1-induced PD-L1 expression. NF-κB also plays a critical role as a mediator between IFN-γ and PD-L1 expression. When PD-L1 expression was stimulated by interferon beta (IFN-β), blocking it required NF-κB inhibitors but not MAPK, PI3K, or STAT3 inhibitors.[119] Also, Peng et al. have found that ovarian cancer treatment causes local immune suppression by upregulating PD-L1 through the NF-κB pathway.[120]

6.1.2 Role of PD-1/PD-L1 integration in different cancers

In a brief, the PD-1/PD-L1 axis is primarily responsible for cancer immune evasion and has major implications for cancer therapy. To improve the specificity and efficacy of the immune system's capacity to combat cancer, it is now essential to get a comprehension of the PD-1/PD-L1 axis.

6.1.2.1 Breast cancer

Recent studies have shown a relationship between PD-L1 expression and EMT in human breast cancer stem cells. According to recent pathological research, PD-L1 expression is notably greater in numerous subtypes of estrogen receptor (ER) α-negative breast cancer. Average PD-L1 mRNA levels were significantly lower in ER α-positive than in ER α-negative breast cancer cell lines, as reported by Lu et al.[121] Remarkably, new studies show that PD-L1 may slow the expansion of cancer cells, and that silencing of PD-L1 can promote both spontaneous and drug-induced apoptotic cell death in breast cancer cells.

6.1.2.2 Lung cancer

It has been proposed that PD-1's capacity to inhibit lung cancer growth can be improved by anti-C5a therapy.[122] PD-L1 has been shown to have a significant correlation with a lung cancer that includes squamous cell carcinoma and a wild-type mutation in the gene that codes for epidermal growth factor receptor.[123] The presence of vascular invasion, smoking, and later stages of cancer were also linked to a higher likelihood of death in males. Patients who tested positive for PD-L1 had a worse prognosis than those who tested negative, as shown by both univariate and multivariate survival analyses.

6.1.2.3 Colorectal cancer

Worsen clinical outcomes have been linked to the expression of the PD-L1 ligand in colorectal cancers and lower differentiation and right colon CRCs were associated with a dismal outcome. The synthesis of PD-L1 is controlled by a wide variety of inflammatory cytokines; when PD-1 and PD-L1 interact, it triggers active T cell apoptosis and the release of IL-10 as a counteracting mechanism.[124] Metastatic CRC is more likely to express PD-L1, which can aid tumor cells in evading immune surveillance and improve Treg function in these cancers.

6.1.2.4 Bladder cancer

Bladder cancer mortality is correlated with the PD-L1 expression. Previous studies shown tumors with a high PD-1 expression have been linked to a greater risk of metastasis and worse prognosis.[125] Additionally, enhanced resistance to BCG treatment is also induced by PD-L1 expression on tumor cells, which is believed to be connected with immune system inactivation.[112] According to Zhu et al., autophagy protein ATG 7 regulates PD-1 expression in bladder cancer (BC) and displayed that overexpression of ATG7 boosted levels of PD-L1, mostly promoting autophagic degradation of FOXO3a.[126] These findings demonstrate the potential of combination treatment through PD-1/PD-L1 immune checkpoint inhibitors and autophagy inhibitors to enhance the efficacy of BCG in humans.

6.1.2.5 Pancreatic cancer

Individuals with pancreatic cancer, PD-L1 has emerged as a promising new prognostic indicator. Some studies have shown that blocking PD-L1 efficiently suppresses established pancreatic cancer in a rat model by increasing IFN-γ and decreasing IL-10 levels.[127] Furthermore, in PD-L1 positive cancers, there are more tumor-infiltrating Treg than in PD-L1 negative tumors. These findings support future studies aiming to target the PD-1 and PD-L1 pathways in pancreatic cancer.[128] It was also shown that patients with pancreatic ductal adenocarcinoma (PDAC) had considerably higher levels of PD-1 expression on peripheral CD8+ T lymphocytes compared to patients with intraductal papillary mucinous neoplasm (IPMN) or healthy doners. By using ROC analysis, we were able to further confirm PD-1 expression's diagnostic value in PDAC.[129]

6.2 CTLA 4

CTLA4, belong to the B7/CD28 family of transmembrane glycoproteins, suppresses T cells activities. Since it is raised on activated T cells (including memory T cells) and generated constitutively on regulatory T cells, it serves a critical function in preserving cell-intrinsic immune control and peripheral tolerance. CTLA4, which belongs to the immunoglobin superfamily (IgSF), was found by means of differential screening of a cDNA library encoding genes for murine cytolytic T cells. CTLA4 and CD28 share structural similarities with the T cell but differ markedly in their functions. A halt signal brought on by TCR ligation that would typically permit the development of an immunological synapse between T lymphocytes and APCs is overridden by CTLA4. CTLA4+ T cells, as shown by *in-vivo* imaging studies, are unable to maintain prolonged interaction with APCs and consequently attenuate T cell activation. With the help of anti-CTLA4 mAb, Treg cells expressing CTLA4 have been demonstrated to suppress CD80/86 on APCs, which is another way T cell activation is constrained. Additionally, it has been proposed that CTLA4 uses a trans-endocytosis method to bind and remove these ligands.[130]

6.2.1 Role of CTLA-4 in suppressing cancer immunity

Initial studies in mouse melanoma models have demonstrated that CTLA4 inhibition results in tumor shrinkage, indicating the potential of inhibitory receptor modification as an immunotherapeutic approach. Ipilimumab, a fully humanized mAb IgG1, was the first mAb to be used in a clinical setting to target CTLA4 for cancer treatment. CTLA4 is activated upon TCR stimulation and inhibits cytokine production and T cell proliferation by binding to B7 than T lymphocyte receptor CD28. At the earliest phases of tumor development, CTLA4 suppresses T cell activation to dampen the immune response. In addition, CTLA4 induces B7-dependent signal reversal and the tryptophan-degrading enzyme indoleamine-2, 3-dioxygenase (IDO), both of which suppress T cell growth.

Additionally, it has been shown that CTLA4 may both upregulate the expression of the Casitas-B-lineage lymphoma (Cbl)-b protein and reduce the synthesis of the zeta-associated protein of 70 kDa (ZAP70), both of which are required to restrict the activation of T cells. Recent research has shown that CTLA4 inhibits NF-kB, PI3K/Akt pathways, cyclin D3, and cyclin-dependent kinases such as cdk4 and cdk6, among other cyclin-dependent kinases and pathways. Even though CTLA4 has been the focus of substantial research, it is still necessary to better investigate how it interacts with its ligands or downstream targets (Fig. 4).[131]

6.2.2 Effect of signaling Wnt/β-catenin pathways on expression of CTLA4 in cancer

Wnt induces the expression of CTLA4 gene is melanoma and Wnt/β-catenin signaling plays a significant role in melanoma development, Wnts are a group of highly conserved cysteine-rich proteins that bind to Frizzled receptors and co-receptors LRP-5/6.[132] Upon binding they activate a cascade of events that results in the accumulation and translocation of β-catenin to the nucleus. Both at the mRNA and protein levels, CTLA4 is greatly increased in human melanoma patients and functions as a direct transcriptional target for Wnt/β-catenin signaling. The activity of regulatory T cells may be modulated by CTLA4. The main function of Tregs, a subset of T cells, is to decrease host immune responses in order

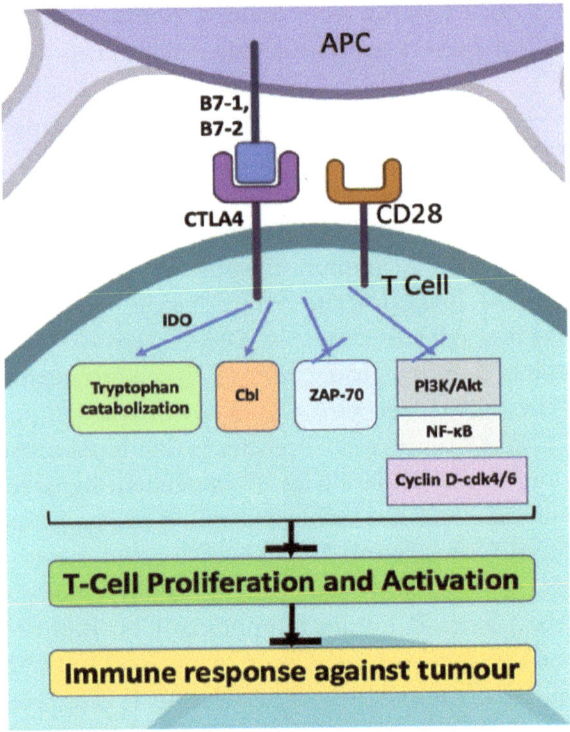

Fig. 4 CTLA4 model establishes its role in the development of cancer. CTLA4 and CD28 share the B7 ligands B7-1 (CD80) and B7-2 (CD86), both of which inhibit T cell activation. CTLA4 is up-regulated in response to T cell receptor (TCR) activation; this in turn stimulates the enzyme IDO, Cbl-b protein, and decreases ZAP 70 production. In addition, CTLA4 generates inhibitory signals that restrict T cell activation by downregulating PI3K/Akt, cyclin D3, cdk4/cdk6, and NF-B. This dampens the anti-tumor immune response.[131] *Adapted from Zhao Y, Yang W, Huang Y, Cui R, Li X, Li B. Evolving roles for targeting CTLA-4 in cancer immunotherapy.* Cell Physiol Biochem. *2018;47(2):721–734. https://doi.org/10.1159/000490025.*

to prevent autoimmune disorders and support tumor immunity.[133] This discovery may shed light on the function of CTLA4 induction in melanoma cells mediated by the Wnt/β-catenin pathway.

6.2.3 Role and function of CTLA4 in melanoma

One of the deadliest forms of cancer, melanoma accounts for just 5% of all skin malignancies yet is responsible for the vast majority of deaths from this disease worldwide.[105] Melanoma stem cells (MSCs) are thought to have the ability to self-renew, differentiate, resist chemotherapeutics, and evade the immune system.[134] In addition to its pro-tumorigenic effect, CTLA4 also induces tumorigenesis and prevents apoptosis in melanoma. The current understanding of the mechanism behind melanoma progression is significantly improved by the discovery of CTLA4 driven carcinogenesis in that it may help CTLA4 targeted therapy by explaining the long-term prognosis in advanced melanoma patients.[131]

6.3 LAG3/CD223

Lymphocyte activation gene 3, also known as LAG3/CD223 LAG3, an inhibitory receptor that has a vague resemblance to CD4, has four Ig-like domains and binds to MHC class II molecules with a greater affinity than CD4. LAG3 also has a comparable function. In a manner similar to that of PD-1, LAG3 exhibits a broad variety of expression patterns. It is expressed constitutively on Treg cells, in addition to being increased on activated T cells and natural killer (NK) cells.[135,136] Plasmacytoid DCs are the ones that express LAG3, as opposed to myeloid DCs, which show PD-1 expression.[137] LAG3 has been demonstrated to adversely influence mouse memory T cell pool size, homeostatic expansion, and T cell activation and proliferation.[136,138] Its exact mode of operation is still unknown. Although it seems that the presence of a KIEELE motif in the cytoplasmic tail is required for its inhibitory function, LAG3 does not include the normal inhibitory motifs.[138] Although soluble LAG3 does not seem to have any biological role in mice, LAG3 function appears to be controlled by cell surface cleavage by ADAM10 and ADAM17 disintegrin/metalloproteases.[139] Last but not least, LAG3 is necessary for both peripherally generated and thymus-derived Treg cells to exert their maximum suppressive function.[135,136] LAG3 is a marker for a severely fatigued T cell phenotype and has been shown to co-express on TILs with other IRs, such as PD-1.[140] Since CD8+ TIL malfunction in patient tumor samples was also discovered to be correlated with increased expression of LAG3, co-expressed with PD-1, clinical studies with LAG3 mAb (BMS-986016) began in 2013. They could offer some preliminary understanding into the potential efficacy of LAG3 inhibition.[141]

7. Concluding remark

In the current era, cancers are associated with high mortality rates and majority of the cancers are highly resistant to radio and chemotherapy. With the emergence of these circumstances, it is becoming imperative to come up with better therapeutic interventions that aim to enhance anti-tumor immune responses in cancer patients. Since invertebrates without B and T lymphocytes have a low prevalence of carcinogenesis, innate immune cells are also crucial in preventing the onset and progression of cancer. Here macrophages can play an imperative role in being used as a target for immunotherapy for cancer by various ways. Primary: The recruitment of macrophages can be inhibited by utilizing inhibitors or antibodies against them or some of their ligands (such as CCL2, CCL5, VEGF, and CSF1) and reduced angiogenesis by targeting Nrp1 and ANG 2 can also have the same effect. Secondary: decrease in macrophage survival. Because CSF1 sends a signal to differentiate macrophages, CSF1 inhibitors can prevent the development of new macrophages. Trabectedin's binding to TRAIL-2/3 receptors, immunotoxins that target SR-A or folate receptor B (FR), and other substances can deplete TAMs and cause apoptosis, reducing the number of macrophages.[133] Inhibition of angiogenesis, which prevents tumor-promoting processes. Tertiary: Dismantling the macrophage blockade involves preventing connections between tumor-associated CD47 and macrophage-associated SIRP-α, which aids tumor cells in avoiding phagocytosis. Quaternary: Repolarization induction: M1 polarization is linked to pro-tumor actions. Numerous factors can cause M1 polarization, Quinary: Targeting receptors like CD19, HER2, and mesothelin to modify effector cells.[134]

Currently, a number of human tumors are being successfully treated using immunotherapy. After becoming activated, effector T cells (Teffs) release CTLA4, a protein that prevents T cells from proliferating, progressing through the cell cycle, and producing cytokines, including IL-2 and IFN-γ. Numerous studies suggest that both hematological and solid cancers benefit from persistent tumor expression of CTLA4 and PD-1, which generates inhibitory signals to dampen the immune response and promotes tumor growth. However, PD-1 is an important barrier to immunological responses and self-tolerance because it controls T cell activity, induces the death of antigen-specific T cells, and inhibits the apoptosis of regulatory T cells.

Two inhibitory checkpoints that are frequently seen on activated T cells, CTLA4 and PD-1, can prove to be the most potential targets for the treatment of several cancer. Numerous human malignancies can be treated by inhibiting these inhibitory receptors on T cells with different antibodies. Anti-CTLA4 immunomodulating monoclonal antibodies (mAb) like ipilimumab and tremelimumab (Table 2) are used alone or in combination with chemotherapy, vaccination, or other antibodies in preclinical and

Table 2 Clinical trials using CTLA4-blocking antibodies alone or with additional antibodies in human cancer.[131]

Treatment	Trial phase	Type of cancer	Patient number	Doses (mg/kg)
Ipilimumab	II	Advance melanoma	217	0.3, 3, 10
Ipilimumab	II	Advance melanoma	155	10
Ipilimumab	III	Advance melanoma	951	10
Ipilimumab	II	Metastatic renal cell carcinoma	61	3 mg/kg followed by 1 mg/kg, 3 mg/kg
Ipilimumab	III	Prostate cancer (castration resistant)	799	10
Ipilimumab	I	Non-Hodgkin lymphoma	18	1, 3
Tremelimumab	II	Malignant mesothelioma	29	15
Tremelimumab	II	Melanoma	251	15
Tremelimumab	II	Refractor metastatic colorectal cancer	47	15
Ipilimumab + Etoposide/carboplatin	II	Extensive-stage small cell lung cancer	42	10
Ipilimumab + Paclitaxel/carboplatin	I	Non-small cell lung cancer	15	3, 10
Ipilimumab + Radiation therapy	I/II	Prostate cancer (castration resistant)	83	3, 5, 10
	I	Solid tumors	35	3
Ipilimumab alone/+nivolumab (PD-1 antibody)	II	Advance melanoma	142	3

Adapted from Zhao Y, Yang W, Huang Y, Cui R, Li X, Li B. Evolving roles for targeting CTLA-4 in cancer immunotherapy. *Cell Physiol Biochem*. 2018;47(2):721–734. https://doi.org/10.1159/000490025.

clinical studies for cancer (anti-PD-1 or anti-OX40). Patient survival has been demonstrated to improve with the use of CTLA4 and PD-1 checkpoint inhibitors in studies of melanoma, renal cell carcinoma, squamous cell carcinoma, and breast cancer.[131]

Although immunomodulatory targeting therapies have increased the average life expectancy in cancer patients, mortality rates are still higher in patients with advanced cancer stage. Therefore, a better understanding of the mechanism of the currently available drugs, cancer formation, their markers and functioning can pave the way for further innovation in the field. These avenues can then open up ways to specifically target those patients that respond well to appropriate biomarker screening procedures and also to combine these checkpoint inhibitors with other available complimentary drugs for the treatment of advanced stage patients thus ultimately increasing the life expectancy.

References

1. Gonzalez H, Hagerling C, Werb Z. Roles of the immune system in cancer: from tumor initiation to metastatic progression. *Genes Dev*. 2018;32(19–20):1267–1284. https://doi.org/10.1101/gad.314617.118.
2. Van Overmeire E, Laoui D, Keirsse J, Van Ginderachter JA, Sarukhan A. Mechanisms driving macrophage diversity and specialization in distinct tumor microenvironments and parallelisms with other tissues. *Front Immunol*. 2014;5. https://doi.org/10.3389/fimmu.2014.00127.
3. Coussens LM, Werb Z. Inflammation and cancer. *Nature*. 2002;420(6917):860–867. https://doi.org/10.1038/nature01322.
4. Teng MWL, Galon J, Fridman W-H, Smyth MJ. From mice to humans: developments in cancer immunoediting. *J Clin Invest*. 2015;125(9):3338–3346. https://doi.org/10.1172/JCI80004.
5. Malgulwar PB, et al. Transcriptional co-expression regulatory network analysis for snail and slug identifies *IL1R1*, an inflammatory cytokine receptor, to be preferentially expressed in ST-EPN-*RELA* and PF-EPN-A molecular subgroups of intracranial ependymomas. *Oncotarget*. 2018;9(84):35480–35492. https://doi.org/10.18632/oncotarget.26211.
6. Qian B-Z, Pollard JW. Macrophage diversity enhances tumor progression and metastasis. *Cell*. 2010;141(1):39–51. https://doi.org/10.1016/j.cell.2010.03.014.
7. Bingle L, Brown NJ, Lewis CE. The role of tumour-associated macrophages in tumour progression: implications for new anticancer therapies. *J Pathol*. 2002;196(3):254–265. https://doi.org/10.1002/path.1027.
8. Kelly P, Davison R, Bliss E, McGee J. Macrophages in human breast disease: a quantitative immunohistochemical study. *Br J Cancer*. 1988;57(2):174–177. https://doi.org/10.1038/bjc.1988.36.
9. Zhang Q, et al. Prognostic significance of tumor-associated macrophages in solid tumor: a meta-analysis of the literature. *PLoS One*. 2012;7(12), e50946. https://doi.org/10.1371/journal.pone.0050946.
10. Murray PJ, Wynn TA. Protective and pathogenic functions of macrophage subsets. *Nat Rev Immunol*. 2011;11(11):723–737. https://doi.org/10.1038/nri3073.

11. Rohan TE, et al. Tumor microenvironment of metastasis and risk of distant metastasis of breast cancer. *J Natl Cancer Inst*. 2014;106(8). https://doi.org/10.1093/jnci/dju136.
12. Mantovani A, Allavena P, Sica A, Balkwill F. Cancer-related inflammation. *Nature*. 2008;454(7203):436–444. https://doi.org/10.1038/nature07205.
13. Mantovani A, Sozzani S, Locati M, Allavena P, Sica A. Macrophage polarization: tumor-associated macrophages as a paradigm for polarized M2 mononuclear phagocytes. *Trends Immunol*. 2002;23(11):549–555. https://doi.org/10.1016/S1471-4906(02)02302-5.
14. Gordon S, Taylor PR. Monocyte and macrophage heterogeneity. *Nat Rev Immunol*. 2005;5(12):953–964. https://doi.org/10.1038/nri1733.
15. Mantovani A, Sica A, Sozzani S, Allavena P, Vecchi A, Locati M. The chemokine system in diverse forms of macrophage activation and polarization. *Trends Immunol*. 2004;25(12):677–686. https://doi.org/10.1016/j.it.2004.09.015.
16. Wang Y-C, et al. Notch signaling determines the M1 versus M2 polarization of macrophages in antitumor immune responses. *Cancer Res*. 2010;70(12):4840–4849. https://doi.org/10.1158/0008-5472.CAN-10-0269.
17. Murray PJ, et al. Macrophage activation and polarization: nomenclature and experimental guidelines. *Immunity*. 2014;41(1):14–20. https://doi.org/10.1016/j.immuni.2014.06.008.
18. Moussai D, et al. The human cutaneous squamous cell carcinoma microenvironment is characterized by increased lymphatic density and enhanced expression of macrophage-derived VEGF-C. *J Invest Dermatol*. 2011;131(1):229–236. https://doi.org/10.1038/jid.2010.266.
19. Aspord C, et al. Breast cancer instructs dendritic cells to prime interleukin 13–secreting CD4+ T cells that facilitate tumor development. *J Exp Med*. 2007;204(5):1037–1047. https://doi.org/10.1084/jem.20061120.
20. Fukuda A, et al. Stat3 and MMP7 contribute to pancreatic ductal adenocarcinoma initiation and progression. *Cancer Cell*. 2011;19(4):441–455. https://doi.org/10.1016/j.ccr.2011.03.002.
21. Mantovani A. Macrophages, neutrophils, and cancer: a double edged sword. *New J Sci*. 2014;2014:1–14. https://doi.org/10.1155/2014/271940.
22. Wang M, et al. Role of tumor microenvironment in tumorigenesis. *J Cancer*. 2017;8(5):761–773. https://doi.org/10.7150/jca.17648.
23. Sica A, Mantovani A. Macrophage plasticity and polarization: in vivo veritas. *J Clin Invest*. 2012;122(3):787–795. https://doi.org/10.1172/JCI59643.
24. Xue J, et al. Transcriptome-based network analysis reveals a spectrum model of human macrophage activation. *Immunity*. 2014;40(2):274–288. https://doi.org/10.1016/j.immuni.2014.01.006.
25. Singh A, et al. Bcl2 negatively regulates protective immune responses during *Mycobacterial* infection. *Biomol Concepts*. 2021;12(1):94–109. https://doi.org/10.1515/bmc-2021-0010.
26. Laoui D, et al. Mononuclear phagocyte heterogeneity in cancer: different subsets and activation states reaching out at the tumor site. *Immunobiology*. 2011;216(11):1192–1202. https://doi.org/10.1016/j.imbio.2011.06.007.
27. Schouppe E, De Baetselier P, Van Ginderachter JA, Sarukhan A. Instruction of myeloid cells by the tumor microenvironment: open questions on the dynamics and plasticity of different tumor-associated myeloid cell populations. *OncoImmunology*. 2012;1(7):1135–1145. https://doi.org/10.4161/onci.21566.
28. Dirkx AEM, Egbrink MGAO, Wagstaff J, Griffioen AW. Monocyte/macrophage infiltration in tumors: modulators of angiogenesis. *J Leukoc Biol*. 2006;80(6):1183–1196. https://doi.org/10.1189/jlb.0905495.

29. Kim J, Bae J-S. Tumor-associated macrophages and neutrophils in tumor microenvironment. *Mediators Inflamm.* 2016;2016:6058147. https://doi.org/10.1155/2016/6058147.
30. Duluc D, et al. Interferon-γ reverses the immunosuppressive and protumoral properties and prevents the generation of human tumor-associated macrophages. *Int J Cancer.* 2009;125(2):367–373. https://doi.org/10.1002/ijc.24401.
31. De Palma M, et al. Tumor-targeted interferon-α delivery by Tie2-expressing monocytes inhibits tumor growth and metastasis. *Cancer Cell.* 2008;14(4):299–311. https://doi.org/10.1016/j.ccr.2008.09.004.
32. Belgiovine C, Digifico E, Anfray C, Ummarino A, Torres Andón F. Targeting tumor-associated macrophages in anti-cancer therapies: convincing the traitors to do the right thing. *J Clin Med.* 2020;9(10):3226. https://doi.org/10.3390/jcm9103226.
33. Metcalf S, Pandha HS, Morgan R. Antiangiogenic effects of zoledronate on cancer neovasculature. *Future Oncol Lond Engl.* 2011;7(11):1325–1333. https://doi.org/10.2217/fon.11.113.
34. Lin EY, et al. Macrophages regulate the angiogenic switch in a mouse model of breast cancer. *Cancer Res.* 2006;66(23):11238–11246. https://doi.org/10.1158/0008-5472.CAN-06-1278.
35. Lin EY, Pollard JW. Tumor-associated macrophages press the angiogenic switch in breast cancer. *Cancer Res.* 2007;67(11):5064–5066. https://doi.org/10.1158/0008-5472.CAN-07-0912.
36. Lin Y, Xu J, Lan H. Tumor-associated macrophages in tumor metastasis: biological roles and clinical therapeutic applications. *J Hematol Oncol.* 2019;12(1):76. https://doi.org/10.1186/s13045-019-0760-3.
37. Jaynes JM, et al. Mannose receptor (CD206) activation in tumor-associated macrophages enhances adaptive and innate antitumor immune responses. *Sci Transl Med.* 2020;12, eaax6337. https://doi.org/10.1126/scitranslmed.aax6337.
38. Dangaj D, et al. Mannose receptor (MR) engagement by mesothelin GPI anchor polarizes tumor-associated macrophages and is blocked by anti-MR human recombinant antibody. *PLoS One.* 2011;6(12), e28386. https://doi.org/10.1371/journal.pone.0028386.
39. Marttila-Ichihara F, et al. Macrophage mannose receptor on lymphatics controls cell trafficking. *Blood.* 2008;112(1):64–72. https://doi.org/10.1182/blood-2007-10-118984.
40. Allavena P, et al. Engagement of the mannose receptor by tumoral mucins activates an immune suppressive phenotype in human tumor-associated macrophages. *Clin Dev Immunol.* 2010;2010:1–10. https://doi.org/10.1155/2010/547179.
41. Minami K, et al. Prognostic significance of CD68, CD163 and folate receptor-β positive macrophages in hepatocellular carcinoma. *Exp Ther Med.* 2018. https://doi.org/10.3892/etm.2018.5959.
42. Komohara Y, Ohnishi K, Kuratsu J, Takeya M. Possible involvement of the M2 anti-inflammatory macrophage phenotype in growth of human gliomas. *J Pathol.* 2008;216(1):15–24. https://doi.org/10.1002/path.2370.
43. Plüddemann A, et al. SR-A, MARCO and TLRs differentially recognise selected surface proteins from *Neisseria meningitidis*: an example of fine specificity in microbial ligand recognition by innate immune receptors. *J Innate Immun.* 2009;1(2):153–163. https://doi.org/10.1159/000155227.
44. La Fleur L, et al. Targeting MARCO and IL37R on immunosuppressive macrophages in lung cancer blocks regulatory T cells and supports cytotoxic lymphocyte function. *Cancer Res.* 2021;81(4):956–967. https://doi.org/10.1158/0008-5472.CAN-20-1885.
45. Shi B, et al. The scavenger receptor MARCO expressed by tumor-associated macrophages are highly associated with poor pancreatic cancer prognosis. *Front Oncol.* 2021;11, 771488. https://doi.org/10.3389/fonc.2021.771488.

46. Lai C, Lemke G. An extended family of protein-tyrosine kinase genes differentially expressed in the vertebrate nervous system. *Neuron*. 1991;6(5):691–704. https://doi.org/10.1016/0896-6273(91)90167-X.
47. Linger RMA, et al. Mer or Axl receptor tyrosine kinase inhibition promotes apoptosis, blocks growth and enhances chemosensitivity of human non-small cell lung cancer. *Oncogene*. 2013;32(29):3420–3431. https://doi.org/10.1038/onc.2012.355.
48. Png KJ, Halberg N, Yoshida M, Tavazoie SF. A microRNA regulon that mediates endothelial recruitment and metastasis by cancer cells. *Nature*. 2012;481(7380): 190–194. https://doi.org/10.1038/nature10661.
49. Quinn JM, et al. Therapeutic inhibition of the receptor tyrosine kinase AXL improves sensitivity to platinum and taxane in ovarian cancer. *Mol Cancer Ther*. 2019;18 (2):389–398. https://doi.org/10.1158/1535-7163.MCT-18-0537.
50. Tworkoski K, et al. Phosphoproteomic screen identifies potential therapeutic targets in melanoma. *Mol Cancer Res*. 2011;9(6):801–812. https://doi.org/10.1158/1541-7786. MCR-10-0512.
51. Bosurgi L, et al. Paradoxical role of the proto-oncogene Axl and Mer receptor tyrosine kinases in colon cancer. *Proc Natl Acad Sci USA*. 2013;110(32):13091–13096. https:// doi.org/10.1073/pnas.1302507110.
52. Lee-Sherick AB, et al. Aberrant Mer receptor tyrosine kinase expression contributes to leukemogenesis in acute myeloid leukemia. *Oncogene*. 2013;32(46):5359–5368. https:// doi.org/10.1038/onc.2013.40.
53. Lew ED, et al. Differential TAM receptor–ligand–phospholipid interactions delimit differential TAM bioactivities. *eLife*. 2014;3, e03385. https://doi.org/10.7554/eLife. 03385.
54. Graham DK, DeRyckere D, Davies KD, Earp HS. The TAM family: phosphatidylserine-sensing receptor tyrosine kinases gone awry in cancer. *Nat Rev Cancer*. 2014;14(12):769–785. https://doi.org/10.1038/nrc3847.
55. Aehnlich P, Powell RM, Peeters MJW, Rahbech A, Thor Straten P. TAM receptor inhibition-implications for cancer and the immune system. *Cancer*. 2021;13(6):1195. https://doi.org/10.3390/cancers13061195.
56. Colotta F, Allavena P, Sica A, Garlanda C, Mantovani A. Cancer-related inflammation, the seventh hallmark of cancer: links to genetic instability. *Carcinogenesis*. 2009;30 (7):1073–1081. https://doi.org/10.1093/carcin/bgp127.
57. Clark CE, Hingorani SR, Mick R, Combs C, Tuveson DA, Vonderheide RH. Dynamics of the immune reaction to pancreatic cancer from inception to invasion. *Cancer Res*. 2007;67(19):9518–9527. https://doi.org/10.1158/0008-5472.CAN-07-0175.
58. Balkwill F, Charles KA, Mantovani A. Smoldering and polarized inflammation in the initiation and promotion of malignant disease. *Cancer Cell*. 2005;7(3):211–217. https:// doi.org/10.1016/j.ccr.2005.02.013.
59. Ruffell B, DeNardo DG, Affara NI, Coussens LM. Lymphocytes in cancer development: Polarization towards pro-tumor immunity. *Cytokine Growth Factor Rev*. 2010;21(1):3–10. https://doi.org/10.1016/j.cytogfr.2009.11.002.
60. Singh A, et al. Deciphering the role of calcium homeostasis in T cells functions during mycobacterial infection. *Cell Immunol*. 2020;357, 104198. https://doi.org/10.1016/j. cellimm.2020.104198.
61. Chaplin DD. Overview of the immune response. *J Allergy Clin Immunol*. 2010;125(2): S3–S23. https://doi.org/10.1016/j.jaci.2009.12.980.
62. Godfrey DI, Berzins SP. Control points in NKT-cell development. *Nat Rev Immunol*. 2007;7(7):505–518. https://doi.org/10.1038/nri2116.
63. Carding SR, Egan PJ. γδ T cells: functional plasticity and heterogeneity. *Nat Rev Immunol*. 2002;2(5):336–345. https://doi.org/10.1038/nri797.

64. Dunn GP, Old LJ, Schreiber RD. The three Es of cancer immunoediting. *Annu Rev Immunol.* 2004;22(1):329–360. https://doi.org/10.1146/annurev.immunol.22.012703.104803.
65. de Visser KE, Korets LV, Coussens LM. De novo carcinogenesis promoted by chronic inflammation is B lymphocyte dependent. *Cancer Cell.* 2005;7(5):411–423. https://doi.org/10.1016/j.ccr.2005.04.014.
66. Patel P, et al. Incidence of types of cancer among HIV-infected persons compared with the general population in the United States, 1992–2003. *Ann Intern Med.* 2008;148(10):728. https://doi.org/10.7326/0003-4819-148-10-200805200-00005.
67. Verma C, et al. Tuberculosis and tuberculosis-associated immune reconstitution inflammatory syndrome in HIV: immunological review of manifestation and immunopathogenesis. *HIV AIDS Rev Int J HIV-Relat Probl.* 2020;19(2):67–73. https://doi.org/10.5114/hivar.2020.96486.
68. Boshoff C, Weiss R. Aids-related malignancies. *Nat Rev Cancer.* 2002;2(5):373–382. https://doi.org/10.1038/nrc797.
69. Fung J. De novo malignancies after liver transplantation: a major cause of late death. *Liver Transpl.* 2001;7(11):S109–S118. https://doi.org/10.1053/jlts.2001.28645.
70. Herrero JI, et al. De novo neoplasia after liver transplantation: an analysis of risk factors and influence on survival. *Liver Transpl.* 2005;11(1):89–97. https://doi.org/10.1002/lt.20319.
71. Vivier E, Tomasello E, Baratin M, Walzer T, Ugolini S. Functions of natural killer cells. *Nat Immunol.* 2008;9(5):503–510. https://doi.org/10.1038/ni1582.
72. Ouyang W, Kolls JK, Zheng Y. The biological functions of T helper 17 cell effector cytokines in inflammation. *Immunity.* 2008;28(4):454–467. https://doi.org/10.1016/j.immuni.2008.03.004.
73. Bettelli E, Korn T, Oukka M, Kuchroo VK. Induction and effector functions of TH17 cells. *Nature.* 2008;453(7198):1051–1057. https://doi.org/10.1038/nature07036.
74. Kryczek I, Wei S, Szeliga W, Vatan L, Zou W. Endogenous IL-17 contributes to reduced tumor growth and metastasis. *Blood.* 2009;114(2):357–359. https://doi.org/10.1182/blood-2008-09-177360.
75. Muranski P, et al. Tumor-specific Th17-polarized cells eradicate large established melanoma. *Blood.* 2008;112(2):362–373. https://doi.org/10.1182/blood-2007-11-120998.
76. Spolski R, Leonard WJ. Interleukin-21: basic biology and implications for cancer and autoimmunity. *Annu Rev Immunol.* 2008;26(1):57–79. https://doi.org/10.1146/annurev.immunol.26.021607.090316.
77. DeNardo DG, et al. CD4+ T cells regulate pulmonary metastasis of mammary carcinomas by enhancing protumor properties of macrophages. *Cancer Cell.* 2009;16(2):91–102. https://doi.org/10.1016/j.ccr.2009.06.018.
78. Johansson M, DeNardo DG, Coussens LM. Polarized immune responses differentially regulate cancer development. *Immunol Rev.* 2008;222(1):145–154. https://doi.org/10.1111/j.1600-065X.2008.00600.x.
79. Langowski JL, et al. IL-23 promotes tumour incidence and growth. *Nature.* 2006;442(7101):461–465. https://doi.org/10.1038/nature04808.
80. Gao Y, et al. γδ T cells provide an early source of interferon γ in tumor immunity. *J Exp Med.* 2003;198(3):433–442. https://doi.org/10.1084/jem.20030584.
81. Kacha AK, Fallarino F, Markiewicz MA, Gajewski TF. Spontaneous rejection of poorly immunogenic P1.HTR tumors by Stat6-deficient mice. *J Immunol.* 2000;165(11):6024–6028. https://doi.org/10.4049/jimmunol.165.11.6024.
82. Saha P, Sharma P, Singh B, Verma C. Role of $CXCR3^+$ $CCR5^+$ Th1 cells in pulmonary tuberculosis patients: at pathogenic site. *Biomed Biotechnol Res J BBRJ.* 2019;3(2):80. https://doi.org/10.4103/bbrj.bbrj_78_19.
83. Guillerey C, Huntington ND, Smyth MJ. Targeting natural killer cells in cancer immunotherapy. *Nat Immunol.* 2016;17(9):1025–1036. https://doi.org/10.1038/ni.3518.

84. Luetke-Eversloh M, et al. NK cells gain higher IFN-γ competence during terminal differentiation: innate immunity. *Eur J Immunol.* 2014;44(7):2074–2084. https://doi.org/10.1002/eji.201344072.
85. Tupin E, Kinjo Y, Kronenberg M. The unique role of natural killer T cells in the response to microorganisms. *Nat Rev Microbiol.* 2007;5(6):405–417. https://doi.org/10.1038/nrmicro1657.
86. Volpedo G, et al. Centrin-deficient Leishmania mexicana confers protection against new world cutaneous leishmaniasis. *NPJ Vaccines.* 2022;7(1):32. https://doi.org/10.1038/s41541-022-00449-1.
87. Varikuti S, et al. MicroRNA-21 deficiency promotes the early Th1 immune response and resistance toward visceral Leishmaniasis. *J Immunol.* 2021;207(5):1322–1332. https://doi.org/10.4049/jimmunol.2001099.
88. Verma C, et al. Calcium dynamics regulate protective responses and growth of *Staphylococcus aureus* in macrophages. *Biomol Concepts.* 2020;11(1):230–239. https://doi.org/10.1515/bmc-2020-0021.
89. Lee PT, Benlagha K, Teyton L, Bendelac A. Distinct functional lineages of human Vα24 natural killer T cells. *J Exp Med.* 2002;195(5):637–641. https://doi.org/10.1084/jem.20011908.
90. Gumperz JE, Miyake S, Yamamura T, Brenner MB. Functionally distinct subsets of CD1d-restricted natural killer T cells revealed by CD1d tetramer staining. *J Exp Med.* 2002;195(5):625–636. https://doi.org/10.1084/jem.20011786.
91. Song L, et al. Vα24-invariant NKT cells mediate antitumor activity via killing of tumor-associated macrophages. *J Clin Invest.* 2009;119(6):1524–1536. https://doi.org/10.1172/JCI37869.
92. Fleming C, Morrissey S, Cai Y, Yan J. γδ T cells: unexpected regulators of cancer development and progression. *Trends Cancer.* 2017;3(8):561–570. https://doi.org/10.1016/j.trecan.2017.06.003.
93. Seo N, Tokura Y, Takigawa M, Egawa K. Depletion of IL-10- and TGF-beta--producing regulatory gamma delta T cells by administering a daunomycin-conjugated specific monoclonal antibody in early tumor lesions augments the activity of CTLs and NK cells. *J Immunol.* 1999;163(1):242–249.
94. Ke Y, Kapp LM, Kapp JA. Inhibition of tumor rejection by gammadelta T cells and IL-10. *Cell Immunol.* 2003;221(2):107–114. https://doi.org/10.1016/s0008-8749(03)00066-2.
95. Hu G, et al. Tumor-infiltrating CD39+γδTregs are novel immunosuppressive T cells in human colorectal cancer. *Oncoimmunology.* 2017;6(2), e1277305. https://doi.org/10.1080/2162402X.2016.1277305.
96. Yuen GJ, Demissie E, Pillai S. B lymphocytes and cancer: a love–hate relationship. *Trends Cancer.* 2016;2(12):747–757. https://doi.org/10.1016/j.trecan.2016.10.010.
97. Pucci F, et al. SCS macrophages suppress melanoma by restricting tumor-derived vesicle–B cell interactions. *Science.* 2016;352(6282):242–246. https://doi.org/10.1126/science.aaf1328.
98. Wejksza K, et al. Cancer-produced metabolites of 5-lipoxygenase induce tumor-evoked regulatory B cells via peroxisome proliferator-activated receptor α. *J Immunol.* 2013;190(6):2575–2584. https://doi.org/10.4049/jimmunol.1201920.
99. Bodogai M, et al. Immunosuppressive and prometastatic functions of myeloid-derived suppressive cells rely upon education from tumor-associated B cells. *Cancer Res.* 2015;75(17):3456–3465. https://doi.org/10.1158/0008-5472.CAN-14-3077.
100. Lindner S, et al. Interleukin 21-induced granzyme B-expressing B cells infiltrate tumors and regulate T cells. *Cancer Res.* 2013;73(8):2468–2479. https://doi.org/10.1158/0008-5472.CAN-12-3450.
101. Ganti SN, Albershardt TC, Iritani BM, Ruddell A. Regulatory B cells preferentially accumulate in tumor-draining lymph nodes and promote tumor growth. *Sci Rep.* 2015;5:12255. https://doi.org/10.1038/srep12255.

102. Kim S, et al. B-cell depletion using an Anti-CD20 antibody augments antitumor immune responses and immunotherapy in nonhematopoetic murine tumor models. *J Immunother.* 2008;31(5):446–457. https://doi.org/10.1097/CJI.0b013e31816d1d6a.
103. Candolfi M, et al. B cells are critical to T-cell—mediated antitumor immunity induced by a combined immune-stimulatory/conditionally cytotoxic therapy for glioblastoma. *Neoplasia.* 2011;13(10):947. https://doi.org/10.1593/neo.11024.
104. Olkhanud PB, et al. Tumor-evoked regulatory B cells promote breast cancer metastasis by converting resting $CD4^+$ T cells to T-regulatory cells. *Cancer Res.* 2011;71(10): 3505–3515. https://doi.org/10.1158/0008-5472.CAN-10-4316.
105. Bodogai M, et al. Anti-CD20 antibody promotes cancer escape via enrichment of tumor-evoked regulatory B cells expressing low levels of CD20 and CD137L. *Cancer Res.* 2013;73(7):2127–2138. https://doi.org/10.1158/0008-5472.CAN-12-4184.
106. Varikuti S, et al. Ibrutinib treatment inhibits breast cancer progression and metastasis by inducing conversion of myeloid-derived suppressor cells to dendritic cells. *Br J Cancer.* 2020;122(7):1005–1013. https://doi.org/10.1038/s41416-020-0743-8.
107. Ohaegbulam KC, Assal A, Lazar-Molnar E, Yao Y, Zang X. Human cancer immunotherapy with antibodies to the PD-1 and PD-L1 pathway. *Trends Mol Med.* 2015;21 (1):24–33. https://doi.org/10.1016/j.molmed.2014.10.009.
108. Sharpe AH, Wherry EJ, Ahmed R, Freeman GJ. The function of programmed cell death 1 and its ligands in regulating autoimmunity and infection. *Nat Immunol.* 2007;8(3):239–245. https://doi.org/10.1038/ni1443.
109. Nunes-Xavier CE, Angulo JC, Pulido R, López JI. A critical insight into the clinical translation of PD-1/PD-L1 blockade therapy in clear cell renal cell carcinoma. *Curr Urol Rep.* 2019;20(1):1. https://doi.org/10.1007/s11934-019-0866-8.
110. Abiko K, et al. IFN-γ from lymphocytes induces PD-L1 expression and promotes progression of ovarian cancer. *Br J Cancer.* 2015;112(9):1501–1509. https://doi.org/10.1038/bjc.2015.101.
111. Bellucci R, et al. Interferon-γ-induced activation of JAK1 and JAK2 suppresses tumor cell susceptibility to NK cells through upregulation of PD-L1 expression. *OncoImmunology.* 2015;4(6), e1008824. https://doi.org/10.1080/2162402X.2015.1008824.
112. Banerjee S, Biehl A, Gadina M, Hasni S, Schwartz DM. JAK–STAT signaling as a target for inflammatory and autoimmune diseases: current and future prospects. *Drugs.* 2017;77(5):521–546. https://doi.org/10.1007/s40265-017-0701-9.
113. Groner B, von Manstein V. Jak Stat signaling and cancer: opportunities, benefits and side effects of targeted inhibition. *Mol Cell Endocrinol.* 2017;451:1–14. https://doi.org/10.1016/j.mce.2017.05.033.
114. Doi T, et al. The JAK/STAT pathway is involved in the upregulation of PD-L1 expression in pancreatic cancer cell lines. *Oncol Rep.* 2017;37(3):1545–1554. https://doi.org/10.3892/or.2017.5399.
115. Peng Q, Deng Z, Pan H, Gu L, Liu O, Tang Z. Mitogen-activated protein kinase signaling pathway in oral cancer (Review). *Oncol Lett.* 2017. https://doi.org/10.3892/ol.2017.7491.
116. Pradhan R, Singhvi G, Dubey SK, Gupta G, Dua K. MAPK pathway: a potential target for the treatment of non-small-cell lung carcinoma. *Future Med Chem.* 2019;11(8):793–795. https://doi.org/10.4155/fmc-2018-0468.
117. Jalali S, et al. Reverse signaling via PD-L1 supports malignant cell growth and survival in classical Hodgkin lymphoma. *Blood Cancer J.* 2019;9(3):22. https://doi.org/10.1038/s41408-019-0185-9.
118. Lim W, Jeong M, Bazer FW, Song G. Curcumin suppresses proliferation and migration and induces apoptosis on human placental choriocarcinoma cells via ERK1/2 and SAPK/JNK MAPK signaling pathways. *Biol Reprod.* 2016;95(4):83. https://doi.org/10.1095/biolreprod.116.141630.

119. Bi X, et al. PD-L1 is upregulated by EBV-driven LMP1 through NF-κB pathway and correlates with poor prognosis in natural killer/T-cell lymphoma. *J Hematol Oncol.* 2016;9(1):109. https://doi.org/10.1186/s13045-016-0341-7.
120. Peng J, et al. Chemotherapy induces programmed cell death-ligand 1 overexpression via the nuclear factor-κB to foster an immunosuppressive tumor microenvironment in ovarian cancer. *Cancer Res.* 2015;75(23):5034–5045. https://doi.org/10.1158/0008-5472.CAN-14-3098.
121. Liu L, et al. ERα is a negative regulator of PD-L1 gene transcription in breast cancer. *Biochem Biophys Res Commun.* 2018;505(1):157–161. https://doi.org/10.1016/j.bbrc.2018.09.005.
122. Ajona D, et al. A combined PD-1/C5a blockade synergistically protects against lung cancer growth and metastasis. *Cancer Discov.* 2017;7(7):694–703. https://doi.org/10.1158/2159-8290.CD-16-1184.
123. Takada K, et al. A comprehensive analysis of programmed cell death ligand-1 expression with the clone SP142 antibody in non–small-cell lung cancer patients. *Clin Lung Cancer.* 2017;18(5):572–582.e1. https://doi.org/10.1016/j.cllc.2017.02.004.
124. Li Y, et al. The prognostic and clinicopathological roles of PD-L1 expression in colorectal cancer: a systematic review and meta-analysis. *Front Pharmacol.* 2019;10:139. https://doi.org/10.3389/fphar.2019.00139.
125. Nakanishi J, Wada Y, Matsumoto K, Azuma M, Kikuchi K, Ueda S. Overexpression of B7-H1 (PD-L1) significantly associates with tumor grade and postoperative prognosis in human urothelial cancers. *Cancer Immunol Immunother.* 2007;56(8):1173–1182. https://doi.org/10.1007/s00262-006-0266-z.
126. Zhu J, et al. A feedback loop formed by ATG7/autophagy, FOXO3a/miR-145 and PD-L1 regulates stem-like properties and invasion in human bladder cancer. *Cancer.* 2019;11(3):349. https://doi.org/10.3390/cancers11030349.
127. Miura. Blockade of B7-H1 or B7-DC induces an anti-tumor effect in a mouse pancreatic cancer model. *Int J Oncol.* 2009;35(4). https://doi.org/10.3892/ijo_00000387.
128. Loos M, et al. Clinical significance and regulation of the costimulatory molecule B7-H1 in pancreatic cancer. *Cancer Lett.* 2008;268(1):98–109. https://doi.org/10.1016/j.canlet.2008.03.056.
129. Masugi Y, et al. Characterization of spatial distribution of tumor-infiltrating CD8+ T cells refines their prognostic utility for pancreatic cancer survival. *Mod Pathol.* 2019;32(10):1495–1507. https://doi.org/10.1038/s41379-019-0291-z.
130. Qureshi OS, et al. Trans-endocytosis of CD80 and CD86: a molecular basis for the cell-extrinsic function of CTLA-4. *Science.* 2011;332(6029):600–603. https://doi.org/10.1126/science.1202947.
131. Zhao Y, Yang W, Huang Y, Cui R, Li X, Li B. Evolving roles for targeting CTLA-4 in cancer immunotherapy. *Cell Physiol Biochem.* 2018;47(2):721–734. https://doi.org/10.1159/000490025.
132. Logan CY, Nusse R. The Wnt signaling pathway in development and disease. *Annu Rev Cell Dev Biol.* 2004;20:781–810. https://doi.org/10.1146/annurev.cellbio.20.010403.113126.
133. Kottarath SK, et al. Folate receptor-β targeted cholesterol-chitosan nanocarrier for treatment of rheumatoid arthritis: an animal study. *J Drug Deliv Sci Technol.* 2020;60, 101946. https://doi.org/10.1016/j.jddst.2020.101946.
134. Duan Z, Luo Y. Targeting macrophages in cancer immunotherapy. *Signal Transduct Target Ther.* 2021;6(1):127. https://doi.org/10.1038/s41392-021-00506-6.
135. Huang C-T, et al. Role of LAG-3 in regulatory T cells. *Immunity.* 2004;21(4):503–513. https://doi.org/10.1016/j.immuni.2004.08.010.
136. Workman CJ, Vignali DAA. Negative regulation of T cell homeostasis by lymphocyte activation gene-3 (CD223). *J Immunol.* 2005;174(2):688–695. https://doi.org/10.4049/jimmunol.174.2.688.

137. Workman CJ, et al. LAG-3 regulates plasmacytoid dendritic cell homeostasis. *J Immunol*. 2009;182(4):1885–1891. https://doi.org/10.4049/jimmunol.0800185.
138. Workman CJ, Dugger KJ, Vignali DAA. Cutting edge: molecular analysis of the negative regulatory function of lymphocyte activation gene-3. *J Immunol*. 2002;169 (10):5392–5395. https://doi.org/10.4049/jimmunol.169.10.5392.
139. Clayton KL, et al. Soluble T cell immunoglobulin mucin domain 3 is shed from CD8 + T cells by the sheddase ADAM10, is increased in plasma during untreated HIV infection, and correlates with HIV disease progression. *J Virol*. 2015;89(7):3723–3736. https://doi.org/10.1128/JVI.00006-15.
140. Woo S-R, et al. Immune inhibitory molecules LAG-3 and PD-1 synergistically regulate T-cell function to promote tumoral immune escape. *Cancer Res*. 2012;72 (4):917–927. https://doi.org/10.1158/0008-5472.CAN-11-1620.
141. Gros A, et al. PD-1 identifies the patient-specific $CD8^+$ tumor-reactive repertoire infiltrating human tumors. *J Clin Invest*. 2014;124(5):2246–2259. https://doi.org/10.1172/JCI73639.

CHAPTER TWELVE

Adrenoceptors and SCD1 in adipocytes/adipose tissues: The expression and variation in health and obesity

Dinh-Toi Chu[a,b,*], Nhat-Le Bui[a,b], and Ngoc Hoan Le[c]

[a]Center for Biomedicine and Community Health, International School, Vietnam National University, Hanoi, Vietnam
[b]Faculty of Applied Sciences, International School, Vietnam National University, Hanoi, Vietnam
[c]Faculty of Biology, Hanoi National University of Education, Hanoi, Vietnam
[*]Corresponding author: e-mail addresses: chudinhtoi.hnue@mail.com; toicd@vnu.edu.vn

Contents

Abstract

Obesity, considered a metabolic disorder, is one of the most significant health issues that the community has to cope with today. A rising number of studies have been conducted to find out promising genetic targets for obese treatment. The sympathetic nervous system was proven to possess remarkable roles in energy metabolism, including the stimulation of lipolysis as well as thermogenesis, via distinct adrenoceptors appearing on the membrane of adipocyte. A decrease of β-adrenoceptor expression has

Progress in Molecular Biology and Translational Science, Volume 194
ISSN 1877-1173
https://doi.org/10.1016/bs.pmbts.2022.06.024

been observed in obese individuals, which is related to reducing energy expenditure and developing obesity. While that the deficiency of stearoyl-CoA desaturase-1 (SCD1), which is a promising target for treatments of metabolic diseases, decreases oxidation and promotes the synthesis of fatty acids. Here, we emphasized several differences between distinct adrenoceptor subtypes, including their mRNA expression level and function in white adipose tissue and brown adipose tissue. We also highlighted SCD1's roles related to the progression of adipocytes and its changing expression under the obese condition in both rodents and humans, and furthermore, tried to figure out the interaction between adrenoceptors and SCD1 in adipose tissue.

Abbreviations

ADSCs	adipose tissue-derived mesenchymal stem cells
C/EBP	CCAAT/enhancer-binding protein
GPCR	G protein-coupled receptor
HFD	high-fat diet
MUFAs	monounsaturated fatty acids
NAFLD	non-alcoholic fatty liver disease
PPARγ	peroxisome proliferator-activated receptor-γ
qRT-PCR	quantitative real-time polymerase chain reaction
SCD1	stearoyl-CoA desaturase 1
SREBP	sterol regulatory element-binding protein
T2D	type 2 diabetes
UCP1	uncoupling protein 1
WAT/BAT/OAT/SAT/VAT	White/Brown/Omental/Subcutaneous/ Visceral Adipose tissue
α/β-AR	alpha/beta-adrenergic receptor

1. Introduction

Obesity is defined as abundant body fat inside the body, harming the health and causing some related metabolic disorders.[1] It is also considered to be one of the major public health problems facing industrialized countries.[2] According to the World Health Organization (WHO), the prevalence of obesity around the world has been tripled since 1975.[3] In the population of 7.5 billion worldwide, there are 0.774 billion people diagnosed with obesity, out of which 1.9 million people are overweight, while 650 million people are diagnosed with obesity.[4] By the year 2030, in the United States, 20% of adults over 65 years old will be obese. In Europe, this number will become 25% in 2035.[5] The three countries with the most obese patients are the United States (over 73 million), China (over 65 million), and India (over 44 million).[4]

Obesity and overweight significantly affect the human quality of life, and furthermore, lead to other metabolic disorders, such as type 2 diabetes

(T2D), cardiovascular diseases, non-alcoholic fatty liver disease (NAFLD), and several certain cancers.[6,7] In 2017, around 426 million people worldwide were suffered from T2D, accounting for 6.28% of the global population.[8] NAFLD's prevalence is 80–90% in obese adults and 30–35% in diabetes patients.[9] Among severely obese patients who undergo bariatric surgery, more than 95% will suffer from NAFLD.[10] Furthermore, it has been shown that obese individuals will have a higher risk of cardiovascular disease and cancers.[11] In fact, about 418,000 cancers worldwide were supposed to be related to obesity and overweight in 2012.[12]

Adipose tissue in the body consists of two main types, white adipose tissue (WAT) and brown adipose tissue (BAT),[13] sometimes also mentioned beige/brite adipose tissue, each type plays a distinct role in controlling obesity and metabolism.[14] WAT is a dynamic and modifiable component that consists of about 3–70% of body mass.[15] The primary function of WAT is described as storing and releasing fatty acids that provided fuel during fasting period.[16] Furthermore, WAT plays an essential part in whole-body energy balance as it stores energy under the form of triglycerides.[17] Nevertheless, excess WAT can cause ectopic fat deposition, including lipid accumulation in hepatocytes and myocytes, which substantially increase the risk of insulin resistance and T2D.[18]

On the other hand, BAT plays a vital part in thermogenesis in newborns[19] through burning triglycerides in various mitochondria to generate heat.[20] BAT mass was most abundant during birth and decreased during the early weeks of life.[21] When activated, BAT can help promote weight loss and reduce body fat mass because of increased energy expenditure and increased local and net glucose disposal.[22,23] Some findings also mention beige or brite adipose tissue, which is transformed from white to brown fat,[24] presence in the human body. The morphology and function of brite adipose tissue are somewhat similar to BAT[25] and are also being researched on its mechanism of turning WAT to BAT.[14]

Research nowadays mainly focuses on the molecular level of how adipose tissue can be related to thermogenesis and metabolic diseases. It has been proven that adipocyte mitochondria can regulate adipose tissue differentiation (Peroxisome proliferator-activated receptor-γ—PPARγ),[26] lipid hemostasis (Adipose triglyceride lipase—ATGL, Perilipin—PLIN), oxidative capacity (oxidative phosphorylation—OXPHOS)[26,27] insulin sensitivity (glucose transporter type 4—GLUT4),[26,28] thermogenesis (uncoupling protein 1—UCP1) and WAT browning (UCP1).[26,29]

With its unique characteristic, BAT has been widely investigated as a promising tool for obesity treatment.[14] Increasing BAT amount can promote the activation of UCP1 and thermogenesis as well as increase total energy expenditure.[30] Furthermore, the increase of BAT activity will uptake the blood lipid and glucose, increasing the oxidation and storage of lipid and glucose. As a result, a considerable reduction in blood glucose levels and insulin resistance was observed, resulting in reducing obesity and metabolic disorder syndromes.[31]

Adrenoceptors and stearoyl-CoA desaturase-1 (SCD1) have been recently concerned as promising therapeutic targets for obesity and related metabolic disorders due to their effect on different tissues and organs, including adipose tissue as well as heart, muscle, and liver.[32–34] Because adipose tissue plays a crucial part in the whole-body energy homeostasis, it is necessary to comprehend how the tissue is modulated by adrenoceptors and SCD1. In this review, we will clarify the roles and expression of adrenoceptors and SCD1, as well as their mutual interaction, in adipocytes.

2. Methods to study the expression and variation of adipose adrenoceptors and SCD1

Gene expression is a two-step process, including transcription and translation, respectively, to produce mRNAs and proteins for cellular function (Fig. 1). The copy numbers of these gene products are diverse in different cell types and may be changed under abnormal health conditions.[35] Analyzing the gene expression of adrenoceptors and SCD1 in adipocytes significantly elucidates their roles on adipose tissue as well as their therapeutic potential of obesity. Recently, cutting-edge technology in the "omic" field allows scientists to study gene expression at a variety of levels.[36] At the genomic level, one of the most popular methods to evaluate mRNA expression level is quantitative real-time polymerase chain reaction (qRT-PCR). To perform qRT-PCR, total RNA from biopsied adipocytes is extracted and then mRNA is converted into complementary DNA (cDNA) via reverse transcription. cDNA is utilized as the template for followed PCR, and specifically targeting mRNA is quantified through indirectly measuring the fluorescence signal by a spectrophotometer.[37] Cao et al. applied qRT-PCR to compare the mRNA expression levels of β_3-adrenergic receptor (β_3-AR) in mature visceral adipocytes and stromal vascular cells between normal-weight and overweight individuals, while Merlin et al. used qRT-PCR results to prove the impact of rosiglitazone treatment on β-AR mRNA levels in white adipocytes.[38,39]

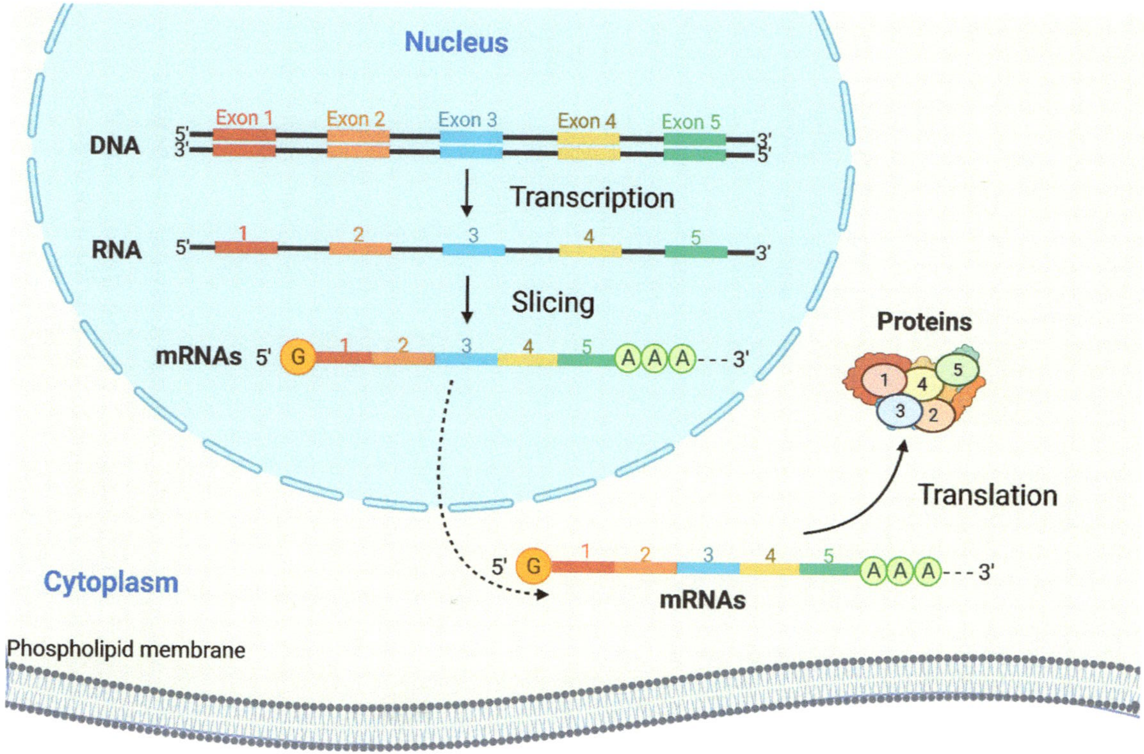

Fig. 1 Gene expression concept. Gene expression is a process that genetic information is converted into products, such as proteins. Eukaryote DNA consists of coding (exon) and non-coding (intron) regions. When a gene is expressed, the base sequence (genetic information) of this gene is initially transcripted into primary RNA. Then mRNA is formed by slicing exons together. mRNA with a poly tail A and an endcap is moved from nucleus to cytoplasm. In the cytoplasm, mRNA is translated into functional proteins with the support of specific ribosomes. (Created with Biorender.com).

Another method for mRNA expression analysis is microarray, a chip where radiolabeled cDNA obtained from samples is hybridized to the complementary sequence of probes on its surface. Data from microarray provide gene expression profiles of a large number of genes, thus investigating the change of expression levels in response to an abnormal condition.[40] Hulver et al. used microarray analysis to detect the relationship between obesity and SCD1 mRNA level, while another study by Rivero et al. combined several public microarray datasets to detect the increased expression level of α_2-AR on tumor cells.[41,42] Burl et al. performed RNA-Sequencing to provide transcriptomes of more than 33,000 epididymial and inguinal white adipocytes during β_3-AR activation.[43] Besides, RNA-Sequencing indicated NF-κB pathway's role through its change in expression level under SCD1-deficient models.[44] Compared to the microarray method, RNA-Sequencing technique diminishes background noises and provides high-throughput sequences to investigate novel RNA species, gene expression, and other applications.[40]

At the proteomic level, Western Blotting is commonly used to quantify the copy numbers of a specific protein. Western Blotting was first utilized to figure out a specific target of protein from complex mixtures via separating by size, transferring protein to a solid base, and then marking target by antibodies to visualize, but now quantitative interpretion is also required.[45] Fold changes in protein expression levels between different samples are calculated based on the associated fluorescent or chemiluminescent signals.[46] Cao et al. used Western Blot Analysis to explore that the mRNA, as well as protein, expression levels of β_3-AR in stromal vascular cells and mature adipocytes in normal-weight were higher than in overweight participants. Thus, to point out the roles of adrenoceptors as well as SCD1, in addition to mRNA level, their protein concentration in different types of adipocyte could be compared.[38]

3. Adrenoceptors and adipose tissues/adipocytes

3.1 Roles of adrenoceptors on the development and function of adipose tissues/adipocytes

Adrenoceptors belong to G protein-coupled receptors (GPCR) superfamily and are broadly allocated throughout the human body with their well-known function on glucose homeostasis and energy metabolism.[47] Adrenoceptors are divided into α-AR and β-AR subtypes. α-AR is further subclassified into α_1-AR and α_2-AR, while β-AR has three siblings, including β_1-, β_2-, and β_3-AR. Adrenergic receptors are specifically activated by noradrenaline and adrenaline, which are both endogenous catecholamines. α_2-AR and β-AR couple to Gi and Gs subunit, respectively, and initiate cAMP-dependent pathway through adenylyl cyclase with opposite effects, while α_1-AR regulates phospholipase C through coupling to the Go protein (Fig. 2).[47–49] Both subtypes of adrenoceptors are apparently found on the adipocyte's membrane.[50,51] However, due to distinct characteristics of WAT and BAT,[14,52] the expression and function of adrenoceptor subtypes are not similar in two major types of adipose tissue (Table 1).

Lipolysis is a hydrolysis process of triglycerides, the main form of energy storage, into fatty acids and glycerol.[53] In WAT, β_3-AR is the predominant β-adrenoceptor which is expressed and responsible for the stimulation of lipolysis in rodents.[54] In WAT of the β_3-AR knockout mice, functions of β-ARs were seriously impaired despite a minor compensatory from β_1-AR.[55,56] Another study indicated that without the expression of β-ARs, white adipose depots were significantly increased in size, and mice susceptibility to diet-induced obesity was increased.[57–59] However, β_3-AR is not that important in humanity, their mRNA level of β_3-AR is even lower than that of

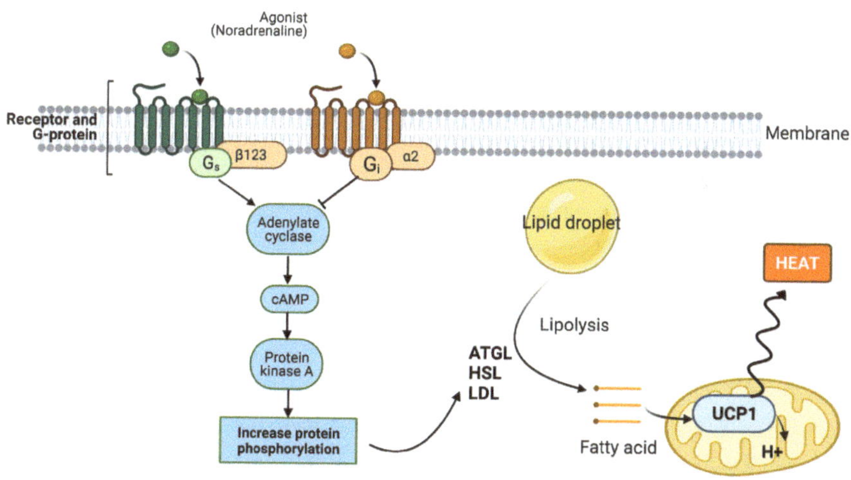

Fig. 2 Mechanism of adrenoceptors effect on adipocyte. Catecholamine (noradrenaline) activates α_2-AR and β-AR on the membrane of adipocytes via Gi/Gs-cAMP-PKA pathway with opposite effects. An increase in protein phosphorylation leads to activate lipases, including lipoprotein lipase (LDL), hormone-sensitive lipase (HSL), and adipose triglyceride lipase (ATGL), which encourage the lipolysis process of triglycerides inside fat droplets into fatty acids. These fatty acids provide protons for UCP1 to activate non-shivering thermogenesis and release heat. (Created with Biorender.com).

Table 1 Differences in characteristics and adrenoceptor expression of WAT and BAT.

	WAT	BAT
Lipid droplet	Unilocular	Multilocular
UCP1	UCP1 (−)	UCP1 (+)
Mitochondrion	Low density	High density
Function	Lipid storage Endocrine	Non-shivering thermogenesis
Adrenoceptor expression in rodent	β_1-AR, β_2-AR, β_3-AR	α_1-AR, α_2-AR β_1-AR, β_2-AR, β_3-AR
Adrenoceptor expression in human	α_1-AR, α_2-AR β_1-AR, β_2-AR, β_3-AR (low)	α_1-AR, α_2-AR β_1-AR, β_2-AR, β_3-AR
Adrenoceptor function	Lipolysis (Induced by β-AR, suppressed by α_2-AR)	Lipolysis Thermogenesis (Both are induced by β-AR)

β_1-AR and β_2-AR.[60] Besides, it has to mention the anti-lipolytic function of α_2-AR. In adipocyte tissue where α_2-AR is more abundant than three subtypes of β-ARs, such as human subcutaneous fat, the inhibition effect of α_2-AR is over the stimulation effect. This contrast action is observed in humans, rabbits, and dogs, but not in rodents.[61] Additionally, there is accumulated evidence suggesting that adrenoceptors also partly modulate the secretion of WAT's adipokines, which is attributed to their effect on lipid metabolism.[62]

Thermogenesis refers to the production of heat to increase energy expenditure and is considerably managed via the sympathetic nervous system.[23,32,63] Via responding to noradrenaline, β-ARs acutely stimulate the increase of UCP1 activity, thus encourage both non-shivering thermogenesis and lipolysis in BAT.[32] Several well-performed research proved the abundance and undeniable function of β_3-AR in the BAT of rodents, however in constrast to rodents, β_1-ARs and β_2-ARs are supposed to be functionally predominant over β_3-ARs in human brown adipocyte. A clinical trial using pharmacological agonists and antagonists of β_2-ARs and evaluating β-adrenergic knockout model reported that in humans, BAT's lipolysis and thermogenesis occurred via β_2-AR signaling.[32] Another study on the human brown adipocyte model suggested that β_1-ARs may play a role as a regulatory of UCP1 expression,[64] while roles of β_3-ARs on human thermogenesis are still controversial topics.[61,65] Additionally, there are growing concerns about the regulatory effect of β-ARs and BAT on glucose homeostasis[66,67] and the expression levels of several BAT's adipokines, such as VEGF.[68]

3.2 The expression and variation of adrenoceptors on adipose tissues/adipocytes

Due to their function on lipolysis and thermogenesis, adrenoceptors, especially β_3-ARs, are considered promising targets for obesity and other metabolic disorders' therapies. Although all types were found on adipocytes, the mRNA level of β_3-ARs were confirmed to be lower than those of β_1-ARs and β_2-ARs in human fat cells, while β_3-ARs were obviously the most dominant subtypes in rodents.[60,69] β_3-AR mARN is mainly expressed in supraclavicular BAT of mature individuals and still increases in response to cold exposure. Although β_3-AR mARN is present in human WAT, its signal is significantly lower than that in infant or perirenal BAT depots.[51] Additionally, in a human cell model of multipotent adipose-derived stem cells, β-AR subtypes expressed in low-level, while only β_1-ARs and β_3-ARs retained their function on UCP1 stimulation.[51,70]

Obesity can be accompanied by dysfunction and deficient expression level of β-AR.[71] When evaluating the expression level and function of three types of β-AR in adipocytes, a dramatically reduction was observed in β_1-ARs and β_3-ARs in genetically obese mice comparing to lean mice. This reduction led to a deficient response when stimulating adenyl cyclase with beta-agonists and suggested that lower expression levels of β-ARs are associated with reducing energy expenditure and obese development.[69,71,72] Another study on obese and control rats also reported consistent results that β_1-ARs were downregulated, which was related to higher leptin concentration and hyperlipidemia in the obese group.[73]

In humans, a study performed by Kurylowicz et al. reported that the gene expression level of β-ARs and several thermogenic genes were considerably lower in adipocyte depots of obese patients compared to the lean group. Moreover, in obese individuals, a lower expression level of β_2-AR and β_3-AR in visceral adipocytes than those in subcutaneous fat was observed. The authors supposed this significant decrease in the expression of β-AR subtypes was related to the reduced response of β-AR to its stimuli, thus, discouraging the thermogenesis activation.[74] A similar study in Chinese people also showed consistent results and the authors even deduced that the reduction in β_3-AR expression level occurs before the onset of obesity, which suggested the downregulation of β_3-AR partly ameliorated the obese condition.[38]

There are several factors affecting the expression of the adrenoceptor and its function on lipolysis induction. Stich et al. indicated that in obese women, a hypocaloric diet reduced α_2-AR mRNA level in the adipose depot, thus lessened its antilipolytic effect, especially during exercise.[75] However, after long-term exercise, the increase in lipolytic activity in obese individuals, which is also partly contributed by β-ARs, was still lower than that in normal-weight subjects.[76,77] Due to differences between men and women in positions of adipose depots, α_2-AR effect was significantly higher in men than in women while β-AR effect was more apparent in females among obese individuals.[76]

4. Stearoyl-CoA desaturase-1 and adipose tissues/adipocytes

4.1 Roles of stearoyl-CoA desaturase-1 (SCD1) on the development and function of adipose tissues/adipocytes

Stearoyl-CoA desaturase (SCD) is a vital protein that is anchored in the endoplasmic reticulum. Such protein is a potential target to control lipogenesis by catalyzing the biosynthesis of monounsaturated fatty acids, mainly including oleate and palmitoleate, from synthesized saturated fatty acids, such as palmitate, and stearate, respectively. These acids are indispensable components of triglycerides, cholesterol ester, wax ester, and especially membrane phospholipids, which determine membrane fluidity and influence robustly lipoprotein metabolism and adiposity. When SCD1 is deficient, saturated fatty acids are accumulated while unsaturated fatty acids are impaired. Biosynthesis is achieved by introducing a cis double bond between the carbons at the Δ9 position of the acyl-CoA substrates. Preferred substrates are stearoyl-CoA and palmitoyl-CoA, which are, respectively, converted into oleoyl-CoA (18:1) and palmitoleoyl-CoA (16:1).[78,79]

SCD1 is considered a short-life protein, with a half-life of just a few hours, due to its fast degradation rate in the microsome fraction. SCD1 transcription could be strictly modulated, so it is well suitable for the regulation of SCD1 expression.[80] Heinemann et al. used a biochemical approach and identified plasminogen as a protease to degrade SCD1 protein in microsome. They also proved that SCD1 expression level in liver microsome is dropped in plasminogen-deficient mice.[81,82]

Elevated SCD expression has been related to a wide range of health issues, including cancers, atherosclerosis, and viral infection.[83] SCD1 has been proved to be involved in the progression of numerous different types of cancer and is being explored as a promising therapeutic target against such deadly diseases.[80]

SCD expression is modulated in transcription by sterol regulatory element-binding protein (SREBP), liver X receptor, insulin signaling, and carbohydrate-responsive element-binding protein (ChREBP).[80] In mammals, expression levels of SCD1 are affected by dietary factors such as vitamin A, cholesterol, and polyunsaturated fatty acids, hormone changes including insulin and glucagon, temperature changes, developmental processes, peroxisomal proliferator, thiazolinediones, alcohol, metals, and phenolic compounds.[83]

A study found that treatment using Dexmedetomidine (DEX) which is a selective α_2-AR agonist suppressed SCD1. An explanation had been presented that the activation of α_2-AR caused by DEX decreased the levels of plasma insulin, so the suppression of SCD1 was partly due to the reduction of insulin concentration in plasma.[84] In addition, the gene expression level of SCD1 is increased in skeletal muscles, liver, and WAT of mice lacking β-adrenoreceptor.[85]

4.2 The expression and variation of SCD1 on adipose tissues/adipocytes

A previous research unveiled that SCD1 expression in mice was implicated in environmental stimuli, with upward regulation in obesity status.[85] Other studies with SCD1 deletions in skin or liver showed that the deficiency of SCD1 functioned precisely in obesity (Fig. 3).[86,87] While an increase in SCD1 expression when feeding mice with an HFD was found in some research,[88,89] others reported an opposite trend toward the expression level of SCD1.[90] Furthermore, SCD1 expression in the liver was found substantially weaker from animals in the HFD-fed group compared to animals fed with HCD for 18 weeks (HFD consisted of 60% kcal from fat, 20% kcal from carbohydrates; HCD consisted of 70% kcal from carbohydrates, 30% kcal from fat).[91] Monounsaturated fatty acids (MUFAs) produced by SCD1, can modulate general metabolism and regulate chronic metabolic diseases. Therefore, high MUFA levels in obese animals and humans indicate an

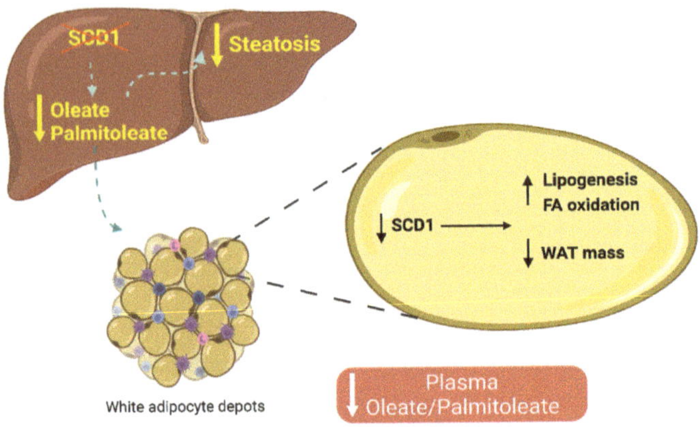

Fig. 3 The role of SCD1 in steatosis. In the SCD1-deficient liver of mice model, reduction in the synthesis process of monounsaturated fatty acids, mainly oleate and palmitoleate, decreases the accumulation of triglycerides, thus decrease steatosis. Besides, it leads to reduced oleate/palmitoleate concentration in plasma, and less lipoprotein transported to adipocytes, which encourages lipogenesis and fatty acid oxidation while suppresses WAT mass. (Created with Biorender.com).

essential role of SCD1 in the development of obesity-related metabolic diseases, such as insulin resistance, hyperlipidemia, and NAFLD.[92–94] SCD1 mRNA was fivefold and protein expression twofold lower than usual in type 2 diabetes.[95] Studies proved that increased SCD1 activity affects fatty acid composition by decreasing oxidation and encouraging the synthesis process of fatty acids.[41,96]

On the contrary, deficient SCD1 mice were lean, hypermetabolic, insulin sensitive, and resistant to diet-induced obesity.[83,93] Mice with mutations in SCD1 were extremely less obese than the control group, significantly increased total energy expenditure, and markedly decreased triglyceride storage and VLDL reproduction.[97] In adipose tissue and liver from SCD −/− mice, the gene absence not only decreased triacylglycerol synthesis but also increased metabolic rate and fatty acid catabolism, which are associated with significant induction of various PPAR targeting genes.[41] Another study performed in SCD1 −/− mice showed that the shortage of SCD reduced the expression of certain lipogenesis-related genes in the liver, including fatty acid synthase (FAS), SREBP-1, and glycerol phosphate acyl-CoA transferase (GPAT).[83]

In previous research, SCD1 was identified as a key enzyme to control beige adipogenesis. Its deficiency in adipose tissue-derived mesenchymal stem cells (ADSCs) provided the capability to differentiate from stem cells into beige adipocytes. A detailed analysis reported that a succinate accumulation caused by SCD1 deficiency enhanced mitochondrial complex II activity and subsequently promoted ADSCs with a tendency toward beige adipogenesis. Moreover, ablation of SCD1 was essential to prohibit ADSCs from differentiating into white adipocytes and maintain a phenotype of beige podocytes.[79]

SCD1 expression was elevated in omental adipose tissue (OAT) of the obesity group, which may show lipid storage activity in OAT.[98] Besides, the correlation of SCD1 expression with BMI in visceral adipose tissue (VAT) and subcutaneous adipose tissue (SAT) has been established in women with obesity.[99] When comparing the lipogenic potential of SAT in the young group and older group, one study reported that SAT mRNA transcripts of SCD1 were less expressed in the older, but it was not statistically significant.[100]

Researchers also pointed out the equivalent abundance of SCD1 mRNA expression levels in human mature myotubes, preconfluent myoblasts, confluent myoblasts, and fibroblasts. Therefore, gene expression did not seem to be modulated during myocyte differentiation and was not enriched in myoblasts. On the contrary, SCD1 expression levels were low in human pre-adipocytes while significantly induced during differentiation.[41]

5. Mechanisms underlying interaction between adrenoceptors and SCD1 in adipocytes/adipose tissues

SCD1 is known as a catalyst that actively supports the synthesis of monounsaturated fatty acids, controlling β-adrenergic thermogenesis. One of the key roles of monounsaturated fatty acids is to mediate the inhibition of thermogenesis by signaling to peripheral tissues.[85] In mice lacking β-ARs, thermogenesis was impaired, leading to an increased likelihood of obesity. Nevertheless, the mRNA level of SCD1 in WAT, not BAT among β-AR—lacking mice was higher than among wildtypes (Fig. 4). WAT, skeletal muscle, and liver are the most prominent sites of SCD1 expression in β-AR-deficient mice. This expression was assessed to be due to SCD1 promoting the synthesis of triglycerides, causing thermogenesis to decrease. Besides, the mice evaluated for SCD1 deficiency promote the synthesis of free fatty acids, resulting in high energy consumption and thermogenesis, which helps to combat the risk of obesity. This result confirmed the inhibition of SCD1 on thermogenesis of β-ARs through an indirect link.[85] Excessive fat accumulation in the body increases, leading to metabolic

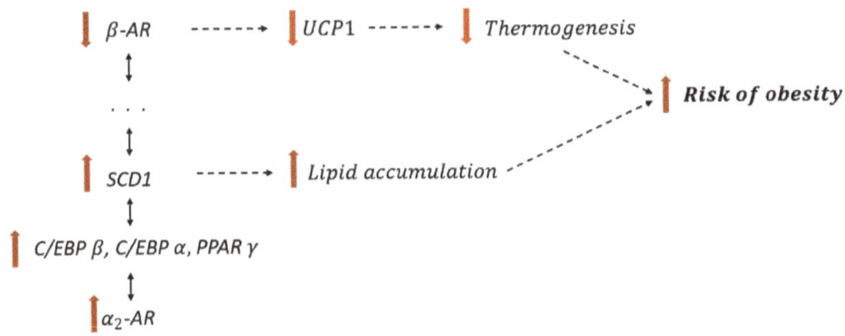

Fig. 4 The interaction between adrenoceptors and SCD1. In response to cold stimuli as well as dietary, the activation of β-AR leads to the elevation of UCP1, which stimulates thermogenesis. Thus, in mice lack of β-AR, thermogenesis is impaired and these mice are more susceptible to obesity. While that the mRNA expression of SCD1 is increased in β-AR—less mice compared to wildtypes.85 Therefore, there is an indirect link between β-AR and SCD1. Besides, the decrease in expression level of SCD1 was associated with the activation of α2-AR via negative feedback with the presence of PPARγ, C/EBPα and C/EBPβ.103

disorders such as insulin resistance, hyperglycemia, hypertension, and cancers. These metabolic disorders have been found to be responsible for the higher expression level of SCD1 in hepatocytes, skeletal muscles, and WAT, as observed in both obese rodent and human model designs. β-ARs controls SCD1 during thermogenesis, altering adaptations to high-fat diets or low-temperature.[85]

In a rat model fed with a high-fat diet (HFD), Tao et al. used DEX to attenuate insulin resistance (IR) levels in rat hepatocytes. The study results reported a change in mouse body mass, lower accumulation of fat in the liver, and a substantially reduced plasma level of triglycerides, and increased promotion of liver damage recovery. SCD1 is considered one of the PPARγ's downstream genes, while PPARγ can be deactivated by suppressing C/EBP α and C/EBP β, which associated with α_2-AR. Thus, the decrease in mRNA level of SCD1 was associated with the activation of α_2-AR via negative feedback, which resulted in restricting lipid accumulation (Fig. 4). Therefore, DEX is viewed as a potential source of drug development for therapeutic strategies of inflammatory responses, IR, and NAFLD through inhibiting lipid accumulation associated with SCD1.[101]

Another study of Maria Iwase et al. focused on the mRNA expression of glycerol kinase (Gyk), which inhibits the thermogenesis of adipocytes. Gyk-knockout interferes with the expression of UCP1 through the β-adrenoceptor agonist isoproterenol pathway in vitro, causing respiratory depression. In addition, increased expression of SCD1 and accumulation of the monounsaturated fatty acid (palmitoleic acid) were found in isoproterenol-treated Gyk-knockout adipocytes.[102]

Seong-Ho Lee et al. have a research paper on SCD1 deficiency in thermogenesis and cold environment. Increased expression of UCP1 in BAT was found in SCD1 homologous mice (SCD1 −/−), indicating a decrease in body fat mass and consumption energy levels rise rapidly. Besides, the expression of β_3-ARs, lipolysis, oxidation of MUFAs, including oleate and palmitoleate, increased simultaneously. With cold exposure, mice showed signs of hypothermia and hypoglycemia, along with significantly low levels of glycogen in the liver. Thus, thermogenesis induced by SCD1 deficiency via controlling the expression of the β_3-AR, followed by an increase in lipolysis and fatty acid oxidation.[103] SCD1 deficiency in adipose tissue and liver of SCD −/− mice inhibits triglyceride synthesis, promotes metabolic processes and fatty acid breakdown, and significantly aids in the induction of different PPAR.[41] Ntambi et al. indicated that deficiency of SCD1 reduced the expression of several hepatic lipogenesis-related genes.[83] In another study, mice with SCD1 deficiency reported reduced body mass, significantly increased metabolism, insulin sensitivity, and reduced risk of obesity.[83,93,97]

A study on β_2-AR function in NAFLD in HFD showed a marked increase in liver mass of rats fed an HFD when β_2-AR was removed. In addition, genetic loss of β_2-AR can cause significant lipid accumulation in the liver and consequently lead to liver damage.[104] SCD1 is capable of producing MUFAs, which partly contribute to the control of chronic metabolic diseases. In obese humans and animals, elevated levels of MUFAs have been observed, leading to hyperlipidemia, insulin resistance, and NAFLD.[92,94]

6. Conclusion

In conclusion, both adrenoceptors and SCD1 have been implicated as essential regulators of energy metabolism. Adrenoceptors stimulate the lipolysis in WAT and thermogenesis in BAT (except for α_2-AR with opposite effect) through coupling to G proteins in both rodent and human models. β_3-AR is of greatest important adrenoceptor in rodent adipose tissue. Based on scientists' observation among rodents, many efforts have been made to develop pharmaceutical products adjusting β_3-AR expression. However, studies about the function of β_2-AR in man is still limited and their results are still controversial. It is indicated that β-AR expression level and thermogenesis activity are suppressed under obese conditions, as well as under the inhibition effect of SCD1. An increase in SCD1 expression reduces oxidation and promotes the synthesis of fatty acids, while SCD1-deficient leads to enhance insulin sensitivity, hypermetabolism, and increase resistance to diet-induced obesity. SCD deficiency encourages the differentiation process of adipose tissue-derived mesenchymal stem cells into beige adipocytes, thus, it can be considered as a key enzyme to control beige adipogenesis. Although adrenoceptors and SCD1 are promising targets for metabolic disorders, further studies need to be conducted to evaluate their therapeutic results in clinical context.

Acknowledgments

This research is funded by Vietnam National Foundation for Science and Technology Development (NAFOSTED) under grant number 106.02-2019.314 (to Dinh Toi Chu).

We would like to thank Suong Mai Vu Ngoc and Thuy Tien Ho Thi (Center for Biomedicine and Community Health, International School, Vietnam National University, Hanoi, Vietnam) and Duong Vu Thuy (Vietnam Military Medical University, Hanoi, Vietnam) for partially collecting of references, and critical reading and checking to improve the manuscript. Dr. Le Bui Minh (NTT Hi-tech Institute, Nguyen Tat Thanh University, 300A Nguyen Tat Thanh St., Ward 13, District 4, Ho Chi Minh City, Vietnam) for the license to use BioRender to create the figures in this work.

References

1. Piché M-E, Tchernof A, Després J-P. Obesity phenotypes, diabetes, and cardiovascular diseases. *Circ Res*. 2020;126(11):1477–1500. https://doi.org/10.1161/CIRCRESAHA.120.316101.
2. Parsons MJ, Moffitt TE, Gregory AM, et al. Social jetlag, obesity and metabolic disorder: Investigation in a cohort study. *Int J Obes (Lond)*. 2015;39(5):842–848. https://doi.org/10.1038/ijo.2014.201.
3. WHO. *Obesity and Overweight*; 2021. https://www.who.int/news-room/fact-sheets/detail/obesity-and-overweight.
4. Karri S, Sharma S, Hatware K, Patil K. Natural anti-obesity agents and their therapeutic role in management of obesity: A future trend perspective. *Biomed Pharmacother*. 2019;110:224–238. https://doi.org/10.1016/j.biopha.2018.11.076.
5. Kalish VB. Obesity in Older Adults. *Prim Care - Clin Off Pract*. 2016;43(1):137–144. https://doi.org/10.1016/j.pop.2015.10.002.
6. Brahe LK, Astrup A, Larsen LH. Can we prevent obesity-related metabolic diseases by dietary modulation of the gut microbiota? *Adv Nutr*. 2016;7(1):90–101. https://doi.org/10.3945/an.115.010587.
7. Chu D-T, Minh Nguyet NT, Nga VT, et al. An update on obesity: Mental consequences and psychological interventions. *Diabetes Metab Syndr Clin Res Rev*. 2019;13(1):155–160. https://doi.org/10.1016/j.dsx.2018.07.015.
8. Khan MAB, Hashim MJ, King JK, Govender RD, Mustafa H, Al KJ. Epidemiology of type 2 diabetes - global burden of disease and forecasted trends. *J Epidemiol Glob Health*. 2020;10(1):107–111. https://doi.org/10.2991/jegh.k.191028.001.
9. Bellentani S, Scaglioni F, Marino M, Bedogni G. Epidemiology of non-alcoholic fatty liver disease. *Dig Dis*. 2010;28(1):155–161. https://doi.org/10.1159/000282080.

10. Mitra S, De A, Chowdhury A. Epidemiology of non-alcoholic and alcoholic fatty liver diseases. *Transl Gastroenterol Hepatol.* 2020;5:16. https://doi.org/10.21037/tgh.2019.09.08.
11. Akil L, Ahmad HA. Relationships between obesity and cardiovascular diseases in four southern states and Colorado. *J Health Care Poor Underserved.* 2011;22 (4 Suppl):61–72. https://doi.org/10.1353/hpu.2011.0166.
12. Colditz GA, Peterson LL. Obesity and cancer: Evidence, impact, and future directions. *Clin Chem.* 2018;64(1):154–162. https://doi.org/10.1373/clinchem.2017.277376.
13. Richard AJ, White U, Elks CM, Stephens JM. *Adipose Tissue: Physiology to Metabolic Dysfunction.* endotext.org; 2020.
14. Chu D-T, Tao Y. Human thermogenic adipocytes: A reflection on types of adipocyte, developmental origin, and potential application. *J Physiol Biochem.* 2017;73(1):1–4. https://doi.org/10.1007/s13105-016-0536-y.
15. Parlee SD, Lentz SI, Mori H, MacDougald OA. Chapter six - quantifying size and number of adipocytes in adipose tissue. In: Macdougald OA, ed. *Methods in Enzymology.* Academic Press; 2014:93–122.
16. Torres N, Vargas-Castillo AE, Tovar AR. Adipose tissue: White adipose tissue structure and function. In: Caballero B, Finglas PM, Toldrá F, eds. *Encyclopedia of Food and Health.* Academic Press; 2016:35–42.
17. Zhong W, Sun Q, Zhou Z. Chapter 12 - role of zinc in alcoholic liver disease. In: Patel VB, ed. *Molecular Aspects of Alcohol and Nutrition.* Academic Press; 2016:143–156.
18. Ning T, Wang QA. Adipose tissue remodeling and adipose precursors. In: Reis RL, ed. *Encyclopedia of Tissue Engineering and Regenerative Medicine.* Academic Press; 2019:457–462.
19. Wickramasinghe M, Weaver JU. Chapter 10 - lipid disorders in obesity. In: Weaver JU, ed. *Practical Guide to Obesity Medicine.* Elsevier; 2018:99–108.
20. Nedergaard J, Cannon B. 35 - Brown adipose tissue: Development and function. In: Polin RA, Abman SH, Rowitch DH, Benitz WE, Fox WW, eds. *Fetal and Neonatal Physiology (Fifth Edition).* Elsevier; 2017:354–363.
21. Dattani MT, Gevers EF. Chapter 22 - endocrinology of fetal development. In: Melmed S, Polonsky KS, Larsen PR, Kronenberg HM, eds. *Williams Textbook of Endocrinology (Thirteenth Edition).* Elsevier; 2016:849–892.
22. Martínez-Fernández L, Fernández-Galilea M, Felix-Soriano E, Escoté X, González-Muniesa P, Moreno-Aliaga MJ. Chapter 4 - inflammation and oxidative stress in adipose tissue: Nutritional regulation. In: del Moral AM, Aguilera García CM, eds. *Obesity.* Academic Press; 2018:63–92.
23. Chu D-T, Tao Y, Son LH, Le D-H. Cell source, differentiation, functional stimulation, and potential application of human thermogenic adipocytes in vitro. *J Physiol Biochem.* 2016;73(3):315–321. https://doi.org/10.1007/s13105-017-0567-z.
24. Zoico E, Rubele S, De Caro A, et al. Brown and Beige adipose tissue and aging. *Review.* 2019;10(368). https://doi.org/10.3389/fendo.2019.00368.
25. Chen Y, Pan R, Pfeifer A. Fat tissues, the brite and the dark sides. *Pflugers Arch.* 2016;468(11 – 12):1803–1807. https://doi.org/10.1007/s00424-016-1884-8.
26. Lee JH, Park A, Oh K-J, Lee SC, Kim WK, Bae K-H. The role of adipose tissue mitochondria: Regulation of mitochondrial function for the treatment of metabolic diseases. *Int J Mol Sci.* 2019;20(19). https://doi.org/10.3390/ijms20194924.
27. Forni MF, Peloggia J, Trudeau K, Shirihai O, Kowaltowski AJ. Murine mesenchymal stem cell commitment to differentiation is regulated by mitochondrial dynamics. *Stem Cells.* 2016;34(3):743–755. https://doi.org/10.1002/stem.2248.
28. Wang W, Seale P. Control of brown and beige fat development. *Nat Rev Mol Cell Biol.* 2016;17(11):691–702. https://doi.org/10.1038/nrm.2016.96.

29. Heinonen S, Buzkova J, Muniandy M, et al. Impaired mitochondrial biogenesis in adipose tissue in acquired obesity. *Diabetes*. 2015;64(9):3135. https://doi.org/10.2337/db14-1937.
30. Kim SH, Plutzky J. Brown fat and Browning for the treatment of obesity and related metabolic disorders. *Diabetes Metab J*. 2016;40(1):12–21. https://doi.org/10.4093/dmj.2016.40.1.12.
31. Bartelt A, Bruns OT, Reimer R, et al. Brown adipose tissue activity controls triglyceride clearance. *Nat Med*. 2011;17(2):200–205. https://doi.org/10.1038/nm.2297.
32. Chen KY, Brychta RJ, Abdul Sater Z, et al. Opportunities and challenges in the therapeutic activation of human energy expenditure and thermogenesis to manage obesity. *J Biol Chem*. 2020;295(7):1926–1942. https://doi.org/10.1074/jbc.REV119.007363.
33. Zhang Z, Dales NA, Winther MD. Opportunities and challenges in developing stearoyl-coenzyme a Desaturase-1 inhibitors as novel therapeutics for human disease. *J Med Chem*. 2014;57(12):5039–5056. https://doi.org/10.1021/jm401516c.
34. Peláez R, Pariente A, Pérez-Sala Á, Larráyoz IM. Sterculic acid: The mechanisms of action beyond stearoyl-CoA desaturase inhibition and therapeutic opportunities in human diseases. *Cell*. 2020;9(1):140. https://doi.org/10.3390/cells9010140.
35. Ohno M, Karagiannis P, Taniguchi Y. Protein expression analyses at the single cell level. *Molecules*. 2014;19(9):13932–13947. https://doi.org/10.3390/molecules190913932.
36. Buccitelli C, Selbach M. mRNAs, proteins and the emerging principles of gene expression control. *Nat Rev Genet*. 2020;21(10):630–644. https://doi.org/10.1038/s41576-020-0258-4.
37. Kuang J, Yan X, Genders AJ, Granata C, Bishop DJ. An overview of technical considerations when using quantitative real-time PCR analysis of gene expression in human exercise research. *PLoS One*. 2018;13(5), e0196438. https://doi.org/10.1371/journal.pone.0196438.
38. Cao WY, Liu Z, Guo F, Yu J, Li H, Yin X. Adipocyte ADRB3 Down-regulated in Chinese overweight individuals adipocyte ADRB3 in overweight. *Obes Facts*. 2018;11(6):524–533. https://doi.org/10.1159/000495116.
39. Merlin J, Sato M, Nowell C, et al. The PPARγ agonist rosiglitazone promotes the induction of brite adipocytes, increasing β-adrenoceptor-mediated mitochondrial function and glucose uptake. *Cell Signal*. 2018;42:54–66. https://doi.org/10.1016/j.cellsig.2017.09.023.
40. San Segundo-Val I, Sanz-Lozano CS. Introduction to the gene expression analysis. In: Isidoro García M, ed. *Molecular Genetics of Asthma*. Springer: New York; 2016:29–43.
41. Hulver MW, Berggren JR, Carper MJ, et al. Elevated stearoyl-CoA desaturase-1 expression in skeletal muscle contributes to abnormal fatty acid partitioning in obese humans. *Cell Metab*. 2005;2(4):251–261. https://doi.org/10.1016/j.cmet.2005.09.002.
42. Rivero EM, Martinez LM, Bruque CD, Gargiulo L, Bruzzone A, Lüthy IA. Prognostic significance of α- and β2-adrenoceptor gene expression in breast cancer patients. *Br J Clin Pharmacol*. 2019;85(9):2143–2154. https://doi.org/10.1111/bcp.14030.
43. Burl RB, Ramseyer VD, Rondini EA, Pique-Regi R, Lee Y-H, Granneman JG. Deconstructing adipogenesis induced by β3-adrenergic receptor activation with single-cell expression profiling. *Cell Metab*. 2018;28(2):300–309. https://doi.org/10.1016/j.cmet.2018.05.025.
44. Zheng R-H, Zhang Y-B, Qiu F-N, et al. NF-κB pathway play a role in SCD1 deficiency-induced ceramide de novo synthesis. *Cancer Biol Ther*. 2021;22(2):164–174. https://doi.org/10.1080/15384047.2021.1883414.
45. Mahmood T, Yang P-C. Western blot: Technique, theory, and trouble shooting. *N Am J Med Sci*. 2012;4(9):429–434. https://doi.org/10.4103/1947-2714.100998.
46. Taylor SC, Posch A. The design of a quantitative western blot experiment. *Biomed Res Int*. 2014;2014:361590. https://doi.org/10.1155/2014/361590.

47. Strosberg AD. Structure, function, and regulation of adrenergic receptors. *Protein Sci.* 1993;2(8):1198–1209. https://doi.org/10.1002/pro.5560020802.
48. Molinoff PB. α- and β-adrenergic receptor subtypes. *Drugs.* 1984;28(2):1–15. https://doi.org/10.2165/00003495-198400282-00002.
49. Vatner DE, Asai K, Iwase M, et al. Beta-adrenergic receptor G protein & adenylyl cyclase signal transduction in the failing heart. *Am J Cardiol.* 1999;83 (12):80–85. https://doi.org/10.1016/S0002-9149(99)00266-0.
50. Robidoux J, Martin TL, Collins S. β-Adrenergic receptors and regulation of energy expenditure: A family affair. *Annu Rev Pharmacol Toxicol.* 2004;44(1):297–323. https://doi.org/10.1146/annurev.pharmtox.44.101802.121659.
51. Evans BA, Merlin J, Bengtsson T, Hutchinson DS. Adrenoceptors in white, brown, and brite adipocytes. *Br J Pharmacol.* 2019;176(14):2416–2432. https://doi.org/10.1111/bph.14631.
52. Chu D-T, Nguyen Thi Phuong T, Tien NL, et al. The effects of adipocytes on the regulation of breast cancer in the tumor microenvironment: An update. *Cell.* 2019;8 (8). https://doi.org/10.3390/cells8080857.
53. Bolsoni-Lopes A, Alonso-Vale MI. Lipolysis and lipases in white adipose tissue - an update. *Arch Endocrinol Metab.* 2015;59(4):2359–4292 [Electronic].
54. Grujic D, Susulic VS, Harper M-E, et al. Adrenergic receptors on White and Brown adipocytes mediate - selective agonist-induced effects on energy expenditure, insulin secretion, and food intake: A STUDY USING TRANSGENIC AND GENE KNOCKOUT MICE *. *J Biol Chem.* 1997;272(28):17686–17693. https://doi.org/10.1074/jbc.272.28.17686.
55. Revelli JP, Preitner F, Samec S, et al. Targeted gene disruption reveals a leptin-independent role for the mouse beta3-adrenoceptor in the regulation of body composition. *J Clin Invest.* 1997;100(5):1098–1106. https://doi.org/10.1172/JCI119620.
56. Germack R, Starzec AB, Vassy R, Perret GY. β-Adrenoceptor subtype expression and function in rat white adipocytes. *Br J Pharmacol.* 1997;120(2):201–210. https://doi.org/10.1038/sj.bjp.0700885.
57. Jimenez M, Léger B, Canola K, et al. β1/β2/β3-adrenoceptor knockout mice are obese and cold-sensitive but have normal lipolytic responses to fasting. *FEBS Lett.* 2002;530 (1):37–40. https://doi.org/10.1016/S0014-5793(02)03387-2.
58. Preite NZ, Nascimento BPP, Muller CR, et al. Disruption of beta3 adrenergic receptor increases susceptibility to DIO in mouse. *J Endocrinol.* 2016;231(3):259–269. https://doi.org/10.1530/JOE-16-0199.
59. Bachman ES, Dhillon H, Zhang C-Y, et al. βAR signaling required for diet-induced thermogenesis and obesity resistance. *Science.* 2002;297(5582):843. https://doi.org/10.1126/science.1073160.
60. Revelli JP, Muzzin P, Paoloni A, Moinat M, Giacobino JP. Expression of the β3-adrenergic receptor in human white adipose tissue. *J Mol Endocrinol.* 1993;10 (2):193–197. https://doi.org/10.1677/jme.0.0100193.
61. Braun K, Oeckl J, Westermeier J, Li Y, Klingenspor M. Non-adrenergic control of lipolysis and thermogenesis in adipose tissues. *J Exp Biol.* 2018;221(Suppl_1). https://doi.org/10.1242/jeb.165381.
62. Evans BA, Agar L, Summers RJ. The role of the sympathetic nervous system in the regulation of leptin synthesis in C57BL/6 mice. *FEBS Lett.* 1999;444(2–3):149–154. https://doi.org/10.1016/S0014-5793(99)00049-6.
63. Chouchani ET, Kazak L, Spiegelman BM. New advances in adaptive thermogenesis: UCP1 and beyond. *Cell Metab.* 2019;29(1):27–37. https://doi.org/10.1016/j.cmet.2018.11.002.
64. Riis-Vestergaard MJ, Richelsen B, Bruun JM, Li W, Hansen JB, Pedersen SB. Beta-1 and not Beta-3 adrenergic receptors may be the primary regulator of human Brown adipocyte metabolism. *J Clin Endocrinol Metabol.* 2020;105(4):e994–e1005. https://doi.org/10.1210/clinem/dgz298.

65. Cero C, Lea HJ, Zhu KY, Shamsi F, Tseng Y-H, Cypess AM. β3-adrenergic receptors regulate human brown/beige adipocyte lipolysis and thermogenesis. *JCI Insight*. 2021;6 (11), e139160. https://doi.org/10.1172/jci.insight.139160.
66. Fernandes GW, Ueta CB, Fonseca TL, et al. Inactivation of the adrenergic receptor β2 disrupts glucose homeostasis in mice. *J Endocrinol*. 2014;221(3):381–390. https://doi.org/10.1530/JOE-13-0526.
67. Finlin BS, Memetimin H, Zhu B, et al. The β3-adrenergic receptor agonist mirabegron improves glucose homeostasis in obese humans. *J Clin Invest*. 2020;130(5):2319–2331. https://doi.org/10.1172/JCI134892.
68. Fredriksson JM, Lindquist JM, Bronnikov GE, Nedergaard J. Norepinephrine induces vascular endothelial growth factor gene expression in Brown adipocytes through a adrenoreceptor/cAMP/protein kinase a pathway involving Src but independently of Erk1/2. *J Biol Chem*. 2000;275(18):13802–13811. https://doi.org/10.1074/jbc.275.18.13802.
69. Lowell MDPBB, Flier MJS. Brown adipose tissue, β3-adrenergic receptors, and obesity. *Annu Rev Med*. 1997;48(1):307–316. https://doi.org/10.1146/annurev.med.48.1.307.
70. Mattsson CL, Csikasz RI, Chernogubova E, et al. β1-adrenergic receptors increase UCP1 in human MADS brown adipocytes and rescue cold-acclimated β3-adrenergic receptor-knockout mice via nonshivering thermogenesis. *Am J Physiol Endocrinol Metab*. 2011;301(6):E1108–E1118. https://doi.org/10.1152/ajpendo.00085.2011.
71. Collins S, Surwit RS. The beta-adrenergic receptors and the control of adipose tissue metabolism and thermogenesis. *Recent Prog Horm Res*. 2001;56(0079–9963 (Print)):309–328.
72. Collins S, Daniel KW, Rohlfs EM, Ramkumar V, Taylor IL, Gettys TW. Impaired expression and functional activity of the beta 3- and beta 1-adrenergic receptors in adipose tissue of congenitally obese (C57BL/6J Ob/Ob) mice. *Mol Endocrinol*. 1994;8 (4):518–527. https://doi.org/10.1210/mend.8.4.7914350.
73. Jiang C, Carillion A, Na N, et al. Modification of the β-adrenoceptor stimulation pathway in Zucker obese and obese diabetic rat myocardium*. *Crit Care Med*. 2015;43(7).
74. Kurylowicz A, Jonas M, Lisik W, et al. Obesity is associated with a decrease in expression but not with the hypermethylation of thermogenesis-related genes in adipose tissues. *J Transl Med*. 2015;13:31. https://doi.org/10.1186/s12967-015-0395-2.
75. Stich V, Marion-Latard F, Hejnova J, et al. Hypocaloric diet reduces exercise-induced α2-adrenergic antilipolytic effect and α2-adrenergic receptor mRNA levels in adipose tissue of obese women. *J Clin Endocrinol Metabol*. 2002;87(3):1274–1281. https://doi.org/10.1210/jcem.87.3.8349.
76. Zouhal H, Lemoine-Morel S, Mathieu M-E, Casazza GA, Jabbour G. Catecholamines and obesity: Effects of exercise and training. *Sports Med*. 2013;43(7):591–600. https://doi.org/10.1007/s40279-013-0039-8.
77. Harant I, Marion-Latard F, Crampes F, et al. Effect of a long-duration physical exercise on fat cell lipolytic responsiveness to adrenergic agents and insulin in obese men. *Int J Obes (Lond)*. 2002;26(10):1373–1378. https://doi.org/10.1038/sj.ijo.0802072.
78. AL AM, Syed DN, Ntambi JM. Insights into stearoyl-CoA Desaturase-1 regulation of systemic metabolism. *Trends Endocrinol Metab*. Dec 2017;28(12):831–842. https://doi.org/10.1016/j.tem.2017.10.003.
79. Liu K, Lin L, Li Q. Scd1 controls de novo beige fat biogenesis through succinate-dependent regulation of mitochondrial complex II. *Proc Natl Acad Sci U S A*. 2020;117(5):2462–2472.
80. Nagao K, Murakami A, Umeda M. Structure and function of Δ9-fatty acid desaturase. *Chem Pharm Bull(Tokyo)*. 2019;67(4):327–332.
81. Heinemann FS, Korza G, Ozols J. A plasminogen-like protein selectively degrades stearoyl-CoA desaturase in liver microsomes. *J Biol Chem*. 2003;278(44):42966–42975.

82. Heinemann FS, Mziaut H, Korza G. A microsomal endopeptidase from liver that preferentially degrades stearoyl-CoA desaturase. *Biochemistry*. 2003;42(33):6929–6937.
83. Ntambi JM, Miyazaki M, Stoehr JP. Loss of stearoyl-CoA desaturase-1 function protects mice against adiposity. *Proc Natl Acad Sci U S A*. 2002;99(17):11482–11486.
84. Mauvoisin D, Mounier C. Hormonal and nutritional regulation of SCD1 gene expression. *Biochimie*. 2011;93(1):78–86. https://doi.org/10.1016/j.biochi.2010.08.001.
85. Mainieri D, Montani JP, Seydoux J, Giacobino JP, Boss O, Dulloo AG. Beta-adrenergic control of stearoyl-CoA desaturase 1 repression in relation to sympathoadrenal regulation of thermogenesis. *Int J Obes (Lond)*. 2007;31(2):378–381. https://doi.org/10.1038/sj.ijo.0803424.
86. Miyazaki M, Flowers MT, Sampath H, et al. Hepatic stearoyl-CoA desaturase-1 deficiency protects mice from carbohydrate-induced adiposity and hepatic steatosis. *Cell Metab*. 2007;6(6):484–496. https://doi.org/10.1016/j.cmet.2007.10.014.
87. Sampath H, Flowers MT, Liu X, et al. Skin-specific deletion of stearoyl-CoA desaturase-1 alters skin lipid composition and protects mice from high fat diet-induced obesity. *J Biol Chem*. 2009;284(30):19961–19973. https://doi.org/10.1074/jbc.M109.014225.
88. Biddinger SB, Miyazaki M, Boucher J, Ntambi JM, Kahn CR. Leptin suppresses stearoyl-CoA desaturase 1 by mechanisms independent of insulin and sterol regulatory element-binding protein-1c. *Diabetes*. 2006;55(7):2032–2041. https://doi.org/10.2337/db05-0742.
89. de Fourmestraux V, Neubauer H, Poussin C, et al. Transcript profiling suggests that differential metabolic adaptation of mice to a high fat diet is associated with changes in liver to muscle lipid fluxes. *J Biol Chem*. 2004;279(49):50743–50753. https://doi.org/10.1074/jbc.M408014200.
90. Toye AA, Dumas ME, Blancher C, et al. Subtle metabolic and liver gene transcriptional changes underlie diet-induced fatty liver susceptibility in insulin-resistant mice. *Diabetologia*. 2007;50(9):1867–1879. https://doi.org/10.1007/s00125-007-0738-5.
91. Schwenk RW, Jonas W, Ernst SB, Kammel A, Jahnert M, Schurmann A. Diet-dependent alterations of hepatic Scd1 expression are accompanied by differences in promoter methylation. *Horm Metab Res*. 2013;45(11):786–794. https://doi.org/10.1055/s-0033-1348263.
92. Puri P, Wiest MM, Cheung O, et al. The plasma lipidomic signature of nonalcoholic steatohepatitis. *Hepatology*. 2009;50(6):1827–1838. https://doi.org/10.1002/hep.23229.
93. Sjogren P, Sierra-Johnson J, Gertow K, et al. Fatty acid desaturases in human adipose tissue: Relationships between gene expression, desaturation indexes and insulin resistance. *Diabetologia*. 2008;51(2):328–335. https://doi.org/10.1007/s00125-007-0876-9.
94. Warensjo E, Riserus U, Vessby B. Fatty acid composition of serum lipids predicts the development of the metabolic syndrome in men. *Diabetologia*. 2005;48(10):1999–2005. https://doi.org/10.1007/s00125-005-1897-x.
95. Bodis K, Kahl S, Simon MC, et al. Reduced expression of stearoyl-CoA desaturase-1, but not free fatty acid receptor 2 or 4 in subcutaneous adipose tissue of patients with newly diagnosed type 2 diabetes mellitus. *Nutr Diabetes*. 2018;8(1):49. https://doi.org/10.1038/s41387-018-0054-9.
96. Sampath H, Miyazaki M, Dobrzyn A, Ntambi JM. Stearoyl-CoA desaturase-1 mediates the pro-lipogenic effects of dietary saturated fat. *J Biol Chem*. 2007;282(4):2483–2493. https://doi.org/10.1074/jbc.M610158200.
97. Cohen P, Miyazaki M, Socci ND, et al. Role for stearoyl-CoA desaturase-1 in leptin-mediated weight loss. *Science*. 2002;297(5579):240–243. https://doi.org/10.1126/science.1071527.

98. Cruz-Color L, Hernandez-Nazara ZH, Maldonado-Gonzalez M, et al. Association of the PNPLA2, SCD1 and leptin expression with fat distribution in liver and adipose tissue from obese subjects. *Exp Clin Endocrinol Diabetes*. 2020;128(11):715–722. https://doi.org/10.1055/a-0829-6324.
99. Mika A, Korczynska J, al e. Visceral and subcutaneous adipose tissue stearoyl-CoA desaturase-1 mRNA levels and fatty acid desaturation index positively correlate with BMI in morbidly obese women. *Eur J Lipid Sci Technol*. 2015;117:926–932.
100. Sramkova V, Koc M, Krauzova E, et al. Expression of lipogenic markers is decreased in subcutaneous adipose tissue and adipocytes of older women and is negatively linked to GDF15 expression. *J Physiol Biochem*. 2019;75(3):253–262. https://doi.org/10.1007/s13105-019-00676-6.
101. Tao L, Guo X, Xu M, et al. Dexmedetomidine ameliorates high-fat diet-induced nonalcoholic fatty liver disease by targeting SCD1 in obesity mice. *Pharmacol Res Perspect*. 2021;9(1), e00700. https://doi.org/10.1002/prp2.700.
102. Iwase M, Tokiwa S, Seno S, et al. Glycerol kinase stimulates uncoupling protein 1 expression by regulating fatty acid metabolism in beige adipocytes. *J Biol Chem*. 2020;295(20):7033–7045. https://doi.org/10.1074/jbc.RA119.011658.
103. Lee SH, Dobrzyn A, Dobrzyn P, Rahman SM, Miyazaki M, Ntambi JM. Lack of stearoyl-CoA desaturase 1 upregulates basal thermogenesis but causes hypothermia in a cold environment. *J Lipid Res*. 2004;45(9):1674–1682. https://doi.org/10.1194/jlr.M400039-JLR200.
104. Tao X, Hu Y, Li L, et al. Genetic deletion of beta2 adrenergic receptors exacerbates hepatocellular lipid accumulation in high-fat diet mice. *Biochem Biophys Res Commun*. 2019;511(1):73–78. https://doi.org/10.1016/j.bbrc.2019.02.037.

CHAPTER THIRTEEN

Low density lipoprotein receptor endocytosis in cardiovascular disease and the factors affecting LDL levels

Rupesh Maurya[a], Gargi Bhattacharjee[a], Nisarg Gohil[a], Khushal Khambhati[a], Shreya Shakhreliya[b], Aaradhya Bhatnagar[a], Priyanka Singh[a], Suresh Ramakrishna[c,d], and Vijai Singh[a,*]

[a]Department of Biosciences, School of Science, Indrashil University, Rajpur, Mehsana, Gujarat, India
[b]Department of Science and Technology, Nottingham Trent University, Nottingham, United Kingdom
[c]College of Medicine, Hanyang University, Seoul, South Korea
[d]Graduate School of Biomedical Science and Engineering, Hanyang University, Seoul, South Korea
*Corresponding author: e-mail addresses: vijaisingh15@gmail.com; vijai.singh@indrashiluniversity.edu.in

Contents

Abstract

Cardiovascular disease (CVD) is the one of major global health issues with approximately 30% of the mortality reported in the mid-income population. Low-density lipoprotein (LDL) plays a crucial role in development of CVD. High LDL along with others forms a plaque and blocks arteries, resulting in CVD. The present chapter deals with the mechanism of receptor-mediated endocytosis of LDL and its management by drugs such as statins and PCSK9 inhibitors along with dietary supplementation for health improvements.

Progress in Molecular Biology and Translational Science, Volume 194
ISSN 1877-1173
https://doi.org/10.1016/bs.pmbts.2022.09.010

1. Introduction

In the past few decades, cardiovascular disease (CVD) is one of the reasons which prevalent and contributes to high rate of mortality. CVD is contributing to about 30% of total mortality in the world of low and middle-income populations.[1] CVD is significantly increasing because of many factors such as smoking,[2] lifestyles,[3–6] food habits,[7,8] nutrition,[9] environmental factors,[10] ageing,[11] stress,[12] and other socioeconomic factors[13]. The total expenditure per patient for the treatment of CVD is reached as high as about 3842 USD[14]. Low-density lipoprotein (LDL) plays a major role in the development of CVD. LDL along with others forms a plaque which blocks arteries, such condition is known as atherosclerosis. LDL level should be less than 70 mg/dL in heart patients. Higher LDL levels can block blood supply in arteries and leads to a heart attack.

In the 1970s, Michael Brown and Joseph Goldstein discovered the activity of LDL as well as a new cell surface receptor called the LDL receptor (LDLR) in familial hypercholesterolemia disease. LDLR helps in the metabolism of cholesterol through feedback inhibition of the hydroxymethylglutaryl CoA reductase enzyme,[15] whereas, LDLR plays a vital role in lipid metabolism, cholesterol homeostasis, receptor-mediated endocytosis, receptor recycling, and feedback regulation of receptors.[16] LDLR also plays an important role in drug delivery,[17] LDLR-related protein-1 plays a significant role in inflammation in injuries to the nervous system, cancer, and atherosclerosis.[18] *LDLR* gene is a mosaic protein of 839 amino acids which is located on chromosome 19 (19p13. 1–13.3), with a length of 45 kb and 18 exons and 17 intervening introns.[19,20] The molecular weight of mature LDLR protein is 160 kDa[21] and it is identified in almost all tissues, but the liver is especially essential for the absorption of plasma LDL-c, which is eliminated in the hepatocytes up to 70%. Transcription of the *LDLR* gene is carried out in all tissues and modulated by a sterol negative feedback mechanism. The promoter region comprises of two 7-base pair segments that include TATA and CT sequences. The promoter's upstream (50) region is significant for gene transcription.[22,23] There are three non-identical 16-bp long regions that are located in between three repetitive regions, transcription enhancer, sterol regulatory element-1 (SRE-1-membrane-bound transcription factor) which are responsible for gene expression and mediated by sterol metabolites. While in the lack of signaling molecules, SRE-1 helps to increase transcription.[23,24] However, in extreme conditions (due to high cholesterol levels), the two persistent repetitive regions are in charge of binding sites for enhancer factors Sp1 and together with SREBP-1. The inhibition of transcription is regulated by enhancing cholesterol in cells that is recognized by SRE-binding protein. In this condition, sterols act as an end product which stimulates the positive enhancers that bind to the promoter region.[25] The most evident member of the LDLR family is first class- very LDLR and apolipoprotein E receptor 2 (LRP 8), second class- giant LDL-related protein 1(LRP 1) and megalin (LRP 2), third class- LRP5 and LRP6, and the fourth class includes LR11, LRP4, ST7 and LRP 3.[26,27] The LDLR is composed of five structurally varied domains and each performs an intended function.

An N-terminal ligand-binding domain has ~40 amino acids which include seven cysteine-rich repeating units (R1-R7 or LDLR class A repeats). It encodes with 2–6 exon codes. The three disulfide-bonds maintain the stability of the two-loop conformation of R1, R2, R6 and R5. The appearance of R2-R7 and EGF repeat A unit helps to bind LDL and R5 unit helps to interact with very-low-density lipoprotein (VLDL).[28,29] The ectodomain in the receptor is an epidermal growth factor precursor homology region which consists of ~ 40 amino acids and three EGF- like repetitive units (A, B and six YWTD motifs (LBL-B repeats) bladed b-propeller bound by a third-repeat C).[30] After the knock-out of the second region of LDL-R, it helps to prevent the acid-dependent dissociation of ligands in endosomes. It seems that the EGF-precursor homology region is responsible for releasing ligand molecules at low pH in the lysosomes and recycled receptors to the cell surface. It is regulated by 7–14 exon codes.[31,32] The O-linked glycan region has 50 amino acids and is glycosylated with 18 serine and threonine, and is encoded by 15 exon codes.[33] There are three primary functions of the third domain: deletion of this region is reasonable for ligand binding, endocytosis, and receptors degradation; O-glycosylation prevents from denaturation and maintains the rate of proteolytic cleavage at the cell surface.[34] The membrane-spanning hydrophobic domain has 24 amino acids, and is attached to the phospholipid bilayer.[35] The transmembrane region of LDLR is transcribed by exon 16 and some regions of exon 17. The remaining region of 17 exons and exon 18 is coded for cytoplasmic elements. The cytosolic domain has 50 amino acids and alanine at C- the terminal.[36,37] It plays a vital role in locating clathrin-coated pits (FDNPVY sequence), endocytosis and intracellular transport.[38]

2. Receptor-mediated endocytosis of LDL

The amount of lipoprotein and cholesterol in the blood mediates the receptor molecule. The increasing or decreasing number of receptor molecules depends upon the level of cellular cholesterol in the body. This helps to avoid the cell from producing excessive intracellular LDL cholesterol.[39] LDLR complexes are further covered by clathrin-coated pits, which engulf the cell membrane and convert the LDLR complexes to endosomes. However, due to the acidic medium, the receptor-lipoprotein bond is broken, and ATP-driven proton pumps dissociate LDL. LDLR recycles back to the cell surface and LDL molecules are broken down by lysosomes into cholesterol and further hydrolysis of cholesterol esters takes place, which stimulates the expression of LDLRs.[40] Endoplasmic reticulum (ER) is associated with the proteins aster (A, B, C). These proteins bind with phosphatidylserine and help aster proteins for the contact of ER in response to the storage of cholesterol in the plasma membrane.[41] There are various kinds of endocytosis. Cells use the process of endocytosis for the internalization of molecules or cargo. Clathrin-mediated endocytosis,[42] caveolae-mediated endocytosis,[43,44] macropinocytosis,[45] transcytosis[46] are the major endocytosis process that is followed by the cell for endocytosis of lipoproteins.

3. Factors influencing LDL levels

3.1 Dietary improvements for LDL management

Several factors can affect LDL levels such as smoking, physical activity, alcohol, diet and obesity.[47] Alginate oligosaccharide lowers plasma LDL-cholesterol levels through the regulation of LDLR expression.[48] The choice of saturated fatty acids instead of carbohydrates or unsaturated fats has enhanced the LDL levels. The enhancement of small size of LDL has been observed due to intake of refined sugar or added sugars. The intake of refined sugar or added sugar has an adverse effect on high-density lipoprotein. The replacement of saturated fatty acid becomes really tricky and important question. Clinical and epidemiologic studies have answered the question. The replacement of saturated fatty acid with polyunsaturated fat is helpful for coronary heart disease. Minimum processed food grains and polyunsaturated fat are recommended for dietary intake. Plant-based proteins can be used for the substitution of saturated fatty acids. Dairy fat and meat in the diet are considered as one the factors for increasing LDL levels. Milk cheese is associated with comparatively lesser cardiovascular risk.[49,50]

Plant-based sterols,[51] stanols,[52] sitosterol[53] play a significant role in lowering LDL cholesterol. Plants can be used for the management of LDL cholesterol such as beans,[54–56] soluble dietary fiber,[57] pulses,[58] high-selenium lentil,[59] *Brassica oleracea*, cabbage[60] and *Aronia melanocarpa* fruit juice.[61] Fish oil,[62] sunflower oil,[63] high-oleic canola oil,[64] a blend of sesame oil and rice bran oil (20:80),[65] sesame oil[66] reduce cardiovascular risk.[67]

3.2 Proprotein convertase subtilisin/kexin type 9 (PCSK9)

PCSK9 is an enzyme that plays an important role in cholesterol metabolism. It can be one of the important targets for the treatment of hypercholesterolemia,[68] *PCSK* gene mutation leads to the reduction in LDLRs which causes an increase in LDL levels due to gain of function.[69] On the other hand, mutation is causing a loss of function where the mutation is causing an enhancement in the LDLR results in a decrease in LDL level.[70,71] Statin and fibrate are being used to enhance the level of PCSK9.[72] It has been earlier reported that about 10–20% of patients are statin intolerant.[73] AMG145 is a monoclonal antibody that is used for PCSK9 for the reduction of LDL levels.[74] PCSK9 binds to the receptor and promotes the degradation of LDLRs. By preventing LDLRs, LDL cholesterol can be lowered up to 50%. Alirocumab and evolocumab are USFDA-approved drugs used for lowering lipids.[75]

3.3 Statins

Statins can significantly reduce the LDL and they are very efficient in enhancing HDL. Statins are a specific, competitive inhibitor of hydroxymethylglutaryl-CoA (HMG-CoA) reductase. During the 1970s, Akira Endo, a Japanese microbiologist during the exploration of antimicrobial agents, first found an incredible inhibitory impact on HMG-CoA reductase(statin), including ML236B (compactin) in a fermentation stock of *Penicillium citrinum*.[76] It has several effects such as calming,[77] antithrombotic,[78] non-lipid lowering,[79] antiplatelet,[80] anticoagulant,[81] in COVID treatment,[82] pleiotropic,[83] pharmacokinetic,[84] neuroprotective,[85] anti-inflammatory, immunomodulatory,[86] anti-ischemic anti-anginal[87] and tumor suppression.[88] As given in Table 1, currently, statins are available in different drugs such as simvastatin, rosuvastatin, atorvastatin, pravastatin, atorvastatin, and fluvastatin.[99] Atorvastatin, simvastatin, and lovastatin show a significant effect on cholesterol

Table 1 Drugs for lowering cholesterol levels.

Drugs	Percentage cholesterol lowering	References
Pitavastatin	33.3–54.7	89
Etofibrate	15–18	90
Gemfibrozil	13	91
Hibiscus sabdariffa extract	12	92
Statin	30	93
Ezetimibe	12–15	94
Alirocumab (PCSK9 inhibitor)	40	95
Clofibrate	9	96
Rosuvastatin	45	97
Atorvastatin	25–60	98

metabolism,[100] statins inhibit HMG-CoA reductase by restricting the enzyme's active site and stopping the substrate-product change condition of the enzyme.[101] Rosuvastatin helps to improve atherogenic and atheroprotective lipid profiles which is an additional benefit.[102] Total 40 mg/kg dose of rosuvastatin, a synthetic drug, has shown a reduction of glucometabolic, changes of which are induced by high-fat-diet (HFD) with amelioration of pancreatic β-cell.[103] Gemfibrozil, *Hibiscus sabdariffa* extract, Ezetimibe and Etofibrate are used to lower cholesterol levels up to 12–18%. Statins are being used for lowering 30% ofLDL levels. Pitavastatin and Alirocumab (PCSK9 inhibitor are used to lower LDL with the rate of 33.3–54.7% and 40%, respectively.

3.4 CRISPR-Cas9 for gene editing for downregulating LDL level

CRISPR (clustered regularly interspaced short palindromic repeats) and CRISPR-associated protein (Cas) is an RNA-mediated immunity of bacteria and archaea.[104–106] It is used for genome editing of a wide range of organisms.[107–112] CRISPR-Cas9 system has been used for repairing defective gene mutation for restoring function.[113,114] CRISPR has been explored for several applications such as the development of disease-resistant crops,[115–117] therapeutic[118–120] and diagnostic.[121–123] In a study, Ding et al.[124] used the CRISPR-*Pcsk9* virus and the GFP virus for 2–11-week-old male mice. CRISPR-*Pcsk9* mice have shown substantial ≈50% mutagenesis which altered half of the *Pcsk9* alleles in the liver. Genome editing with the CRISPR–Cas9 system had been used significantly to disrupt the *Pcsk9* gene *in vivo* and decreased blood cholesterol levels in mice. Such an approach can also be used to protect heart disease in humans.

Very interesting work has been done by Zhang et al.,[125] in which, they have developed nuclear localization signal tagged with Cas9 along with *Pcsk9* that can be targeted by sgRNA. It was modified with cationic HIV-1- transactivating transcriptor (TAT) peptide. Subsequently, it was encapsulated in galactose modified lipid layer for targeting nanoclusters in the liver. They have found 60 % *Pcsk9*-editing efficiency using nanoclusters *in vitro* and also decrease 30% of plasma LDL in mice. In another study, Guo et al.[126] used Syrian golden hamsters because they have a similar metabolic profile as humans. Triglyceride levels have been reduced after the deletion of *Apoc3* genes by using CRISPR-Cas9.

4. Conclusion and future perspectives

CVDs are one of the major causes of mortality globally and it is estimated that 17.9 million people died from CVDs in 2019, representing 32% of all global deaths. LDL plays a vital role for promoting CVD and drugs such as statins and PCSK9 inhibitors along with dietary supplementation help to improve overall health by reducing LDL levels. Based on the mechanism of endocytosis of LDLRs and its binding to LDL proteins have helped the scientific community to design various medication ways for protection against CVD. Advanced biotechnological techniques also have the potential for the future development of diagnostic and therapeutic markers of CVDs. Genome-engineering technology for the editing of cholesterol genes can be an advantage for future health and therapy.

Acknowledgment

N.G. and G.B. acknowledge the Indian Council of Medical Research of the Government of India for financial assistance as Senior Research Fellowship File No. 5/3/8/63/ITR-F/2020 and File No. 5/3/8/41/ITR-F/2022-ITR, respectively. P.S. gratefully acknowledges the Gujarat State Biotechnology Mission for financial assistance as Junior Research Fellowship (Project ID: 5LY45F). The financial assistance from National Research Foundation of Korea (2021M3A9H3015389) and the Korean Fund for Regenerative Medicine (KFRM) grant funded by the Korea government (the Ministry of Science and ICT, the Ministry of Health & Welfare) (22A0304L1-01) to S.R. is acknowledged.

References

1. Murray CJ, Lopez AD, Organization WH. *The global burden of disease: A comprehensive assessment of mortality and disability from diseases, injuries, and risk factors in 1990 and projected to 2020: Summary*. World Health Organization; 1996.

2. Chen Z, Boreham J. Smoking and cardiovascular disease. *Semin Vasc Med.* 2002;2(3):243–252.
3. Barbaresko J, Rienks J, Nothlings U. Lifestyle indices and cardiovascular disease risk: A Meta-analysis. *Am J Prev Med.* 2018;55(4):555–564.
4. Iso H. Lifestyle and cardiovascular disease in Japan. *J Atheroscler Thromb.* 2011;18(2):83–88.
5. Kelishadi R, Poursafa P. A review on the genetic, environmental, and lifestyle aspects of the early-life origins of cardiovascular disease. *Curr Probl Pediatr Adolesc Health Care.* 2014;44(3):54–72.
6. Mozaffarian D, Wilson PW, Kannel WB. Beyond established and novel risk factors: Lifestyle risk factors for cardiovascular disease. *Circulation.* 2008;117(23):3031–3038.
7. Anand SS, Hawkes C, de Souza RJ, et al. Food consumption and its impact on cardiovascular disease: Importance of solutions focused on the globalized food system: A report from the Workshop convened by the world heart federation. *J Am Coll Cardiol.* 2015;66(14):1590–1614.
8. Kaikkonen JE, Mikkila V, Raitakari OT. Role of childhood food patterns on adult cardiovascular disease risk. *Curr Atheroscler Rep.* 2014;16(10):443.
9. Getz GS, Reardon CA. Nutrition and cardiovascular disease. *Arterioscler Thromb Vasc Biol.* 2007;27(12):2499–2506.
10. Cosselman KE, Navas-Acien A, Kaufman JD. Environmental factors in cardiovascular disease. *Nat Rev Cardiol.* 2015;12(11):627–642.
11. Samani NJ, van der Harst P. Biological ageing and cardiovascular disease. *Heart.* 2008;94(5):537–539.
12. Kivimaki M, Kawachi I. Work stress as a risk factor for cardiovascular disease. *Curr Cardiol Rep.* 2015;17(9):630.
13. Kaplan GA, Keil JE. Socioeconomic factors and cardiovascular disease: A review of the literature. *Circulation.* 1993;88(4 Pt 1):1973–1998.
14. Kumar A, Siddharth V, Singh SI, Narang R. Cost analysis of treating cardiovascular diseases in a super-specialty hospital. *PLoS One.* 2022;17(1), e0262190.
15. Goldstein JL, Brown MS. The LDL receptor. *Arterioscler Thromb Vasc Biol.* 2009;29(4):431–438.
16. Brown MS, Goldstein JL. A receptor-mediated pathway for cholesterol homeostasis. *Science.* 1986;232(4746):34–47.
17. Chung NS, Wasan KM. Potential role of the low-density lipoprotein receptor family as mediators of cellular drug uptake. *Adv Drug Deliv Rev.* 2004;56(9):1315–1334.
18. Gonias SL, Campana WM. LDL receptor-related protein-1: A regulator of inflammation in atherosclerosis, cancer, and injury to the nervous system. *Am J Pathol.* 2014;184(1):18–27.
19. Al-Khateeb A, Zahri MK, Mohamed MS, et al. Analysis of sequence variations in low-density lipoprotein receptor gene among Malaysian patients with familial hypercholesterolemia. *BMC Med Genet.* 2011;12:40.
20. Galicia-Garcia U, Benito-Vicente A, Uribe KB, et al. Mutation type classification and pathogenicity assignment of sixteen missense variants located in the EGF-precursor homology domain of the LDLR. *Sci Rep.* 2020;10(1):1727.
21. Tolleshaug H, Goldstein JL, Schneider WJ, Brown MS. Posttranslational processing of the LDL receptor and its genetic disruption in familial hypercholesterolemia. *Cell.* 1982;30(3):715–724.
22. Lindgren V, Luskey KL, Russell DW, Francke U. Human genes involved in cholesterol metabolism: Chromosomal mapping of the loci for the low density lipoprotein receptor and 3-hydroxy-3-methylglutaryl-coenzyme A reductase with cDNA probes. *Proc Natl Acad Sci U S A.* 1985;82(24):8567–8571.

23. Sudhof TC, Van der Westhuyzen DR, Goldstein JL, Brown MS, Russell DW. Three direct repeats and a TATA-like sequence are required for regulated expression of the human low density lipoprotein receptor gene. *J Biol Chem*. 1987;262(22):10773–10779.
24. Goldstein JL, Brown MS. Regulation of the mevalonate pathway. *Nature*. 1990;343 (6257):425–430.
25. Rajavashisth TB, Taylor AK, Andalibi A, Svenson KL, Lusis AJ. Identification of a zinc finger protein that binds to the sterol regulatory element. *Science*. 1989;245(4918): 640–643.
26. May P, Bock HH, Herz J. Integration of endocytosis and signal transduction by lipoprotein receptors. *Sci STKE*. 2003;2003(176):PE12.
27. Schneider WJ, Nimpf J. LDL receptor relatives at the crossroad of endocytosis and signaling. *Cell Mol Life Sci*. 2003;60(5):892–903.
28. Clayton D, Brereton IM, Kroon PA, Smith R. Three-dimensional NMR structure of the sixth ligand-binding module of the human LDL receptor: Comparison of two adjacent modules with different ligand binding specificities. *FEBS Lett*. 2000;479 (3):118–122.
29. Daly NL, Djordjevic JT, Kroon PA, Smith R. Three-dimensional structure of the second cysteine-rich repeat from the human low-density lipoprotein receptor. *Biochemistry*. 1995;34(44):14474–14481.
30. Springer TA. An extracellular beta-propeller module predicted in lipoprotein and scavenger receptors, tyrosine kinases, epidermal growth factor precursor, and extracellular matrix components. *J Mol Biol*. 1998;283(4):837–862.
31. Davis CG, Goldstein JL, Sudhof TC, Anderson RG, Russell DW, Brown MS. Acid-dependent ligand dissociation and recycling of LDL receptor mediated by growth factor homology region. *Nature*. 1987;326(6115):760–765.
32. Rudenko G, Henry L, Henderson K, et al. Structure of the LDL receptor extracellular domain at endosomal pH. *Science*. 2002;298(5602):2353–2358.
33. Defesche JC. Low-density lipoprotein receptor- -its structure, function, and mutations. *Semin Vasc Med*. 2004;4(1):5–11.
34. Magrane J, Casaroli-Marano RP, Reina M, Gafvels M, Vilaro S. The role of O-linked sugars in determining the very low density lipoprotein receptor stability or release from the cell. *FEBS Lett*. 1999;451(1):56–62.
35. Gent J, Braakman I. Low-density lipoprotein receptor structure and folding. *Cell Mol Life Sci*. 2004;61(19–20):2461–2470.
36. Patthy L. Evolution of the proteases of blood coagulation and fibrinolysis by assembly from modules. *Cell*. 1985;41(3):657–663.
37. Russell DW, Schneider WJ, Yamamoto T, Luskey KL, Brown MS, Goldstein JL. Domain map of the LDL receptor: Sequence homology with the epidermal growth factor precursor. *Cell*. 1984;37(2):577–585.
38. Davis CG, van Driel IR, Russell DW, Brown MS, Goldstein JL. The low density lipoprotein receptor. Identification of amino acids in cytoplasmic domain required for rapid endocytosis. *J Biol Chem*. 1987;262(9):4075–4082.
39. Innerarity TL, Mahley RW. Enhanced binding by cultured human fibroblasts of apo-E-containing lipoproteins as compared with low density lipoproteins. *Biochemistry*. 1978;17(8):1440–1447.
40. Jeon H, Blacklow SC. Structure and physiologic function of the low-density lipoprotein receptor. *Annu Rev Biochem*. 2005;74:535–562.
41. Sandhu J, Li S, Fairall L, et al. Aster proteins facilitate nonvesicular plasma membrane to ER cholesterol transport in mammalian cells. *Cell*. 2018;175(2):514–529 e20.
42. Boucrot E, Saffarian S, Zhang R, Kirchhausen T. Roles of AP-2 in clathrin-mediated endocytosis. *PLoS One*. 2010;5(5), e10597.

43. Kim HR, Gil S, Andrieux K, et al. Low-density lipoprotein receptor-mediated endocytosis of PEGylated nanoparticles in rat brain endothelial cells. *Cell Mol Life Sci*. 2007;64(3):356–364.
44. Sverdlov M, Shajahan AN, Minshall RD. Tyrosine phosphorylation-dependence of caveolae-mediated endocytosis. *J Cell Mol Med*. 2007;11(6):1239–1250.
45. Kruth HS, Jones NL, Huang W, et al. Macropinocytosis is the endocytic pathway that mediates macrophage foam cell formation with native low density lipoprotein. *J Biol Chem*. 2005;280(3):2352–2360.
46. Zhang X, Sessa WC, Fernandez-Hernando C. Endothelial transcytosis of lipoproteins in atherosclerosis. *Front Cardiovasc Med*. 2018;5:130.
47. Siekmeier R, Wulfroth P, Wieland H, Gross W, Marz W. Low-density lipoprotein susceptibility to in vitro oxidation in healthy smokers and nonsmokers. *Clin Chem*. 1996;42(4):524–530.
48. Yang JH, Bang MA, Jang CH, Jo GH, Jung SK, Ki SH. Alginate oligosaccharide enhances LDL uptake via regulation of LDLR and PCSK9 expression. *J Nutr Biochem*. 2015;26(11):1393–1400.
49. Briggs MA, Petersen KS, Kris-Etherton PM. Saturated fatty acids and cardiovascular disease: Replacements for saturated fat to reduce cardiovascular risk. *Healthcare (Basel)*. 2017;5(2).
50. Siri-Tarino PW, Sun Q, Hu FB, Krauss RM. Saturated fatty acids and risk of coronary heart disease: Modulation by replacement nutrients. *Curr Atheroscler Rep*. 2010;12 (6):384–390.
51. Lau VW, Journoud M, Jones PJ. Plant sterols are efficacious in lowering plasma LDL and non-HDL cholesterol in hypercholesterolemic type 2 diabetic and nondiabetic persons. *Am J Clin Nutr*. 2005;81(6):1351–1358.
52. Katan MB, Grundy SM, Jones P, et al. Efficacy and safety of plant stanols and sterols in the management of blood cholesterol levels. *Mayo Clin Proc*. 2003;78(8):965–978.
53. Fernandez ML, Vega-Lopez S. Efficacy and safety of sitosterol in the management of blood cholesterol levels. *Cardiovasc Drug Rev*. 2005;23(1):57–70.
54. Doma KM, Dolinar KF, Dan Ramdath D, Wolever TMS, Duncan AM. Canned beans decrease serum Total and LDL cholesterol in adults with elevated LDL cholesterol in a 4-wk multicenter, randomized. *Crossover Study J Nutr*. 2021;151(12):3701–3709.
55. Han KH, Fukushima M, Kato T, et al. Enzyme-resistant fractions of beans lowered serum cholesterol and increased sterol excretions and hepatic mRNA levels in rats. *Lipids*. 2003;38(9):919–924.
56. Loader TB, Zahradka P, Ahmadi S, Taylor CG. Processing method modulates the effectiveness of black beans for lowering blood cholesterol in spontaneously hypertensive rats. *J Sci Food Agric*. 2021;101(2):449–458.
57. Bazzano LA. Effects of soluble dietary fiber on low-density lipoprotein cholesterol and coronary heart disease risk. *Curr Atheroscler Rep*. 2008;10(6):473–477.
58. Abeysekara S, Chilibeck PD, Vatanparast H, Zello GA. A pulse-based diet is effective for reducing total and LDL-cholesterol in older adults. *Br J Nutr*. 2012;108(Suppl 1): S103–S110.
59. Krohn RM, Lemaire M, Negro Silva LF, et al. High-selenium lentil diet protects against arsenic-induced atherosclerosis in a mouse model. *J Nutr Biochem*. 2016; 27:9–15.
60. Suido H, Tanaka T, Tabei T, et al. A mixed green vegetable and fruit beverage decreased the serum level of low-density lipoprotein cholesterol in hypercholesterolemic patients. *J Agric Food Chem*. 2002;50(11):3346–3350.
61. Valcheva-Kuzmanova S, Kuzmanov K, Tsanova-Savova S, et al. Lipid-lowering effects of Aronia melanocarpa fruit juice in rats fed cholesterol-containing diets. *J Food Biochem*. 2007;31(5):589–602.

62. Herold PM, Kinsella JE. Fish oil consumption and decreased risk of cardiovascular disease: A comparison of findings from animal and human feeding trials. *Am J Clin Nutr.* 1986;43(4):566–598.
63. Vazquez-Velasco M, Esperanza Diaz L, Lucas R, et al. Effects of hydroxytyrosol-enriched sunflower oil consumption on CVD risk factors. *Br J Nutr.* 2011;105 (10):1448–1452.
64. Jones PJ, MacKay DS, Senanayake VK, et al. High-oleic canola oil consumption enriches LDL particle cholesteryl oleate content and reduces LDL proteoglycan binding in humans. *Atherosclerosis.* 2015;238(2):231–238.
65. Devarajan S, Singh R, Chatterjee B, Zhang B, Ali A. A blend of sesame oil and rice bran oil lowers blood pressure and improves the lipid profile in mild-to-moderate hypertensive patients. *J Clin Lipidol.* 2016;10(2):339–349.
66. Vennila LJWJOPR. Cardioprotective effect of sesame oil: A review. *World J Pharma Res.* 2017;6(11):331–341.
67. Sun Y, Neelakantan N, Wu Y, Lote-Oke R, Pan A, van Dam RM. Palm oil consumption increases LDL cholesterol compared with vegetable oils low in saturated fat in a Meta-analysis of clinical trials. *J Nutr.* 2015;145(7):1549–1558.
68. Benjannet S, Rhainds D, Essalmani R, et al. NARC-1/PCSK9 and its natural mutants: Zymogen cleavage and effects on the low density lipoprotein (LDL) receptor and LDL cholesterol. *J Biol Chem.* 2004;279(47):48865–48875.
69. Abifadel M, Varret M, Rabes JP, et al. Mutations in PCSK9 cause autosomal dominant hypercholesterolemia. *Nat Genet.* 2003;34(2):154–156.
70. Cohen J, Pertsemlidis A, Kotowski IK, Graham R, Garcia CK, Hobbs HH. Low LDL cholesterol in individuals of African descent resulting from frequent nonsense mutations in PCSK9. *Nat Genet.* 2005;37(2):161–165.
71. Zhao Z, Tuakli-Wosornu Y, Lagace TA, et al. Molecular characterization of loss-of-function mutations in PCSK9 and identification of a compound heterozygote. *Am J Hum Genet.* 2006;79(3):514–523.
72. Mayne J, Dewpura T, Raymond A, et al. Plasma PCSK9 levels are significantly modified by statins and fibrates in humans. *Lipids Health Dis.* 2008;7:22.
73. Bruckert E, Hayem G, Dejager S, Yau C, Begaud B. Mild to moderate muscular symptoms with high-dosage statin therapy in hyperlipidemic patients- -the PRIMO study. *Cardiovasc Drugs Ther.* 2005;19(6):403–414.
74. Sullivan D, Olsson AG, Scott R, et al. Effect of a monoclonal antibody to PCSK9 on low-density lipoprotein cholesterol levels in statin-intolerant patients: The GAUSS randomized trial. *JAMA.* 2012;308(23):2497–2506.
75. Chaudhary R, Garg J, Shah N, Sumner A. PCSK9 inhibitors: A new era of lipid lowering therapy. *World J Cardiol.* 2017;9(2):76–91.
76. Hajar R. Statins: Past and present. *Heart Views.* 2011;12(3):121–127.
77. Koh KK, Han SH, Sakuma I, Zhao D. Calming down chaos regarding redefining blood pressure targets-the importance of statin-based therapy. *Int J Cardiol.* 2016;221: 572–574.
78. Owens 3rd AP, Mackman N. The antithrombotic effects of statins. *Annu Rev Med.* 2014;65:433–445.
79. Sorrentino S, Landmesser U. Nonlipid-lowering effects of statins. *Curr Treat Options Cardiovasc Med.* 2005;7(6):459–466.
80. Mitsios JV, Papathanasiou AI, Goudevenos JA, Tselepis AD. The antiplatelet and antithrombotic actions of statins. *Curr Pharm Des.* 2010;16(34):3808–3814.
81. Undas A, Brummel-Ziedins KE, Mann KG. Anticoagulant effects of statins and their clinical implications. *Thromb Haemost.* 2014;111(3):392–400.
82. Lee KCH, Sewa DW, Phua GC. Potential role of statins in COVID-19. *Int J Infect Dis.* 2020;96:615–617.

83. Pedersen TR. Pleiotropic effects of statins. *Am J Cardiovasc Dis.* 2010;10(2):10–17.
84. Corsini A, Bellosta S, Baetta R, Fumagalli R, Paoletti R, Bernini F. New insights into the pharmacodynamic and pharmacokinetic properties of statins. *Pharmacol Ther.* 1999;84(3):413–428.
85. Vaughan CJ, Delanty N. Neuroprotective properties of statins in cerebral ischemia and stroke. *Stroke.* 1999;30(9):1969–1973.
86. Arnaud C, Braunersreuther V, Mach F. Toward immunomodulatory and anti-inflammatory properties of statins. *Trends Cardiovasc Med.* 2005;15(6):202–206.
87. Lardizabal JA, Deedwania PC. The anti-ischemic and anti-anginal properties of statins. *Curr Atheroscler Rep.* 2011;13(1):43–50.
88. Sassano A, Platanias LC. Statins in tumor suppression. *Cancer Lett.* 2008;260(1–2): 11–19.
89. Adams SP, Alaeiilkhchi N, Wright JM. Pitavastatin for lowering lipids. *Cochrane Database Syst Rev.* 2020;6:CD012735.
90. Degenring FH, Schatton W, Hotz W. Atherosclerosis treatment with etofibrate retard. New perspectives. *Fortschr Med.* 1983;101(30):1391–1396.
91. Groggel GC, Cheung AK, Ellis-Benigni K, Wilson DE. Treatment of nephrotic hyperlipoproteinemia with gemfibrozil. *Kidney Int.* 1989;36(2):266–271.
92. Lin T-L, Lin H-H, Chen C-C, Lin M-C, Chou M-C, Wang C-J. Hibiscus sabdariffa extract reduces serum cholesterol in men and women. *Nutrition Res.* 2007;27(3): 140–145.
93. Scott R. Lipid modifying agents: Mechanisms of action and reduction of cardiovascular disease. *Clin Exp Pharmacol Physiol.* 1997;24(5):A26–A28.
94. Nutescu EA, Shapiro NL. Ezetimibe: A selective cholesterol absorption inhibitor. *Pharmacotherapy.* 2003;23(11):1463–1474.
95. Manniello M, Pisano M. Alirocumab (Praluent): First in the new class of PCSK9 inhibitors. *P T.* 2016;41(1):28–53.
96. Pichardo R, Boulet L, Davignon J. Pharmacokinetics of clofibrate in familial hypercholesterolemia. *Atherosclerosis.* 1977;26(4):573–582.
97. Kastelein JJ. Should rosuvastatin be withdrawn from the market? *Lancet.* 2004;364 (9445):1577–1578 [author reply 1578-9].
98. Nawrocki JW, Weiss SR, Davidson MH, et al. Reduction of LDL cholesterol by 25% to 60% in patients with primary hypercholesterolemia by atorvastatin, a new HMG-CoA reductase inhibitor. *Arterioscler Thromb Vasc Biol.* 1995;15(5):678–682.
99. Pahan K. Lipid-lowering drugs. *CMLS.* 2006;63(10):1165–1178.
100. Scharnagl H, Schinker R, Gierens H, Nauck M, Wieland H, Marz W. Effect of atorvastatin, simvastatin, and lovastatin on the metabolism of cholesterol and triacylglycerides in HepG2 cells. *Biochem Pharmacol.* 2001;62(11):1545–1555.
101. Zhou Q, Liao JK. Statins and cardiovascular diseases: From cholesterol lowering to pleiotropy. *Curr Pharm Des.* 2009;15(5):467–478.
102. Hunninghake DB, Stein EA, Bays HE, et al. Rosuvastatin improves the atherogenic and atheroprotective lipid profiles in patients with hypertriglyceridemia. *Coron Artery Dis.* 2004;15(2):115–123.
103. Al-Kuraishy HM, Al-Gareeb AI. Effects of rosuvastatin on metabolic profile: Versatility of dose-dependent effect. *J Adv Pharm Technol Res.* 2019;10(1):33–38.
104. Barrangou R. The roles of CRISPR-Cas systems in adaptive immunity and beyond. *Curr Opin Immunol.* 2015;32:36–41.
105. Hille F, Richter H, Wong SP, Bratovic M, Ressel S, Charpentier E. The biology of CRISPR-Cas: Backward and forward. *Cell.* 2018;172(6):1239–1259.
106. Gohil N, Bhattacharjee G, Lam NL, Perli SD, Singh V. CRISPR-Cas systems: Challenges and future prospects. *Prog Mol Biol Transl Sci.* 2021;180:141–151.

107. Bikard D, Euler CW, Jiang W, et al. Exploiting CRISPR-Cas nucleases to produce sequence-specific antimicrobials. *Nat Biotechnol.* 2014;32(11):1146–1150.
108. Cong L, Ran FA, Cox D, et al. Multiplex genome engineering using CRISPR/Cas systems. *Science.* 2013;339(6121):819–823.
109. Mali P, Esvelt KM, Church GM. Cas9 as a versatile tool for engineering biology. *Nat Methods.* 2013;10(10):957–963.
110. Singh V, Braddick D, Dhar PK. Exploring the potential of genome editing CRISPR-Cas9 technology. *Gene.* 2017;599:1–18.
111. Singh V, Gohil N, Ramirez Garcia R, Braddick D, Fofié CK. Recent advances in CRISPR-Cas9 genome editing technology for biological and biomedical investigations. *J Cell Biochem.* 2018;119(1):81–94.
112. Khambhati K, Gohil N, Bhattacharjee G, Singh V. Development and challenges of using CRISPR-Cas9 system in mammalians. In: Singh V, Dhar PK, eds. *Genome Engineering via CRISPR-Cas9 System.* London: Academic Press; 2020:83–93.
113. Long C, McAnally JR, Shelton JM, Mireault AA, Bassel-Duby R, Olson EN. Prevention of muscular dystrophy in mice by CRISPR/Cas9-mediated editing of germline DNA. *Science.* 2014;345(6201):1184–1188.
114. Nelson CE, Hakim CH, Ousterout DG, et al. In vivo genome editing improves muscle function in a mouse model of Duchenne muscular dystrophy. *Science.* 2016;351 (6271):403–407.
115. Borrelli VMG, Brambilla V, Rogowsky P, Marocco A, Lanubile A. The enhancement of plant disease resistance using CRISPR/Cas9 technology. *Front Plant Sci.* 2018; 9:1245.
116. Tripathi L, Ntui VO, Tripathi JN. CRISPR/Cas9-based genome editing of banana for disease resistance. *Curr Opin Plant Biol.* 2020;56:118–126.
117. Tyagi S, Kumar R, Kumar V, Won SY, Shukla P. Engineering disease resistant plants through CRISPR-Cas9 technology. *GM Crops Food.* 2021;12(1):125–144.
118. Dey A, Nandy S. CRISPER/Cas in plant natural product research: Therapeutics as anticancer and other drug candidates and recent patents. *Recent Pat Anticancer Drug Discov.* 2021;16(4):460–468.
119. Bhattacharjee G, Mani I, Gohil N, et al. CRISPR technology for genome editing. In: Faintuch J, Faintuch S, eds. *Precision Medicine for Investigators, Practitioners and Providers.* London: Academic Press; 2020:59–69.
120. Bhattacharjee G, Gohil N, Khambhati K, et al. Current approaches in CRISPR-Cas9 mediated gene editing for biomedical and therapeutic applications. *J Control Release.* 2022;343:703–723.
121. Kaminski MM, Abudayyeh OO, Gootenberg JS, Zhang F, Collins JJ. CRISPR-based diagnostics. *Nat. Biomed Eng.* 2021;5(7):643–656.
122. Khambhati K, Bhattacharjee G, Gohil N, et al. Phage engineering and phage-assisted CRISPR-Cas delivery to combat multidrug-resistant pathogens. *Bioeng Transl Med.* 2022:e10381.
123. Bhattacharjee G, Khambhati K, Gohil N, Singh V. Programmable removal of bacterial pathogens using CRISPR-Cas9 system. In: Singh V, Dhar PK, eds. *Genome Engineering Via CRISPR-Cas9 System.* London: Academic Press; 2020:39–44.
124. Ding Q, Strong A, Patel KM, et al. Permanent alteration of PCSK9 with in vivo CRISPR-Cas9 genome editing. *Circ Res.* 2014;115(5):488–492.
125. Zhang L, Wang L, Xie Y, et al. Triple-targeting delivery of CRISPR/Cas9 to reduce the risk of cardiovascular diseases. *Angew Chem Int Ed Engl.* 2019;58(36):12404–12408.
126. Guo M, Xu Y, Dong Z, et al. Inactivation of ApoC3 by CRISPR/Cas9 protects against atherosclerosis in hamsters. *Circ Res.* 2020;127(11):1456–1458.

CHAPTER FOURTEEN

Endocytosis of LXRs: Signaling in liver and disease

Mettle Brahma[a], Soumika Ghosal[b], Mulaka Maruthi[a,*], and Suresh K. Kalangi[c,*]

[a]Department of Biochemistry, Central University of Haryana, Jant-Pali, Mahendragarh, Haryana, India
[b]Amity Stem cell Institute, Amity Medical School, Amity University Haryana, Amity Education Valley Pachgaon, Manesar, Gurugram, Haryana, India
[c]Stem Cell and Drug Discovery Division, INOVAUGMET Research Pvt Ltd., Bengaluru, Karnataka, India
*Corresponding authors: e-mail address: mulaka@cuh.ac.in; suresh.kalangi@inovaugmetresearch.com

Contents

Abstract

Nuclear receptors are among one of the major transcriptional factors that induces gene regulation in the nucleus. Liver X receptor (LXR) is a transcription factor which regulates essential lipid homeostasis in the body including fatty acid, cholesterol and phospholipid synthesis. Liver X receptor-retinoid X receptor (LXR-RXR) heterodimer is activated

Progress in Molecular Biology and Translational Science, Volume 194
ISSN 1877-1173
https://doi.org/10.1016/bs.pmbts.2022.07.008

by either of the ligand binding on LXR or RXR. The promoter region of the gene which is targeted by LXR is bound to the response element of LXR. The activators bind to the heterodimer once the corepressor is dissociated. The cellular process such as endocytosis aids in intracellular trafficking and endosomal formation in transportation of molecules for essential signaling within the cell. LXR isotypes play a crucial role in maintaining lipid homeostasis by regulating the level of cholesterol. In the liver, the deficiency of LXRα can alter the normal physiological conditions depicting the symptoms of various cardiovascular and liver diseases. LXR can degrade low density lipoprotein receptors (LDLR) by the signaling of LXR-IDOL through endocytic trafficking in lipoprotein uptake. Various gene expressions associated with cholesterol level and lipid synthesis are regulated by LXR transcription factor. With its known diversified ligand binding, LXR is capable of regulating expression of various specific genes responsible for the progression of autoimmune diseases. The agonists and antagonists of LXR stand to be an important factor in transcription of the ABC family, essential for high density lipoprotein (HDL) formation. Endocytosis and signaling mechanism of the LXR family is broad and complex despite their involvement in cellular growth and proliferation. Here in this chapter, we aimed to emphasize the master regulation of LXR activation, regulators, and their implications in various metabolic activities especially in lipid homeostasis. Furthermore, we also briefed the significant role of LXR endocytosis in T cell immune regulation and a variety of human diseases including cardiovascular and neuroadaptive.

Abbreviations

CETP cholesteryl ester transfer protein
FAS fatty acid synthase
HDL high density lipoprotein
LXR liver X receptor
RCT reverse cholesterol transporter

1. Introduction

Liver is a vital organ that plays a crucial role as endocrine gland for maintaining various physiological functions in the body. The hepatocytes of the liver play numerous metabolic and anabolic roles that are essential to maintain homeostasis and lipid metabolism. Cholesterol is an important component synthesized by the liver that is essential for production of steroids. Further, liver cells are well known for their various immunological functions, fluid balance and excretion of toxic substances out of the body. These cells produce insulin-like growth factors that can regulate proliferation, growth, and development of the liver cells. The food particles, rich fat intake and drug is absorbed by gastrointestinal tract which reaches to liver via the circulation where it gets stored in adipose tissue present in fat cells.

This stored fat is an important source for the regulation of heat and energy.[1] Since liver cells are important sites commonly associated with various diseases and drug metabolism, the alteration or dysfunction of liver can cause mild to chronic diseases and ultimately leads to multiorgan failure which can be fatal.[2] Hence, understanding the role of liver cell receptors and their metabolism of action in the body is of prime importance in controlling diseases.

Liver X receptor (LXR) is a transcription factor of the nuclear receptor family that plays a crucial role in lipid metabolism, maintaining the cholesterol level and immune regulation. This ligand activated transcription factor can regulate the lipid homeostasis in normal and pathophysiological conditions by activation of the genes and hence is regarded as the master regulator of lipogenic pathway.[3] LXR, a nuclear receptor, was first identified as RLD-1 consisting of 445-amino-acid long proteins by Apfel et al. in 1994. The receptor, LXR, is highly expressed in the organs of liver, spleen, lungs, pituitary and fat.[4] Later, Willy et al. in 1995 in their study found out the similarity between RLD-1 and LXR-α; and their ability to regulate responses of retinoid X-receptors (RXR). The two isotypes of LXR, i.e., LXR-α and LXR-β are encoded by Nr1h3 and Nr1h2, respectively, and hence named as NR1H3 and NR1H2, though their presence is also found in all different types of cells but LXR-α is mostly abundant in liver, adipose tissue, spleen, and kidney.[5]

The signaling of LXR requires the activation of LXR which is governed by the bindings of a ligand or an agonist. The various target genes of LXR have the LXR response element in their conserved region on which LXR remains bound until activated by the agonists. The expression of target genes of LXR requires fork-head box A2 (Foxa2), and the binding of agonists to LXR can increase the occupancy of Foxa2 in ligand-dependent manner.[6] RXR belongs to the receptor family which forms a heterodimer structure with the LXR, later gets activated on binding with specific ligands followed by the action of co-activators. These bindings transduce the process of gene transcription and expression on binding with the LXR elements (LXRE) on the promoter region of target genes (Fig. 1).[7] Since the heterodimer formation of LXR-RXR is directly bound to the repeats of response elements on the conserved sequence which remains activated until the bindings of ligands to it and hence it belongs to the family of type 2 nuclear receptors.[8] The LXRs are also regarded as oxysterol receptors since they consist of oxysterols or derivatives of oxidized cholesterol as their specific ligands.[9] The hydroxyl group of side chains present in cholesterol can activate the

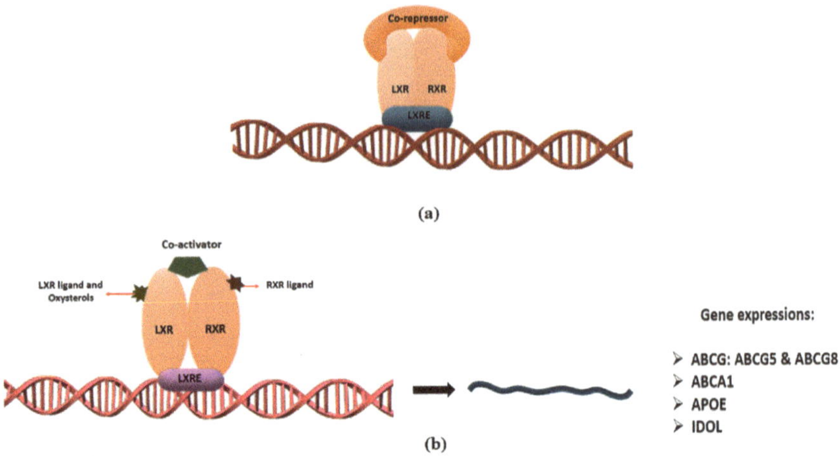

Fig. 1 Inactive and active stages of LXR-RXR heterodimer. The nuclear factor LXR-RXR heterodimer binds to response element, LXRE present on the target genes. (A) Inactivated form of LXR-RXR heterodimer on binding with co-repressor. The LXR-RXR heterodimer complex remains inactive until the binding of either RXR or LXR with their ligands. (B) Replacement of co-repressor by co-activators on LXR-RXR complex can activate the complex LXR-RXR followed by the activation of target genes and their gene expression. The expression of specific genes ATP-binding cassette super-family G (ABCG), ATP-binding cassette (ABCA1), apolipoprotein E (APOE) and inducible degrader of the LDLR (IDOL) are responsible for the various lipid metabolisms which are regulated by LXR signaling. *Redrawn from Hong C, Tontonoz P. Liver X receptors in lipid metabolism: Opportunities for drug discovery.* Nat Rev Drug Discov. *2014;13:433–444.*

LXR which then alters the heterodimer structure and removes the co-repressor of LXR.[5] Oxysterols, the physiological ligand of LXR, can bind and activate the LXRs signaling and promotes the lipid metabolism and cholesterol homeostasis.[10]

Endocytosis is a cellular process by which the extracellular materials or cargo are transported into intracellular compartments by a series of pathways followed by the formation of vesicles.[11] The process of endocytosis aids in the internalization of various surface proteins and macromolecules by trafficking them to form an endosomal network.[12] Clathrin-mediated endocytosis and Caveolin-mediated endocytosis are the two processes of receptor-mediated endocytosis that transport the cargo depending upon various sizes of particles by the endocytic pathway.

Clathrin-mediated endocytosis is crucial for cellular surface receptor internalization, transmembrane transporter, intracellular signaling, cellular interactions, and homeostasis.[13] This process of endocytosis further aids in the recycling of receptors for signal transduction, as the internalization of proteins or ligands is returned back to the cell surface by the receptor-mediated endocytosis.[14] The regulators of endocytic pathways depend upon the specific ligand binding to the receptors such as low-density lipoproteins, proteins, etc. LXRs must bind particular ligands in order to carry out the intracellular signaling pathway, and these ligand binds activate the LXRs and their pathways.

2. Liver X receptor: Master regulator for lipid metabolism and homeostasis

LXRs are expressed in two isoforms, which form heterodimers with RXR. The signaling of LXRs in the liver, intestine, and kidney plays a basic role in cholesterol regulation and lipid metabolism. LXR signaling begins with the activation of LXR by its ligands, activation leads to the formation of LXR complex with RXR and coactivators, which further translocate to the nucleus and bind to LXR response elements to regulate the expression of genes for metabolic pathways of cholesterol and other lipids.[7] LXRs are involved in various mechanisms in physiology by regulating various metabolic products. LXRs can be termed as a master regulator, as it consists of a cholesterol sensing response which aids in the continuous track on its level and can regulate hormone production indirectly. LXRs can regulate reverse cholesterol transport (RCT) to maintain the intracellular cholesterol level.[15]

2.1 LXRs in lipid metabolism

Lipid is an essential source to produce energy in cells. Lipid is stored as fat in adipose tissue via various metabolic pathways while producing ATP as a source of energy. LXRs play various direct and indirect roles in fatty acid synthesis contributing to lipid metabolism.[16] Further, the interconnection between the inflammation and the metabolism of lipids is established due to LXRs signaling.[3] In hepatocytes, the agonists of LXRs can induce lipogenesis and upregulates essential enzymes including acetyl CoA carboxylase, fatty acid synthase (FAS), and stearoyl-CoA desaturase-1c (SREBP-1c).[17,18] LXRs can bind to the LXR response element on the regulatory region of fatty acid synthase, acetyl CoA carboxylase, and stearoyl-CoA desaturase-1 genes and can regulate the lipogenic pathway.[19–21] The expression of SREBP-1c is upregulated due to LXR signaling that can bind the sterol response element of the promoter region of the target genes of enzymes essential for the lipogenesis pathway.[22] LXRs can induce several apolipoproteins, phospholipid transfer protein, and cholesterol ester transporting protein (CETP).[9]

Hepatic triglyceride accumulation can lead to severe liver steatosis and dysfunction. The treatment with the agonist T0901317 can increase the triglyceride content of hepatic and plasma cells in mice.[23,24] The level of plasma triglycerides reduces when the signaling of LXRs is inhibited in LXR double knockout mice. Another glucose-sensitive transcription factor known as carbohydrate response element-binding protein (ChREBP) induces various essential lipogenic enzymes on higher levels of glucose converting glucose to lipids.[9] LXR and ChREBP share transcriptional targets for many genes and the expression of ChREBP can be regulated by LXR signaling. LXRα and ChREBPα can coactivate the target genes of ChREBP independent of LXRE.[25]

2.2 LXRs in cholesterol homeostasis

Cholesterol is an important bio-molecule essential for cells, required to maintain the normal physiology of the body. It is a major precursor for the synthesis of steroid hormones. The accumulation of cholesterol, however, can alter the normal functioning of various organs including the hepatocyte cells in liver function. The process of reverse cholesterol transport (RCT) transports excess cholesterol in the peripheral to plasma, which is utilized by liver cells to synthesize bile.[26] Polyunsaturated fatty acids that can induce lipid peroxidation can damage the liver cells.[27] In the liver and small intestine, the transport of cholesterol is regularly kept on track by LXRs for the regulation of cholesterol homeostasis (Fig. 2).[28]

The important target genes of LXR for the regulation of cholesterol are ATP-binding cassette subfamily A type 1 (ABCA1) and ATP-binding cassette subfamily G member 1 (ABCG1). ABCA1 protein is an integral membrane transporter that plays a major role in HDL formation.[29] The expression of these genes is upregulated on the activation of LXRs along with carbohydrate-response element-binding protein (ChREBP) and sterol regulatory element-binding protein 1c (SREBP1c) playing a crucial role in the lipogenic pathway.[3] The increase in the level of HDL has greater benefits in reducing cardiovascular disease due to the process of reverse cholesterol transport. The inhibition in the degradation of ABCA1 expression upregulates the level of circulating HDL which then regulates RCT which reduces the risk of the disease, atherosclerosis.[30] ABCA1 cholesterol efflux

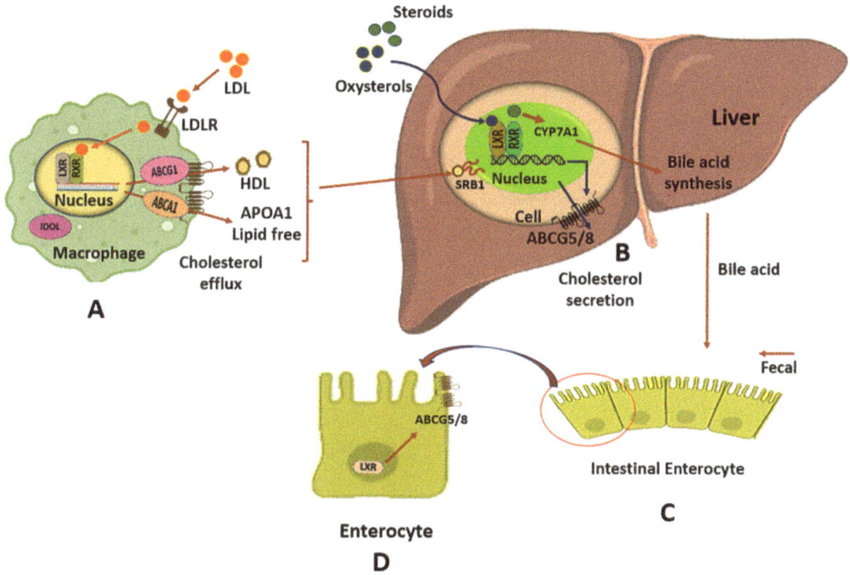

Fig. 2 Liver X receptors regulate cholesterol homeostasis. (A) In macrophages, LDL binds to LDL receptor (LDLR) that activates the cascade reactions including activation of LXR-RXR heterodimer and enables the expression of specific ABC genes for cholesterol efflux. HDL and lipid free APOA1 are produced on the receptors ABCG1 and ABCA1, respectively. (B) In the liver, steroids and derivatives of cholesterol activate the various genes associated with lipid metabolism. Oxysterols are the specific ligands/agonists that bind to the LXR nuclear factor and can activate the heterodimers followed by the transcription of specific genes and expression on cell surfaces such as ABCG and ABCA1. CYP7A1 is an enzyme that can synthesize bile acid from cholesterol in the liver. (C) Bile acid is released on the intestine where the fecal materials are transported. (D) The enterocytes present in the intestine express the ABCG5 and ABCG8 on the surface which is regulated by the signaling of LXR for a further role in the absorption of cholesterol. *Redrawn from Schulman IG. Liver X receptors link lipid metabolism and inflammation.* FEBS Lett. *2017;591:2978–2991.*

capacity is regulated and promoted by apolipoprotein A1 (ApoA1) associated with HDL, which acts as key mediator.[31] Further, APOA1 can also regulate the HDL-remodeling protein.[32] ATP-binding cassette subfamily G member 5 (ABCG5) and ATP-binding cassette subfamily G member 8 (ABCG8) also play an effective role in cholesterol metabolism. ABCG5 ABCG8 has a common promoter for the transcription and encodes the transporter and mutation on this gene can cause various diseases.[33] LXRs can also regulate ABCG5 and ABCG8 heterodimers that are expressed in enterocytes as they have an LXRα response element in their conserved region.[34]

LXR plays an essential role in association with ApoE in immunosuppressive innate cell populations. This ApoE gene is a target for LXR in macrophages and adipocytes and has a crucial role in cholesterol efflux.[35] These apolipoproteins are the carrier proteins of lipid in lipid metabolism.[36] LXR/ApoE pathway can regulate innate immunity and elicit adaptive-like immunity in immunosuppressed patients.[37] The genes of various types of Apos have LXR responsive elements in their conserved region such as ApoC1, ApoC2, ApoC4 and ApoD.[35,36,38,39] ApoE is synthesized on adipocytes, smooth muscle, macrophages, kidneys, and the brain. ApoE can decrease the level of low-density lipoprotein due to its response of inhibition of cholesteryl ester transfer protein which enhances the RCT inducing LXR and PPARα signaling.[40]

LXR signaling can regulate the expression of various genes such as Scavenger receptor, class B type 1 (SR-B1), Inducible degrader of the LDLR (IDOL), CYP7A1, and Cholesteryl ester transfer protein (CETP) along with ABC Family.[41] SR-B1 is a multiligand lipoprotein membrane receptor that influx HDL-derived cholesteryl esters into cells.[42] In mature adipocytes, the expression of SR-B1 mRNA gets elevated. Further, SR-B1 protein level is regulated by the expression of LXR in posttranscriptional regulation in microRNA level.[43] SR-B1 plays an essential role in lipid metabolism which can uptake the lipids and involve in the delivery of cholesterol esters. Moreover, LXRα signaling can regulate the mRNA expression of cholesterol 7-α-hydroxylase (CYP7A1), a cytochrome P450 protein family.[41,44] The alteration in the polymorphism of CYP7A1 gene can alter the risk factor for coronary artery diseases.[45] Further, the CYP7A1 protein can regulate the bile synthesis by the rate-limiting activity producing 7-α-hydroxycholesterol.

Cholesteryl ester transfer protein (CETP) is a glycoprotein that plays a basic role in lipid homeostasis modulating the level of HDL-C.[46] CETP can control the level of lipoprotein and helps in transferring triglycerides from VLDL into LDL and HDL.[47] These proteins are synthesized in the liver and their expression is significantly regulated by the expression of LXR nuclear receptors.[48] The expression of CETP is limited to certain mammalian species excluding humans and monkeys where CETP plays a role in lipid metabolism. Another E3 ubiquitin ligase enzyme, the inducible degrader of LDLR (IDOL) is a cholesterol regulator that can regulate the signaling of LDLR. The posttranscriptional regulation SUMOylation can prevent autoubiquitination and elevate the level of IDOL proteins and their modulation further decreases the level of LDLR.[49] The LXR-IDOL pathway plays an active role in lipid metabolism and regulates physiological homeostasis.[50] This IDOL can degrade the LDLR and its expression is controlled at the transcriptional level by LXRs.

2.3 LXRs in immune regulation

In the liver, many immune cells including macrophages and natural killer T (NKT) cells exist for the body's immune defense. These cells collectively protect the liver cells and different parts of the body. The LXR expression and their signaling depict its important role in immunometabolism. LXRs signaling can show its response to various inflammatory stimuli which helps in innate and adaptive immune regulation. LXRs have emerged as important regulators of innate immunity and several studies have shown that LXRs can contribute to liver fibrosis via hepatic stellate cells (HSC) activation. The involvement of LXRs in various chronic inflammatory diseases and other immune disorders has been understood in various studies. These studies of LXRs have been widely performed in macrophages along with lipid metabolism. However, signaling and its complete role in immune regulation is much more complex and yet to be understood. There are various modulators that show response upon LXR signaling and are used as therapeutic targets against several inflammatory diseases.

The LXR can perform dual roles in the regulation and differentiation of CD4+ T cells. Herold et al. in their study found that its activation can suppress the pro-inflammatory T cells, however, it enhanced the T regulatory cell production.[51] The LXR knocked down decreases the endocytosis of LDL cholesterol in mice and reduces the activity of phagocytosis by the action of Kupffer cells. However, the level of $CD36^+$ has elevated in LXR knockdown mice.[52] The activation of LXR nuclear transcription factors can decrease the macrophage receptor independent, fluid-phase pinocytosis of LDL.[53] The role of LXRs is also involved in the release of cytokines, lymphocyte proliferation, migration and chemotaxis.[54] The reprogramming of cholesterol metabolic pathways can influence innate immunity, where LXR plays an active role. Their activation can induce the long-term pro-inflammatory memory (trained) immunity phenotype in monocytes involving Acetyl-CoA and IL-1β.[15]

The role of macrophages involved in lipid metabolism is well known. The attenuation of inflammatory response and reverse cholesterol transport is upregulated in macrophages. Further, the downregulation in NF-κB (nuclear factor kappa-light-chain-enhancer of activated B-cell) synthesis and reduction in inflammatory response were found in the murine study during LXR activation.[55] LXRs further, can negatively regulate the pro-inflammatory genes NF-κB and activator protein 1 (AP-1) through SUMO2/3 SUMOylation-dependent mechanism to bind to the co-repressor complex.[56] The activation of LXR can alter the function of the toll-like receptor (TLR) leading to immunoregulation. TLR9 is inhibited in the activation of LXR however, TLR4 expression is upregulated in human monocyte-derived macrophages (MDMs).[57] The agonists of LXR can further increase the production of IL-1β independent of TLR activity.[58] The LXR/ApoE pathway induces innate immunity in immunocompetent cells regulating anti-tumor response and eliciting T cell response against tumors.[37]

2.4 LXR in glucose level maintenance

The liver acts as an important organ for the synthesis of various metabolic products and is involved in glucose homeostasis. Though the study of LXR in association with glucose metabolism and various transporters is limited, the binding response of LXR and glucose provides important evidence of LXR in glucose pathways. LXRs can be activated on binding with glucose and can act as the LXR ligand. The activation of LXR by glucose can transcript the target genes and induce transcription.[59] Despite the significant knowledge of the direct role of LXR signaling in lipid and cholesterol metabolism, the involvement of LXR in glucose metabolism is yet to be understood. Genes that are involved in glycolysis such as GLUT1 and Hexokinase-2 (HK2) are induced by LXRs with the help of hypoxia-inducible factor 1α (HIF 1α) in human macrophages.[58] The inhibition of LXR signaling by the binding of ligand antagonists can decrease the rate of glycolysis by decreasing the expression of various protein factors of the glycolytic pathway.[60] Furthermore, the activation of LXR in macrophages can induce HIF-1α-dependent pathway glycolysis.[58] In liver, peroxisome proliferator-activated receptor α (PPAR α) can regulate the fibroblast growth factor essential for bile acid synthesis.[61,62] Fibroblast growth factor-21 (FGF21) is a regulator of glucose and lipid metabolism through (the ERK1/2)/PPARγ/LXRα pathway in macrophages.[63,64] This regulator can promote cholesterol efflux by upregulating the level of ABCA1 and ABCG1 mRNA and Apo A1 which promote the CRT system.[65]

3. Liver X receptor agonists and antagonists

Agonists or ligands of LXR are either natural or synthetic. The rise in importance of LXR led to the synthesis and development of various synthetic agonists for specific diseases. Agonists can activate the LXR-RXR and either upregulate or downregulate the gene function in association with LXR heterodimer. LXR agonists can bind to LXR, activating it to upregulate or downregulate the target genes. The binding of ligands to the RXR, a heterodimer component of LXR, also activates the LXR. Agonists and their co-signaling with LXR heterodimer can inhibit certain diseases such as atherosclerosis. The function of LXR agonists and their effectiveness in diseases firmly confirms the LXR as a target for therapeutics.

Antagonists of LXR can inhibit the LXR-RXR signaling on binding with heterodimer. Various exogenous proteins or metabolic products can act as antagonists of LXR and LXR signaling by inhibiting the heterodimer activation of LXR-RXR. The complete process of downregulation of LXR signaling by antagonists playing an important role in lipid metabolism regulation is a complex process.

3.1 LXR agonists

In the early stages, LXRs was regarded as orphan family receptor with unknown ligands. Recent developments in biotechniques have helped to identify the various ligands for LXRs. The natural ligands are oxysterols that can bind to species specifically followed by the signaling process of LXRs. The activation of LXR nuclear receptor and expression of LXR target genes by agonists is ligand-dependent and requires coactivators.[6] The oxygenated cholesterol and bile acids are natural ligands of LXRs. The natural ligands hydroxycholesterols such as 20(*S*)-hydroxy-cholesterol, 22(*R*)-hydroxy-cholesterol, and 24(*S*)-hydroxy-cholesterol can activate the signaling of LXR. Their ability to activate the LXRs shows that 27-OHC has the highest affinity among all. There are naturally existing ligands or agonists for LXRs that help in regulating the signaling of LXRS in various metabolisms. Paxilline, an indole alkaloid found naturally in *Penicillium paxillin, is a* toxic non-oxysterol LXR agonist. This agonist can activate both LXR α and LXRβ heterodimer and enhances the transcription of ABCA1 and SREBP genes.[66]

Lehman et al. studied the oxysterols and hormonal regulation inducing the transcription of LXR target genes essential for the regulation of cholesterol levels.[67] Various synthetic agonists of LXRs were previously used to study the activity of LXRs association with immunological and cardiovascular diseases. The agonists are species-specific for animals with its isotypes. The synthetic agonists are well designed to activate the specific LXRs for signaling and to transcribe their target genes. LXR agonists can induce the IDOL-dependent pathway which regulates the concentration of LDL cholesterol.[50]

Most of synthetic LXR agonists can induce the mechanism of reverse cholesterol transport (RCT) and the mechanism is well known as risk-reducing in various inflammatory and cardiovascular diseases.[38,50,68] In the RCT mechanism, the liver synthesizes apolipoprotein A1 (Apo A1) which helps in the efflux of cholesterol by involving the ATP binding cassette family, where the formation of HDL takes place. This mechanism is induced by the specific agonists which upregulate the expression of ABC family.[68] These agonists can also induce IDOL expression in hepatocytes which reduce LDLR level and raise plasma levels.[50] However, in mice, the IDOL expression due to agonists occurs peripheral and not in the liver. Synthetic agonists of LXR agonists are RCT and TICE that can prevent the development of atherosclerosis in various physiological condition. Furthermore, LXR agonists can activate the lipogenesis by stimulating the regulation of sterol regulatory element-binding protein-1c (SREBP-1c), a lipogenic transcription factor expression leading to hypertriglyceridemia and liver steatosis. Other effects of LXRs include stimulation of several apolipoproteins (apo-E, apo-D, apo-AIV) as well as of cholesterol ester transporting protein (CETP) and phospholipid transfer protein (PLTP).[9] The inverse agonists of LXR and their configurations H3, H11 and H12 can downregulate the target gene expression for LXR and inhibit the process of lipogenesis.[18]

The synthetic agonist of LXR is well known for inducing various metabolic activities regulating the expression of various genes. LXR agonists such as T091317, AZ876 and GW3965 are specific ligands of LXR that can activate the transcription of essential genes associated with lipid metabolism. Additionally, these agonists play an influential role in LXR-related diseases. T091317 is an LXR agonist that can regulate the gene and protein expression of ApoD, ABCA1, and CYP27A1. The binding of this agonist to LXR modulates or controls the amount of protein expression through corresponding posttranslational regulation of the mRNAs of those genes.[38] T0901317 can upregulate the expression of ABCA1, ABCG5, and ABCG8 in the small intestine with LXRα.[60] Any alteration on genes ABCG5 and ABCG8 can lead to sitosterolemia due to sterol accumulation and cholesterol circulation. The agonist, T0901317 can further upregulate the level of mRNA expression of IL-1β in macrophages.[58] This agonist can play a basic role in inhibiting the fibroblast-like synoviocyte (FLS) inhibition induced by IL-1β and IL-6-induced.[69] Moreover, T0901317 can upregulate the expression of ApoD to aid in the suppression of angiogenesis and tube formation in human umbilical vein endothelial cells.[38] GW3965 is an LXR-specific agonist used in the study of LXR signaling in the immune system and allergies. This agonist on binding with LXR can activate the LXR heterodimer which produces IL3 and IL4 increasing the response of Th2 cytokines.[70] AZ876 agonists in low doses can inhibit the progress of atherosclerosis promoting RCT but without inducing anti-inflammatory or hypertriglyceridemia.[71] This agonist reduces the risk of atherosclerosis without the alteration in the level of triglycerides in the liver or plasma. AZ876 and GW3965 both are well known for their anti-atherosclerosis response targeting LXR signaling. However, AZ876 has lesser side effects or even no side effects and a more stable lesion phenotype when compared with GW3965 agonist.

3.2 LXR antagonists

The involvement of LXR signaling can upregulate the genes associated with the disease, elevating the risk. Antagonists of LXR can either prevent or inhibit the function of LXR by regulating coactivators and co-repressor. Antagonists-like substances exist in nature in the form of dust in air such as triphenyl phosphate (TPHP) and 2-ethylhexyl diphenyl phosphate (EHDPP), their capability of antagonizing the LXR can inhibit the efflux of cholesterol.[72] Despite the role of reducing the progression of diseases by various agonists, the role of antagonists is also known for reducing the risk of diseases. Liver disease such as non-alcoholic fatty liver disease is treated with LXR antagonists.[73] Various intermediate products of endogenous and metabolic pathways can regulate the LXR signaling as antagonists.

A mevalonate cascade product, Geranylgeranyl pyrophosphate (GGPP) is an LXR antagonist that can inhibit LXR/RXR signaling by preventing the interaction of LXRα with nuclear coactivators.[9,74] GSK2033 is a specific antagonist of LXR that can suppress the expression of LXR target genes such as fatty acid synthase (FAS) and sterol regulatory element-binding protein 1c (SREBP 1c) by recruiting co-repressor of nuclear receptor.[75] This agonist further activates various nuclear receptors in the liver such as glucocorticoid receptor and farnesoid X receptor. Another antagonist, SR9243 antagonist can increase the level of reactive oxygen species (ROS) and can reduce Warburg effect-associated enzymes which are regulated on inhibition of LXR signaling in cancer cells. This antagonist further promotes the apoptosis, suppresses various metabolic enzymes and clonogenicity of cancer stem cells.[60] In macrophages, the LXR signaling can regulate the gene expression for efferocytosis receptors and signaling.[76] In efferocytosis, inhibition of LXR signaling by antagonists can suppress the expression of ABCA1 cholesterol efflux from efferocytosis cells.[77] The functions of antagonists altering various gene expressions at the molecular level become more significant in terms of therapeutics, which further promotes the importance of developing more synthetic antagonists.

4. Diet on LXR signaling

Diet is an important source of energy production in cells. Various biomolecules present in the diet plays an influential role in normal body functioning by metabolic and anabolic pathways. The alteration in the levels of end and intermediate products of synthetic pathways ultimately alters the physiology in the body. Cholesterol is an important precursor required for hormone synthesis. Cholesterol rich food contains high quantity oxysterols that are retained for a longer period. High fat diet can induce the upregulation in the expression of SR-B1. The diet source with high LDL or HDL or high fat content can alter the signaling of LXR in lipid metabolism and cholesterol level maintenance. Low-density lipoprotein receptors (LDLR) can bind to hepatitis B virus-associated apoE which plays a crucial role in HBV viral infection.[78] An isoflavone compound, BiochaniniA found in cabbage and red clover can regulate the expression of ABCA1 and ABCG1 that helps to improve the plasma lipid by enhancing the RCT essential for cholesterol efflux through the pathway peroxisome proliferator-activated receptor γ (PPARγ) and LXRα.[79] This isoflavone can inhibit the inflammatory response and reduce the activity of proinflammatory cytokines improving the risks of atherosclerosis.

5. Endocytosis of liver X receptor

The process of endocytosis is important for the receptors for their renewal and continuation of their metabolic signaling. The process helps in the internalization of defective or inactive forms of receptors that are already involved in signaling pathways. Despite different types of endocytosis, the Clatherin mediated endocytosis is widely understood in higher animals for surface receptors. However, LXR is a nuclear receptor found inside the nucleus and binds with its heterodimeric form in their target genes. The process of LXR endocytosis is not well understood in liver cells. However, their renewal and degradation might involve ubiquitin proteasome pathway. The C-N terminal of the protein domain plays a crucial role for various endocytic pathways. Though the mechanism of endocytosis of LXRs nuclear receptor is completely unknown, its involvement in receptor recycling and internalization is like other NRs families. The endocytosis and internalization of various surface receptors are driven by the conserved family of C-terminals in the domain.[80] This domain has the endocytic motif that governs the process of endocytosis which is conserved in all receptor proteins. The process of endocytosis begins with the binding of ligands or agonists to the receptors followed by signal transduction. The end process of the endocytic pathway of receptors leads to the degradation or recycling of the receptor proteins that are regulated by the formation of endosomes.[80]

LXR, a transcription factor, remains bound to the promoter region of genes involved in cholesterol transport. LXRs require bindings of ligands or agonists to either of the heterodimer forms of LXR/RXR. This binding activates the heterodimer of LXR and regulates the transcription of target genes. The conserved region, the DR4 response element, is present in the DNA of the LXR target genes.[81] In absence of ligands, the LXR heterodimer remains bound to the target genes along with the signal inhibitor known as nuclear receptor co-repressor (NCoR). The binding of agonists or ligands activates the heterodimer form of LXR-RXR which then releases the NCoR after alteration in its complex of LXR-RXR-NCoR.[82] This process further gets assisted by the recruitment of nuclear receptor coactivators. The process is similar for both the isotypes, LXRα and LXRβ despite their dissimilarity in molecular structure.

The degradation of LXRs and its associated signaling protein factors by the ubiquitin-proteasome pathway is well known. The proteolysis targeting chimera (PROTAC) can degrade the LXRβ in a ubiquitin-proteasome-dependent manner involving VHL E3 ligase.[61] Internalization of Liver X receptors due to endocytic factors is important for the renewal of transcription factors.

The LXR-RXR signaling is closely associated and may be parallel to caveolin-mediated endocytosis. Caveolin-1 is an important protein of

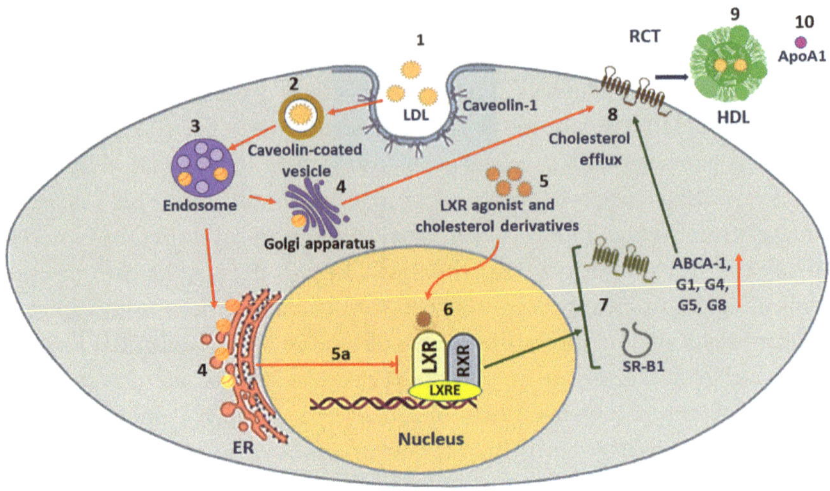

Fig. 3 Involvement of transcription factors in the regulation of inflammatory response and genes involved in reverse cholesterol transport by caveolae and Cav-1. LDL are endocytosed by the caveolin with the help of caveolin-1 (1) which forms the vesicle coated with caveolin (2). The endosomes are transported to the endoplasmic reticulum (RE) (3, 4), where cholesterol accumulates and downregulates the expression of LXR function inhibiting the expression of genes involved in cholesterol efflux (5a). The endosomes that reach the Golgi apparatus have higher chances for function in cholesterol efflux. The agonists of LXR or RXR can bind to LXR and activate the heterodimer by replacing the co-repressor with coactivators (5,6). Like PPARγ, the activation of LXR-RXR can upregulate the expression of specific ABC family receptors and SR-B1 genes (7). These receptors can further help in cholesterol efflux, and HDL production and can regulate cholesterol reverse transport mechanisms. *Redrawn from Qin L, Zhu N, Ao BX, Liu C, Shi YN, Du, K, Chen JX, Zheng, XL, Liao DF Caveolae and Caveolin-1 integrate reverse cholesterol transport and inflammation in atherosclerosis.* Int J Mol Sci. *2016;17:429.*

caveolae that plays an important role in endocytosis (Fig. 3). The vesicle formation of the endocytosed LDL or cholesterol takes place, where it is coated with a layer and known as a caveolin-coated vesicle or caveolar vesicle. This later is transported as caveosome as in endosome. The transport of the caveosome to the endoplasmic reticulum (ER) can help in further signaling. However, the accumulation of cholesterol in ER can downregulate the functions of nuclear factor LXR and its heterodimer LXR-RXR. The caveosome that undergoes towards the Golgi reaches the late endosome stage followed by the lysosomal action as in the Clatherin-mediated endocytosis. The cholesterol that are agonist to LXR-RXR, such as oxysterols, can bind to either LXR or RXR bound to LXRE and activates the genes and produce HDL and provide RCT mechanism.

6. Role of LXRs in human disease

LXR is a major transcription factor that helps in signaling and sensing the cholesterol level in the liver and many other organs. The role of LXR is biased in different types of tissues due to the involvement of their two different isotypes. Cholesterol and glucose are the most important molecule known for the association with LXR in the liver and small intestine, respectively. The understanding of LXR signaling is more important due to its involvement with various types of diseases associated with lipids and the immune system. The alteration in the signaling of LXR and its heterodimer can lead to various cardiovascular diseases along with autoimmune and immune disorders. Since, LXRs are nuclear receptors and transcription factors, understanding their complete role in diseases is important for therapeutic approaches. LXR signaling along with proliferator-activated receptor γ (PPARγ) responses are well known for their involvement in various metabolic disorders including non-alcoholic fatty liver disease, diabetes, and obesity.[83]

6.1 Hyperlipidemia

Hyperlipidemia is a state of disease associated with various lipid disorders like hypercholesterolemia and hypertriglyceridemia. This disease can be caused due to atherosclerosis. This disease is characterized by the elevation of the levels of triglyceride and serum cholesterol contributing risk factors for various cardiovascular diseases.[84] This disease is either inherited or can be acquired as it can develop secondarily.[85] This disease is more common in Western hemisphere with more than 3 million people in the United States and Europe being diagnosed with hyperlipidemia.[86] Various inverse agonists are effective against the hyperlipidemia by inhibitory process of lipogenesis.[18] Since, the diseases are associated with the metabolism of lipids, the signaling of LXR is well associated with it. The anti-hyperlipidemia property of compounds is regulated by the involvement of RCT which can elevate the expression of various ABC family genes including ABCA1, ABCG8, and SR-BI', PPARγ/α. Further, study performed by Liu et al. provided significant evidence of the involvement of LXRs, as LXRα protein expression was upregulated in the liver.[87]

6.2 Inflammatory diseases

LXRs is well known for their association with various inflammatory diseases. LXRs signaling is involved in chronic inflammatory autoimmune disease such as rheumatoid arthritis and atherosclerosis. As the signaling of LXR can regulate various proinflammatory and anti-inflammatory cytokines its involvement in inflammatory diseases is significant.[88] The downregulation of LXR expression is well known in autoimmune disorders as well as in certain inflammatory diseases. The deficiency in the Liver X receptor, the LXR α can cause the inflammatory diseases, however, LXRβ expression may act vice versa.[89] The downregulation of LXRs promotes chronic inflammatory disease, Atherosclerosis which is caused due to the accumulation of lipid content and formation of plaque in arteries. LXRs signaling are also an important factor in neuroinflammation and synaptic plasticity in the liver. Sleep-deprived cognitive impairment disease is associated with a deficiency in LXRβ expression. The downregulation of LXRβ expression can upregulate the inflammatory pathway by the expression of high mobility group box 1 (HMGB1)/TLR4/NF-κB p65. This expression can alter the immune system by inducing inflammatory responses. Further, the activation of LXRβ can inhibit neuroinflammation and improve sleep-deprived diseases.[8]

6.2.1 Atherosclerosis

The accumulation of lipids can cause chronic inflammatory disease and plaque formation leading to atherosclerosis within the arteries. The decrease in the activity of pro-inflammatory cytokines by LXR signaling indicates the effective activity of LXR in atherosclerosis. Various factors can influence atherosclerosis development including endoplasmic reticulum stress and prolonged inflammation. The inhibition of ER stress can significantly inhibit the development of atherosclerosis by upregulating the expression of genes, ABCA1, and ABCG1.[68] Various synthetic agonists of LXRs are significantly effective against anti-atherosclerosis. The risk of atherosclerosis has also been found higher when there is a dysfunction or deactivation of LXR isotypes. Further, the activation of LXRs reduced the progression of inflammatory diseases. The progress of foam cell accumulation takes place in the mice with the knocked-out genes of LXRs.[90] CM3II (isolated from *Cordyceps militaris*) is an anti-atherosclerotic which can act by lowering the plasma lipid in mice deficient with ApoE. This compound can promote the activity of LXRα and downregulate the SREBP-2 signaling in the liver.[91] The agonists of LXRs, T0901317 and GW3965 are well known for their anti-atherosclerosis activity.[92]

6.2.2 *Rheumatoid arthritis*

Rheumatoid arthritis is a chronic inflammation and an autoimmune disease. This disease can damage the joint tissues of different parts in the body making it severe pain and lack of balance. Homeostatic immune regulation requires a balance between proinflammatory and anti-inflammation factors. Various upregulation in the levels of pro-inflammatory cytokines can damage its own cells in the body unknowingly as it fails to discriminate the cells. Rheumatoid arthritis can be treated by various NSAIDs despite their adverse effects on the body. Antigen-presenting cells such as dendritic cells (DCs), and macrophages induce the production of various cytokines and chemokines which can either activate the immune processes or generate the maintenance of immune tolerance while presenting to T cells.[93] The Liver X receptors signaling pathway is associated with rheumatoid arthritis since the level of LXR expression is upregulated and their activation induces cytokine secretion driven by TLR signaling.[94] However, the LXRs signaling plays a significant role in associated with fibroblast-like synoviocyte invasion. The agonist of LXR can inhibit the signaling of IL-1β-induced and IL-6-induced fibroblast-like synoviocytes in the rheumatoid arthritis patient.[69]

6.2.3 *Respiratory disease*

Various factors such as inflammation, allergy in the respiratory tract can cause mild to chronic respiratory diseases. LXR signaling is responsive in various respiratory syndromes such as allergic asthma. Unlike the effective role of LXR expression in many diseases as a therapeutic approach, the LXRs can increase airway inflammation and hyperresponsiveness in chronic asthma.[70] The immune cells and inflammatory cytokines level in various allergic reactions and induce the production of antibodies. The reduction in LXR expression can decrease the allergy and hyperresponsiveness in asthma which further decreases the airway remodeling. This influential role of LXR in associated with asthma makes it a critical target for asthma therapy.

6.3 LXR in cancer cell progression

Cancer cell's progression completely depends upon various extrinsic and intrinsic factors, mutation and epigenetics. Immune cells play an important role in inhibiting or prohibiting cancer cells under various physiological conditions. Various transcription factors can induce the signal transduction pathways targeting the specific genes important in the development of cancer cells. In cancer cells, the metabolic uncoupling of LXR nuclear receptors can alter the intracellular cholesterol level due to the reduction in oxysterol synthesis by cells.[95] LXR nuclear receptors can induce various immune regulations either by downregulating or upregulating the immune cells. The signaling role of LXR in different cancer types such as breast cancer, melanoma, colon cancer, prostate, and glioblastoma is little known. The activation of LXR and signaling is associated with the expression of major histocompatibility complex class I chain-related molecule A (MICA) and MICB in multiple myeloma cancer cells. This expression of the NK group 2D receptor can enhance the activity of NK cell-mediated cytotoxicity in multiple myeloma.[3] Cholesterol plays an important role in cancer cell progression. The cholesterol metabolite acts as a ligand of LXR which can induce an immune suppressive program in cancer cell development.[96] The activation of LXR by specific agonists can induce the hypo-osmotic stress and inhibit the proliferation, migration and invasion of cancer stem cells.[60] The signaling nuclear translocation of androgen receptors is inhibited by the activation of LXRβ.[97] The LXR activation and upregulation associated with immune inhibitor treatment provide attainable anti-tumor responses.[3] Wong et al. studied the response of LXR/RXR signaling in oral squamous cell carcinoma and found out this signaling is essential for development and progress.[98] LXRβ activation can suppress the growth of tumor cells and reduce the progress of myeloid-derived suppressor cells in immunocompetent patients. This activation of LXRβ by ligand agonists can induce tumor regression in the lung, colon, glioblastoma, melanoma, and ovary.[37]

6.4 In liver diseases

In recent studies, there has been strong evidence of involvement of correlation between cholesterol efflux of HDL and cardiovascular diseases.[99] Cardiovascular diseases are well associated with the signal regulation of LXR isotypes. The transport of cholesterol is regularly kept on track by LXRs for the regulation of cholesterol homeostasis.[28] Various genes associated with LXRs for the regulation of cholesterol transport are ABCA1, ABCG1, ABCG5 and ABCG8, ApoE, SR-B1, CYP7A1, Cholesteryl ester transfer protein (CETP), and inducible degrader of the LDLR (IDOL).

6.4.1 Non-alcoholic fatty liver disease (NAFLD)

The alteration in the lipid metabolism can lead to the accumulation of lipids in liver cells. This in turn can form non-alcoholic fatty liver in association with obesity, insulin resistance, and hyperlipidemia.[73] NAFLD is a chronic liver disease linked with atherosclerosis, coronary artery disease, hypertension, and cardiovascular diseases.[100] LXR signaling has a specific response in association with NAFLD. The signaling of LXR and its heterodimer is the main regulator of lipid metabolism in NAFLD, this regulation is induced by the expression of SREBP-1c.[101] Hydroxysteroid sulfotransferase 2B1b (SULT 2B1b) induced LXRα signaling suppression can regenerate the fatty liver in NAFLD mice.[102] Blnc1, a long non-coding RNA can induce lipogenesis by regulating the pathway LXR/SREBP 1c and regulating NAFLD pathogenesis.[103]

6.4.2 Diabetic cardiomyopathy (DCM)

DCM disease is altered in myocardial metabolic, inflammation, oxidative stress, cardiomyocyte apoptosis, ventricular dysfunction, and structural remodeling which is developed due to diabetes-induced pathologic abnormalities.[104] The mRNA expression of LXRα and target genes of LXR is elevated in diabetic rats indicating the potential role of LXR in diabetes and the progression of diabetic cardiomyopathy.[105] The role of LXR/RXR pathway in diabetic cardiomyopathy has been reported in various studies.[106] However, LXRα is upregulated by hyperglycemia in diabetic mice. The role of LXR agonists with effective antidiabetic properties relates to the involvement of LXR signaling. These agonists can reduce hyperglycemia and improve insulin sensitivity.[107]

6.5 LXRs in hepatitis

Hepatitis is a liver disease caused by a virus. The disease is a major health concern despite the presence of vaccines. The death rate caused by hepatitis is higher than the death rate of HIV. Low-density lipoprotein receptors (LDLR) can bind to hepatitis B virus-associated apoE which plays crucial role in HBV viral infection.[78] LXRs signaling plays an effective role in autoimmune hepatitis disease.[108] The downregulation of LXRα along with its target genes has been reported in autoimmune hepatitis, however, upregulation of LXRβ has been notified by Gao et al. in 2020. Furthermore, the expression of LXRα is also associated with the disease hepatitis C which is induced by the transcription of various target genes.[109]

6.6 Alzheimer's disease

Alzheimer's disease (AD) is a complex progressive neurodegenerative disease that is caused due to alteration in protein structure and aggregation such as deposition of amyloid plaque and alters in neuroinflammatory responses. AD is more common in aged people and hence regarded as an age-associated disease. Alzheimer's patients have many neurological and psychological symptoms in the brain including dementia, loss of memory, disorientation, aggressiveness, and hallucination. Altered in the innate immune system and formation of amyloid plaques cause prominent development of AD. Amyloid-β peptide (Aβ) and tau tangle regulate the neuroinflammatory processes in neurodegenerative diseases that worsen the condition of AD.

The regulation of inflammatory mediators in AD patients is induced by the intracellular signaling of liver X receptors.[110] The role of the ABCA1 cholesterol transporter in AD is highly influenced yet poorly understood.[111] However, the LXR present in ABC is well known for its regulation in the immune system which may play a key role in AD. As the ligands of LXR can regulate the LXR signaling, agonists and modulators of LXR play an inducible role in Alzheimer's disease.[112] Apolipoprotein E (ApoE) is a cholesterol carrier in the brain that influences lipid metabolism and also act as a neuroprotective that plays a significant role in Alzheimer's disease.[113] Apolipoprotein E4 (ApoE4) induces inflammatory responses in Aβ deposition and tau-mediated mechanisms in AD.[110] The basic role of LXR in lipid metabolism associated with AD makes it more precise for the drug targets by specific agonists by inhibiting the abnormal neuromodulation.

6.7 Human immunodeficiency virus (HIV)

HIV infection is governed by an infectious virus. ApoE plays an essential role in pathogenesis of HIV infection. The downregulation in the expression of ApoE increases the infectivity of HIV. This enzyme acts as an anti-HIV-1 by associating with envelope protein, gp160 Env precursor that later inhibits viral production and infectivity of HIV-1.[114] However, limited studies have been performed with the evidence of direct involvement of LXR signaling in associated with HIV.

7. Future prospects of LXR endocytosis in therapeutics

Cholesterol and lipids are important biomolecules that play crucial roles in liver cell metabolism. In hepatocytes, the LXR nuclear receptor has a significant role in lipid metabolism and cholesterol level to maintain metabolic functions of the liver. LXR has been targeted for the therapy of rheumatoid arthritis due to its role in lipid homeostasis.[88] The understanding of LXR and its signaling at the molecular level is important. The role of LXR associated with various diseases and cardiovascular disease is a potential source of target therapy which can be treated by many agonists and antagonists. The upregulation and downregulation of LXR by various synthetic ligands can influence and alter the progress of diseases. Hence, understanding the role of LXR endocytosis becomes more significant for various metabolic diseases to develop various drugs in future.

8. Conclusion

The recent techniques used to study the nuclear receptors has enhanced the understanding of various ligands involved with the nuclear receptor signaling, regulations of physiological pathways, and their association with various types of diseases. Despite the multiple types of endocytosis, receptor-mediated endocytic pathway is highly studied and crucial in animal cells for cell–cell, cell–ligand interactions. LXRs are an important nuclear receptor and transcriptional factor for the regulation of lipid metabolism. In hepatocytes, it plays a variety of roles in maintaining the homeostatic condition of cells. The endocytosis of many receptors has been studied to understand the importance of receptor signaling. Further, the LXRs have been found to be associated with various metabolic and genetic diseases such as cardiovascular diseases, neurological disorders, autoimmune diseases, and cancer. The study of LXRs and endocytosis will provide an idea of its importance in treating diseases with unique targets related to it.

References

1. Leiskau C, Baumann U. Structure, function, and repair of the liver. In: Kelly DA, ed. *Diseases of the Liver and Biliary System in Children*. Wiley-Blackwell; 2017:1–17.
2. Beckwitt CH, Clark AM, Wheeler S, et al. Liver 'organ on a chip'. *Exp Cell Res*. 2018;363:15–25.

3. Bilotta MT, Petillo S, Santoni A, Cippitelli M. Liver X receptors: Regulators of cholesterol metabolism, inflammation, autoimmunity, and cancer. *Front Immunol.* 2020;11:584303.
4. Apfel R, Benbrook D, Lernhardt E, Ortiz MA, Salbert G, Pfahl M. A novel orphan receptor specific for a subset of thyroid hormone-responsive elements and its interaction with the retinoid/thyroid hormone receptor subfamily. *Mol Cell Biol.* 1994;14:7025–7035.
5. Willy PJ, Umesono K, Ong ES, Evans RM, Heyman RA, Mangelsdorf DJ. LXR, a nuclear receptor that defines a distinct retinoid response pathway. *Genes Dev.* 1995;9:1033–1045.
6. Kain J, Wei X, Reddy NA, Price AJ, Woods C, Bochkis IM. Pioneer factor Foxa2 enables ligand-dependent activation of type II nuclear receptors FXR and LXRalpha. *Mol Metab.* 2021;53:101291.
7. Baranowski M. Biological role of liver X receptors. *J Physiol Pharmacol.* 2008;59:31–55.
8. Qiu C, Wang M, Yu W, et al. Activation of the hippocampal LXRbeta improves sleep-deprived cognitive impairment by inhibiting neuroinflammation. *Mol Neurobiol.* 2021;58:5272–5288.
9. Wójcicka G, Jamroz-Wiśniewska A, Horoszewicz K, Bełtowski J. Liver X receptors (LXRs). Part I: Structure, function, regulation of activity, and role in lipid metabolism Receptory wątrobowe X (LXR). Część I: Budowa, funkcja, regulacja aktywności i znaczenie w metabolizmie lipidów. *Postepy Hig Med Dosw (Online).* 2007;61:736–759.
10. Janowski BA, Grogan MJ, Jones SA, et al. Structural requirements of ligands for the oxysterol liver X receptors LXRalpha and LXRbeta. *Proc Natl Acad Sci U S A.* 1999;96:266–271.
11. Khan I, Steeg PS. Endocytosis: A pivotal pathway for regulating metastasis. *Br J Cancer.* 2021;124:66–75.
12. Elkin SR, Lakoduk AM, Schmid SL. Endocytic pathways and endosomal trafficking: A primer. *Wien Med Wochenschr.* 2016;166:196–204.
13. Mettlen M, Chen PH, Srinivasan S, Danuser G, Schmid SL. Regulation of Clathrin-mediated endocytosis. *Annu Rev Biochem.* 2018;87:871–896.
14. Takei K, Haucke V. Clathrin-mediated endocytosis: Membrane factors pull the trigger. *Trends Cell Biol.* 2001;11:385–391.
15. Sohrabi Y, Sonntag GVH, Braun LC, et al. LXR activation induces a proinflammatory trained innate immunity-phenotype in human monocytes. *Front Immunol.* 2020;11:353.
16. Pineda-Torra I, Siddique S, Waddington KE, Farrell R, Jury EC. Disrupted lipid metabolism in multiple sclerosis: A role for liver X receptors? *Front Endocrinol (Lausanne).* 2021;12:639757.
17. Wang Y, Viscarra J, Kim SJ, Sul HS. Transcriptional regulation of hepatic lipogenesis. *Nat Rev Mol Cell Biol.* 2015;16:678–689.
18. Chen Z, Chen H, Zhang Z, et al. Discovery of novel liver X receptor inverse agonists as lipogenesis inhibitors. *Eur J Med Chem.* 2020;206:112793.
19. Joseph SB, Laffitte BA, Patel PH, et al. Direct and indirect mechanisms for regulation of fatty acid synthase gene expression by liver X receptors. *J Biol Chem.* 2002;277:11019–11025.
20. Talukdar S, Hillgartner FB. The mechanism mediating the activation of acetyl-coenzyme a carboxylase-alpha gene transcription by the liver X receptor agonist T0-901317. *J Lipid Res.* 2006;47:2451–2461.
21. Wang Y, Kurdi-Haidar B, Oram JF. LXR-mediated activation of macrophage stearoyl-CoA desaturase generates unsaturated fatty acids that destabilize ABCA1. *J Lipid Res.* 2004;45:972–980.

22. Repa JJ, Liang G, Ou J, et al. Regulation of mouse sterol regulatory element-binding protein-1c gene (SREBP-1c) by oxysterol receptors, LXRalpha and LXRbeta. *Genes Dev.* 2000;14:2819–2830.
23. Schultz JR, Tu H, Luk A, et al. Role of LXRs in control of lipogenesis. *Genes Dev.* 2000;14:2831–2838.
24. Ress C, Kaser S. Mechanisms of intrahepatic triglyceride accumulation. *World J Gastroenterol.* 2016;22:1664–1673.
25. Fan Q, Norgaard RC, Grytten I, et al. LXRalpha regulates ChREBPalpha Transactivity in a target gene-specific manner through an agonist-modulated LBD-LID interaction. *Cell.* 2020;9.
26. Ouimet M, Barrett TJ, Fisher EA. HDL and reverse cholesterol transport. *Circ Res.* 2019;124:1505–1518.
27. Puschel GP, Henkel J. Dietary cholesterol does not break your heart but kills your liver. *Porto Biomed J.* 2018;3:e12.
28. Ma Z, Deng C, Hu W, et al. Liver X receptors and their agonists: Targeting for cholesterol homeostasis and cardiovascular diseases. *Curr Issues Mol Biol.* 2017;22:41–64.
29. Phillips MC. Is ABCA1 a lipid transfer protein? *J Lipid Res.* 2018;59:749–763.
30. Huang L, Fan B, Ma A, Shaul PW, Zhu H. Inhibition of ABCA1 protein degradation promotes HDL cholesterol efflux capacity and RCT and reduces atherosclerosis in mice. *J Lipid Res.* 2015;56:986–997.
31. Pamir N, Hutchins PM, Ronsein GE, et al. Plasminogen promotes cholesterol efflux by the ABCA1 pathway. *JCI. Insight.* 2017;2.
32. Cooke AL, Morris J, Melchior JT, et al. A thumbwheel mechanism for APOA1 activation of LCAT activity in HDL. *J Lipid Res.* 2018;59:1244–1255.
33. Williams K, Segard A, Graf GA. Sitosterolemia: Twenty years of discovery of the function of ABCG5ABCG8. *Int J Mol Sci.* 2021;22.
34. Back SS, Kim J, Choi D, Lee ES, Choi SY, Han K. Cooperative transcriptional activation of ATP-binding cassette sterol transporters ABCG5 and ABCG8 genes by nuclear receptors including liver-X-receptor. *BMB Rep.* 2013;46:322–327.
35. Laffitte BA, Repa JJ, Joseph SB, et al. LXRs control lipid-inducible expression of the apolipoprotein E gene in macrophages and adipocytes. *Proc Natl Acad Sci U S A.* 2001;98:507–512.
36. Zhou Y, Luo G. Apolipoproteins, as the carrier proteins for lipids, are involved in the development of breast cancer. *Clin Transl Oncol.* 2020;22:1952–1962.
37. Tavazoie MF, Pollack I, Tanqueco R, et al. LXR/ApoE activation restricts innate immune suppression in cancer. *Cell.* 2018;172:825–840 [e818].
38. Lai CJ, Cheng HC, Lin CY, et al. Activation of liver X receptor suppresses angiogenesis via induction of ApoD. *FASEB J.* 2017;31:5568–5576.
39. Wang B, Tontonoz P. Liver X receptors in lipid signalling and membrane homeostasis. *Nat Rev Endocrinol.* 2018;14:452–463.
40. Liu Y, Morton RE. Apolipoprotein F: A natural inhibitor of cholesteryl ester transfer protein and a key regulator of lipoprotein metabolism. *Curr Opin Lipidol.* 2020;31:194–199.
41. Yang Z, Yin J, Wang Y, et al. The fucoidan A3 from the seaweed Ascophyllum nodosum enhances RCT-related genes expression in hyperlipidemic C57BL/6J mice. *Int J Biol Macromol.* 2019;134:759–769.
42. Shen WJ, Azhar S, Kraemer FB. SR-B1: A unique multifunctional receptor for cholesterol influx and efflux. *Annu Rev Physiol.* 2018;80:95–116.
43. Briand O, Touche V, Colin S, et al. Liver X receptor regulates triglyceride absorption through intestinal Down-regulation of scavenger receptor class B, type 1. *Gastroenterology.* 2016;150:650–658.

44. Chiang JY, Ferrell JM. Up to date on cholesterol 7 alpha-hydroxylase (CYP7A1) in bile acid synthesis. *Liver Res.* 2020;4:47–63.
45. Iwanicki T, Balcerzyk A, Niemiec P, et al. CYP7A1 gene polymorphism located in the 5′ upstream region modifies the risk of coronary artery disease. *Dis Markers.* 2015;2015:185969.
46. Chapman MJ, Le Goff W, Guerin M, Kontush A. Cholesteryl ester transfer protein: At the heart of the action of lipid-modulating therapy with statins, fibrates, niacin, and cholesteryl ester transfer protein inhibitors. *Eur Heart J.* 2010;31:149–164.
47. Morton RE, Mihna D, Liu Y. The lipid substrate preference of CETP controls the biochemical properties of HDL in fat/cholesterol-fed hamsters. *J Lipid Res.* 2021;62:100027.
48. Shimada A, Kimura H, Oida K, et al. Serum CETP status is independently associated with reduction rates in LDL-C in pitavastatin-treated diabetic patients and possible involvement of LXR in its association. *Lipids Health Dis.* 2016;15:57.
49. Wang JQ, Lin ZC, Li LL, et al. SUMOylation of the ubiquitin ligase IDOL decreases LDL receptor levels and is reversed by SENP1. *J Biol Chem.* 2021;296:100032.
50. Hong C, Tontonoz P. Liver X receptors in lipid metabolism: Opportunities for drug discovery. *Nat Rev Drug Discov.* 2014;13:433–444.
51. Herold M, Breuer J, Hucke S, et al. Liver X receptor activation promotes differentiation of regulatory T cells. *PLoS One.* 2017;12:e0184985.
52. Ishikiriyama T, Nakashima H, Endo-Umeda K, et al. Contrasting functional responses of resident Kupffer cells and recruited liver macrophages to irradiation and liver X receptor stimulation. *PLoS One.* 2021;16:e0254886.
53. Buono C, Li Y, Waldo SW, Kruth HS. Liver X receptors inhibit human monocyte-derived macrophage foam cell formation by inhibiting fluid-phase pinocytosis of LDL. *J Lipid Res.* 2007;48:2411–2418.
54. Waddington KE, Jury EC, Pineda-Torra I. Liver X receptors in immune cell function in humans. *Biochem Soc Trans.* 2015;43:752–757.
55. Joseph SB, Castrillo A, Laffitte BA, Mangelsdorf DJ, Tontonoz P. Reciprocal regulation of inflammation and lipid metabolism by liver X receptors. *Nat Med.* 2003;9:213–219.
56. Ghisletti S, Huang W, Ogawa S, et al. Parallel SUMOylation-dependent pathways mediate gene- and signal-specific transrepression by LXRs and PPARgamma. *Mol Cell.* 2007;25:57–70.
57. Fontaine C, Rigamonti E, Nohara A, et al. Liver X receptor activation potentiates the lipopolysaccharide response in human macrophages. *Circ Res.* 2007;101:40–49.
58. Menegaut L, Thomas C, Jalil A, et al. Interplay between liver X receptor and hypoxia inducible factor 1alpha potentiates interleukin-1beta production in human macrophages. *Cell Rep.* 2020;31:107665.
59. Mitro N, Mak PA, Vargas L, et al. The nuclear receptor LXR is a glucose sensor. *Nature.* 2007;445:219–223.
60. Dianat-Moghadam H, Khalili M, Keshavarz M, et al. Modulation of LXR signaling altered the dynamic activity of human colon adenocarcinoma cancer stem cells in vitro. *Cancer Cell Int.* 2021;21:100.
61. Xu H, Ohoka N, Yokoo H, et al. Development of agonist-based PROTACs targeting liver X receptor. *Front Chem.* 2021;9:674967.
62. Christidis G, Karatayli E, Hall RA, et al. Fibroblast growth factor 21 response in a preclinical alcohol model of acute-on-chronic liver injury. *Int J Mol Sci.* 2021;22.
63. Lin XL, He XL, Zeng JF, et al. FGF21 increases cholesterol efflux by upregulating ABCA1 through the ERK1/2-PPARgamma-LXRalpha pathway in THP1 macrophage-derived foam cells. *DNA Cell Biol.* 2014;33:514–521.

64. Xu Y, Denning KL, Lu Y. PPARalpha agonist WY-14,643 induces adipose atrophy and fails to blunt chronic ethanol-induced hepatic fat accumulation in mice lacking adipose FGFR1. *Biochem Pharmacol.* 2021;192:114678.
65. Shang W, Yu X, Wang H, et al. Fibroblast growth factor 21 enhances cholesterol efflux in THP-1 macrophage-derived foam cells. *Mol Med Rep.* 2015;11:503–508.
66. Bramlett KS, Houck KA, Borchert KM, et al. A natural product ligand of the oxysterol receptor, liver X receptor. *J Pharmacol Exp Ther.* 2003;307:291–296.
67. Lehmann JM, Kliewer SA, Moore LB, et al. Activation of the nuclear receptor LXR by oxysterols defines a new hormone response pathway. *J Biol Chem.* 1997;272: 3137–3140.
68. Wang T, Zhao Y, You Z, et al. Endoplasmic reticulum stress affects cholesterol homeostasis by inhibiting LXRalpha expression in hepatocytes and macrophages. *Nutrients.* 2020;12.
69. Laragione T, Gulko PS. Liver X receptor regulates rheumatoid arthritis fibroblast-like synoviocyte invasiveness, matrix metalloproteinase 2 activation, interleukin-6 and CXCL10. *Mol Med.* 2012;18:1009–1017.
70. Zhang J, Wu Z, Yu F, et al. Role of liver-X-receptors in airway remodeling in mice with chronic allergic asthma. *Exp Ther Med.* 2021;22:920.
71. van der Hoorn J, Linden D, Lindahl U, et al. Low dose of the liver X receptor agonist, AZ876, reduces atherosclerosis in APOE*3Leiden mice without affecting liver or plasma triglyceride levels. *Br J Pharmacol.* 2011;162:1553–1563.
72. Hu W, Jia Y, Kang Q, et al. Screening of house dust from Chinese homes for chemicals with liver X receptors binding activities and characterization of atherosclerotic activity using an in vitro macrophage cell line and ApoE −/− mice. *Environ Health Perspect.* 2019;127:117003.
73. Fan S, Zhang H, Wang Y, et al. LXRalpha/beta antagonism protects against lipid accumulation in the liver but increases plasma cholesterol in rhesus macaques. *Chem Res Toxicol.* 2021;34:833–838.
74. Forman BM, Ruan B, Chen J, Schroepfer Jr GJ, Evans RM. The orphan nuclear receptor LXRalpha is positively and negatively regulated by distinct products of mevalonate metabolism. *Proc Natl Acad Sci U S A.* 1997;94:10588–10593.
75. Griffett K, Burris TP. Promiscuous activity of the LXR antagonist GSK2033 in a mouse model of fatty liver disease. *Biochem Biophys Res Commun.* 2016;479:424–428.
76. Gage MC. Measuring apoptotic cell engulfment (efferocytosis) efficiency. *Methods Mol Biol.* 2019;1951:143–152.
77. Mota AC, Dominguez M, Weigert A, Snodgrass RG, Namgaladze D, Brune B. Lysosome-dependent LXR and PPARdelta activation upon efferocytosis in human macrophages. *Front Immunol.* 2021;12:637778.
78. Li Y, Luo G. Human low-density lipoprotein receptor plays an important role in hepatitis B virus infection. *PLoS Pathog.* 2021;17:e1009722.
79. Yu XH, Chen JJ, Deng WY, et al. Biochanin a mitigates atherosclerosis by inhibiting lipid accumulation and inflammatory response. *Oxid Med Cell Longev.* 2020;2020 [8965047].
80. Scott DB, Michailidis I, Mu Y, Logothetis D, Ehlers MD. Endocytosis and degradative sorting of NMDA receptors by conserved membrane-proximal signals. *J Neurosci.* 2004;24:7096–7109.
81. Quack M, Frank C, Carlberg C. Differential nuclear receptor signalling from DR4-type response elements. *J Cell Biochem.* 2002;86:601–612.
82. Kim J, Castellano JM, Jiang H, et al. Overexpression of low-density lipoprotein receptor in the brain markedly inhibits amyloid deposition and increases extracellular a beta clearance. *Neuron.* 2009;64:632–644.

83. Dixon ED, Nardo AD, Claudel T, Trauner M. The role of lipid sensing nuclear receptors (PPARs and LXR) and metabolic lipases in obesity, diabetes and NAFLD. *Genes (Basel)*. 2021;12.
84. Andreadou I, Schulz R, Badimon L, et al. Hyperlipidaemia and cardioprotection: Animal models for translational studies. *Br J Pharmacol*. 2020;177:5287–5311.
85. Handhle A, Park A. Hyperlipidaemia. *Medicine*. 2021;49:587–591.
86. Hill MF, Bordoni B. *Hyperlipidemia*. Treasure Island (FL): StatPearls; 2022.
87. Liu HF, Huang JB, Huang MC, et al. Study on anti-hyperlipidemia effect of Linderae radix via regulating reverse cholesterol transport. *Zhongguo Zhong Yao Za Zhi*. 2021;46:1795–1802.
88. Mai CT, Zheng DC, Li XZ, Zhou H, Xie Y. Liver X receptors conserve the therapeutic target potential for the treatment of rheumatoid arthritis. *Pharmacol Res*. 2021;170, 105747.
89. Bogie JFJ, Vanmierlo T, Vanmol J, et al. Liver X receptor beta deficiency attenuates autoimmune-associated neuroinflammation in a T cell-dependent manner. *J Autoimmun*. 2021;124:102723.
90. Zierfuss B, Weinhofer I, Buda A, et al. Targeting foam cell formation in inflammatory brain diseases by the histone modifier MS-275. *Ann Clin Transl Neurol*. 2020;7: 2161–2177.
91. Yang X, Lin P, Wang J, et al. Purification, characterization and anti-atherosclerotic effects of the polysaccharides from the fruiting body of cordyceps militaris. *Int J Biol Macromol*. 2021;181:890–904.
92. Collins JL, Fivush AM, Watson MA, et al. Identification of a nonsteroidal liver X receptor agonist through parallel array synthesis of tertiary amines. *J Med Chem*. 2002;45:1963–1966.
93. Lin YJ, Anzaghe M, Schulke S. Update on the Pathomechanism, diagnosis, and treatment options for rheumatoid arthritis. *Cell*. 2020;9.
94. Asquith DL, Ballantine LE, Nijjar JS, et al. The liver X receptor pathway is highly upregulated in rheumatoid arthritis synovial macrophages and potentiates TLR-driven cytokine release. *Ann Rheum Dis*. 2013;72:2024–2031.
95. Bovenga F, Sabba C, Moschetta A. Uncoupling nuclear receptor LXR and cholesterol metabolism in cancer. *Cell Metab*. 2015;21:517–526.
96. Baek AE, Krawczynska N, Das Gupta A, et al. The cholesterol metabolite 27HC increases secretion of extracellular vesicles which promote breast cancer progression. *Endocrinology*. 2021;162.
97. El-Kenawi A, Dominguez-Viqueira W, Liu M, et al. Macrophage-derived cholesterol contributes to therapeutic resistance in prostate cancer. *Cancer Res*. 2021;81:5477–5490.
98. Wong YL, Ramanathan A, Yuen KM, et al. Comparative sera proteomics analysis of differentially expressed proteins in oral squamous cell carcinoma. *PeerJ*. 2021;9:e11548.
99. Groenen AG, Halmos B, Tall AR, Westerterp M. Cholesterol efflux pathways, inflammation, and atherosclerosis. *Crit Rev Biochem Mol Biol*. 2021;56:426–439.
100. Polyzos SA, Kechagias S, Tsochatzis EA. Review article: Non-alcoholic fatty liver disease and cardiovascular diseases: Associations and treatment considerations. *Aliment Pharmacol Ther*. 2021;54:1013–1025.
101. Higuchi N, Kato M, Shundo Y, et al. Liver X receptor in cooperation with SREBP-1c is a major lipid synthesis regulator in nonalcoholic fatty liver disease. *Hepatol Res*. 2008;38:1122–1129.
102. Zhang X, Xu Y, Bai Q, et al. Inhibition of LXR signaling by SULT2B1b promotes liver regeneration after partial hepatectomy in mouse models of nonalcoholic fatty liver disease. *Am J Physiol Gastrointest Liver Physiol*. 2020;319:G87–G96.

103. Zhao XY, Xiong X, Liu T, et al. Long noncoding RNA licensing of obesity-linked hepatic lipogenesis and NAFLD pathogenesis. *Nat Commun*. 2018;9:2986.
104. Huang J-P, Cheng M-L, Wang C-H, et al. Therapeutic potential of cPLA2 inhibitor to counteract dilated-cardiomyopathy in cholesterol-treated H9C2 cardiomyocyte and MUNO rat. *Pharmacol Res*. 2020;160:105201.
105. Cheng Y, Liu G, Pan Q, Guo S, Yang X. Elevated expression of liver X receptor alpha (LXRalpha) in myocardium of streptozotocin-induced diabetic rats. *Inflammation*. 2011;34:698–706.
106. He Q, Pu J, Yuan A, et al. Liver X receptor agonist treatment attenuates cardiac dysfunction in type 2 diabetic db/db mice. *Cardiovasc Diabetol*. 2014;13:149.
107. Schulman IG. Liver X receptors link lipid metabolism and inflammation. *FEBS Lett*. 2017;591:2978–2991.
108. Gao L, Li B, Wang J, et al. Activation of liver X receptor alpha sensitizes mice to T-cell mediated hepatitis. *Hepatol Commun*. 2020;4:1664–1679.
109. Lima-Cabello E, Garcia-Mediavilla MV, Miquilena-Colina ME, et al. Enhanced expression of pro-inflammatory mediators and liver X-receptor-regulated lipogenic genes in non-alcoholic fatty liver disease and hepatitis C. *Clin Sci*. 2011;120:239–250.
110. Kloske CM, Wilcock DM. The important Interface between apolipoprotein E and neuroinflammation in Alzheimer's disease. *Front Immunol*. 2020;11:754.
111. Zelcer N, Khanlou N, Clare R, et al. Attenuation of neuroinflammation and Alzheimer's disease pathology by liver x receptors. *Proc Natl Acad Sci U S A*. 2007;104:10601–10606.
112. Goher SS, Griffett K, Hegazy L, et al. Development of novel liverXreceptor modulators based on a 1,2,4-triazole scaffold. *Bioorg Med Chem Lett*. 2019;29:449–453.
113. Lanfranco MF, Ng CA, Rebeck GW. ApoE lipidation as a therapeutic target in Alzheimer's disease. *Int J Mol Sci*. 2020;21.
114. Siddiqui R, Suzu S, Ueno M, et al. Apolipoprotein E is an HIV-1-inducible inhibitor of viral production and infectivity in macrophages. *PLoS Pathog*. 2018;14:e1007372.

CHAPTER FIFTEEN

Frizzled receptors and SFRP5 in lipid metabolism: Current findings and potential applications

Dinh-Toi Chu[a,b,*] and Thanh-Lam Nguyen[a]

[a]Center for Biomedicine and Community Health, International School, Vietnam National University, Hanoi, Vietnam

[b]Faculty of Applied Sciences, International School, Vietnam National University, Hanoi, Vietnam

*Corresponding author: e-mail addresses: chudinhtoi.hnue@mail.com; toicd@vnu.edu.vn

Contents

Abstract

Lipid metabolism plays a very important role as the central metabolic process of the body. Lipid metabolism interruptions may cause many chronic diseases, for example, non-alcoholic fatty liver disease (NAFLD), diabetes, and obesity. Secreted Frizzled Related Protein 5 (SFRP5) and Frizzled receptors (FZD) are two newly discovered adipokines that are involved in lipid metabolism as well as lipogenesis. Both of these adipokines affect lipid metabolism and adipogenesis through three WNT signaling pathways (WNTSP): WNT/β-catenin, WNT/Ca^{2+}, and WNT/JNK. FZD consists of 10 species, which have a cysteine-rich domain (CRD) to bind to the WNT protein for signal transduction. Depending on the type of ligand or co-receptor, they can stimulate or inhibit

Progress in Molecular Biology and Translational Science, Volume 194
ISSN 1877-1173
https://doi.org/10.1016/bs.pmbts.2022.06.025

adipogenesis. In lipid metabolism, they play a role in recognizing fatty acids. In obesity, gene expression of the WNT/FZD receptors is significantly increased. In contrast, SFPR5 serves as an antagonist that can compete with FZD for inhibition of WNTSP. It is believed to have anti-inflammatory potential in obesity and diseases related to abnormal lipid metabolism. In these cases, the expression of SFRP5 is found to be very low leading to the promoted production of proinflammatory cytokines (PICS). Some methods that include using recombinant SFRP5 to improve non-alcoholic steatohepatitis (NASH), using secreted Ly-6/uPAR-related protein 1 (Slurp1) to regulate fat accumulation in the liver through SFRP5, and dietary and lifestyle interventions to improve overweight/obesity have been studied. However, understandings of the molecular mechanisms of these two adipokines and their interactions are very limited. Therefore, more in-depth studies are needed in the future.

Abbreviations

AT	adipose tissue
CEBP	transcription factor CCAAT/enhancer binding protein
CRD	cysteine-rich domain
FAs	fatty acids
FZD	frizzled receptors
GPCRs	G-protein coupled receptors
IL	interleukin
IR	insulin resistance
JNK	c-Jun N-terminal kinase
LDL-C	low density lipoprotein cholesterol
lncRNAs	long non-coding RNAs
LP	lipoprotein production
LRP	low-density lipoprotein receptor-related protein
MCP	monocyte chemoattractant protein
NAFLD	non-alcoholic fatty liver disease
NASH	non-alcoholic steatohepatitis
NEFA	non-esterified fatty acids
PICs	proinflammatory cytokines
PPARγ	peroxisome proliferator- activated receptor gamma
RAC1	Ras-related C3 botulinum toxin substrate 1
SCD	stearoyl-CoA desaturase
SFRP5	secreted frizzled-related protein 5
Slurp1	secreted Ly-6/uPAR-related protein 1
T2D	type 2 diabetes
TAGs	triacylglycerols
TNFα	tumor necrosis factor α
USP17	ubiquitin-specific protease 17
VAT	visceral adipose tissue
VLDL	very low density lipoprotein
WNTSP	WNT signaling pathways

1. Introduction

Lipid metabolism is the process of synthesizing, breaking down, and storing fat in cells for energy storage. It is a complex process taking place in eukaryotic cells that are involved in many metabolic reactions.[1] It is also the basis for forming chemical compounds.[1] The lipids are derived from the precursor acetyl-CoA and divided into five main classes: fatty acids (FAs) are intermediates in lipid biosynthesis; free sterols are components of the membrane structure; sterol esters act as lipid storage compounds made up of sterols and FAs; triacylglycerols (TAGs) serve as a lipid reserve formed from glycerol and FAs; phospholipids are structural compounds in membranes made up of FAs, glycerol, and an alcohol base.[1] These classes may vary in composition depending on age. Lipid biosynthesis has been presented since the fetal stage. At this stage, the development and growth of organs (especially the heart and small intestine) of the fetus are closely linked with the development of lipid metabolism and adipogenesis.[2] Differences in lipid composition between maternal and fetal organs will affect the physiological properties of cell membranes.[2] This means that molecules such as receptors, enzymes and transporters that reside in their cell membranes will change as lipid composition changes.[2] Therefore, maternal health status, in particular, gestational diabetes may lead to different fetal development through impact on lipid composition. In addition, lipid metabolism and its products participate in gene regulation. With aging, lipid breakdown and utilization will decrease, causing many health problems in the elderly.[3] The difference in lipid metabolism between men and women also directly affects the different health statuses between the two sexes. The ratio of lipids and lipolysis in women may be higher than in men, leading to women being more susceptible to liver-related diseases such as fatty liver than men.[4] In general, lipid metabolism plays an important role in cell survival, so interference with abnormalities in this process is essential.

Lipid metabolism disorder is associated with many chronic diseases: NAFLD, obesity, diabetes, etc. These are common chronic diseases and seriously affect the quality of life of patients. Lipids are metabolized mainly in the liver and adipocytes, so an abnormality in this process can lead to NAFLD.[5] NAFLD is the result of an imbalance between lipid accumulation and lipolysis in the liver.[6] It can develop into NASH when liver cells are damaged and die through activation of c-Jun N-terminal kinase (JNK).[6,7] NASH appears to be accompanied by symptoms of sustained inflammation.[6] This symptom, along with fibrillation in NASH, is driven by chronic elevation of leptin.[6] More seriously, NASH can progress to hepatocellular carcinoma through fibroblast proliferation leading to death.[6] On the other hand, NAFLD is associated with cognitive decline in middle age. An experiment in rats had shown that NAFLD causes cerebral hypoperfusion and unbalanced metabolism of monounsaturated fatty acids and polyunsaturated fatty acids.[8] Abnormalities of lipid metabolism also contribute to the mechanism of obesity. A study in China demonstrated that obesity in adults is related to total cholesterol and protein cholesterol levels.[9] Several other studies have also found an association between certain genes involved in obesity and abnormalities in lipid metabolism.[10–12] This confirms that the development of obesity may be related to impaired lipid metabolism. Furthermore, lipid metabolism disorder is associated with NAFLD in obese individuals.[13] In addition, it also contributes to the development of complications in diabetes.[14,15] It contributes to determining the fetal development of mothers with gestational diabetes.[2,16]

FAs and TAGs are two important components in lipid metabolism. FAs are the main storage and source of energy. TAGs is the most common form of non-toxic fatty acid. Non-esterified fatty acids (NEFA) are formed by hydrolysis of complex fats or by hydrolysis of fatty acids-CoA.[5] The liver then absorbs NEFAs from the blood in proportion to their concentrations.[5] In hepatocytes, acyl-CoA synthetases or fatty acid-binding proteins activate them.[5] After activation, NEFA and acyl-CoA are transported to the intracellular compartment for metabolism or to the nucleus to catalyze transcription.[5] Exogenous fatty acids are assimilated to neutral and polar lipids, some of which are oxidized.[5] As a result, fatty acyl-CoA and intracellular NEFA are insignificant. De novo lipogenesis is an important metabolic pathway in higher animals that helps with energy homeostasis. FAs are synthesized by lipoprotein production (LP) in the liver to provide energy and materials to build membranes.[5] De novo synthesis helps to store energy in adipose tissue (AT). De novo synthesis takes place in the cytosol via a series of decarboxylation condensation reactions.[5] Palmitic acid is the major product of this process through the catalysis of the palmitate.[5] The contribution of AT and liver to de novo lipogenesis varies from animal to species. Triacylglycerol synthesis is controlled by transcription factors and nuclear receptors [sterol regulatory element-binding transcription factor 1, carbohydrate response element-binding protein, peroxisome proliferator-activated receptor gamma (PPARγ), and liver X receptors and their ligands].[5,17] The NEFAs are esterified to glycerol, which exists as glycerolipid.[5] Triacylglycerol will be kept in cytosolic storage so that it can be mobilized through lipolysis (re-esterification as needed).[5] Phospholipids will be transferred to build membranes. Unesterified acyl-CoAs can be completely or incompletely oxidized in mitochondria or peroxisomes. Factors affecting the degree of fat oxidation in the liver include the supply of FAs to the liver through lipolysis and the split between oxidation and esterification in the microsomes.[5] For incomplete oxidation, ketone bodies will be produced.[5]

Measures for ameliorating diseases associated with lipid metabolism disorder are widely studied. Regulatory mechanisms based on long noncoding RNAs (lncRNAs) play a critical role in the development of lipid-associated diseases.[18] Therefore, lncRNAs are considered biomarkers for the examination, and treatment of lipid-associated diseases.[18] In the future, more in-depth studies are needed. Low density lipoprotein cholesterol (LDL-C) has always been the target of treatment for diseases related to hyperlipidemia.[19] However, currently used drugs have many side effects so phytochemicals and whole foods of animal origin are considered a promising alternative strategy.[19] NAFLD could be considered the most severe disease associated with this abnormality. It progresses silently, making it difficult to detect early using only conventional clinical measures. Therefore, lipidomics analysis is considered a breakthrough to detect specific fatty acid profiles that are associated with metabolic diseases (NAFLD, diabetes mellitus) and disease staging.[20,21] Its development also allows the identification of products of the lipid metabolism pathway involved in the pathogenesis of NAFLD.[20] In addition, the suggestion of combining lipidomics analysis with machine learning also opens the future of predicting possible risks depending on the disease condition.[21] Current treatments only focus on reducing fat accumulation and liver damage to improve the disease.[22] There are four main types of drugs: PPAR agonists, bile acid metabolism regulators, thyroid hormone receptor co-transporters, and type 2 diabetes (T2D) drugs.[22] Fibroblast growth factor 21 is being tested for the treatment of NAFLD and gets positive results.[22] However, it has some undesirable effects. Besides advanced medications, herbal therapy and lifestyle modifications are also suggested as measures to treat and improve the condition of this disease.[20,23] Herbal therapy affects this disease by causing changes to the intestinal microflora.[23] In addition, methods to directly influence factors associated with NAFLD should also be considered.[24,25] Obesity is also a severe disease and may be associated with T2D. Traditional Chinese medicines are used to increase lipids in obesity through modulation of absorption, synthesis, breakdown, and transport of lipid.[26] In the future, the proposed methods need to be studied more to effectively treat and remove the chronic disease burden related to lipid metabolism disorders.

Because of the importance and severity of lipid metabolism, we conducted this chapter to synthesize knowledge about the Frizzled receptor and SFRP5 adipokines that play an important role in the mechanism of lipid metabolism disorders and their potential applications. This knowledge can provide important insights for the development of new studies and effective treatment of lipid disorders through these two adipokines.

2. Frizzled receptors and lipid metabolism

2.1 The expression of frizzled receptors on adipose tissues/adipocytes

FZD are a family of G-protein coupled receptors (GPCRs). There are 10 types of FZD divided into main groups based on similarity in protein sequence: FZD1/2/7, FZD5/8, FZD3/6, FZD4/9/10.[27] They are known as an unusual form of GPCRs because there is no evidence of their interaction with other G proteins.[27] The FZD receptor acts through three distinct pathways: canonical/β-catenin, non-canonical/non-β-catenin, and the WNT/Ca^{2+} pathways. In which, FZD1,2,3 and 7 are abundant in visceral adipose tissue (VAT).[28] FZD mediates WNT signaling leading to the promotion or inhibition of adipogenesis depending on the type of WNT.[28] The activity of FZD in adipocytes is dependent on their ligands and co-receptors. FZD1,2 binds to low-density lipoprotein receptor-related protein (LRP) 5 or LRP 6 it causes a standard WNT signal (WNT1, WNT10b).[28,29] Among them, WNT10b is an endogenous activator of WNT signaling that strongly inhibits adipogenesis.[30] Thus, the β-catenin pathway will be activated to inhibit adipocyte differentiation.[28] When FZD1 and the receptor tyrosine kinase like orphan receptor 1/2 co-receptor are present, the WNT-FZD/LRP pathway is inhibited and the adipogenesis is promoted.[28] During adipogenic differentiation, FZD1, FZD7, and β-catenin decrease.[31] (Table 1).

In obese patients, gene expression of the WNT5 receptors FZD2 and FZD5 increase significantly.[32] At this time, the sensitivity of the arteries to non-standard Wnt signals also increases, leading to increased vascular oxidative stress.[32] Ras-related C3 botulinum toxin substrate 1 (RAC1) is ubiquitin-specific protease 17 (*USP17*)-mediated by FZD2-induced

Table 1 Characteristics of frizzled receptors on adipose tissue and in lipid metabolism.

Ref.	Important messages
27	FZD include 10 types, divided into 4 groups based on similarity in protein sequences: FZD1/2/7, FZD5/8, FZD3/6, FZD4/9/10
28	FZD inhibit or promote adipocyte through three WNTSP: canonical/β-catenin, non-canonical/non-β-catenin, and the WNT/Ca^{2+} pathways
32	In obesity, gene expression of the WNT5 receptors FZD2 and FZD5 is significantly increased
33	In lipid metabolism, FZD may have a function to recognize fatty acids

noncanonical activation.[32] Currently, studies on the expression of FZD in obesity are limited. Therefore, more in-depth studies are needed in the future. (Table 1).

2.2 Roles of frizzled receptors in lipid metabolism

FZD typically interact with ligands and co-receptors either grouped or ungrouped for WNT signaling.[27] From the transmitted WNT signals, one of three pathways is activated to influence adipogenesis. Depending on WNT signaling, inhibition or promotion of adipogenesis will take place. WNT plays an important role in adipocyte differentiation. If WNT is disrupted, spontaneous adipocyte differentiation occurs.[30] In lipid metabolism, FZD contributes to the recognition of FAs. WNT8a binds to the C14 fatty acid to occupy the hydrophobic groove of the mFZD8 CRD.[33] The TcdB toxin binds to the FZd1/2/7 to enter the toxin. FZD2 CRD binds to TcdB forming a T-shape.[33] The interaction between FZD2 CRD and FAs is mainly mediated by hydrophobic interactions.[33] In TcdB toxin mutations, the FZD2 CRD complex and FAs can be destroyed.[33] FZD5 and TcdB interactions may be promoted by fatty acid addition. All types of FZD are dimerized. FZD dimerization is associated with WNT signaling.[33,34] WNT5 stimulates dimerization of FZD4 CRD. WNT8a and WNT3a stimulate dimerization of the FZD8 CRD.[33] Free unsaturated fatty acids of C14 or more stimulate dimerization of FZD7 CRD.[33] This represents a reciprocal interaction between fatty acids, FZD CRD, and WNT in lipid metabolism. The fatty acid recognition mechanism of the FZD CRD has four supporting factors: position of the fatty acyl chain in the lipid compartment, aggregation of residues forming the lipid cavity, flexibility in the side chains, and inherent flexibility of the chain.[33,34] (Table 1).

3. Secreted frizzled-related protein 5 (SFRP5) in lipid metabolism

3.1 The expression SFRP5 on adipose tissues/adipocytes and other places associated abnormal adipose tissues

SFRP5 is an adipokine. It is the only SFRP that can interact with both standard and non-standard WNT signaling pathways to influence adipocyte differentiation. It inhibits the WNT signaling pathway thereby influencing adipocyte differentiation.[35] Furthermore, one study found much lower levels of SFRP5 in adipocyte lysates than in supernatants, demonstrating that SFRP5 is primarily present as a secretory protein in adipocytes.[36] Therefore, the expression of SFRP5 is closely associated with the size of the AT,[37] and the expansion of AT.[38,39] PPARγ may be a transcription factor that regulates its expression because adenovirus transfection activated by PPARγ can increase the expression of SFRP5.[40] During adipocyte differentiation, expression of SFRP5 gradually increased and reached a maximum on day 8 of differentiation through 3T3L1 cell differentiation.[41] As for the marker of "brown adipogenesis" in the white fat store, its expression was maximal in 21-day-old rats and was inhibited from 21 days of age after weaning to 56 days of age.[39] During white adipose tissue and brown adipose tissue differentiation, SFRP5 mRNA was formed.[41] In contrast to Mest expression, SFRP5 mRNA was found to be homologous and rather low in both interscapular brown adipose tissue, Inguinal fat.[39] Furthermore, SFRP5 overexpression also did not increase the expression of adipocyte-specific marker genes.[41] Therefore, SFRP5 does not affect white and brown fat production but may be involved only in AT enlargement (Table 2).

SFRP5 is the adipokine involved in the development of obesity. When combined with WNT5a, it may contribute to the mechanism of obesity by the WNT/Ca^{2+} pathway. In contrast to normal adipose tissue, SFRP5 in AT during weight gain was very low.[42] Gradual growth in fat percentage is related to decreased expression of SFRP5 in AT.[42] Expression of SFRP5 decreased for mice fed a 24-week high-fat/sucrose-rich diet, while the WNT5a expression was increased and the ratio of WNT5a/SFRP5 was also high.[35,42] In addition, people who consumed a huge amount of sugary drinks while ignoring fruits and vegetables had lower levels of SFPR5.[42] In contrast, for the HF/HS diet in obese rodents, the expression of SFPR5 was significantly improved.[42] Furthermore, some studies have demonstrated that circulating SFPR5 plasma concentrations are significantly reduced in obese patients compared to normal patients, and this concentration is inversely

Table 2 Characteristics of SFRP5 on adipose tissue and in lipid metabolism.

Ref.	Important messages
35	SFRP5 inhibits the WNT signaling pathway
42	Increased fat percentage over time is related to decreased expression of SFRP5 in AT
35,42	Circulating SFRP5 plasma concentrations are reduced in obese subjects and are inversely correlated with body fat expression indices.
43	Mothers who gain too much weight during pregnancy have low levels of SFRP5 in the umbilical cord
44–46	When SFRP5 expression is decreased, the production of PICs (TNFα, IL-6, and MCP-1) is promoted leading to increased lipolysis and free fatty acid secretion
47	SFRP5 helps increase lipid accumulation to expand AT
47,48	SFRP5 reduced the expression of two genes relevant to lipid production, PPARγ, and CEBPα, through the action of WNT10b
49	The hypothalamus and liver promote SFRP5 for LP through inhibition of SCD-1 in the liver
48	In the SFRP5 group with low plasma, the percentage of total cholesterol increased was much higher than in the high plasma SFRP5 group

correlated with body fat expression indices.[35] The same is true for overweight/obese children.[35] For mothers with excessive weight gain during pregnancy, the SFRP5 concentration in the umbilical cord of the infant is determined by maternal serum SFRP5 concentration, hemoglobin A1c, several other indices, and anthropometric measurements.[43] In these cases, SFRP5 in both umbilical cord and maternal serum was much lower than the standard.[43] Therefore, weight management during pregnancy also needs attention. In addition, when SFRP5 is decreased, latent cytokines, for example, *tumor necrosis factor α* (TNFα), *monocyte chemoattractant protein (MCP)-1*, and interleukin (IL)-6 are increased leading to inflammation and insulin resistance (IR).[44] When measuring the expression of SFRP5 in biopsies of VAT, it had been shown that obese patients with AT inflammation had lower SFRP5 expression than those without it.[35] Overall, obesity is associated with upregulating the expression of SFRP5. (Table 2).

3.2 Roles of SFRP5 in lipid metabolism

SFRP5 is implicated in many metabolic syndromes, for example, lipid metabolism disorders, IR, T2D, obesity, and cardiovascular disease.[45]

Moreover, it also helps increase lipid accumulation in fat cells to open up wide AT.[47] In the mechanism of lipid metabolism, SFRP5 may induce abnormalities through interference with the WNTSP.[50] The increase in mitochondrial respiration would help SFRP5 inhibit oxidation by preventing the receptors from binding to WNT.[47,48] From there, the WNT signal is suppressed. SFRP5 also reduced the expression of two genes relevant to lipid production, PPARγ, and transcription factor CCAAT/enhancer binding protein (CEBP) α, through the action of WNT10b.[47,48] This action causes fibroblasts and myogenicity to be produced leading to decreased adipogenesis.[47] The PPARγ receptor enhances proteasome-dependent degradation of β-catenin to regulate the early stages of pre-adipocyte differentiation, lipid accumulation, and suppression of WNT/β-catenin signaling.[47] In addition, when SFPR5 levels decreased, PICs including TNF-α, IL-6, and IL-1β increased.[44,45] From there, they inhibit lipoprotein lipase, which increases lipolysis and free fatty acid secretion, leading to insulin resistance.[46] The mechanism that causes SFRP5 to suppress WNT signaling may also be related to the neutralization of WNT5a by SFRP5.[47] Mechanisms involving intracerebroventricular SFRP5 may also be involved in lipid metabolism disorders. Stearoyl CoA desaturase-1 in the liver is inhibited so that lipids cannot be converted to apoB-lipoprotein leading to decreased production of very low density lipoprotein (VLDL)-triglycerides.[49] Thus, the hypothalamus and liver promote SFRP5 for LP through inhibition of stearoyl-CoA desaturase (SCD)-1 in the liver.[49] When dyslipidemia occurs, SFPR5 levels decrease. A study demonstrated that in the SFRP5 group with low plasma, the percentage of total cholesterol increased was much higher than in the high plasma SFRP5 group.[48] In addition, the risk of metabolic disorders increased about 2 times. When other variables were adjusted, the same results were obtained.[48] (Table 2).

4. Mechanisms underlying interaction between frizzled receptors and SFRP5 in lipid metabolism

In the WNTSP, FZD serves as a stimulatory receptor that transmits WNT signaling in 3 pathways with 1 CRD region and 7 transmembrane and cytoplasmic regions. In contrast, SFRP5 in particular and SFRP in general act as antagonists to modulate WNT signaling by competing with FZDs.[51,52] Because they contain functional domains similar to the CRD domain of FZDs, they may combine with WNT proteins to prevent it from binding to FZDs or they form nonfunctional complexes with frizzled proteins.[50,52] Specifically, SFRP5 can compete with FZD to block both standard and non-standard WNTSP in embryonic and endodermal development (Fig. 1).[53] They bind and compete primarily with the WNT5 and WNT11 proteins. When WNTSP are inhibited, transcription of downstream target genes and cytoskeleton is inhibited.[50] SFRP5 through association with WNT5a can also suppress JNK signaling leading to signaling a decline in the production of PICs and antagonism of IRS-1 Ser

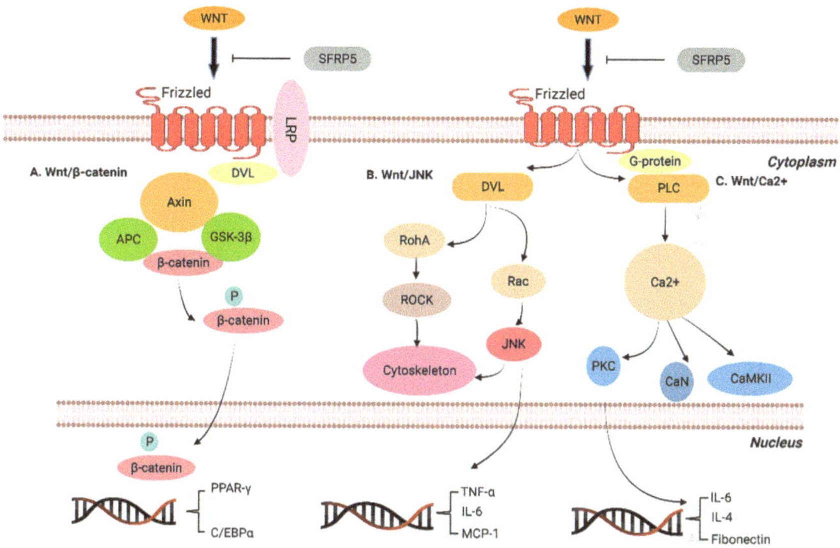

Fig. 1 Interaction between Frizzled receptors and SFRP5 during adipogenesis. In WNTSP, FZD binds to the WNT protein by complexing with ligands and co-receptors for signal transduction into the cell. For the WNT/β-catenin signaling pathway, after the signal is transduced into the cell, the Axin/APC/GSK-3β complex formed in the cell plays a role in activating β-catenin through phosphorylation. The β-catenin activation signal is transmitted to the nucleus to influence the transcriptional activity of target genes (PPARγ, CEBPα). For the WNT/JNK pathway, a signal transduced into the cell activates the downstream Rho/ROCK and Rac/JNK pathways. From there, it induces cytoskeleton involvement and affects the transcription of target genes (TNF-α, IL-6, and MCP-1). For the pathway, WNT/Ca^{2+}, FZD/WNT activates PLC leading to intracellular Ca^{2+} accumulation. Then, activation of CamKII, CaN and PKC take place to signal to the nucleus to interact with the activity of target genes (IL-6, IL-4, Fibonectin). SFRP5 competes with FZD for complexation with the WNT protein. From there, the WNT signal is suppressed. β-Catenin and its target genes are not activated. WNT/SFRP5 also blocks downstream JNK signaling that causes TNF-α, IL-6, and MCP-1 genes to be suppressed leading to reduced proinflammatory but increased insulin resistance. In addition, it also prevents PLC activation and Ca^{2+} accumulation leading to the destruction of target genes IL-6, IL-4, Fibonectin. (Created with BioRender.com).

phosphorylation.[50,54] When the body is overweight/obese, SFRP5 acts as a decoy to isolate WNT5a from FZD leading to suppression of WNT signaling.[47] WNT5a affects adipocyte functional activity because it is produced by macrophages in adipose tissue.[55] In chronic low-grade inflammation, WNT5a signaling is downregulated by SFRP5 activity by inhibiting the formation of a bond between FZD and WNT5a.[47] The level of this activity can be regulated through the association between SFRP5 and LRP1.[56] When bound to WNT5a it can reduce the expression of proinflammatory cytokines.[55] Since then, it is believed to have anti-inflammatory effects and improve the condition of diseases related to lipid metabolism disorders and obesity. Currently, studies on the combination and interaction between SFRP5 and FZD as well as adipokines are very limited. Therefore, more research is needed to understand the association between SFRP5 as well as other adipokines.

5. Potential applications of findings on frizzled receptors and SFRP5 in lipid metabolism

WNT5a and SFRP5 seem to take part in regulating inflammation in obesity, as well as balancing mediators between AT and other important metabolic organs such as the liver. SFRP5 binding to WNT5a through competition with the FZD receptor may be to reduce WNT5a-induced inflammation. The WNT5a/SFRP5 duo also contributes to regulating metabolic homeostasis in prepubertal children.[57] In SFRP5-deficient mice, pancreatic β-cell proliferation was detected through IGFBP-3, a lesion commonly seen in obesity.[58,59] Furthermore, SFRP5 is involved in adipocyte differentiation. In addition, in obese patients, the expression of SFRP5 was also found to be very low.[57,60,61] Thus, SFRP5 has been identified as having anti-inflammatory potential in metabolic disorders as well as obesity. Modification of lifestyle and diet is considered as an effective measure to ameliorate inflammation in obesity through the regulation of SFRP5 expression.[57,62] Oral glucose therapy and liraglutide have also been tested. After treatment with liraglutide, the homeostasis model assessment of β-cell function was significantly increased while BMI was moderately decreased.[63] SFRP5 expression was also significantly increased after liraglutide treatment.[63] It is considered a potential therapy for the treatment of obesity, but further experimental studies are needed. SFRP5 also has a role in lipid metabolism. Therefore, it participates in the mechanism of lipid-related diseases in the liver. SFRP5 is to reduce liver fibrosis and protect the pancreas through WNT5a/FZD2 signaling.[52] SFRP5 mutations induced in animal models of high fat intake cause liver failure, insulin resistance, and glucose intolerance.[64] Slurp1 interacts with SFRP5 to promote its expression in palmitate-induced HepG2 cells.[64] From there, Slurp1 can regulate triglyceride accumulation in hepatic steatosis through the regulation of SFRP5 expression.[64] In addition, it is involved in NASH—a serious disease related to lipid metabolism disorders. SFRP5 level in NASH was significantly lower compared with normal.[65] Recombinant SFRP5 was used for efficacy assays on NASH. After recombinant SFRP5 was administered, inflammation scores in the liver and expression of PICs in the liver were markedly reduced.[65] SFRP5 inhibits NASH by inhibiting Kuffer and inflammation in the liver.[65] In general, based on the changes of SFRP5 levels in pathological manifestations, future studies can develop methods for the early detection and diagnosis of obesity-related diseases and lipid metabolism disorders. Through the change of SFRP5 concentration. At the same time, the new disease intervention methods discovered above also need to be proven and further studied for effectiveness and side effects. The contribution of FZD to the pathogenesis of obesity-related diseases and lipid metabolism, as well as methods of intervention in these diseases, is very limited. Therefore, future studies also need to clarify this mechanism and measures to improve the disease through FZD.

6. Conclusion

Obesity and lipid metabolism disorders are common metabolic disorders. The diseases associated with these disorders are often severe chronic diseases and cause a global burden of disease. In particular, lipid metabolism disorders are also associated with liver diseases. NAFLD can progress to NASH. It is the 1st cause of death globally from liver diseases. SFRP5 and Frizzled receptor are newly discovered adipokines involved in the mechanism of these metabolic disorders. Recently, the role of SFRP5 has been studied. Several methods of improving NASH, hepatic steatosis, and obesity through SFRP5 have been considered. However, the current understanding of the mechanisms of these two adipokines in lipid metabolism disorders and obesity is very limited, especially the role of the Frizzled receptor and its relationship with SFRP5. I suggest that more in-depth studies are needed on the effects of the two adipokines and their interactions in obesity and dyslipidemia. In addition, disease intervention methods and early diagnosis based on their manifestations also need attention to improve the burden of diseases related to lipid disorders and obesity.

Acknowledgments

We would like to thank Dr. Le Bui Minh (NTT Hi-tech Institute, Nguyen Tat Thanh University, 300A Nguyen Tat Thanh St., Ward 13, District 4, Ho Chi Minh City, Vietnam) for the license to use BioRender to create the figures in this work. We would like also want to thank Mr. Nghia Pham (Northeastern University, Boston, MA, USA) for critical reading and checking to improve the manuscript.

References

1. Nielsen J. Systems biology of lipid metabolism: From yeast to human. *FEBS Lett.* 2009;583(24):3905–3913.
2. Desoye G, Herrera E. Adipose tissue development and lipid metabolism in the human fetus: The 2020 perspective focusing on maternal diabetes and obesity. *Prog Lipid Res.* 2020;101082.
3. Chung KW. Advances in understanding of the role of lipid metabolism in aging. *Cell.* 2021;10(4):880.
4. Mittendorfer B. Sexual dimorphism in human lipid metabolism. *J Nutr.* 2005;135 (4):681–686.
5. Nguyen P, Leray V, Diez M, et al. Liver lipid metabolism. *J Anim Physiol Anim Nutr.* 2008;92(3):272–283.
6. Nahon P, Allaire M, Nault J-C, Paradis V. Characterizing the mechanism behind the progression of NAFLD to hepatocellular carcinoma. *Hepatic. Oncology.* 2020;7(4): [HEP36].
7. Hliwa A, Ramos-Molina B, Laski D, Mika A, Sledzinski T. The role of fatty acids in non-alcoholic fatty liver disease progression: An update. *Int J Mol Sci.* 2021;22(13).
8. Pinçon A, De Montgolfier O, Akkoyunlu N, et al. Non-alcoholic fatty liver disease, and the underlying altered fatty acid metabolism, reveals brain hypoperfusion and contributes to the cognitive decline in APP/PS1 mice. *Metabolites.* 2019;9(5):104.
9. Yin T, Zhang J-X, Wang F-X, et al. The association between sarcopenic obesity and hypertension, diabetes, and abnormal lipid metabolism in Chinese adults. *Diabetes Metab Syndr Obes.* 2021;14:1963.
10. Wu Z, Liu J, Chen G, et al. CD146 is a novel ANGPTL2 receptor that promotes obesity by manipulating lipid metabolism and energy expenditure. *Adv Sci.* 2021;8(6):2004032.
11. Czogała W, Czogała M, Strojny W, et al. Methylation and expression of FTO and PLAG1 genes in childhood obesity: Insight into anthropometric parameters and glucose-lipid metabolism. *Nutrients.* 2021;13(5).
12. Lee JH, Jung SB, Lee SE, et al. Expression of LONP1 is high in visceral adipose tissue in obesity, and is associated with glucose and lipid metabolism. *Endocrinol Metab.* 2021;36 (3):661–671.
13. Hazer İ, Kabukçu HO, Yağcı M, Ertürk Z, Yıldırım GK, Kirel B. The association of lipid metabolism and non-alcoholic fatty liver disease in children with obesity. *Turk Pediatri Arsivi.* 2020;55(3):263–269.
14. Suryawanshi N, Bhutey A, Nagdeote A, Jadhav A, Manoorkar G. Study of lipid peroxide and lipid profile in diabetes mellitus. *Indian J Clin Biochem.* 2006;21(1):126.
15. Eid S, Sas KM, Abcouwer SF, et al. New insights into the mechanisms of diabetic complications: Role of lipids and lipid metabolism. *Diabetologia.* 2019;62(9):1539–1549.
16. Furse S, Fernandez-Twinn DS, Chiarugi D, Koulman A, Ozanne SE. Lipid metabolism is dysregulated before, during and after pregnancy in a mouse model of gestational diabetes. *Int J Mol Sci.* 2021;22(14).
17. Sozio M, Crabb DW. Alcohol and lipid metabolism. *Am J Physiol Endocrinol Metab.* 2008;295(1):E10–E16.

18. Huang SF, Peng XF, Jiang L, Hu CY, Ye WC. LncRNAs as therapeutic targets and potential biomarkers for lipid-related diseases. *Front Pharmacol.* 2021;12:729745.
19. Islam SU, Ahmed MB, Ahsan H, Lee YS. Recent molecular mechanisms and beneficial effects of phytochemicals and plant-based whole foods in reducing LDL-C and preventing cardiovascular disease. *Antioxidants (Basel, Switzerland).* 2021;10(5).
20. Notarnicola M, Osella AR, Caruso MG, et al. Nonalcoholic fatty liver disease: Focus on new biomarkers and lifestyle interventions. *Int J Mol Sci.* 2021;22(8):3899.
21. Castañé H, Baiges-Gaya G, Hernández-Aguilera A, et al. Coupling machine learning and Lipidomics as a tool to investigate metabolic dysfunction-associated fatty liver disease. A general overview. *Biomolecules.* 2021;11(3).
22. Dabravolski SA, Bezsonov EE, Baig MS, Popkova TV, Orekhov AN. Mitochondrial lipid homeostasis at the crossroads of liver and heart diseases. *Int J Mol Sci.* 2021;22(13).
23. Yang X-F, Lu M, You L, et al. Herbal therapy for ameliorating nonalcoholic fatty liver disease via rebuilding the intestinal microecology. *Chinese Med.* 2021;16(1):1–16.
24. Shabalala SC, Dludla PV, Mabasa L, et al. The effect of adiponectin in the pathogenesis of non-alcoholic fatty liver disease (NAFLD) and the potential role of polyphenols in the modulation of adiponectin signaling. *Biomed Pharmacother.* 2020;131:110785.
25. Mass Sanchez PB, Krizanac M, Weiskirchen R, Asimakopoulos A. Understanding the role of perilipin 5 in non-alcoholic fatty liver disease and its role in hepatocellular carcinoma: A review of novel insights. *Int J Mol Sci.* 2021;22(10):5284.
26. Fan Q, Xu F, Liang B, Zou X. The anti-obesity effect of traditional Chinese medicine on lipid metabolism. *Front Pharmacol.* 2021;12:696603.
27. Nichols AS, Floyd DH, Bruinsma SP, Narzinski K, Baranski TJ. Frizzled receptors signal through G proteins. *Cell Signal.* 2013;25(6):1468–1475.
28. Im H, Park J-H, Im S, Han J, Kim K, Lee Y-H. Regulatory roles of G-protein coupled receptors in adipose tissue metabolism and their therapeutic potential. *Arch Pharm Res.* 2021;1–13.
29. Jeon M, Rahman N, Kim YS. Wnt/β-catenin signaling plays a distinct role in methyl gallate-mediated inhibition of adipogenesis. *Biochem Biophys Res Commun.* 2016;479 (1):22–27.
30. Bennett CN, Ross SE, Longo KA, et al. Regulation of Wnt signaling during adipogenesis. *J Biol Chem.* 2002;277(34):30998–31004.
31. Park JR, Jung JW, Lee YS, Kang KS. The roles of Wnt antagonists Dkk1 and sFRP4 during adipogenesis of human adipose tissue-derived mesenchymal stem cells. *Cell Prolif.* 2008;41(6):859–874.
32. Akoumianakis I, Sanna F, Margaritis M, et al. Adipose tissue–derived WNT5A regulates vascular redox signaling in obesity via USP17/RAC1-mediated activation of NADPH oxidases. *Sci Transl Med.* 2019;11(510).
33. Nile AH, Hannoush RN. Fatty acid recognition in the frizzled receptor family. *J Biol Chem.* 2019;294(2):726–736.
34. Nile AH, Mukund S, Stanger K, Wang W, Hannoush RN. Unsaturated fatty acyl recognition by frizzled receptors mediates dimerization upon Wnt ligand binding. *Proc Natl Acad Sci U S A.* 2017;114(16):4147–4152.
35. Wang D, Zhang Y, Shen C. Research update on the association between SFRP5, an anti-inflammatory adipokine, with obesity, type 2 diabetes mellitus and coronary heart disease. *J Cell Mol Med.* 2020;24(5):2730–2735.
36. Zhou W, Ye C, Li L, et al. Adipocyte-derived SFRP5 inhibits breast cancer cells migration and invasion through Wnt and epithelial-mesenchymal transition signaling pathways. *Chin J Cancer Res.* 2020;32(3):347.
37. Mori H, Prestwich TC, Reid MA, et al. Secreted frizzled-related protein 5 suppresses adipocyte mitochondrial metabolism through WNT inhibition. *J Clin Invest.* 2012;122(7):2405–2416.

38. Jura M, Jarosławska J, Chu DT, Kozak LP. Mest and Sfrp5 are biomarkers for healthy adipose tissue. *Biochimie*. 2016;124:124–133.
39. Chu D-T, Malinowska E, Gawronska-Kozak B, Kozak LP. Expression of adipocyte biomarkers in a primary cell culture models reflects preweaning Adipobiology. *J Biol Chem*. 2014;289(26):18478–18488.
40. Zeng J, Hu J, Lian Y, Jiang Y, Chen B. SFRP5 is a target gene transcriptionally regulated by PPARγ in 3T3-L1 adipocytes. *Gene*. 2018;641:190–195.
41. Wang R, Hong J, Liu R, et al. SFRP5 acts as a mature adipocyte marker but not as a regulator in adipogenesis. *J Mol Endocrinol*. 2014;53(3):405–415.
42. Koutaki D, Michos A, Bacopoulou F, Charmandari E. The emerging role of Sfrp5 and Wnt5a in the pathogenesis of obesity: Implications for a healthy diet and lifestyle. *Nutrients*. 2021;13(7):2459.
43. Kimber-Trojnar Ż, Patro-Małysza J, Trojnar M, Darmochwał-Kolarz D, Oleszczuk J, Leszczyńska-Gorzelak B. Umbilical cord SFRP5 levels of term newborns in relation to normal and excessive gestational weight gain. *Int J Mol Sci*. 2019;20(3):595.
44. Wang C. Obesity, inflammation, and lung injury (OILI): The good. *Mediators Inflamm*. 2014;2014.
45. Farkhondeh T, Llorens S, Pourbagher-Shahri AM, et al. An overview of the role of adipokines in cardiometabolic diseases. *Molecules*. 2020;25(21):5218.
46. Jager J, Grémeaux T, Gonzalez T, et al. Tpl2 kinase is upregulated in adipose tissue in obesity and may mediate interleukin-1β and tumor necrosis factor-α effects on extracellular signal–regulated kinase activation and lipolysis. *Diabetes*. 2010;59(1):61–70.
47. Liu L-B, Chen X-D, Zhou X-Y, Zhu Q. The Wnt antagonist and secreted frizzled-related protein 5: Implications on lipid metabolism, inflammation, and type 2 diabetes mellitus. *Biosci Rep*. 2018;38(4):[BSR20180011].
48. Bai Y, Du Q, Jiang R, et al. Secreted frizzled-related protein 5 is associated with glucose and lipid metabolism related metabolic syndrome components among adolescents in northeastern China. *Diabetes Metab Syndr Obes*. 2021;14:2735.
49. Li Y, Tian M, Yang M, et al. Central Sfrp5 regulates hepatic glucose flux and VLDL-triglyceride secretion. *Metabolism*. 2020;103:154029.
50. Tong S, Ji Q, Du Y, Zhu X, Zhu C, Zhou Y. Sfrp5/Wnt pathway: A protective regulatory system in atherosclerotic cardiovascular disease. *J Interferon Cytokine Res*. 2019;39 (8):472–482.
51. Maekawa T, Kulwattanaporn P, Hosur K, et al. Differential expression and roles of secreted frizzled-related protein 5 and the wingless homolog Wnt5a in periodontitis. *J Dent Res*. 2017;96(5):571–577.
52. Chatani N, Kamada Y, Kizu T, et al. Secreted frizzled-related protein 5 (Sfrp5) decreases hepatic stellate cell activation and liver fibrosis. *Liver Int*. 2015;35(8):2017–2026.
53. Stuckenholz C, Lu L, Thakur PC, Choi T-Y, Shin D, Bahary N. Sfrp5 modulates both Wnt and BMP signaling and regulates gastrointestinal organogensis in the zebrafish, Danio rerio. *PLoS One*. 2013;8(4):e62470.
54. Li Q, Zuo L-l, Lin Y-q, et al. Cloning and expression of SFRP5 in Tibetan chicken and its relationship with IMF deposition. *Anim Biotechnol*. 2016;27(4):231–237.
55. Catalán V, Gómez-Ambrosi J, Rodríguez A, et al. Activation of noncanonical Wnt signaling through WNT5A in visceral adipose tissue of obese subjects is related to inflammation. *J Clin Endocrinol Metabol*. 2014;99(8):E1407–E1417.
56. Au DT, Migliorini M, Strickland DK, Muratoglu SC. Macrophage LRP1 promotes diet-induced hepatic inflammation and metabolic dysfunction by modulating Wnt signaling. *Mediators Inflamm*. 2018;2018.
57. Tan X, Wang X, Chu H, Liu H, Yi X, Xiao Y. SFRP 5 correlates with obesity and metabolic syndrome and increases after weight loss in children. *Clin Endocrinol (Oxf)*. 2014;81(3):363–369.

58. Rebuffat SA, Oliveira JM, Altirriba J, et al. Downregulation of Sfrp5 promotes beta cell proliferation during obesity in the rat. *Diabetologia*. 2013;56(11):2446–2455.
59. Wang HZ, Ning G. Sfrp5 mediates glucose-induced proliferation in rat pancreatic β-cells. *J Endocrinol*. 2016;229:73–83.
60. Ouchi N, Higuchi A, Ohashi K, et al. Sfrp5 is an anti-inflammatory adipokine that modulates metabolic dysfunction in obesity. *Science*. 2010;329(5990):454–457.
61. Xu Q, Wang H, Li Y, et al. Plasma Sfrp5 levels correlate with determinants of the metabolic syndrome in Chinese adults. *Diabetes Metab Res Rev*. 2017;33(6):e2896.
62. Schulte DM, Müller N, Neumann K, et al. Pro-inflammatory wnt5a and anti-inflammatory sFRP5 are differentially regulated by nutritional factors in obese human subjects. *PLoS One*. 2012;7(2):e32437.
63. Hu W, Li L, Yang M, et al. Circulating Sfrp5 is a signature of obesity-related metabolic disorders and is regulated by glucose and liraglutide in humans. *J Clin Endocrinol Metabol*. 2013;98(1):290–298.
64. Zhao A, Jiang F, Yang G, Liu H, Li L. Sfrp5 interacts with Slurp1 to regulate the accumulation of triglycerides in hepatocyte steatosis model. *Biochem Biophys Res Commun*. 2019;512(2):256–262.
65. Chen L, Zhao X, Liang G, et al. Recombinant SFRP5 protein significantly alleviated intrahepatic inflammation of nonalcoholic steatohepatitis. *Nutr Metab*. 2017;14(1):1–8.

Index

Note: Page numbers followed by "*f*" indicate figures and "*t*" indicate tables.

D

E

F

G

H

O

P

T

U

V

W

Y

Kolibri 360 GmbH
Ferdinand-Jühlke-Straße 7
99095 Erfurt, Deutschland
produktsicherheit@kolibri360.de